C0-DAW-917

# Information Technologies, Methods, and Techniques of Supply Chain Management

John Wang
*Montclair State University, USA*

Managing Director: Lindsay Johnston
Senior Editorial Director: Heather A. Probst
Book Production Manager: Sean Woznicki
Development Manager: Joel Gamon
Development Editor: Heather Probst
Acquisitions Editor: Erika Gallagher
Typesetter: Nicole Sparano
Cover Design: Nick Newcomer, Lisandro Gonzalez

Published in the United States of America by
Business Science Reference (an imprint of IGI Global)
701 E. Chocolate Avenue
Hershey PA 17033
Tel: 717-533-8845
Fax: 717-533-8661
E-mail: cust@igi-global.com
Web site: http://www.igi-global.com



Library of Congress Cataloging-in-Publication Data

Information technologies, methods, and techniques of supply chain management / John Wang, editor.
p. cm.
Includes bibliographical references and index.
Summary: "This book has compiled chapters from experts from around the world in the field of supply chain management and provides a vital compendium of the latest research, case studies, frameworks, methodologies, architectures, and best practices within the field of supply chain management"--Provided by publisher.
ISBN 978-1-4666-0918-1 (hbk.) -- ISBN 978-1-4666-0919-8 (ebook) -- ISBN 978-1-4666-0920-4 (print & perpetual access) 1. Business logistics--Information technology. I. Wang, John, 1955-
HD38.5.I52 2012
658.7--dc23

2012000060

British Cataloguing in Publication Data
A Cataloguing in Publication record for this book is available from the British Library.

All work contributed to this book is new, previously-unpublished material. The views expressed in this book are those of the authors, but not necessarily of the publisher.

# Editorial Advisory Board



# List of Reviewers

# Table of Contents

# Detailed Table of Contents



Although many companies have implemented ERP systems to track and share information across cross functional business processes, they often supplement them with legacy, custom, or best of breed applications to support supply chain execution and management. This article offers a framework for understanding all types of enterprise applications that support the supply chain. In this study, the authors organize these applications, define acronyms, and describe the various types of systems that make up an information infrastructure for supply chain management.



The increasing rate of technology growth has resulted in decrease in cost of information. These technologies are helpful in coordinating the activities resulting in effective management of the supply chain. Literature shows that the use of Information Technology (IT) plays an important role in managing the processes of SCM. This has resulted in increasing use of IT in SCM. The computerization of SCM processes, if implemented in one go may result in failure. IT implementation prioritization in supply chain is a major issue before the planner as there is no clear cut formula to solve this problem. This paper considers components of SCM like material management, purchase management, production management, logistics and distribution and customer interface for IT implementation prioritization. Two multi-criteria decision making methods (MCDM) viz. analytical hierarchy process (AHP) and a technique for order preference by similarity to ideal solution (TOPSIS) are used in this paper. The novelty of the paper lies in integration of AHP and TOPSIS methods for IT implementation prioritization. The weights of the criterions and the alternatives are calculated using AHP method which is used as an input for TOPSIS analysis for prioritization of IT implementation.

In this study, the author examines organizations' perceptions of the importance of absorptive capacity attributes in the deployment of radio frequency identification (RFID) in a supply chain and their relationships with operational efficiency and market knowledge creation as moderated by information technology infrastructure integration and supply chain process integration. Data was collected using a survey questionnaire administered online to members of the Council of Supply Chain Management Professionals (CSCMP). Four proposed hypotheses were partially supported in this study. Both variables, IT infrastructure integration and supply chain process integration, moderate the relationships between three predictor variables, business process modularity, standard electronic business interfaces, and breadth of information exchange and the two dependent variables examined in this study, operational efficiency and market knowledge creation to a considerable extent. This study has clear implications for how decision makers affecting their firm's supply chains should make a business case for robust IT elements that support both IT infrastructure integration and supply chain process integration.



Historically, the growth of the beef industry has been hampered by various entities, i.e., breeders, cow-calf producers, stockers, backgrounders, processors, etc..., within the beef industry's supply chain. The primary obstacles to growth are the large numbers of participants in the upstream side of the supply chain and the lack of coordination between them. Over the last decade significant advances have been made in information and communication technologies, and many new companies have been founded to promote these technical advances. This research looks at both the upstream and downstream participants to determine the degree to which information technologies are currently being utilized and the degree that these new technologies have driven performance improvements in the beef industry's supply chain. Through surveys, the authors find that the beef industry does not use information technologies to their benefit and that the US beef supply chain is not yet strategically poised to enable the use of these technologies.



Amid the turmoil of the current economic crisis, the wild expectations for business-to-business electronic marketplaces or 'e-hubs' as transformative modes of exchange for all industries have subsided. However, e-hubs continue to elicit interest in industries such as car production. Yet, there is little research that investigates firms' strategies for e-procurement in the automotive industry and the potential benefits of e-hubs to them. This research re-examines the transition from bespoke electronic data interchange to generic electronic procurement, and conflicting motivations and complex barriers at firm and industry level are revealed. The article develops a framework that examines the benefits and barriers to firms joining e-hubs, applies the framework to the car industry, and proposes an e-procurement matrix that offers alternative strategies. Six cases from vehicle manufacturers and component suppliers demonstrate a shallow industry structure that lacks supplier integration, where a particular concern is the emergence of consortium e-hubs that combine a transactional approach for reducing price, with a collaborative approach that requires sharing knowledge. While this dispels the myth of simplistic e-commerce models, the governance of e-procurement across collaborative supply chains is still uncertain.


*Stefan Henningsson, Centre for Applied Information and Communication Technology, Copenhagen Business School, Denmark*
*Jonas Hedman, Centre for Applied Information and Communication Technology, Copenhagen Business School, Denmark*

Experiences from enterprise-wide integration initiatives during more than four decades indicate that industry-wide information integration could render substantial benefits. Two ways in which industry-wide integration differs from enterprise-wide integration are that there is no common management level and the economic units in the integration are the constituent units, not the industry. Management involvement has been emphasized as perhaps the most critical success factor for enterprise-wide information integration. The common economic unit enables increased costs in one part of the organization to lower the total cost in the company as a whole. In this article the authors address which consequence these two differences have for the development of information integration in four industry-wide supply chains. The authors find the existing methods for enterprise-wide information integration, such as BPR, virtually impossible to apply on industry-wide information integration and that the disjoint economic responsibility is a hampering aspect in reaching potential benefits of industry-wide information integration.


*Kenneth Saban, Duquesne University, USA*
*John Mawhinney, Duquesne University, USA*

Supply chain performance is often equated with acquiring the best technology or process. However, current studies suggest that supply chain performance also requires human collaboration. To change conventional thinking, this paper proposes a holistic approach to supply chain management (SCM), clarifies the forces that facilitate human collaboration, and identifies the steps management can take to create a more collaborative network.


*Frank Wolf, Nova Southeastern University, USA*
*Lee Pickler, Baldwin Wallace College, USA*

This paper examines how supply chain conflicts across domestic and international jurisdictions arise and become resolved given that conventional conflict resolution tribunals cannot effectively settle fast enough to serve the needs of supply chain partners. Observations from the field should guide practitioners, and in combination with information technology, may lead to best practice rules in dispute resolution. For

Manufacturers in a high-tech durable product industry may have to make operational decisions in the presence of uncertainties associated with product demand and supplier's wholesale price. In this paper, the authors investigate the impact of such uncertainties on the activities of a manufacturer and its supplier and develop an optimization model that describes how the manufacturers should reflect the uncertainty issues in their pricing and order quantity policy to achieve a desirable profit. In the modeling process, three important managerial problems are discussed, i.e., the effect of coordination between a manufacturer and its supplier in dealing with uncertainties on product demand and supplier's wholesale price, strategies for mitigating both errors in demand forecasting and supply risk, and modeling frameworks to determine the optimal solution for price and order quantity based on the varying levels of coordination. To identify best operational decisions under market uncertainty, the authors use the stochastic optimal control theory.


*Ibrahim Al Kattan, American University of Sharjah, UAE*
*Taha Al Khudairi, American University of Sharjah, UAE*

This paper employs a simulation model in a Supply Chain Management (SCM) system. This study is one of the first to present simulation model of inventory control system in supply chain management using barcode and Radio Frequency Identification (RFID). The main objective of this model is to compare two inventory systems in a supply chain, one using RFID, versus the barcode. The model will help company to consider moving from a barcode system to the RFID application. A quantitative analysis based on a simulation model is developed. The model runs for both systems using ARENA simulation software with a comparison between the two systems. Furthermore, the simulation model is tested by applying three different types of demand for both scenarios. The results have shown that regardless of demand distribution pattern and customer order rate, the outcomes of the model are consistent and provide promising RFID technology adoption to improve inventory control of the entire supply chain system. The installation and unit cost of RFID implementation were estimated and considered to be the main barrier. Such model can offer the policymakers insight into how RFID might improve SCM system performance. Additional test has been conducted for demand with normal and triangular distributions using real data provided by ABC-Dubai Company. The results obtained from running the two models for these distributions are consistent with the original results.


*N. Anbazhagan, Algappa University, India*
*B. Vigneshwaran, Thiagarajar College of Engineering, India*

This article examines a two commodity substitutable inventory system—two different brands of super computers under continuous review. The demand points for each commodity are assumed to form independent Poisson processes. The reordering policy is to place orders for both the commodities when the total net inventory level drops to any one of the prefixed levels with prescribed probability distribution. Lost sales are assumed during the stock out period. The lead time for a reorder is exponentially distributed with parameter( , depending on the size of the ordering quantity. The limiting probability distribution for the joint inventory levels is also evaluated. Various operational characteristics and total expected cost rate are derived. Numerical examples are provided to find optimal reorder quantity and band width .

# Preface

## INFORMATION TECHNOLOGIES, METHODS, AND TECHNIQUES OF SUPPLY CHAIN MANAGEMENT

### The Use of Information Technology in Supply Chain Management

Section one consists of four chapters. Chapter 1 entitled "Enterprise Applications for Supply Chain Management" is written by Susan A. Sherer. Coordination of information to effectively operate and manage a supply chain can be a source of competitive advantage today. In order to gather, utilize, and effectively share information, companies continue to invest in information technology, acquiring and integrating many different types of enterprise applications. The foundation of most enterprise information infrastructures today is the enterprise resource planning system (ERP) which has greatly reduced the number of applications required to track and share information across cross functional business processes. However, many companies supplement basic ERP systems with legacy, custom, and best of breed applications to optimize supply chain execution and management.

These systems not only help run and manage the business, but support collaboration among supply chain partners. In addition to systems that primarily support traditional supply chain processes such as source, plan, make, deliver, and return, businesses must integrate information from systems that manage supporting processes such as design and engineering and customer focused applications. Applications such as business intelligence, and productivity, communication, and collaboration tools, must also be part of an enterprise information infrastructure for supply chain management. Without integrating these applications, effective analysis and information sharing cannot be achieved.

Most previous research has focused on subsets of a complete information systems infrastructure, for example, supply chain planning or customer relationship management or logistics or business intelligence systems. The main contribution of this chapter is that it brings together in a single framework all the various applications that are required to build an enterprise information infrastructure for supply chain management. The chapter organizes and describes all the different applications that make up an information infrastructure for supply chain management including ERP; e-procurement, supplier relationship management, and B2B marketplaces; advanced planning and optimization, and collaborative planning, forecasting, and replenishment; computer aided manufacturing; content providers of freight ratings, import/export compliance, and environmental health and safety; manufacturing integration and intelligence; vendor managed inventory systems; warehouse and transportation management systems, and transportation exchanges; reverse logistics tracking and management; computer aided design and collaborative product development and product lifecycle management; customer relationship management; content management, productivity tools; and business intelligence. Chapter 1 includes examples of the use of the different types of applications. And it provides an overview of the different applications to assist in understanding the scope of applications needed to develop a complete information systems infrastructure for competitive supply chain management capabilities.

In Chapter 2, Debendra Mahalik and Gokulananda Patel explore "Information Technology Implementation Prioritization in Supply Chain: An Integrated Multi Criteria Decision Making Approach". In today's competitive business environment information technology (IT) is the key enabler for business survival and success. The role of IT has undergone a sea change from data accusation and processing to the function which supports and drives each components of business. Companies implementing Supply Chain Management (SCM) are looking for more responsiveness, greater flexibility, better supplier relations and improved customer satisfaction, which can be achieved through efficient IT implementation. Application of IT in different process of supply chain as an important and essential ingredient has already been felt by academic researchers as well as practitioners, resulting in concepts like Information Technology enabled SCM (ITeSCM). The real challenge lies in finding a way for the IT implementation in real situation, as SCM has number of process starting from supplier to the end customer through various stages. Some of the strategies taken for IT implementation in SCM have shown results which are more failure or partial successes in nature. The reasons may be many; some of them are implementation in one go, quick implementation, non availability of clear rules etc. Over a period practitioner are adopting a component approach for a success in implementation. In this approach also they are in a dilemma in selection of component in the whole Supply Chain for IT implementation. As the decision of component is a complex due to presence of multi criteria decision making and solving it in a traditional way does not always give better results. This calls for a better and more scientific approach.

Mahalik and Patel consider five components viz. Materials Management, Purchase Management, Production Management, Logistics & Distribution function and Customer interface for prioritization of computerization in SCM. These five components are considered as five alternatives before the company. The study is carried out in a company which is largest manufacturer of aluminum in India having fourteen plants and ten offices spread across the country. The study first identifies fifteen parameters viz. Process Improvement, Speed, Easy, Reliable, Productivity, Information availability, Accurate, Focus on core work, Resources utilization, Secure, Uniform Standard, Cost, Transparency, Corporate image and Environment through literature review as reasons to go for computerization in SCM. A structured questionnaire is prepared based on these fifteen variables and data is collected from the executives of the company. The collected data are subjected to factor analysis and three factors according to the importance like primary attributes, secondary attributes and tertiary attributes are identified.

It is found that there are five primary reasons for IT implementation in supply chain. The five primary reasons identified are Process Improvement, Speed, Easy, Reliable and Productivity. These five identified variables are considered as the criterion for evaluating the five alternatives before the company. Analytic Hierarchy Process (AHP) which deals with complex systems for a choice among several alternatives is used to find weights of criteria and alternatives. The weights so calculated are used as an input to for TOPSIS (Technique for Order Preference by Similarity to Ideal Solution) analysis. TOPSIS is a very effective method in multi attribute decision analysis. It uses normalized matrix to find the superior project and inferior project (that is ideal solution and non-ideal solution), then calculates the distances of other projects from the ideal and the non-ideal solution. Then the relative closeness to the ideal solution is calculated. The relative closeness can be ranked in descending order. The relative closeness can be put in the range of 0-1, if it is close to 1, evaluating projects is said to be closer to the ideal. The alternatives are ranked on the basis of value of closeness coefficient which is calculated for each alternative. According to the closeness coefficient, the ranking order of five alternatives is found to be - first computerize the materials management department followed by computerization of customer interface. The other departments as per the order for computerization shall be purchase, logistics and production department respectively. The management of any similar type of company can use the procedure for their IT implementation prioritization. The novelty of the chapter lies in integration of AHP and TOPSIS method for IT implementation prioritization.

In Chapter 3, Rebecca Angeles's survey study planned radio frequency identification (RFID) implementation by member firms of the Council of Supply Chain Management Professionals (CSCMP) focuses on the moderating role of information technology (IT) infrastructure integration and supply chain process integration between absorptive capacity attributes and two deployment outcomes --- operational efficiency and market knowledge creation. Applying the moderated regression data analysis method, the key finding of this study is that of the two proposed moderator variables, supply chain process integration more strongly moderates the relationship between three key absorptive capacity attributes that turned out to be significant --- business process modularity, standard electronic business interfaces, and breath of information exchanged and the two outcome variables, operational efficiency and market knowledge creation.

Use of standard electronic business interfaces for RFID brings several benefits. First, firms could develop a well-defined, robust, and tested RFID solution that has been tried and tested in the real world. Second, firms could avoid the costs of lengthy development efforts and vendor lock-in through the use off-the-shelf RFID software solutions that conform to strict specifications for solution components. Third, firms can use RFID software solutions that are compatible with a wide range of related software systems. The most recognizable standards are mentioned here. EPCglobal has developed standards for designing, implementing, and adoption electronic product code (EPC) and the EPCglobal Network, which specifies RFID specifications for supply chain operations. There is also the Electronic Product Code Information Service (EPCIS), which is a specification for a standard interface for accessing EPC-related information such as unique serial numbers that allow firms to track objects and collect real-time data about them.

Modularity of business processes is key to managing complexity and flexibility in global supply chains. This is crucial in today's digital marketplace where there is a need to integrate related business processes executed in a loosely coupled environment where messages are exchanged among supply chain trading partners across diverse IT platforms. Models are needed with business process modules that represent important supply chain workflows, rules, practices, and exception handling in a flexible manner.

The use of both standard electronic business interfaces and RFID business process modules is important in generating data of greater breadth than what is provided by routine transactional events. An important goal is to support RFID-generated intelligence that can deliver actionable data that decision makers can use for critical issues such as inventory levels, stock locations, delivery rates, missing or stolen products, etc. In addition to these operational data, though, customer-centric information that can be gathered from an RFID-enabled retail outlet, for instance, that addresses questions like --- "What are the store behaviors of my customers?" "What are they looking at in the store?" "What are they bringing to the fitting rooms?" "What sells and what doesn't sell?" "Which products convert to actual sales?" --- have the potential to extend market knowledge and support higher-level intelligence-based decisions.

This study is only one in a series of RFID research projects the author has conducted which feature the following topic areas: (1) "Effects of Reciprocal Investments and Relational Interaction in Deploying RFID Supply Chain Systems"; (2) "Moderated Regression: Effects of IT Infrastructure Integration and Supply Chain Process Integration on The Relationships Between RFID Critical Success Factors and System Deployment Outcomes"; and (3) "Moderated Regression: Effects of IT Infrastructure Integration and Supply Chain Process Integration on the Relationships between RFID Adoption Attributes and System Deployment Outcomes."

George Kenyon and Brian D. Neureuther make "A Comparison of Information Technology Usage across Supply Chains: A Comparison of the U.S. Beef Industry and the U.S. Food Industry" in Chapter 4. Cattle ranching have long held a nostalgic value in the America consciousness; it also has significant health and monetary value as part of the U.S. food supply. Historically, the growth of the beef industry has been hampered by the various entities (breeders, cow-calf producers, stockers, backgrounders, processors, etc...) within the beef industry's supply chain. Traditionally, the participants of the U.S. beef industry have been very independent of each other; making communication and coordination efforts difficult. Unfortunately, over the last several decades, this independence and the economics of the marketplace has made it increasingly harder for ranchers to make a profit. One of the primary obstacles to growth and profitability in the beef industry is the large number of participants in the upstream side of the supply chain. Another obstacle is the lack of coordination between participants in the beef industry's supply chain.

Since the mid-1980's, significant advances have been made in information and communication technologies. These advances have enabled supply chain members to improve their ability to transfer information and to coordinate activities. In many industries the benefits of this improved capability has resulted in improved profitability. Many new companies have been founded to promote these technical advances in the beef industry; but have the profitability of ranching and processing operations in the beef industry improved?

Kenyon and Neureuther's research used data collected by a survey to analyze the performance of food processes and food manufacturing plants in the US to determine the degree to which information technologies are being utilized and the degree to which these new technologies have driven performance improvements in the beef industry's supply chain. Though there is significant usage of information technologies in the downstream portion of the food manufacturing supply chains, there has not been any significant improvement in plant performance. The usage of online technologies by upstream participants in the beef industry's supply chain has also failed to make any significant impact to the profitability of ranches. These findings raise serious concerns for the future of beef production in the United States. If the United States wants to continue being able to food its population and avoid becoming dependent upon foreign food sources, greater awareness of the infrastructural requirements of our agricultural supply chains needs to occur.

## Supply Chain Collaboration

Section two consists of five chapters. In Chapter 5, Mickey Howard, Richard Vidgen and Philip Powell discover "Strategies for E-Procurement: Auto Industry Hubs Re-Examined". While the wild expectations for business-to-business electronic marketplaces or 'e-hubs' have subsided, they continue to elicit interest in industries such as car production. Technology exists to create electronic marketplaces - 'one stop' procurement and product development hubs - but a key barrier to their development is collaboration amongst a critical mass of industry players. Howard, Vidgen, and Powell re-examine the received wisdom of the transition from supply chain to electronic hub. The initial proliferation of e-hubs by manufacturers and component suppliers is explained by long-term e-commerce developments by some firms, countervailing power exercised by suppliers concerned over their lack of representation and the desire to develop core competence in electronic markets.

The research confirms that a consortium-based approach is adopted if price and control are the prime motivations. If supply chain management, supplier development and product innovation is of greater importance, then a private-based e-hub is adopted. In the auto industry, the introduction of e-commerce resulted in a mess of overlapping networks. This was a shallow structure, lacking in supplier integration and resembles an ad-hoc arrangement of spokes rather than a singular hub design. While suppliers worried over complexity in information exchange of using electronic media, an inter-firm system was introduced that combined a transactional approach for reducing price with a collaborative product development approach that required sharing organizational knowledge. The apparent strategies in the industry were not long-term. Manufacturers and suppliers cannot sustain a system characterized by duplicated services, multiple standards and restrictions on membership.

The chapter develops a framework that examines the benefits and barriers to firms joining e-hubs, applies the framework to the car industry, and proposes an e-procurement matrix that offers alternative strategies. Six cases from vehicle manufacturers and component suppliers demonstrate organizations were taken in by simplistic, technology-driven models professing to enhance competitive and co-operative capability. In moving from bespoke supply chain to electronic hub, this research found multiple barriers to information sharing, overlapping networks competing for membership, and isolated pockets of collaboration. This transitory phase involved conflict in motivation between price, collaboration and membership exclusivity, which prevented e-hubs from becoming fully integrated and the industry realizing full benefit from electronic markets. An e-Procurement strategy matrix highlights the difficulties of implementing e-market strategies that attempt to reconcile the mutually exclusive nature of buyer-supplier relationships: price versus collaboration, and inclusive versus exclusive hub membership.

If management commitment is the most critical success factor for information integration, what then if there is no management level? In Chapter 6, Stefan Henningsson and Jonas Hedman from Copenhagen Business School make an interesting exploration into why many industries are not optimally integrated in their information flows. Experiences from enterprise-wide integration initiatives during more than four decades indicate that industry-wide information integration could render substantial benefits for the industry as a whole. Yet, many industries' supply chains are only marginally integrated across individual actors.

The authors find two fundamental ways in which industry-wide integration differs from enterprise-wide integration: there is no common management level, and the economic units in the integration are the constituent units, not the industry. Management involvement has been emphasized as perhaps the most critical success factor for enterprise-wide information integration. The common economic unit enables increased costs in one part of the organization to lower the total cost in the company as a whole.

In the search to understand how these two particularities affect the industry-wide information integration, Henningsson and Hedman do in-depth investigations of the consequences in four agrifood industries, based on milk, sugar, pea, and pork production. They find that: Information integration was dominated by supply chain captains, dominant companies in the middle to the supply chain that controlled the integration agenda. Information integration that would have been desirable by other actors was hardly ever implemented. To the extent other actors information integration needs were met it was due to legislation or other governmental demand. Asynchronous costs and savings due to perceived or real difference in backward and forward integration worked hampering on the industry-wide integration. As long as the complete chain is within one economic unit this is not an issue, but as companies exists at different places their benefits are not equal and if possible the unbeneficial contest integration. One way information integration was used to ensure availability of key resources and as lock on less powerful actors in the supply chain.

The authors make the eye-opening conclusion that the fundamental differences in the structure of the unit to be integrated, make traditional methods for enterprise-wide information integration, such as BPR, virtually impossible to apply on industry-wide information integration and that the disjoint economic responsibility forms a severe challenge in reaching potential benefits of industry-wide information integration.

Kenneth Saban and John Mawhinney provide a unique view of an ever growing business concern among SCM in Chapter 7 - "The Strategic Role of Human Collaboration in Supply Chain Management". The willingness and ability of the people that comprise the human capital of business organizations to collaborate within and between enterprises is proving to be the next frontier of competitiveness. Many have looked upon this challenge as one focused on the need for more and better technology, yet while technology and systems tools continue to develop at warp speeds, collaboration within and between businesses is slowly progressing. This research uncovers six factors that facilitate human collaboration and provides the foundation for the premise that it is the human element that drives collaborative technology and processes.

The relationships of these factors are depicted in the human collaboration model which provides clarity and logic to each component. The model establishes the framework for the four step managerial enhancement methodology to create a collaborative environment within an enterprise. While it has long been understood that internal collaboration is the first building block to inter-enterprise supply chain collaboration, this work confirms the critical factors that must be addressed to establish a human collaboration culture.

This work establishes the importance of the human element in the collaboration model, identifies the critical factors of a successful collaborative environment, and provides logical process for addressing these issues. The authors have bridged the divide between theoretical research and practical application in this innovative work.

In Chapter 8, Frank Wolf and Lee Pickler present "Supply Chain Dispute Resolution: A Delphi Study". Wolf and Pickler noticed that conflict resolution in the field of SCM is different from conflict resolution in otherareas of business conflict. In a supply chain, there is a pressing need for swift settlement of disputes in order to preserve the larger good of members along the chain. Even less formal forums like mediation and arbitration can be too slow and too costly to be of practical value. Just as business bankruptcies are in many cases preventable, supply chain conflict may be preventable by knowing where to look. The business relationship suffers dearly if the conflict resolution process becomes too legal. With the average supply chain having four tiers, any dispute between any two levels calls for direct settlement between the parties. Only when the nature of the dispute affectsall parties, will the supply chain anchor step in to preserve the greater good.

The subject of supply chain dispute resolution is not well developed, yet enormously important. Nature itself provides many reasons for supply interruptions. The prevention of man-made supply chain failures is a productive research endeavor, and the purpose of this work. Preventive measures include frequent communication between parties, coordination, supply chain re-design and cultural awareness, information technology and freed back. The Delphi Method requires multiple interactions with the same survey participants, plus feedback at each iteration. Getting participants to sit still that long can test their patience, unless the feedback is quick and benefits the parties. Their recommendation is to speed up the process in order to not lose survey members. In this study the members were executives in aerospace, food marketing, defense and materials, import-export.

Supply chain managers should pay close attention to the needs of each participant and orchestrate what are typically 4 levels/tier chains, into a well functioning whole, supported by Information Technology and frequent contact between the parties. Disputes must be settled rapidly and on the spot if possible. Common dispute issues are the day-to-day payments, quality, and pricing of products and services. For issues like accidental patent infringement for example, New York law applies worldwide, as it does to most commerce. Arbitration and mediation, voluntary or contractually mandated are alternative methods for settlement. Governmental help is not usually wanted by the parties. United Nations trade harmonizing efforts like UNCITRAL are not well known.

Chapter 9 entitled "Fair Distribution of Efficiency Gains in Supply Networks from a Cooperative Game Theory Point of View" is written by Stephan Zelewski and Malte L. Peters. One principal aim of supply network management is to realize efficiency gains by coordinating the activities of all actors in a supply network. When efficiency gains are realized in supply networks, a distribution problem arises. The cooperating actors know that they are realizing efficiency gains by mutually coordinating their activities. Each actor is interested in maximizing his own gain probably even at the expense of the other actors in the supply network. Therefore, supply networks suffer from a built-in conflict between cooperation and defection. The problem lies in distributing efficiency gains among the actors in a manner that the actors will accept as fair and thus find it advantageous to cooperate with each other.

The contribution presents a solution concept involving τ-value from cooperative game theory for the problem of distributing efficiency gains in supply networks that aims at distributing efficiency gains in a fair manner. The first main contribution of the work is that the proposed method determines a unique solution of the distribution problem. This unique solution is a recommendation for allocating the efficiency gain among the actors of a supply network in a fair manner. The second main contribution is that the proposed method ensures that the distribution results can be communicated easily. The distribution results are obtained by successively restricting the solution space. This successive procedure facilitates the communication of the distribution results. This is a major advantage over other solution approaches for distribution problems from cooperative game theory. These are the Shapley value and the Nucleolus. These solution approaches suffer from a serious drawback: The fairness of the distribution results is hard to justify, since the 'logic' of these approaches is difficult to communicate. The third main contribution is that the application of the proposed method prevents a supply network from collapsing, since the actors accept the distribution results as fair and thus they remain in the supply network.

Over the past several years, a lot of research has been done on finding ways of realizing efficiency gains by coordinating the activities of all actors in a supply network. However, the problem of how to distribute the realized efficiency gains among the actors in a manner that the actors will accept as fair has been neglected in the specialized literature. Therefore, it has been important to fill this scientific gap. However, the readers should bear in mind that up to now there is no well tried software implementation of the proposed method available.

Other main contributions in this research area have been developed in two directions. On the one hand, the concept of the τ-value has been covered extensively in a monograph and has been developed further to cope with incredible threats as well as with insignificant actors; see Zelewski, S. (2009). *Fair distribution of efficiency gains in supply webs: a game theoretic approach based on the τ-value* (in German). Berlin: Logos. On the other hand, a generalized version of the τ-value, the χ-value, has also been applied on the problem of distributing efficiency gains; see Jene, S., & Zelewski, S. (2011).

Fair Distribution of Profit in Supply Chains. In W. Kersten, T. Blecker, & C. Jahn (Eds.), *International Supply Chain Management and Collaboration Practices* (pp. 299-313), Lohmar et al.: Eul. Jene, S., & Zelewski, S. (2011). Distributive Justice in Supply Chains – Fair Distribution of Collectively Earned Profits in Supply Chains. In E. Sucky, B. Asdecker, A. Dobhan, S. Haas, & J. Wiese (Eds.), *Logistics Management* (in German), Vol. III (pp. 115-132), Bamberg: University of Bamberg Press.

## Production Planning and Inventory Management

Section three has four chapters. In Chapter 10, "Dynamic Price and Quantity Postponement Strategies", Yohanes Kristianto examined why the price and quantity postponement affect oppositely over the supply chain profit and recommend ways of using those two postponement strategies. The research draws attention to the fact that in mass customized manufacturing, the choices of contract type are important to maximize the profit, depending on the product commonality. The dynamic behavior analysis in this chapter helps decision makers to decide their long term postponement policy with regard to their manufacturing types, namely make-to-stock or make-to-order. The analysis results also support both modularity and customization principles in mass customized products, where decision uncertainty can be reduced by making closer customer order decoupling, point to sales point.

Chapter 10 suggests that product developers design common platform products and decide the price according to customer specific requirements. A practical evidence of the proposed model comes from an example of airline pricing strategy retained its market share by postponing the ticket price just after the demands are observed. Further investigations reveal that this market share is also retained by increasing the commonality amongst products which regards the part interchangeability as a main ingredient of product substitutability.

It is recommended: Price postponement is superior to production postponement at many respects. This type of contract guarantees profit stability and at the same time supports the product standardization effort. Price postponement is also a dominant strategy for substitutable products. This conclusion is at odds with the previous literatures on price and production postponements. This discrepancy is caused by the previous literatures perhaps assuming that in Bertrand price-like competition, the quantity setting will avoid both firms having to reduce their production quantity further. On the contrary, this chapter assumes sticky prices and quantities, where it pushes both firms to cooperate at higher levels. By sticky prices and quantities, this chapter is more appropriate for common platform based products instead of two widely differentiated products. Production quantity postponement (make-to-order) is a dominant strategy for highly differentiable products. This conclusion supports the previous literatures, who discuss postponement strategies differentiation according to their applicability.

In Chapter 11, Seong-Hyun Nam, Hisashi Kurata, John Vitton and Jaesun Park show "Determining Optimal Price and Order Quantity under the Uncertainty in Demand and Supplier's Wholesale Price". In the turbulent business environment of today, it is critical for manufacturers in a high-tech durable product industry to understand how to make operational decisions in the presence of unexpected or unknown variables associated with product demand and supplier's wholesale price. Nam et. al. proposed in their chapter a very useful and practical optimization model that helps manufacturers deal with such production related uncertainties. The model developed in Chapter 11 focuses on how the manufacturer should reflect those uncertainties in their pricing and order quantity policy in order to achieve its desirable profit. In the process of modeling, the authors integrated two major variables of forecasting error associated with demand uncertainty and risk incurred from unpredictable changes in supplier's wholesale price. They also showed the stochastic steps to reach optimal decisions on price and order quantity based on the different levels of coordination between a manufacturer and its supplier.

Chapter 11 contains two important managerial implications. First, they found that there may be significant performance improvements of manufacturers from creating a collaborative work relationship with their suppliers. If the use of a coordination mechanism mitigates the uncertainties associated with product demand and supplier's wholesale price, then manufacturers may be able to have increased power to make the right operational decisions. Although high levels of coordination require an investment cost, an important advantage of the close interaction with suppliers may be the manufacturer's superior decision-making capability. Second, the findings of the model enable manufacturers to formulate their optimal pricing and ordering strategies to handle production related uncertainties. If the optimization policy with a quality coordination mechanism can lead to improved operational performance associated with product availability, then manufacturers may be able to offer better customer service, thus improving sales. Recognizing research with a focus on supply-side uncertainty seems to be limited in the SCM field. This chapter contributes to the understanding of the effect of supplier's wholesale price uncertainty on a manufacturer's operational decisions.

Ibrahim Al Kattan and Taha Al Khudairi perform a "Simulation of Inventory Control System in a Supply Chain Using RFID" in Chapter 12. Most companies using SCM are interested in new technologies such as RFID application to track information of assets and its utilization, reducing inventory and labour cost. The cost and the payback period of RFID implementation are one of the most important factors facing any company in adopting this technology. Recently, the cost RFID implementation became more acceptable and has shown significant benefits in cost saving and service enhancement. However, the lack of transparency of costs/benefits, data collision represents some of challenges for RFID.

Chapter 6 does examine the impact on customer service, as expressed in terms of the number of lost customer and its impact on the company's market. This is actually an area that is likely to be of interest to companies seeking an RFID solution. This research used simulation model to compare the saving of labor cost during scanning operations, reduction in inventory cost and loss of customer's good-will. Two scenarios with and without use of RFID were developed using computer simulation modeling to evaluate supply chain system to assess the benefit of RFID. These two models were tested using RFID technology and the current barcode used by ABC Company. Subsequently, several comparisons have been accomplished by measuring both the total inventory cost and customer satisfaction for the entire supply chain storages for both systems.

A quantitative analysis based on a simulation model used to compare two scenarios through the entire inventory systems and customer service for ABC Company. The installation and unit cost of RFID implementation were estimated and considered to be the main barrier. The model can offer the policymakers insight into how RFID might improve SCM system performance. The RFID risk and challenges are in technology, Return on Investment (ROI), privacy/security and implementation. The models make possible for company consider moving from a barcode system to the RFID application. The application showed the RFID solutions face similar problems as they would with adoption of other technology solutions, including general corporate fear of change and unclear implementation costs and processes. Users should also expect to maintain dual systems and processes (existing and new RFID systems) for several years to work out issues and wait for supply chain partner participation.

In Chapter 13, N. Anbazhagan and B. Vigneshwaran demonstrate a "Two-Commodity Markovian Inventory System with Set of Reorders". With the fast expansion of activities in Business and Industrial sectors, many inventory systems are increasingly found to operate with more than single commodity.

These systems unlike those dealing with single commodity, involve more complexities in the reordering procedures. In the modeling of such systems, initially models were proposed with independently established reorder points. But in situations where several products share the same transport facility or are procured from the same source, the above method overlooks potential savings associated with joint ordering and hence may not be optimal. This motivated the researchers to build a multi commodity especially two commodity inventory models. In this aspect the authors developed two commodity inventory control system with coordinated and joint ordering policies.

Recently the authors developed a model for associated commodities like products like specific blood group and its accessories for transfusion of blood, computer and table, television and holding stand. The challenge of managing a production/inventory system in the presence of random demand and lead time has inspired a considerable amount of research effort in recent years. Marketing of consumable goods which has high obsolete rate is a challenging task that has to be dealt carefully. Incentive for bulk purchase of prime commodity together with associated gift article is one strategy to boost sale volume at the same time to avoid the risk of getting obsolete state. This motivated the researchers to build inventory models dealing with two items in stock, of which one is regular with stochastic demand and the other one is a gift item supplied to a customer for the bulk purchase of the regular item. The bulk order scheme is actually equivalent to price break for bulk purchase, but it differs from classical price break models because in their model, the sales promotion for gift article is taken into account.

The significance of modeling such stochastic systems can be directly attributed to the severity of their potential negative effort on operating costs and customer service measures in modern manufacturing and business environments. Previously, Anbazhagan and Vigneshwaran assumed a model with a joint ordering policy which places orders for both commodities whenever the total net inventory level drops to a prefixed level. The demand points for each commodity form independent Poisson processes and the lead time is distributed as negative exponential. Unit demand for both commodities is assumed for efficient use of transaction reporting system control. Here the authors have extended the above model by assuming the set of reorder levels with prescribed probability distribution for reordering. This is a more realistic model in which the vendor or manager has the freedom to reorder the required quantity with probabilistic stint of the random environment. The author has received Young Scientist Award (2004) from DST, New Delhi, India, Young Scientist Fellowship (2005) from TNSCST, Chennai, India and Career Award for Young Teachers (2005) from AICTE, India. He has successfully completed one research project, funded by DST, India.

## Logistics

Section four has three chapters. In Chapter 14, Reza Farzipoor Saen takes "A New Look at Selecting Third-Party Reverse Logistics Providers". Many companies have realized that their core competencies are not in the logistics-field, and have progressively sought to buy logistics services and functions from third-party reverse logistics (3PL) providers. The 3PL providers play a role in helping organizations in closing the loop for products. Traditionally, reverse logistics is an activity within organizations delegated to the customer service function, where customers with warranted or defective products would return them to their supplier. The outsourcing of non-core processes and activities makes it possible to focus on core manufacturing activities, while, at the same time, 3PL providers have specific logistics core competences, and they can manage logistics processes more efficiently than their customers.

The use of Data Envelopment Analysis (DEA) in many fields is based on total flexibility of the weights. However, the problem of allowing total flexibility of the weights is that the values of the weights obtained by solving the unrestricted DEA program are often in contradiction to prior views or additional available information. Also, many applications of DEA assume complete discretionary of decision making criteria. However, they do not assume the conditions that some factors are nondiscretionary. To select the most efficient 3PL provider in the conditions that both weight restrictions and nondiscretionary factors are present, a methodology in the context of DEA is introduced.

Chapter 14 is the first attempt that discusses 3PL provider selection in the presence of both weight restrictions and nondiscretionary factors. The contributions of the chapter are as follows: The proposed model does not demand exact weights from the decision maker. In traditional models of 3PL provider selection, the weights are allocated in a crisp value, while in the proposed model; weights are defined in an interval. It is clear that interval definition of the weights for the decision maker is easier than the crisp weight assignment. The proposed model considers nondiscretionary factors for 3PL provider selection. The proposed model considers weight restrictions for 3PL provider selection. Weights restrictions and nondiscretionary factors are considered simultaneously. A numerical example demonstrates the application of the proposed method.

In Chapter 15, Bjørnar Aas and Stein W. Wallace concentrate on "Management of Logistics Planning". We are all living in a world where information availability is steadily increasing. At the same time, our capability to solve complex problems is increasing more or less at the same pace. This is at least true when we look at the big picture. However, snapshots at a micro level usually show a significant bias in one or the other direction. In a business setting, such a bias typically represents waste of resources, low customer service levels, lost business opportunities and so on. Companies, departments or employees fail to utilize existing information, or the necessary information is lacking.

The general understanding and acceptance of this reality description is not a topic of much discussion, but when addressing specific problems, for instance at a task-level, to obtain an objective conclusion is very often extremely difficult. The analysis would typically be complicated by a lack of data to establish a good picture of the present situation. To make it worse, this is very often accompanied by the presence of a number of more or less well-qualified subjective interpretations of the situation. Further, the problem that is addressed is usually interacting with several other problems with their corresponding, often contradicting, goals, unclear priorities and so on.

Aas and Wallace's experience from working with industry for many years is that the situation described is present in many different facets within all industries and companies. However, for this chapter, experiences and examples from logistics in the Norwegian oil & gas industry have provided the authors with much empirical evidence. In Chapter 15, the particular challenges of logistics management are addressed by exploring how logistics planning should be managed in order to be as efficient as possible. This is done by establishing the availability of information and problem-solving capability as the core parts of logistics planning. Thereafter, a conceptual model for the management of logistics planning is proposed and discussed.

With this research, the authors have managed to successfully establish an innovative conceptual contribution within the field of logistic management which would be extremely useful, particular at operational level. Notwithstanding, the contributed knowledge of this chapter has also been proven to be an excellent contribution when teaching logistics.

In Chapter 16, Duangpun Kritchanchai, Albert Wee Kwan Tan and Peter Hosie make "An Empirical Investigation of Third Party Logistics Providers in Thailand: Barriers, Motivation and Usage of Information Technologies". The authors observed that Third Party Logistics (3PL) in Asia has emerged as an important trend in logistical management. In particular, 3PL in developing countries continue to develop rapidly and gain importance. While a great deal has been written about the usage of information technology (IT) in general, there is still a lack of research specifically in the field of small and medium size 3PLs. In particular, there have been few empirical investigations into the use of IT in relation to 3PLs in Asia. The authors of this chapter have conducted an empirical study designed to investigate the profiles of 3PLs in Thailand and their company strategies for providing logistics service and their use of IT.

Their research has provided a good overview of 3PL companies in Thailand, mainly SME and their markets and IT implementation status. The majority of companies surveyed have implemented the basic IT systems but in order to compete with the larger companies and they will need to consider investing more in this area as well as in IT manpower. These companies according to the authors should develop an IT system that is flexible and able to accommodate legacy data, Internet and related web-based services logistics information systems. Web-based information will help to reduce communications barriers such as complex logistics operational systems.

According to the authors, 3PL companies in Thailand and other developing countries receive little support for training and education. Government support is needed to train and educate employees for these smaller logistics companies. In addition, educating the customers on how to use the logistics information system will certainly help to improve customer satisfaction.

Having conducted the survey in Singapore and Thailand, the authors have replicated the surveys in Malaysia, United Arab Emirates (UAE) and Finland so that comparison can be made among the countries involved. The results gathered from these surveys are largely similar in terms of barriers and motivation. The findings are published in journal papers subsequently.

## Supply Chain Monitoring and Performance Management

Section five has four chapters. In Chapter 17, Dimitris Folinas and Ioannis Manikas conduct "Design and Development of an e-Platform for Supporting Liquid Food Supply Chain Monitoring and Traceability". Developing a traceability system is a very challenging and demanding effort. Such a system must be able to file and communicate information regarding product quality and origin, and consumer safety. Moreover, it must integrate a number of functionalities / features. The main features of such a system include adequate filtering and extracting of information from available databases and harmonization with international codification standards, Internet standards and up to date technologies. But first of all, it requires the existence of a conceptual framework, a model that fulfills all the above features.

The development of such a model requires the identification and analysis of each stage of the supply chain for each one of the product categories under study, from farm to fork, including all factors that affect quality, such as packaging materials, agrochemicals, antibiotics, fertilizers, climate, soil etc. It also requires the modeling of the main entities of the proposed framework in order to design and develop the web-platform in a reliable and effective manner. In this study, the authors develop a reference model and a respective web platform for traceability management for liquid food by establishing and modeling these basic concepts and features. The proposed web platform aims to support efficiently food traceability by monitoring and administering the data gathered from the various production and logistics processes along the supply chain.

Another critical aspect for the development of any information system is also its testing in real life cases. Thus the lessons learned for the implementation of the proposed web platform in a particular dairy production line in Greece is also presented and analyzed. Synoptically, the following outcomes were emerged by the above application; industrial stakeholders - especially production and quality assurance managers - noticed that the main benefits derived from implementing the developed traceability management system were: user friendliness, since it required more business than technical background by the users, pragmatic identification of traceability requirements, and risk reduction. Moreover, the system functionality was based on data already available from databases that support HACCP and ISO standards, while data communication tools (RFID, EPC) are based in EAN-UCC standards. Data processing uses XML technologies and information filtering is achieved by implementing the six elements model (PML) presented and analyzed above. On the other hand, users with technical background observed that the main feature of the proposed platform is the simplicity in use and the ability of communicating information through commonly accessible means such as the Internet, e-mail, and cell phones.

Chapter 18 entitled "A Composite Method to Compare Countries to Ascertain Supply Chain Success: Case of USA and India" is written by Mark Gershon and Jagadeesh Rajashekharaiah. The operation of the global supply chain contributes to the success or failure of all multinational corporations. Most analysis assumes that the structure of that supply chain, its locations and facilities, are in place. But the bigger issue is one of design. In which countries should the main supply chain facilities be located?

To answer this question, this chapter provides a methodology for comparing the required infrastructures, both physical and political, of countries prior to making the location decision. India and the United States are the two countries used in the chapter to demonstrate the approach. Three methods are provided and applied to the example countries. These are the Global Comparative Index (GCI), the Analytical Hierarchy Process (AHP) and Data Envelopment Analysis (DEA).

For each country, supply chains are assessed regarding the contribution that they make to improving business processes, return on investment, and the functionality of the entire supply chain. The first major contribution is the set of metrics that define success, factors to consider in evaluating the supply chain. Both quantitative and qualitative measures are used, with the quantitative measures divided into financial and non-financial measures. The composite method provided is a multi-level approach that combines the three methodologies. The goal is to begin at the global level, and finally refine the analysis to the level of the firm. At the highest level, it uses the GCI to decide about the suitability of the country for establishing a supply chain. Next, AHP allows the decision makers to compare partnering countries using pair wise comparisons. Finally, DEA is used to create a series of inputs and outputs for the comparison of each country.

The application showed the USA to be ranked higher than India. This was due to its advantages in infrastructure, and a weighting of the factors to emphasize the importance of infrastructure. India ranked close to the USA because of its large and highly skilled workforce. The DEA helped to identify key areas for improvement. For example, India needs to improve both reliability and capacity to pass out the USA. The main result is not any conclusive opinion about any particular countries. Instead, it is the methodology for making such comparisons.

In Chapter 19, Firat Kart, Louise E. Moser, and P. M. Melliar-Smith display "An Automated Supply Chain Management System and Its Performance Evaluation". The world of commerce is moving rapidly towards global just-in-time manufacturing and build-to-order products. These business processes reduce inventory carrying costs, but can increase logistic and administrative costs. Manufacturers must be able to respond quickly to requests for quotations, and must be able to negotiate prices, availability, and delivery times with their potential suppliers. Even so, their supply chains might be disrupted unexpectedly, such as by earthquakes in Japan or by floods in Thailand. Human negotiators are too slow, and are too costly because of the highly variable workloads.

SCM involves information flow, product flow, and financial flow. This chapter describes MIDAS, an automated SCM system based on the Service Oriented Architecture that focuses on information flow in the supply chain. At each manufacturer, MIDAS maintains a database of the components and materials required for each product. MIDAS also maintains globally a registry of suppliers. A manufacturer can consult the registry to discover potential suppliers of components or materials, and to discover alternative suppliers. This information can be regarded as a virtual inventory that replaces the traditional physical inventory. MIDAS contacts several potential suppliers to obtain quotations for the prices, availability, and delivery times of a product, and also for packaging, transportation, storage, finance and insurance, before selecting a supplier and placing an order.

The processing of an order from a customer involves two phases, a Waiting phase in which the order can be aggregated with orders from other customers, and a Quotes phase in which quotations are obtained from alternative suppliers for the components and materials required to fill the order. MIDAS is based on the premise that it is the customers' satisfaction that counts. The durations of the Waiting phase and the Quotes phase are investigated to balance order completion time and to obtain the best quotation.

There are many challenges to be faced in establishing an automated SCM system, particularly in agreeing on a standard and in securing its widespread adoption. MIDAS represents a first step in that process, of great interest to those who will design the standard, and also to those awaiting the widespread availability of an automated SCM system.

In Chapter 20, Reza Saen and Mark Gershon describe "Supplier Selection by the Pair of AR-NF-IDEA Models". In a world where companies are trying to reduce the number of suppliers in their supply chains, the problem of how to select the optimal set of those suppliers to be included is an important one. That is the problem addressed in this chapter. A methodology for ranking suppliers is developed and demonstrated. The supplier selection problem is a multi-criteria decision problem. For this type of problem, DEA is a useful approach to measure how well each potential supplier performs on the multiple criteria relative to other suppliers in the same market. Using this approach, it is possible to evaluate each supplier's performance relative to the best suppliers in the market, using the DEA efficiency measures.

This issue is one that has been addressed previously in many forms, including the use of DEA. However, previous models have relied on the availability of cardinal data. But a complete set of cardinal data is not available in most situations. The IDEA (Imprecise DEA) approach described here allows for a combination of cardinal and ordinal data to be used. Previous models have also assumed that the selection criteria are chosen by management and in this way are discretionary and in the control of management. But there are often non-discretionary measures as well, ones that are outside the control of management. The approach presented handles both sets of criteria.

Their model is a pair of AR-NF-IDEA models. It yields a final efficiency score for each Decision Making Unit characterized by an interval. After describing the use of virtual weight restrictions, the chapter describes how to find the virtual assurance regions. These provide the bounds of the interval. A numerical example is provided next, using 18 suppliers to be ranked. Three criteria are used: total cost, distance and supplier reputation, the three most commonly used factors for a problem of this type. The resulting efficiency intervals are provided for all eighteen suppliers as well as a list of peer group suppliers for each.

The main contributions of the chapter are as follows:

- The inclusion of nondiscretionary factors
- The ability to treat weight restrictions
- The ability to use imprecise data
- The expansion of the criteria considered to go well beyond using only cost
- Demonstration that the approach is computationally efficient for practical use

Additional applications could be used for ranking in other situations. Examples include project selection in project portfolio design or in Six Sigma applications.

*John Wang*
*Montclair State University, USA*

*Bin Zhou*
*Kean University, USA*

## Section 1

# The Use of Information Technology in Supply Chain Management

# Chapter 1
# Enterprise Applications for Supply Chain Management

**Susan A. Sherer**
*Lehigh University, USA*

## ABSTRACT

*Although many companies have implemented ERP systems to track and share information across cross functional business processes, they often supplement them with legacy, custom, or best of breed applications to support supply chain execution and management. This article offers a framework for understanding all types of enterprise applications that support the supply chain. In this study, the authors organize these applications, define acronyms, and describe the various types of systems that make up an information infrastructure for supply chain management.*

## INTRODUCTION

Coordination of information to effectively operate and manage a supply chain can be a source of competitive advantage today; a supply chain is only as good as its weakest information link. Companies continue to invest in technology to gather and utilize information, acquiring and integrating various different types of enterprise applications. The foundation of most enterprise information infrastructures today is the enterprise resource planning system (ERP) which has greatly reduced the number of applications required to track and share information across cross functional business processes. Basic ERP vendors such as SAP and Oracle have greatly enhanced their offerings with modules that add on functionality for many supply chain processes. However, many companies supplement basic ERP systems with legacy, custom, and best of breed applications for supply chain execution and management.

Supply chain management information systems have been defined as "systems used to coordinate the movement of products and ser-

DOI: 10.4018/978-1-4666-0918-1.ch001

vices from suppliers to customers" (Wang, Yan, Hollister, & Xing, 2009). Studies of supply chain management information systems primarily focus on logistics and production processes (Rutner, Gibson, & Gustin, 2001), or B2B integration (Chalasani & Sounderpandian, 2004; Dong, Xu, & Zhu, 2009), or a category of systems for sourcing called supplier relationship management systems (Choy, Lee, & Lo, 2004). Some studies focus on specific types of applications, e.g., green SCMS (Ko, Tseng, Yin, & Huang, 2008; Sahay & Ranjan, 2008) or supporting technologies such as RFID (Ozelkan, 2008). No peer-reviewed studies have encompassed the range of functionality of even the most widely-used commercial SCM IS packages (McLaren & Vuong, 2008). And even the broader studies that focus on systems to support supply chain processes (Wang et al., 2009) do not consider all the various applications that need to integrate with these processes in order to create an overall enterprise infrastructure for supply chain management, including, for example, business intelligence for supply chain management (Sahay & Ranjan, 2008).

Our purpose is to bring together all the various applications that are required to build an enterprise information infrastructure for supply chain management. Our main objective is to organize, define acronyms, and describe all the different applications that make up an information infrastructure for supply chain management. Examples of the use of the different applications are provided. Our objective is to provide an overview of the different applications, rather than an in-depth analysis of each one, to assist in understanding the scope of applications related to development of a complete supply chain infrastructure. This article offers a framework for understanding all the various types of information systems that support the supply chain, not just those that support B2B integration or logistics or sourcing.

## BUILDING AN ENTERPRISE INFORMATION INFRASTRUCTURE

Table 1 provides an overview of the various applications supporting supply chain management today. This table was developed after an extensive review of literature, both academic and practitioner, web sites of major vendors, and interviews. Interviews with both supply chain and IT infrastructure managers in several companies in different industries validated our classification of the three objectives of the existing applications: to run the business, manage the business, and support collaboration with business partners. Participants felt that this framework was more compelling for organizing supply chain management applications than traditional classifications of information technologies by organizational structure, functional area, or support provided (Turban, McLean, & Wetherbe, 2001).

Our objective was to provide an overview of all possible types of applications that might be considered to support the supply chain. In addition to applications supporting the key SCOR processes, we therefore included supporting applications, such as communication and productivity tools and applications that support cross business processes, as well as SCM related processes. The integration of ALL these applications today can serve to enhance supply chain management.

Fundamental to all information processing are communications and office productivity tools that enable access, sharing, and interpretation of enterprise information in a timely manner. Enterprise Content Management (ECM) systems capture, store, preserve, and deliver unstructured content, providing an integrated location for supply chain team members to find organizational resources and expertise, manage content and workflow, and collaborate.

ERP and BI (business intelligence) applications support cross functional transaction processing and management. Collaborative work spaces such as Microsoft's Sharepoint support portals, wikis,

*Table 1. Enterprise information applications for supply chain management*

| | **Run the Business** | **Manage the Business** | **Collaborate with partners** |
|---|---|---|---|
| Supports | Transaction processing | Advanced integration, optimization, and intelligence | Internal/external relationships |
| **Communication and Productivity** | Productivity/Communication<br>Enterprise Content Management (ECM) | | |
| **Cross Business Processes** | ERP | Business Intelligence (BI) | Collaborative work spaces |
| **Supply Chain SCOR Processes** | | | |
| Source | e-procurement | Supplier Relationship Management (SRM) | B2B Markets |
| Plan | | Advanced Planning & Optimization (APO) | Collaborative Planning Forecasting & Replenishment |
| Make | Process control, Computer Aided Manufacturing (CAM), Environmental Health & Safety (EHS) | Manufacturing Integration &Intelligence (MII) | CPFR |
| Deliver | Import/export compliance; freight ratings | Transportation Management Systems (TMS), Warehouse Management Systems (WMS) | Transportation exchanges; Vendor Managed Inventory (VMI) |
| Return | Reverse logistics tracking | Reverse logistics management | |
| **SC Related Processes** | | | |
| Engineering and Design | Computer Aided Design (CAD)/Product Data Management (PDM) | Product Lifecycle Management (PLM) | Collaborative Product Development (CPD) |
| Sales/Marketing/ Service | Sales Force Automation (SFA) | Customer Relationship Management (CRM); Field service; spare parts management | B2B Markets |

and workflow for collaboration within companies but also with business partners.

The strength of an ERP is its ability to track, record, and integrate information about all transactions and activities within a corporation. By adopting a single ERP, companies can adopt consistent work processes across all regions. This can bring significant improvement. For example, from 2004-2007 when Air Products and Chemicals, Inc., a $10 billion global supplier of chemicals and industrial gases, implemented SAP, they reduced SGA by 2.5% and inventory by 10%, with a 10% improvement in employee productivity and a 3% increase in return on capital (SAP, 2007). Similar success stores are available at the websites of the key ERP vendors.

However, an ERP alone is often insufficient to manage a complete supply chain. Many corporations integrate additional software applications to run their supply chains because their needs are not covered or handled at the level required. Some companies choose not to use ERP in every business segment because of cost, legacy, or personnel issues or they operate in a dynamic environment involving acquisitions still evolving to the same infrastructure. While ERP vendors have added on many supply chain capabilities, a 2008 survey of business and IT managers indicates that they are not always selected, particularly for managing supply chain processes beyond demand planning and inventory optimization, with best of breed and custom built applications preferred in fulfillment and delivery processes such as warehouse management, international trade compliance and documentation, and transportation management, as well as supply chain network design and collaborative applications that extend beyond the enterprise (Fontanella & Klein, 2008). While Air Products and Chemicals, Inc. was able to retire more than 600 enterprise applications when implementing its ERP, it only cut its IT portfolio by 25% (SAP, 2007).

## SOURCING

Basic ERP greatly improves sourcing particularly by keeping track of all vendors and orders. Additional applications that support sourcing include e-procurement, supplier relationship management systems, and trading community systems in B2B markets.

Self service or e-procurement applications typically support purchase of indirect goods/ services such as office supplies and travel. Here is how they work. The procurement department pre-negotiates contracts. Employees interested in purchasing specific items see only those pre-negotiated items when they access the system on their own computers. For example, if the procurement department standardizes on a particular computer vendor and within that vendor, only certain computer configurations, all employees wishing to purchase computers see only those options. Once the employee decides on a particular purchase, if approvals are required, work flow functionality automates them. Links to vendor sites provide delivery information. These systems decrease procurement costs by reducing maverick buying since the centralized procurement department pre-negotiates contracts with suppliers while simultaneously reducing non value added time spent in the procurement department. This avoids field locations developing vendor contracts based upon local preferences and relationships.

E-Procurement functionality is often acquired as part of a Supplier Relationship Management (SRM) application. Originally introduced and popularized by Ariba, ERP vendors now do provide this functionality. SRM supports the procurement professional in analyzing spend. Spend management functionality provides visibility into what is being bought so that managers can drill down into different spending patterns. For example, a leading industrial manufacturer's *Spend Analysis & Vendor Evaluation System (SAVES)*, captures all expenditures for direct items across all operating units and compares prices paid at each location against baselines (Monczka, Trent, & Petersen, 2006). SRM generally also includes supplier qualification, negotiation tools, contract generation and management, forward and reverse auctions, as well as self service procurement. Additionally, these systems enable procurement managers to analyze performance of their own departments. Air Products and Chemicals have saved 15 to 40 percent with Ariba's sourcing applications. With approximately 40 unique catalogs for centrally negotiated contracts, the system is used to purchase many items at Air Products from business cards to cranes (Ariba, 2008).

ERP vendors have lagged behind in collaborative computing, where information is collected and disseminated within a trading community (Fontanella & Klein, 2008). B2B markets provide discovery and facilitation capabilities by aggregating content from multiple suppliers, supporting vendor search, providing matching capabilities, and facilitating settlement, payment, and logistics. While some organizations build their own marketplaces, others access independent or industry consortia marketplaces. For example, Elemica was founded by 22 leaders of the global chemical industry to facilitate order processing and supply chain management of chemicals buying. Elemica has over 1,800 industry trading partners with 40 of the top 50 chemical companies participating, representing over $50 billion of combined transactional revenue (Elemica, 2008).

## PLANNING

Basic ERP systems include some planning functionality but originally lacked sophisticated optimization tools. Introduced by vendors such as i2 and Manugistics (now part of JDA Software), planning optimization applications are now available in supply chain management add-ons to the basic ERP systems, e.g., APO, Advanced Planning and Optimization, now part of SAP's supply chain management application. However, while more than 70% of 336 respondents to a 2008 AMR survey prefer their ERP vendors for demand planning (Fontanella & Klein, 2008), ERP vendors were preferred by only 55% for manufacturing/distribution planning, 50% for sales and operation planning, 44% for capacity planning, and 41% for network design (Fontanella & Klein, 2008). Consumer goods manufacturer Church & Dwight, maker of Arm & Hammer baking soda, has used Manugistics software for more than 10 years. This application determines forecasting algorithms and parameters and automatically generates customer specific forecasts (J. V. Murphy, 2008).

Collaborative Planning, Forecasting, and Replenishment (CPFR) applications support CPFR processes by tracking and highlighting planning discrepancies. Here is how they work. Both suppliers and customers input their forecasts for all stock keeping units (SKUs) into the system. The system determines if there are substantial discrepancies between these forecasts and alerts partners to discuss and develop a better collaborative forecast. Motorola uses the Manugistics system to support CPFR with its major cell phone retailers. This system automates input of demand signals into Motorola's overall planning system, feedback to the customer of Motorola's supply commitment, analysis of inventory over or under stock, sending signals to responsible planners, and management by exception (Cederlund, Kohli, Sherer, & Yao, 2007).

## FULFILLMENT (MANUFACTURING AND DELIVERY)

Specialized applications support unique manufacturing and delivery capabilities. Process control and computer aided manufacturing (CAM) applications automate manufacturing and integrate design and engineering with production. Applications from specialized content providers provide freight ratings, import/export compliance policies, specialized tax rates, and environmental health and safety regulations that are integrated with transactional processing systems. One newly defined application is the GSCMS or green supply chain management system which insures that products conform to relevant environmental controls before they are exported (Ko et al., 2008). Other applications integrate and support manufacturing and delivery processes including collaboration to meet customer demand. For a comparison of some leading vendors of applications that manage warehouses, transportation, and logistics to coordinate movement of products and services from suppliers to customers, see (Wang et al., 2009). A recent technology that has received much attention for its ability to support logistics is RFID since its tracking capabilities improve inventory visibility and automation in inventory management (Ozelkan, 2008).

Manufacturing management software links shop floor data to enterprise applications. It enables production personnel to monitor, measure, and control process improvement projects (e.g., lean manufacturing and Six Sigma), and aggregate and analyze shop floor data so that management quickly responds to disruptions. SAP's Manufacturing Integration and Intelligence, for example, constantly evaluates machine conditions, assists in root cause analysis of performance deviation, analyzes shift performance, and triggers alerts based on rule violations.

Vendor Managed Inventory (VMI) requires that suppliers have real time access to POS data in buyer organizations. Approximately 60% of respondents to a 2008 survey of more than 300 professionals believe that ERP vendors are not the best source of VMI applications which must scale for use across multiple trading partners (Fontanella & Klein, 2008). Although a small percentage prefer custom applications, most prefer best of breed applications that monitor partner inventory levels, support data transfer, analyze retailer supplied demand information, automatically generate orders, set inventory policies, determine replenishment plans, and support shipment.

Air Products has implemented VMI with some key gas customers. In fact, they now market a technology product called TELALERT, a telemetry system that supports efficient inventory management at gas customer locations using a dynamic target refill system, collecting, storing, and managing data about customer inventory levels and usage patterns. This system remotely monitors gas levels and if levels are below target refill levels, the remote telemetry unit sends out an alarm to a scheduling group at Air Products and a replenishment delivery is automatically scheduled. This supports tonnage level billing. Telemetry information also supports customer service by notifying Air Products if any remote units are not forwarding data (Trent, 2008).

Only ½ of 336 respondents to AMR Research's annual survey of supply chain technology buying intentions prefer ERP vendors for warehouse management systems (WMS) and transportation management systems (TMS) (Fontanella & Klein, 2008). WMS control automated material handling equipment, yard management, and pick replenishment, and facilitate cross docking. They better utilize warehouse layout and space, leading to faster fulfillment and overhead reduction, and lower costs. Pep Boys, a $2.2 billion leader in the automotive aftermarket industry, uses a WMS from Manhattan Associates. The slotting optimization solution enables Pep Boys to determine the best location for every car part and accessory in its five distribution facilities according to each product's physical attributes, pick frequency, and location in the company's stores (Manhattan Associates, 2006).

TMS determine the most efficient and cost-effective way to execute the movement of product(s). They support carrier selection, freight cost analysis, and parcel scheduling, routing, and tracking. TMS systems create more efficient loads within schedule requirements and maximize utilization of transportation assets. TMS systems also support dispatch optimization, continual evaluation and allocation of resources and loading with real time information (Killingsworth, 2008). Church & Dwight have used the JDA transportation management solution since 1996. This software helps them select the best carriers from a spend standpoint and consolidate orders into full truckload quantities while meeting on-time delivery targets (Murphy, 2008).

Air Products has its own custom logistics system that supports efficient deliveries of packaged gases. Since Air Products is not shipping commodity items, they have built a custom application that selects best routes, loads appropriate products, and minimizes fuel costs. This system generates delivery lists, and has a route optimizer that tweaks the schedule. Drivers log in to a computer network to view their scheduled trips and download information to a handheld computer that is plugged into an on-board truck computer used to update the amount of gas delivered and each driver's hours of service (Trent, 2008).

Transportation exchanges facilitate full load two way shipments. Exchanges are available for air cargo (e.g., Global Freight Exchange), trucking (e.g., the Internet Truckstop), and ocean transport (e.g., LevelSeas.com). Databases of available planes, trucks, and ships, loads, routes, and bids are maintained and aggregated to facilitate logistics.

## RETURNS

Generic software designed specifically for reverse logistics, non-existent in 2000, continues to be difficult to find due to the fact that it requires much customization (Bayles, 2000), particularly in determining asset disposition to increase recovery value.

Estee Lauder is an example of a company that developed its own reverse logistics system. The system uses scanned expiration dates to calculate disposition, for example, sold in other markets or at employee stores, given away to charities, or destroyed. It also determines economic shipping plans based upon returned inventory levels. The system increased Estee Lauder's ability to put return goods back on the market and provided better data on why goods were returned (Caldwell, 1999).

Some companies collaborate with third party providers (3PLs) to manage reverse logistics. For example, Sears partnered with GENCO whose reverse logistics application supports recall management, testing and warranty, repair and refurbishment, and product liquidation, and identifies ways to improve return processes. GENCO software provides Sears managers real-time visibility into exact inventory levels in return centers while supporting goods disposition to appropriate channels (Harps, 2003).

## ENGINEERING AND DESIGN

Integrating engineering and design with supply chain information infrastructure is critical especially as more suppliers take an active role in development. Computer aided design (CAD) and product data management (PDM) automate design and engineering and track data that must be linked to the ERP system. Additionally, Collaborative Product Development (CPD) enables partners to collaborate on design. Virtual customer environments enable customer participation in product conceptualization, design, test, support, and marketing. For example, AB Volvo involves customers in virtual product concept tests; two of the concept cars have become production vehicles (Mabisan & Nambisan, 2008). PLM, product lifecycle management, integrates the various computer assisted design applications, including CAD and PDM with product lifecycle and portfolio tools that support resource allocation and innovation during development.

## MARKETING/SALES/SERVICE

The specialized needs of the sales department have been supported by sales force automation (SFA) tools which track all leads, contacts, and opportunities. They are now offered as part of customer relationship management (CRM) applications which integrate data from all customer facing functions: marketing, sales, and service. By tracking all customer touchpoints, and individualizing and customizing customer contacts, CRM applications support marketing campaigns, integrating sales lead data, and customer service contacts. Adapting to changing customer needs is a critical component of agile supply chains.

Product configuration tools are often custom designed to support sales efforts. For example, in the late 1990s, Air Products built a product configurator for a line of products that delivers specialty chemicals to the electronics industry. The configurator not only supported sales efforts, but enabled mass customization without significant engineering cost. This configurator not only serves as a repository for design information and rules, it identifies equipment for various offerings, and creates proposals, specifications, and engineering drawings that can be shared immediately with customers supporting their sales efforts (Chowdhury, Sherer, & Ray, 2001).

In addition to CRM applications that support service contacts with customers, field service and spare parts management applications are available for appropriate industries. Since service requirements are not always predictable, managing service inventory presents unique challenges. Systems that track spares parts inventory (on trucks, in depots, at customer locations) and support field service technicians, allowing them to estimate costs, order parts, fix equipment, bill customers, process payments, and communicate effectively from the field, support after sales service operations.

At PP&L, the automated meter reading system (AMR) not only generates more accurate and timely electric bills, but helps better utilize expensive field service technicians. Service technicians no longer need to visit locations for disconnection or reconnection, eliminating thousands of service calls annually (Trent, 2008). The AMR system is linked to PPL's outage management system so that service crews identify the locations of electrical problems quicker, leading to faster service restoration.

Logistics providers can provide customer service, for example, UPS stores repair Toshiba laptops, eliminating extra transportation, centralizing parts and repairs, and reducing costs. Collaborative applications support this interaction, particularly in sharing performance metrics to ensure that outsourced customer service meets needs and does not damage brand image.

## BUSINESS INTELLIGENCE (BI)

There is a growing need for business intelligence in the supply chain. According to the leader of Accenture's Supply Chain Management service:

Effective supply chain management requires integration across functions and departments. Some of the most useful reporting is around cross-functional department processes such as total cost to serve; product or customer profitability incorporating logistics, ordering, fulfillment, selling and other costs; vendor scorecards; the perfect order; order-to-cash cycles; and variable cost productivity. Each of the reports requires assembling data from multiple source systems, and most companies cannot easily do this (Mulani, 2008).

Traditional transaction systems are not designed to support efficient enterprise reporting and business analytics. BI systems need to draw information from all operational systems, often accomplished by developing and working with a data warehouse. "Hence, it is well understood that SCM alone cannot deliver the expected value at right time in an organization. Clearly a BI system needs to draw information from all operational systems" (Sahay & Ranjan, 2008).

Leading users of BI go beyond functional data analysis; they pool data generated in-house from all areas with data acquired form outsides sources to gain a comprehensive understanding of their customers and business (Davenport, 2006). BI tools support extraction of data at different levels of aggregation and utilize advanced methods for enterprise wide data analysis (Sahay & Ranjan, 2008). Harrah's Entertainment uses analytics to personalize customer experiences and determine what incentives best persuade customers to return, creating individual relationships and differentiating service among 40 million customers (Stanley, 2006).

Coupled with real time event monitoring, business intelligence applications enable management recognition and timely response to critical issues. Continental's Flight Operations department uses a data warehouse to feed a real time application that provides up to the minute airline performance statistics, alerting operators if a flight has been sitting away from a gate for at least two hours (Watson & Wixom, 2007). Sometimes referred to as SCEM (Supply Chain Event Management) or BAM (Business Activity Monitoring) applications, many of these applications not only detect and evaluate unexpected events; but they support the decision maker with alternative solutions. For example, an event management application could alert a supply chain manager of a shipping delay or slow down in manufacturing operations. By accessing and aggregating information from SCM, ERP, and other transactional systems, as well as information from a data warehouse, a business intelligence application can drill into historical information to identify the cause of a problem.

For example, data analysis could suggest whether a delayed shipment results from a typical seasonal slowdown or a new supplier service agreement. The system gives users information to make informed decisions and take action.

## CONCLUSION

Businesses today integrate a variety of different information systems in order to coordinate information to effectively operate and manage their supply chains. These systems not only help run and manage the business, but support collaboration among supply chain partners. Most previous research has focused on subsets of these systems. We have developed a framework that brings together all the different systems, not only systems that primarily support source, plan, make, deliver, and return, but also those that support processes such as design and engineering and customer focused applications, which must be well integrated in order to assure effective supply chain management. Additionally, we suggest that applications such as business intelligence, and productivity, communication, and collaboration tools, must also be part of an enterprise information infrastructure for supply chain management. Without integrating these applications, effective analysis and information sharing cannot be achieved. While ERP and productivity suites often form the foundation for the information infrastructure with wide access within the company, many companies continue to integrate custom and legacy applications. According to AMR Research "The demise of the best of breed supply chain technology vendor is not imminent," .... "In fact, it's likely that gains will be made against ERP vendors in areas that require multi-enterprise integration and leading edge functionality" (Supply Chain Digest, 2008).

The elusive goal of a single system that meets all needs has not yet been reached by most corporations. Perhaps as software as a service evolves, or if grid computing become more prevalent for supply chain support (Vassiliadis, 2009), we will see significant changes in supply chain information infrastructures. Other factors that will contribute to changes in IT applications include greater integration of RFID technology, enhanced tools for simulations and "what-if" analyses connecting with real time business intelligence, and technology and certification programs to insure environmental sustainability in supply chain design and execution (S. Murphy, 2008). For now, most companies integrate many different types of applications even with common ERP infrastructures. To increase market share, ERP vendors will need to continue to evolve their infrastructures, improve support of cross business integration, and provide strong capabilities to support global and green initiatives.

A limitation of this research is the fact that, while we attempted to be as complete as possible in including applications that integrate with supply chain management, new applications evolve and others may include some applications that we may not have considered. However, we believe the we have provided a starting point for attempting to put together as many applications as possible in one location so that supply chain researchers and managers can both have a framework that we hope is as inclusive as possible at this time.

## REFERENCES


Ariba. (2008). *Air Products and Chemicals customer success story*. Retrieved August 2008, from http://www.ariba.com//pdf/Air_Products.pdf

Bayles, D. (2000). Send it back! The role of reverse logistics. *InformIT*. Retrieved from www.informit.com/articles/printerfriendly.aspx?p=164926

Caldwell, B. (1999). Reverse logistics. *Information Week Online*. Retrieved from http://www.informationweek.com/729/logistics.htm

Cederlund, J., Kohli, R., Sherer, S., & Yao, Y. (2007). How Motorola put CPFR into action. *Supply Chain Management Review*.

Chalasani, S., & Sounderpandian, J. (2004). Performance benchmarks and cost sharing models for B2B supply chain information systems. *Benchmarking, 11*(5), 447–464. doi:10.1108/14635770410557690doi:10.1108/14635770410557690

Chowdhury, N., Sherer, S. A., & Ray, M. (2001). Realizing IT value at Air Products and Chemicals. *Communications of the AIS, 7*(23).

Choy, K. L., Lee, W. B., & Lo, V. (2004). Development of a case based intelligent supplier relationship management system - linking supplier rating system and product coding system. *Supply Chain Management, 9*(1), 86–101. doi:10.1108/13598540410517601doi:10.1108/13598540410517601

Davenport, T. (2006). Competing on analytics. *Harvard Business Review, 84*(1), 98–106.

Dong, S., Xu, S. X., & Zhu, K. X. (2009). Information technology in supply chains; the value of IT-enabled resources under competition. *Information Systems Research, 20*(1), 18–32. doi:10.1287/isre.1080.0195doi:10.1287/isre.1080.0195

Fontanella, J., & Klein, E. (2008). Supply chain technology spending outlook. *Supply Chain Management Review, 12*(4), 14.

Harps, L. H. (2003). Revving up returns. *Inbound Logistics*. Retrieved from www.inboundlogistics.com/articles/featurs/1103_feature02.shtml

Killingsworth, K. (2008). Improving fleet management performance. *Inbound Logistics, 32*. Retrieved from www.inboundlogistics.com/articles/itmatters/itmatters0508.shtml

Ko, H.-C., Tseng, F.-C., Yin, C.-P., & Huang, L.-C. (2008). The factors influence suppliers satisfaction of green supply chain management systems in Taiwan. *International Journal of Information Systems and Supply Chain Management, 1*(1), 66–79.

Mabisan, S., & Nambisan, P. (2008). How to profit from a better "virtual customer environment". *MIT Sloan Management Review, 49*(3), 53–61.

Manhattan Associates. (2006). *A fine-tuned slotting strategy keeps Pep Boys rolling*. Retrieved from www.manh.com

McLaren, T. S., & Vuong, D. C. H. (2008). A "genomic" classification scheme for supply chain management information systems. *Journal of Enterprise Information Management, 21*(4), 409–423. doi:10.1108/17410390810888688doi:10.1108/17410390810888688

Monczka, R. M., Trent, R. J., & Petersen, K. J. (2006). *Effective global sourcing and supply for superior results*. Tempe, AZ: CAPS Research.

Mulani, N. (2008). Business intelligence enters the supply chain. *Logistics management, 47*(4), 27.

Murphy, J. V. (2008). CPG maker that bought forecasting software 10 years ago still benefits. *Global logistics & supply chain strategies*. Retrieved from www.suppychainbrain.com

Murphy, S. (2008). The supply chain in 2008. *Supply Chain Management Review, 12*(1).

Ozelkan, E. (2008). When does RFID make business sense for managing supply chains? *International Journal of Information Systems and Supply Chain Management, 1*(1), 15–47.

Rutner, S. M., Gibson, B. J., & Gustin, C. M. (2001). Longitudinal study of supply chain information systems. *Production and Inventory Management Journal, 42*(2), 49–56.

Sahay, B. S., & Ranjan, J. (2008). Real time business intelligence in supply chain analytics. *Information Management & Computer Security, 16*(1), 28–48. doi:10.1108/09685220810862733doi:10.1108/09685220810862733

SAP. (2007). Air Products becomes one company with SAP software. *Business transformation study*. Retrieved from http://www.sap.com/usa/solutions/business-suite/erp/customersuccess/index.epx

Stanley, T. (2006, February 1). High-stakes analytics. *Optimize*.

Supply Chain Digest. (2008). Supply chain software: AMR Research remains bullish on supply chain software spend. *Supply Chain Digest*. Retrieved from http://www.scdigest.com/assets/On_Target/08-05-19-2.php?cid=1688

Trent, R. J. (2008). *End-to-end lean management: a guide to complete supply chain improvement*. Ft. Lauderdale, FL: J. Ross Publishing.

Turban, E., McLean, E., & Wetherbe, J. (2001). *Information technology for management* (2nd ed.). New York: John Wiley and Sons.

Vassiliadis, B. (2009). The grid as a virtual enterprise enabler. *International Journal of Information Systems in the Service Sector*, *1*(1), 78–92.

Wang, Z., Yan, R., Hollister, K., & Xing, R. (2009). A relative comparison of leading supply chain management software packages. *International Journal of Information Systems and Supply Chain Management*, *2*(1), 81–96.

Watson, H., & Wixom, B. (2007). Enterprise agility and mature BI capabilities. *Business Intelligence Journal*, *12*(3), 4–6.

*This work was previously published in International Journal of Information Systems and Supply Chain Management, Volume 3, Issue 3, edited by John Wang, pp. 18-28, copyright 2010 by IGI Publishing (an imprint of IGI Global)*

# Chapter 2
# Information Technology Implementation Prioritization in Supply Chain:
## An Integrated Multi Criteria Decision Making Approach

**Sreekumar**
*Rourkela Institute of Management Studies, India*

**Debendra Kumar Mahalik**
*Sambalpur University, India*

**Gokulananda Patel**
*Birla Institute of Management Technology, India*

## ABSTRACT

*The increasing rate of technology growth has resulted in decrease in cost of information. These technologies are helpful in coordinating the activities resulting in effective management of the supply chain. Literature shows that the use of Information Technology (IT) plays an important role in managing the processes of SCM. This has resulted in increasing use of IT in SCM. The computerization of SCM processes, if implemented in one go may result in failure. IT implementation prioritization in supply chain is a major issue before the planner as there is no clear cut formula to solve this problem. This paper considers components of SCM like material management, purchase management, production management, logistics and distribution and customer interface for IT implementation prioritization. Two multi-criteria decision making methods (MCDM) viz. analytical hierarchy process (AHP) and a technique for order preference by similarity to ideal solution (TOPSIS) are used in this paper. The novelty of the paper lies in integration of AHP and TOPSIS methods for IT implementation prioritization. The weights of the criterions and the alternatives are calculated using AHP method which is used as an input for TOPSIS analysis for prioritization of IT implementation.*

DOI: 10.4018/978-1-4666-0918-1.ch002

## INTRODUCTION

One of the key ingredients for business survival in today's dynamic and eclectic business scenario is information technology (IT). The IT application helps the organizations in becoming more competitive and is an essential ingredient for business survival. Information technology, in any form, has its own inherent tendencies and influences the nature and direction of organizations. It has significantly changed supply chain processes across different organizations and, in many cases, improved competitiveness (Singh et al., 2007). Thompson (1998) discussed physical supply chain and the financial supply chain which exist virtually in all forms of commerce. More recently, Rahman (2004) discussed the information supply chain. The efficiency of supply chain can be improved using the advanced information technology. In the present scenario the ability to effectively manage information within the firm has become critically important because it provides a competitive advantage. It is therefore not surprising that many firms have begun to develop strategies, focusing on using information technology as a resource to facilitate the effective collection and utilization of information (Bharadwaj, 2000). The rapid development of information technology and the internet, together with the emerging trend towards global logistics systems has prompted an increasing emphasis to be placed on the circulation function of the supply chain (Sanders & Premus, 2002; Trappey et al., 2004; Rahman, 2004). Narasimhan and Kim (2001) proposed that Earl's classification not only was applicable to the internal value chain of a firm, but could also be extended to the company's supply chain, linking suppliers, and customers. Several studies have examined the impact of information technology on the supply chain (Byrd & Davidson, 2003; Kearns & Lederer, 2004). Liao et al. (2004) indicated that the effect of implementation of IT functions on a business becomes a critical issue not only theoretically but also in practice. A good control policy must be beneficial for the whole supply chain. The aim of control is to optimize some performance measure, which typically comprises revenue from sales (Saharidis, 2009)

Prior to the 1980s, a significant portion of the information flows between functional areas within an organization and between supply chain member organizations were paper based. In many instances, these papers based transactions and communications were slow, unreliable and error prone. Bowersox and Closs (1996) emphasized on timely and accurate information for American Business. In fact distorted information creates lot of confusion and misbalance the whole chain, form suppliers to end customers. This is in fact boosted the requirement for accurate and timely information.

The paper considers a smelter plant of a company named XYZ which is the largest manufacturers of aluminum in India for the study. Five processes of the plant viz. customer interface, purchase, logistics, production and materials management are considered for the IT implementation prioritization. Two methods multi-criteria decision making methods (MCDM) viz. analytical hierarchy process (AHP) and technique for order preference by similarity to ideal solution (TOPSIS) are used in the paper. The weights of the criterions and the alternatives are calculated using AHP method which is used as an input for TOPSIS analysis. So the paper attempts to demonstrate an integrative method to prioritize various processes for IT implementation.

## NEED FOR INFORMATION AND TECHNOLOGY

Every supply chain has an information chain that parallels the flow of products (Andel, 1997). If the information is not relayed at the right time to the right place, there will not be any purchase order, no shipment messages, no payment, no coordinated marketing and sales efforts, and the supply chain shuts down (Zuckerman, 1998). SCM is based on the exchange of substantial quantities of information among buyers, suppliers, and carriers to increase the efficiency and effectiveness of the supply chain (Carter, Ferrin, & Carter, 1995). At the ultimate level of information integration, all member links in the supply chains are continuously supplied with information in real time (Balsmeier & Voisin, 1996). Information provides better visibility of physical goods as they move within the company (Lewis & Talalayevsky, 1997). Hence, high levels of information enable companies to develop unique capabilities to achieve competitive advantage. A survey conducted by Gustin et al. (1995) found that the successful implementation of the integrated logistics concept was related to high level of information availability. They also found that companies with integrated logistics exhibited enhanced information systems performance compared with non integrated companies. This, in turn leads to better decision making capabilities at the strategic, tactical, and operational level.

(Altekar, 2005) proposed three reasons for the use of information technology in SCM; like Data transfer, Information Retrieval and Process Control. There are also other reasons such as, easy data storage, retrieval, less human error, less paper work, automation of process, more secure and ease of operation.

## LITERATURE REVIEW

Numerous information system have been developed for increasing the efficiency of SCM and support both planning and execution (Shere, 2005). The most typical role of IT in SCM is reducing the friction in transactions between supply chain partners through cost-effective information flow (Auramo et al., 2005). For integration of supply chain there is a requirement for the specifications of the business processes and development of information systems (Bechtel & Jayaram, 1997; Hewitt, 1994). It can be transaction processing systems focusing on day-today operations; operational planning systems, or strategic planning tools used to redesign the supply chain infrastructure (Richmond et al., 1998).

## COMPONENTS OF SCM

A supply chain is an interlinked network of suppliers, manufacturers, distributors and customers where materials or services flow from the suppliers through manufacturers and distributors to the customers (Sevkli, 2007). Supply chain is a process which needs interface and interaction with the supply chain enabler like vendor, customer, carrier, and intermediate parties. Raghuram and Rangraj (2000) have identified the 15 key thrust areas, such as minimizing uncertainty, reducing lead times, minimizing the number of stages, improving flexibility, improving process quality, minimizing variety, managing demand, delaying differentiation, kitting of supplies, focusing on "A" category, planning for multiple supply Chains, modifying performance Measures, competing on services, moving from function to process, taking initiatives at an industry level.

A typical supply chain consists of customers, retailers, whole sellers/distributors, and manufactures, Components/Raw material suppliers (Chopra & Meindl, 2003). These SCM not only include supplier and manufactures, but also transporters, warehouses, retailers and customers themselves.

Basically supply chain consists of upstream, internal, and downstream process, which consists of different process as follows:

- Upstream supply chain processes are mainly consists of Procurement, Transportation, Warehousing, Inventory
- Internal supply chain processes are mainly consists Manufacturing, work on progress
- Downstream supply chain process are mainly delivery, Transportation, Distribution management, Inventory, Warehousing

## INTEGRATION REQUIREMENT

In an organization a typical SCM activity starts with purchasing, supplier selection, Transportation, warehousing, work in progress inventory, distribution. In a conventional supply chain, the information flows in the opposite direction, from the customers through the distributors and manufacturers to the suppliers. Conventional supply chains are therefore characterized by the forward flow of material and the backward flow of information (Sevkli, 2007).

Over the past few years, most well run companies have focused on improving their physical supply chain efficiencies (Carr et al., 2004). In doing so, they have realized such benefits as short time to market, reduced production costs, lower inventory costs, and closer collaboration between the trading partners (Boubekri, 2001). Individually organizations are trying to improve performance. The companies are trying to improve performance thought physical, Information and financial supply chain. But there is a need for a collaborative performance improvement for better SCM. Organizations have made a little progress towards integrating the three. There is opportunity lies to improve coordination and integration between different components. The extent of the opportunities is reflected in the inefficiencies that current exist (Rainbird, 2004). Improper integration are reflected in a study by (Svensson, 2003), this result increases the time to order completion, transaction costs, and reconciliation activities. These examples of uncoordinated physical and financial supply chain activities are a financial detriment to the organization (Lin et al., 2005). Problems at the physical-financial supply chain interface affects other performance metrics as well, such as inventory turns safety stock availability, manual processing costs, and dropped transaction because of inefficiencies (Rahman, 2006).

The success of supply chain depends largely on the better management of co-ordination of various activities and sharing of the information. But how much to share with the business partner is still an issue. Klaus (1993) states that inter-disciplinary research could bring new idea in to SCM discipline. This could facilitate practitioner to address various domain specific issues. Stock (1997) argues that SCM has its roots in theories borrowed from the established discipline.

## CASE ANALYSIS: CASE OF XYZ COMPANY

For our study, we considered a company named XYZ which is the largest manufactures of aluminum in India and also one of the largest manufactures in the world. The company was incorporated in the year 1938 as a subsidiary unit of a Canadian company. The company is pioneer in the production and marketing of alumina, aluminum metal and semi-fabricated products. It has 14 plants and 10 offices spread across the country. Every stage of the company is vertically integrated including bauxite mining, alumina refining, alumina smelting, semi-fabricating (sheets and foils), product development and captive power generation etc... In the year 2000 the company joined ABC group of companies which has dominance in metal and is known for competitive cost structure.

One of the smelter plants of the company is located at western part of Orissa, India is taken for the study. This smelter plant pursues production of primary metal, which is based on electrolytic reduction process through the primitive horizontal stud soderberg (HSS) technology. The major raw materials used for the primary aluminum production are alumina, cryolite, aluminum fluoride and electrical power. The smelter which was started in January 1959 with 10 KTPA capacity has undergone expansion in stages to reach the present level with the capacity of 65 KTPA. Expansion and modernization project for increasing the smelting capacity to 146 KTPA is in the completion stage. For meeting the extra power requirement the captive power plant (CPP) is also expanding from 67.5 MW to 367.5 MW. Presently the CPP capacity has already reached at 267.5 MW and another 100 MW unit is expected to be commissioned by the end of 2007, which increase total capacity to 368 MW (http://www.hindalco.com/about_us/). The company has also taken a coal mine in the nearby area to meet increasing requirement its power plant.

The company has several departments and for our study of prioritization of computerization in Supply Chain Management; we have taken following components as a part of Supply Chain Management namely

- Material Management
- Purchase Management
- Production management
- Logistics and Distribution function
- Customer Interface

According to Raghuram and Rangaraj (2000), we have found that there are 15 reasons to go for computerization of Supply Chain Management. The important reasons can be - Security, Speed, Accuracy, Easy, Reliability, Cost, Process Improvement, Corporate image, Focus on core work, Resources utilization, Information availability, Transparency, Environment, Productivity.

## FACTOR ANALYSIS

As discussed above there may be fifteen variables which may be the reasons for computerization. To find the factors important for computerization a structured questionnaire is prepared containing these fifteen variables. The questionnaire is distributed among one hundred thirty executives of the company, out of which fifty six responded fully. The data collected is used for factor analysis for two purposes, to reduce the number of variables and to detect structure in the relationships between variables i.e. is to classify variables. Therefore, factor analysis is applied as a data reduction or structure detection method (the term *factor analysis* was first introduced by Thurstone, 1931). The rotated component matrix is shown in Table 1.

The variables with factor loading more than 0.50 were considered as significant under each dimension. The eigen values of selected factors were greater than 1. Three factors are extracted on analysis and three variables viz. transparency, environment and corporate image is dropped as their factor loading is less than 0.50. It shows that these factors, such as transparency, environment and corporate image have no significant impact of on computerization system. From the factor analysis we got three sub factors according to the importance like primary attributes, Secondary attributes and Tertiary attributes.

Table 2 represents the three sub-factors which contribute to the computerization system of the organization.

There are five primary reasons for IT implementation in supply chain. We have used two methods viz. "analytic hierarchy process" and "technique for order preference by similarity to ideal solution" for prioritization of IT of five variables under primary reasons.

## ANALYTICAL HIERARCHY PROCESS (AHP) APPLICATION

The Analytic Hierarchy Process (AHP) deals with complex systems for a choice among several alternatives. And at the same time it provides a comparison of the considered options. This method was first presented by Saaty (1980). The decision contains many social and economic factors, which needs to be evaluated by linguistics variables and it, has been found that AHP has been applied in various type of problem (Paulson, 1993). The method is based on the subdivision of the problem in a hierarchical form and helps the analysts to organize the critical aspects of a problem into a hierarchical structure similar to a family tree. The method reduces complex decisions to

*Table 1. Rotated component matrix*

| Items: | Factor | | |
|---|---|---|---|
| | 1 | 2 | 3 |
| Process Improvement | **0.925** | 0.207 | 0.091 |
| Speed | **0.889** | 0.22 | 0.024 |
| Easy | **0.870** | 0.256 | 0.032 |
| Reliable | **0.865** | 0.233 | 0.022 |
| Productivity | **0.857** | 0.207 | -0.02 |
| Information availability | 0.208 | **0.88** | -0.022 |
| Accurate | 0.234 | **0.878** | 0.078 |
| Focus on core work | 0.096 | **0.762** | 0.061 |
| Resources utilization | 0.28 | **0.730** | 0.035 |
| Secure | 0.462 | **0.501** | 0.065 |
| Uniform Standard | 0.001 | 0.044 | **0.952** |
| Cost | -0.036 | 0.016 | **0.655** |
| Transparency | 0.335 | 0.174 | 0.493 |
| Corporate image | 0.274 | -0.025 | 0.477 |
| Environment | -0.132 | 0.015 | 0.460 |
| Extraction Method: Principal Axis Factoring | | | |
| Rotation Method: Varimax with Kaiser Normalization | | | |

a series of simple comparisons and rankings, then synthesizing the results. By doing so, the AHP not only helps the analysts to arrive at the best decision, but also provides a clear rationale for the choices made. The objective of using an analytic hierarchy process (AHP) is to identify the preferred alternative and also determine a ranking of the alternatives when all the decision criteria are considered simultaneously (Saaty, 1980). Recently there is an increase use of AHP in SCM, one such application was done by (Min, 2007). Others (Jing, 2006; Milind, 2007; Jukka, 2001; Felix, 2008; Jing-yuan, 2006) also applied AHP in supply chain evaluation.

The detailed step wise procedure of using AHP is as follows:

**Step 1:** Define decision criteria in the form of a hierarchy of objectives. The hierarchy is structured on different levels: from the top

*Table 2. Factors influencing the IT implementation*

| **Primary Reasons** | **Secondary Reason** | **Tertiary reasons** |
|---|---|---|
| Process Improvement | Information availability | Uniform Standard |
| Speed | Accurate | Cost |
| Easy | Focus on core work | |
| Reliable | Resources utilization | |
| Productivity | Secure | |

(i.e., the goal) through intermediate levels (criteria and sub-criteria on which subsequent levels depend) to the lowest level (i.e., the alternatives);

**Step 2:** Weigh the criteria, sub-criteria and alternatives as a function of their importance for the corresponding element of the higher level. For this purpose, AHP uses simple pair wise comparisons to determine weights and ratings so that the analyst can concentrate on just two factors at one time.

**Step 3:** After a judgment matrix has been developed, a priority vector to weight the elements of the matrix is calculated.

This is the normalized eigenvector of the matrix. The use of AHP instead of another multi-criteria technique is due to the following reasons:

1. Quantitative and qualitative criteria can be included in the decision making.
2. A large quantity of criteria can be considered
3. A flexible hierarchy can be constructed according to the problem. After getting all the relevant data in different table AHP analysis has been used to priorities these computerization projects. Here we have used Expert choice 11.5 version, and the results are as below

*Table 3. Pair wise comparison matrix of criteria by experts*

| | $C_1$ | $C_2$ | $C_3$ | $C_4$ | $C_5$ |
|---|---|---|---|---|---|
| $C_1$ | 1 | 3 | 4 | 3 | 3 |
| $C_2$ | 1/3 | 1 | 2 | 3 | 2 |
| $C_3$ | 1/4 | 1/2 | 1 | 3 | 5 |
| $C_4$ | 1/3 | 1/3 | 1/3 | 1 | 2 |
| $C_5$ | 1/3 | 1/2 | 1/5 | 1/2 | 1 |

The main objective of this analysis was to find out the component, which need to be computerized. In this respect three experts are being contacted from the company XYZ discussed above. These experts have sufficient expertise and involved in the functional areas such as purchase, finance, production, materials management, distribution and related functions. These experts were initially asked about the level of importance of each criterion under primary reasons with respect to each other. The response of experts on each of the criterion is shown in Table 3.

$C_1$: Process Improvement; $C_2$: Speed; $C_3$: Easy; $C_4$: Reliability ; $C_5$: Productivity

Now the issue before us is to prioritize the computerization of various components of SCM. Each computerization is being taken as separate project and now each projects are compared with respect to each (five) different criteria /factors mention above using a Likert type 1-9 scale. The alternatives are presented as follows:

**$A_1$:** IT implementation in Purchase;
**$A_2$** : IT implementation in Logistics;
**$A_3$:** IT implementation in Material management;
**$A_4$:** IT implementation in Production management;
**$A_5$:** IT implementation in Customer Interface

The comparison of each alternatives against each other on various criteria are given in Table 4.

For AHP application we have used Expert Choice 11.5, the various weight results obtained in analysis are shown in Table 5. There may be issues like inconsistency which may require special attention like, if criterion A is just as important as criterion B, then the pair wise judgments for A and B to any other criterion should be identical. When this doesn't happen in the judgment process, inconsistency can arise. Saaty (1994) suggested the error in these measurements is tolerable only when it is of lower order of magnitude (10%) than the actual measurement itself. Consistency ratios (CR) can be calculated and compared to indexes

*Table 4. Comparison matrix of alternatives on each criteria*

| Criteria | Alternatives | $A_1$ | $A_2$ | $A_3$ | $A_4$ | $A_5$ |
|---|---|---|---|---|---|---|
| $C_1$ | $A_1$ | 1 | 3 | 1/3 | 3 | 1/3 |
| | $A_2$ | 1/3 | 1 | 1/3 | 1/2 | 1/3 |
| | $A_3$ | 3 | 3 | 1 | 3 | 3 |
| | $A_4$ | 1/3 | 2 | 1/3 | 1 | 1/3 |
| | $A_5$ | 3 | 3 | 1/3 | 3 | 1 |
| $C_2$ | $A_1$ | 1 | 3 | 1 | 3 | 2 |
| | $A_2$ | 1/3 | 1 | 1/3 | 3 | 1/3 |
| | $A_3$ | 1 | 3 | 1 | 3 | 2 |
| | $A_4$ | 1/3 | 1/3 | 1/3 | 1 | 1/3 |
| | $A_5$ | 1/2 | 3 | 1/2 | 3 | 1 |
| $C_3$ | $A_1$ | 1 | 2 | 1/2 | 2 | 2 |
| | $A_2$ | 1/3 | 1 | 1/2 | 3 | 2 |
| | $A_3$ | 2 | 2 | 1 | 3 | 1 |
| | $A_4$ | 1/2 | 1/3 | 1/3 | 1 | 1/3 |
| | $A_5$ | 1/2 | 1/3 | 1 | 3 | 1 |
| $C_4$ | $A_1$ | 1 | 2 | 1/3 | 2 | 2 |
| | $A_2$ | 1/2 | 1 | 1/3 | 2 | 2 |
| | $A_3$ | 3 | 3 | 1 | 2 | 2 |
| | $A_4$ | 1/2 | 1/2 | 1/2 | 1 | 1/3 |
| | $A_5$ | 1/2 | 1/2 | 1/2 | 3 | 1 |
| $C_5$ | $A_1$ | 1 | 2 | 1/3 | 3 | 2 |
| | $A_2$ | 1/2 | 1 | 1/2 | 4 | 3 |
| | $A_3$ | 3 | 2 | 1 | 3 | 3 |
| | $A_4$ | 1/3 | 1/4 | 1/3 | 1 | ½ |
| | $A_5$ | 1/2 | 1/3 | 1/3 | 2 | 1 |

*Table 5. Weights of criteria and alternatives*

| | $A_1$ | $A_2$ | $A_3$ | $A_4$ | $A_5$ | Inconsistency |
|---|---|---|---|---|---|---|
| $C_1$ (0.444) | 0.166 | 0.074 | 0.40 | 0.096 | 0.259 | 0.09 |
| $C_2$ (0.225) | 0.303 | 0.116 | 0.303 | 0.073 | 0.206 | 0.05 |
| $C_3$ (0.166) | 0.244 | 0.199 | 0.300 | 0.080 | 0.176 | 0.08 |
| $C_4$ (0 .075) | 0.215 | 0.165 | 0.372 | 0.097 | 0.151 | 0.09 |
| $C_5$ (0.090) | 0.338 | 0.148 | 0.239 | 0.172 | 0.103 | 0.08 |

derived from random judgments. As long as the CR is less than 0.10, analysis can proceed. Saaty also emphasized that greater consistency does not imply greater accuracy. In all our analysis the inconsistency ratio are found to be less than specified level.

The bracketed value shows the weights of the criterion and the cell values shows the weight of alternatives against each criterion. This table is used as the input for TOPSIS analysis in the following section.

## TOPSIS APPLICATION

TOPSIS stands for technique for order preference by similarity to ideal solution and was developed by Hwang and Yoon (1981). The TOPSIS reduces the influence of experts' subjective factors to the decision of scheme and at the same time this method avoids the complex calculation. It is a very effective method in multi attribute decision analysis. It uses normalized matrix to find the superior project and inferior project (that is ideal solution and non-ideal solution), then calculates the distances of other projects from the ideal and the non-ideal solution. Then the relative closeness to the ideal solution is calculated. The relative closeness can be ranked in descending order. The relative closeness can be put in the range of 0-1, if it is close to 1, evaluating projects is said to be closer to the ideal.

The algorithm of TOPSIS applications is as follows:

- Consider Table -5 as an input.
- In a multi-criteria decision making process we may have 'm' criterions $C_1, C_2, \ldots, C_m$ which has to be evaluated on 'n' properties and performance indices $X_1, X_2, \ldots, X_n$ . This evaluation will give a m x n matrix.

$$r_{ij} = \frac{x_{ij}}{\sqrt{\sum_{i=1}^{m} x_{ij}^2}}$$

The original data can be normalized to get the normalized rating $r_{ij}$ ; which can be calculated using the formulae:

Table 6 gives the weights of five alternatives against the five criteria and calculates the sum of squares of column.

$$D = \begin{array}{c} A_1 \\ A_2 \\ A_3 \\ \cdot \\ \cdot \\ \cdot \\ A_m \end{array} \begin{bmatrix} x_{11} & x_{12} & x_{13} & \cdot & \cdot & \cdot & x_{1n} \\ x_{21} & x_{22} & x_{23} & \cdot & \cdot & \cdot & x_{2n} \\ x_{31} & x_{32} & x_{33} & \cdot & \cdot & \cdot & x_{3n} \\ \cdot & \cdot & \cdot & & & & \cdot \\ \cdot & \cdot & \cdot & & & & \cdot \\ \cdot & \cdot & \cdot & & & & \cdot \\ x_{m1} & x_{m2} & x_{m3} & \cdot & \cdot & \cdot & x_{mn} \end{bmatrix}$$

- This step calculates the weighted normalized ratings $v_{ij}$. The weight of j[th] attribute or criteria is taken $w_j$ . The weighted normalized ratings matrix $V$ is obtained by $v_{ij} = w_i$ *(.)* $r_{ij}$ shown in Table 7.

*Table 6. Weights of alternatives against criteria*

| | $C_1$(0.444) | $C_2$(0.225) | $C_3$(0.166) | $C_4$(0 .075) | $C_5$(0.090) |
|---|---|---|---|---|---|
| $A_1$ | 0.166 | 0.303 | 0.244 | 0.215 | 0.338 |
| $A_2$ | 0.074 | 0.116 | 0.199 | 0.165 | 0.148 |
| $A_3$ | 0.40 | 0.303 | 0.300 | 0.372 | 0.239 |
| $A_4$ | 0.096 | 0.073 | 0.08 | 0.097 | 0.172 |
| $A_5$ | 0.259 | 0.206 | 0.176 | 0.151 | 0.103 |
| Sum of square of column | 0.52283 | 0.49481 | 0.47593 | 0.494 | 0.48317 |

*Table 7. Weighted normalized rating matrix*

| | $C_1$(0.444) | $C_2$(0.225) | $C_3$(0.166) | $C_4$(0 .075) | $C_5$(0.090) |
|---|---|---|---|---|---|
| $A_1$ | 0.3175 | 0.6124 | 0.5127 | 0.4352 | 0.6995 |
| $A_2$ | 0.1415 | 0.2344 | 0.4181 | 0.3340 | 0.3063 |
| $A_3$ | 0.7746 | 0.6124 | 0.6303 | 0.7530 | 0.4946 |
| $A_4$ | 0.1836 | 0.1475 | 0.1680 | 0.1964 | 0.3559 |
| $A_5$ | 0.4954 | 0.4163 | 0.3698 | 0.3057 | 0.2132 |

- In this step the positive –ideal solution A* and negative –ideal solution A⁻ is calculated.

$$A^{+} = \{(\max_i v_{ij} | j \in J), (\min_i v_{ij} | j \in J') | i = 1,2,...m\} = \{v_1^{+}, v_2^{+}, ..., v_j^{+}, ..., v_n^{+}\}$$

$$A^{-} = \{(\min_i v_{ij} | j \in J), (\max_i v_{ij} | j \in J') | i = 1,2,...m\} = \{v_1^{-}, v_2^{-}, ..., v_j^{-}, ..., v_n^{-}\}$$

$$\text{where } J = \{j = 1,2,...,n | j \text{ associated with benefit criteria}\}$$
$$J' = \{j = 1,2,...,n | j \text{ associated with cost criteria}\}$$

Then we calculate the closeness coefficient ( $CC_i$ ) for each alternatives as,

$$CC_i = \frac{d_i^-}{d_i^- + d_i^*}, \quad i = 1,2,...,m$$

Where,

$$d_i^* = \sqrt{\sum_{j=1}^{n}\left(v_{ij} - v_j^{+}\right)^2} \qquad i = 1,2,...,m$$

It can be noted that

$$CC_i = \begin{cases} 1 & \text{if } A_i = A^* \\ 0 & \text{if } A_i = A^- \end{cases}$$

The result of above calculation is shown in Table 8.

$$d_i^- = \sqrt{\sum_{j=1}^{n}\left(v_{ij} - v_j^{-}\right)^2} \qquad i = 1,2,...,m$$

Based on the value of $CC_i$, $A_3 > A_5 > A_1 > A_2 > A_4$, It shows that Computerization of Materials management leads over other (Table 9). The computerization priorities can be seen in Table 10.

So, the company shall first computerize the materials management department followed by computerization of customer interface. Next departments as per the order for computerization shall be purchase, logistics and production department respectively.

*Table 8. Positive and negative ideal solutions*

| | $C_1$ | $C_2$ | $C_3$ | $C_4$ | $C_5$ |
|---|---|---|---|---|---|
| $A_1$ | 0.14097 | 0.13778 | 0.09381 | 0.07224 | 0.06295 |
| $A_2$ | 0.06284 | 0.05274 | 0.07651 | 0.05544 | 0.02756 |
| $A_3$ | 0.34393 | 0.13778 | 0.11535 | 0.12500 | 0.04451 |
| $A_4$ | 0.08152 | 0.03319 | 0.03076 | 0.03259 | 0.03230 |
| $A_5$ | 0.21994 | 0.09367 | 0.06767 | 0.05074 | 0.01918 |
| $A^*$:*Positive ideal solution* | 0.34393 | 0.13778 | 0.11535 | 0.12500 | 0.06295 |
| $A^-$:*Negative ideal solution* | 0.06284 | 0.03319 | 0.03076 | 0.03259 | 0.01918 |
| $d^*$:*Positive separation measure* | 0.20792 | 0.18155 | 0.09423 | 0.29649 | 0.14777 |
| $d^-$: *Positive separation measure* | 0.16037 | 0.08593 | 0.21589 | 0.09515 | 0.18766 |

*Table 9. Computation of* $d_i^-, d_i^*$ and $CC_i$

| | $d_i^*$ | $d_i^-$ | $d_i^* + d_i^-$ | $CC_i = \frac{d_i^-}{d_i^- + d_i^*}$ |
|---|---|---|---|---|
| $A_1$ | 0.20792 | 0.16037 | 0.36829 | 0.435445 |
| $A_2$ | 0.18155 | 0.08593 | 0.26748 | 0.321258 |
| $A_3$ | 0.09423 | 0.21589 | 0.31012 | 0.69615 |
| $A_4$ | 0.29649 | 0.09515 | 0.39164 | 0.242953 |
| $A_5$ | 0.14777 | 0.18766 | 0.33543 | 0.559461 |

## CONCLUSION

In today's dynamic and challenging marketplace the companies are facing acute competition from the competitors. This has compelled the organizations to be more cost effective and be more responsive to the market. More and more companies are now competing on their supply chain to have cutting edge over others. The challenge for successful supply chain management lies in developing strategies which will minimize the cost and maximize the flexibility. Hence the implementation of information technology is one of the important factors. This paper proposes an integrated multiple criteria decision model for IT implementation prioritization in a supply chain. This is considered as one of the critical decision making process for improving the efficiency of the supply chain. A novel integrated technique using two methods AHP and TOPSIS, is used to prioritize IT implementation in an organisation. Five department viz. materials management, customer interface, purchase, logistics and production department are considered for IT implementation prioritization. The method has been elicited with the help of real life case study. The case study problem deals with five criteria and five alternatives. The method prioritizes the five alternatives based on five criteria. The output of AHP method is used as an input to the second method TOPSIS. The method uses the concept of positive-ideal solution and negative-ideal solution for solving the problem. The alternatives are ranked on the basis of value of closeness coefficient which is

*Table 10.*

| Priority Order | Process | Code |
|---|---|---|
| 1 | IT implementation in Material Management | $A_3$ |
| 2 | IT implementation in Customer Interface | $A_5$ |
| 3 | IT implementation in Purchase Department | $A_1$ |
| 4 | IT implementation in Logistics Department | $A_2$ |
| 5 | IT implementation in Production Management | $A_4$ |

calculated for each alternative. According to the closeness coefficient, the ranking order of five alternatives has been determined as $A_3 > A_5 > A_1 > A_2 > A_4$. So, the company shall first computerize the materials management department followed by computerization of customer interface. The other departments as per the order for computerization shall be purchase, logistics and production department respectively. The management of any similar type of company can use the procedure for their IT implementation prioritization.

## ACKNOWLEDGMENT

The author acknowledges the Editor-in-Chief of the journal, Professor John Wang, and the anonymous reviewers for their indispensable input that improved the paper significantly.

## REFERENCES

Agrawal, D. K. (2001). 7 Ms to make supply chain relationships harmonious. In *Proceeding of National Conference on People, Processes and Organizations: Emerging Realities.* New Delhi, India: Excel books.

Altekar, R. V. (2005). *Supply chain management concept and cases*. Washington, DC: IEEE.

Andel, T. (1997). Information supply chain: set and get your goals. *Transportation and Distribution, 38*(2), 33.

Auramo, J., Kauremaa, J., & Tanskanen, K. (2005). Benefits of IT in supply chain management: an explorative study of progressive companies. *International Journal of Physical Distribution & Logistics Management, 35*(2), 82–100. doi:10.1108/09600030510590282

Balsmeier, P. W., & Voisin, W. J. (1996). Supply chain management: a time based strategy. *Industrial Management (Des Plaines), 38*(5), 24–27.

Bechtel, C., & Jayaram, J. (1997). Supply chain management: a strategic perspective. *The International Journal of Logistics Management*, 15-34.

Bharadwaj, A. S. (2000). A resource-based perspective on information technology capability and firm performance: an empirical investigation. *Management Information Systems Quarterly, 24*(1), 169–196. doi:10.2307/3250983

Boubekri, N. (2001). Technology enablers for supply chain. *Integrated Manufacturing Systems, 12*(6), 394–399. doi:10.1108/EUM0000000006104

Bowersox, D. J., & Closs, D. C. (1996). *Logistical management: the integrated supply chain process*. New York: McGraw-Hill.

Byrd, T. A., & Davidson, N. W. (2003). Examining possible antecedents of IT impact on the supply chain and its effect on firm performance. *Information & Management, 41*(3), 243–255. doi:10.1016/S0378-7206(03)00051-X

Carr, P., Rainbird, M., & Walters, D. (2004). Measuring the implications of virtual integration in the new economy: a process-led approach. *International Journal of Physical Distribution & Logistics Management, 34*(3), 358–372. doi:10.1108/09600030410533646

Carter, J. R., Ferrin, B. G., & Carter, C. R. (1995). The effect of less-than-truckload rates on the purchase order lot size decision. *Transportation Journal, 34*(4), 35–44.

Chopra, S., & Meindl, P. (2003). *Supply chain management – strategic planning and operation* (2nd ed.). Upper Saddle River, NJ: Prentice Hall.

Felix, T. S., Chan, K. N., Tiwari, M. K., Lau, H. C. W., & Choy, K. L. (2008). Global supplier selection: a fuzzy-AHP approach. *International Journal of Production Research, 46*(14), 3825-3857.

Guo, J.-y., Liu,, J., & Qiu, L. (2006). Research on supply chain performance evaluation based on DEA/AHP model. In *Proceedings of the 2006 IEEE Asia-Pacific Conference on Services Computing* (pp. 609-612).

Gustin, C. M., Daughery, P. J., & Stank, T. P. (1995). The effect of information availability on logistics integration. *Journal of Business Logistics, 16*(1), 1–21.

Hewitt, F. (1994). Supply chain redesign. *The International Journal of Logistics Management*, 1-9.

Hwang, C. L., & Yoon, K. (1981). *Multiple attribute decision making methods and applications*. Berlin: Springer Verlag.

Kearns, G. S., & Lederer, A. L. (2004). The impact of industry contextual factors on IT focus and the use of IT for competitive advantage. *Information & Management, 41*(2), 899–919. doi:10.1016/j.im.2003.08.018

Klaus, P, Henning, H., Muller-Steinfahr, U., & Stein. (1993). The promise of interdisciplinary research in logistics. In *Proceedings of the Twenty second Annual Transportation and Logistics Educators Conference* (pp. 161-87).

Korpela, J., Lehmusvaara, A., & Tuominen, M. (2001). An analytic approach to supply chain development. *International Journal of Production Economics, 71*(1-3), 145–155. doi:10.1016/S0925-5273(00)00114-6

Lewis, I., & Talalayevsky, A. (1997). Logistics and information technology: a coordination perspective. *Journal of Business Logistics, 18*(1).

Liao, S. H., Chem, Y. M., & Liu, F. H. (2004). Information technology and relationship management: a case study of Taiwan's small manufacturing firm. *Technovation, 24*(2), 97–108. doi:10.1016/S0166-4972(02)00037-8

Lin, F., Sheng, O., & Wu, S. (2005). An integrated framework for e-chain bank accounting systems. *Industrial Management & Data Systems, 105*(3), 291–306. doi:10.1108/02635570510590129

Lombardo, S. (2001). *AHP reference listing*. Retrieved January 2, 2003, from http://www.expertchoice.com/ahp/default.htm

Min, W. (2007). Topsis-AHP simulation model and its application to supply chain management. *World Journal of Modelling and Simulation, 3*(3), 196–201.

Narasimhan, R., & Kim, S. W. (2001). Information system utilization strategy for supply chain integration. *Journal of Business Logistics, 22*(2), 51–75.

Raghuram, G., & Rangraj, N. (2000). *Logistics and supply chain management, concepts and cases*. New Delhi, India: Macmillan India Ltd.

Rahman, Z. (2004). Use of internet in supply chain management: a study of Indian companies. *Industrial Management & Data Systems*, *104*(1), 31–41. doi:10.1108/02635570410514070

Rahman, Z. (2006). Integrating the physical, information and financial flows - the next corporate paradigm. *The ICFAI Journal of Supply Chain Management*, *3*(3), 12–22.

Rainbird, M. (2004). Demand and supply chain: the value catalyst. *International Journal of Physical Distribution & Logistic Management*, *34*(3), 230–250. doi:10.1108/09600030410533565

Richmond, B., Burns, A., Maybe, J., Nutthsll, L., & Toole, R. (1998). Supply chain management tools: minimizing the benefits. In Gattoma, J. (Ed.), *Strategic Supply Chain Alignment: Best Practices in Supply Chain Management* (pp. 509–520). Aldershot, UK: Gower.

Saaty, T. L. (1980). *The analytic hierarchy process*. New York: McGraw-Hill.

Saharidis, K. D., Georgios, K., Vassilis, S., & Dallery, Y. (2009). Centralized and decentralized control polices for a two-stage stochastic supply chain with subcontracting. *International Journal of Production Economics;*,(n.d) 117–126. doi:10.1016/j.ijpe.2008.10.001

Sanders, N. R., & Premus, R. (2002). IT applications in supply chain organizations: a link between competitive priorities and organizational benefits. *Journal of Business Logistics*, *23*(1), 65–83.

Sevkli, M., Koh, S. C., Zaim, L., Selim, D. M., & Tatoglu, E. (2007). An application of data envelopment analytic hierarchy process for supplier selection: a case study of BEKO in Turkey. *International Journal of Production Research*, *45*(9), 1973–2003. doi:10.1080/00207540600957399

Sharma, M. K., & Rajat, B. (2007). An integrated BSC-AHP approach for supply chain, management evaluation. *Measuring Business Excellence*, *11*(3), 57–68. doi:10.1108/13683040710820755

Shere, S. A. (2005). From supply chain management to value network advocacy: implication for e-supply chains. *Supply Chain Management: An International Journal*, *10*(2), 77–83. doi:10.1108/13598540510589151

Singh, N., Lai, K.-H., & Cheng, T. C. E. (2007). Intra-organizational perspectives on IT-enabled supply chains. *Communications of the ACM*, *50*(1), 59–65. doi:10.1145/1188913.1188918

Stock, J. R. (1997). Applying theories from other disciplines of logistics. *International Journal of Physical Distribution & Logistics Management*, *27*(9/10), 515–539. doi:10.1108/09600039710188576

Svensson, G. (2003). Holistic and cross-disciplinary deficiencies in the theory generation of supply chain management. *Supply Chain Management: An International Journal*, *8*(4), 303–316. doi:10.1108/13598540310490062

Thompson, P. (1998). Bank lending and the environment: policies and opportunities. *International Journal of Bank Marketing*, *16*(6), 243–252. doi:10.1108/02652329810241384

Thurstone, L. L. (1931, September). Multiple factor analysis. *Psychological Review*, *38*(5), 406–427. doi:10.1037/h0069792

Trappey, A. J. C., Trappey, C. V., Hou, J. L., & Chen, B. J. G. (2004). Mobile agent technology and application for online global logistic services. *Industrial Management & Data Systems*, *104*(1/2), 169–184. doi:10.1108/02635570410522143


*This work was previously published in International Journal of Information Systems and Supply Chain Management, Volume 3, Issue 4, edited by John Wang, pp. 83-96, copyright 2010 by IGI Publishing (an imprint of IGI Global)*

# Chapter 3
# Moderated Multiple Regression of Absorptive Capacity Attributes and Deployment Outcomes:
## The Importance of RFID IT Infrastructure Integration and Supply Chain Process Integration

**Rebecca Angeles**
*University of New Brunswick Fredericton, Canada*

## ABSTRACT

*In this study, the author examines organizations' perceptions of the importance of absorptive capacity attributes in the deployment of radio frequency identification (RFID) in a supply chain and their relationships with operational efficiency and market knowledge creation as moderated by information technology infrastructure integration and supply chain process integration. Data was collected using a survey questionnaire administered online to members of the Council of Supply Chain Management Professionals (CSCMP). Four proposed hypotheses were partially supported in this study. Both variables, IT infrastructure integration and supply chain process integration, moderate the relationships between three predictor variables, business process modularity, standard electronic business interfaces, and breadth of information exchange and the two dependent variables examined in this study, operational efficiency and market knowledge creation to a considerable extent. This study has clear implications for how decision makers affecting their firm's supply chains should make a business case for robust IT elements that support both IT infrastructure integration and supply chain process integration.*

DOI: 10.4018/978-1-4666-0918-1.ch003

## INTRODUCTION

Radio frequency identification (RFID) studies conducted in the last couple of years indicate the reluctance of firms to implement it due to a variety of reasons: scepticism concerning the ability of RFID to deliver cost savings or positive return on investment in the near future; lack of knowledge about RFID implementation; inability to make a business case supporting RFID; concern over the level of financial resources required for RFID implementation; and lack of clarity about the protection and security of the information that needs to be embedded in the RFID tags (Dos Santos & Smith, 2008; (Reyes et al., 2007; Godon, Visich, & Li, 2007; Vijayaraman & Osyk, 2006). On the more positive side, RFID studies on the use of the tags for inventory management have also been coming out demonstrating potential and actual benefits in terms of operational efficiencies and reduction in losses due to shrinkage or theft (Lee, Cheng, & Leung, 2009; Rekik, Sahin, & Dallery, 2009; De Kok, Van Donselaar, & Van Woensel, 2008; Szmerekovsky & Zhang, 2008; Moon & Ngai, 2008). Practical guidelines on RFID implementation have also surfaced to help firms seriously considering it plan ahead and create a business case for an earnest initiative (Angeles, 2005; Reyes et al., 2007).

This study approaches RFID implementation from a different angle, focusing on the organizational learning aspects of its deployment, thus, contributing a new exploration into the technology's future prospects within the firm. Organizational learning within the supply chain context using RFID-enabled systems is of great interest nowadays when mandates for RFID used are being issued by powerful channel masters (i.e., hub firms). RFID initiatives are interenterprisewide system applications that require mutual buy in and learning experiences between and among value chain participants. As trading partners seek to pursue initiatives of this scale, they would be embarking in knowledge gaining experiences, with the "conscripted" trading partners following the lead of and learning from the hub firm introducing the use of RFID such as the case of Wal-Mart in order to gain operational efficiency and/or market knowledge creation. This study uses the concepts of absorptive capacity and the dynamic capabilities perspective (DCP) to frame this major current challenge to supply chains implementing RFID. Thus, this study looks at firms' perceptions of the importance of absorptive capacity attributes in the deployment of RFID in a supply chain context and their relationships with two RFID system outcomes—operational efficiency and market knowledge creation. More importantly, this study investigates the ability of information technology infrastructure integration and supply chain process integration to moderate the relationships between the absorptive capacity attributes and these RFID deployment outcomes.

Zahra and George (2002) define "absorptive capacity" as "...a set of organizational routines and processes by which firms acquire, assimilate, transform, and exploit knowledge to produce a dynamic organizational capability. (2002, p. 186)" Within the context of the RFID supply chain, a firm's ability to value, assimilate, and apply knowledge received from external sources such as suppliers, customers, competitors, and alliance partners can boost its future chances of gaining new knowledge and sensitivity to the significance of new information emerging in the marketplace (Cohen & Levinthal, 1990; Lindsay & Norman, 1977). The absorptive capacity capability involves the use of valued organizational resources such as socially complex routines to support an "information architecture" (Knudsen & Madsen, 2002) that enables the transfer of knowledge and facilitate communication among parties involved and a large and active network of internal and external relationships, supported by cross-functional interfaces (Collis, 1994; Hall, 1992; Cohen & Levinthal, 1990).

The dynamic capabilities perspective, on the other hand, refers to the capability of firms to renew its competencies in terms of organizational resources needed to align themselves with environmental business demands (Teece, Pisano, & Shuen, 1997; Eisenhardt & Martin, 2000). Firms with this capability continually build, adapt, and reconfigure both internal and external resources to meet challenges such as keeping up with rapid technological changes and unpredictable customer demands. Cultivating competitive advantage depends, to a great extent, on the firm's ability to convert knowledge into key capabilities in responding to environmental demands (Lane & Lubatkin, 1998). This challenge becomes even more onerous in turbulent environments where a lack of organizational learning capacity could account for a firm's inability to adopt important emerging technologies (Huber, 1996).

The absorptive capacity concept is appropriate here and is relevant to the dynamic capabilities perspective (DCP) in assessing a firm's current state of "fitness" as it tries to meet marketplace challenges.

A few examples of using absorptive capacity attributes within the RFID-enabled value chains illustrate the point being made. Wal-Mart is a world-class retailer that has mandated its top suppliers to use RFID in cases and pallets of products delivered to designated distribution centers.

Beaver Street Fisheries, a Jacksonville, Florida frozen seafood dealer, is among Wal-Mart's top 300 suppliers required to use RFID by 2006. Although the firm responded to this mandate, its executives made its learning curve steeper by embarking on a three-phased approach to RFID adoption in order to increase its volume of business not only with Wal-Mart but also with other major customers as well (Wasserman, 2007). In the first phase of implementation, the firm tried to overcome technical problems in using RFID tags on account of the fact that seafood products contain a high concentration of ice which interferes with read rates. Various products in different pallet configurations were tested until read rates reached at least 90 percent. In the second phase, the firm's warehouse management system was configured to electronically communicate retail orders for RFID-tagged seafood products directly to the assembly line. In the third phase, a number of things were intended to be accomplished. The firm's receiving operations will be automated through the use of RFID interrogators that will read RFID-tagged cases and pallets upon arrival, thus, replacing the current tedious manual scanning of bar codes. Furthermore, the following business processes will be automated: shipping and receiving; issuing electronic billing to customers; and sending customers advanced ship notices.

Lego Systems, the U.S. division of building-block toy manufacturer LEGO Group, also complied with the RFID mandates of Wal-Mart and Target as well. Lego Systems' Enfield, Connecticut distribution center implemented an RFID system which was integrated with its existing shipping and order verification business processes (O'Connor, 2006). The firm's sales department specifically benefited from tracking promotions by analyzing read points of RFID-tagged cases in the retailers' stores and compared the readings with sales data. By doing so, Lego Systems determined if the products shipped out to these retailers during product promotion periods were actually put on display in the shelves and if product replenishments arrived on time.

Canus, a Canadian manufacturer of skin care products based on goat's milk and a supplier of Wal-Mart, was different in that the firm proactively anticipated Wal-Mart's RFID mandate and embarked on experimentation prior to the actual issuance of the mandate (Collins, 2004). Working with Canadian vendor Ship2Save, Canus implemented a multiphase project to test, configure, and use RFID in its three distribution centers and two manufacturing sites. Canus will focus on RFID's ability to monitor the temperature of the products in real-time en route from the warehouses to the customers. The firm's personal care products ranging from soaps to fragrances have been made from plant extracts and oils and goat's milk that make it imperative that the products be kept within the temperature range of 25 and 104 degrees Fahrenheit to avoid spoilage.

## LITERATURE REVIEW

### Absorptive Capacity: Four Dimensions of the Concept of "Absorptive Capacity"

The following are the four dimensions of the concept of "absorptive capacity:"

1. Acquisition: the firm's capability to identify and acquire knowledge critical to its operations from sources external to the firm (Zahra & George, 2002)
2. Assimilation: the use of routines and processes that support the analysis, interpretation, and comprehension of the external knowledge obtained by the firm (Kim, 1997a; Kim, 1997b; Szulanski, 1996)
3. Transformation: the firm's capability to combine the firm's existing knowledge base with the newly acquired and assimilated information; and
4. Exploitation: the firm's capability to refine, extend, and leverage its existing competencies by incorporating acquired external knowledge and using the combination for the benefit of its operations

### Concepts supporting absorptive capacity used in this study

#### Routinization

By routinizing tasks, the firm is able to spend just enough time to the process of transforming inputs into outputs (Galunic & Rodan, 1998; Perrow, 1967). Repetitive and structured tasks are ideal for routinization (Hage & Aiken, 1967; Perrow, 1967; Withey, Daft, & Cooper, 1983). In this study, routinization is expressed in a number of ways:

a. Use of interorganizational business process modularity
b. Use of standard electronic business interfaces, and
c. The exchange of coordination information

#### Interorganizational Business Process Modularity

"Modularity" allows the structuring of interorganizational business processes so that those who support them need conduct only the minimum amount of coordination communication while maximizing rich information exchange (Malhotra, Gosain, & El Sawy, 2005). By breaking up interrelated business processes into subprocesses, trading partners can undertake their respective tasks independently and simultaneously, thus, enhancing expeditious performance across the chain.

#### Standard Electronic Business Interfaces

The exchange of information among firms is facilitated by the use of standard electronic business interfaces to handle the interoperability of both the data and business processes. A supplier, for instance, may need to broadcast changes in product data such as price changes or the discontinuation of a certain model or color, to all its valued customers. In terms of the data, standards form the foundation of the data architecture needed to define the structure of the data and the relationships among data entities in order to achieve data consistency, a key requirement for interorganizational data sharing (Van Den Hoven, 2004).

#### Coordination Information Exchanged

A classical problem in the supply chain involves the "bull whip effect," or the amplification of the variability of order information communicated in the supply chain causing problems (Moyaux & Chaib-draa, 2007). To reduce the "bullwhip effect," the customer firm needs to communicate both the original demand information and subsequent revisions to it to the supplier firm. This pervasive supply chain problem clearly demonstrates the importance of exchanging coordination information.

## Interpretation Systems

After collecting a considerable amount of information across trading partners, there is a need to organize, rearrange, process, and interpret this information. "Data mining" or the process of analyzing data to reveal useful patterns and relationships hidden in the data could help here (Rupnick, Kukar, & Krisper, 2007). Data mining makes use of statistical, pattern recognition, and machine learning methods to enable data analysis and the discovery of insights embedded in the data (Spangler, Gal-Or, & May, 2003). The ability to gain insights from the data using information technology-supported tools has contributed to knowledge creation (Cooper et al., 2000) and important customer behaviour predictions that are a key focus of customer relationship management systems (Padmanabhan & Tuzhilin, 2003).

## Memory Systems for Interorganizational Activities

"Organizational memory" refers to the saving, representation, and sharing of corporate knowledge (Croasdell, 2001) that can be used by members of the firm in carrying on regular operations and responding to environmental challenges as well (Stein, 1995; Huber, 1991; Walsh & Ungson, 1991; Pralahad & Hamel, 1990). In the context of today's complex supply chain activities, organizational memory embedded in electronic datawarehouses, databases, filing systems, and manuals, could support multiple interrelated tasks spanning diverse corporate environments (Ackerman, 1996).

## Partner Interaction

"Partner interaction" is defined as the extent to which the partnering firms interact with each other in terms of trust, adjustment, and conflict (Chen, 2004). Prior studies have recognized the importance of trust to the alliance performance during the interfirm cooperation period (Casson, 1991; Buckley & Casson, 1988; Larson, 1991). In this study, "partner interaction" will be operationalized in terms of: joint decision making, exchange of privileged information, quality of information, and breadth of information.

### Joint Decision Making

Jansen, Van Den Bosch, and Volberda (2005) articulate the need for organizational mechanisms associated with coordination capabilities for absorptive capacity that include the importance of participation in decision making. This study extends this concept within the context of supply chain partnerships. The focal firm's efforts to invite key supply chain partners participating in the RFID system implementation in the decision making process allows the focal firm to receive knowledge from these partners (i.e., external sources) and enhance the performance of the system.

### Privileged Information Exchanged

As trading partners move closer to each other, the nature of the information exchanged also changes and they are far more willing to share "privileged" information that is specific to the trading partner (Malhotra, Gosain, & El Sawy, 2005). This usually include proprietary and confidential information that provides the receiving firm far more insightful information that will enable it to innovate alongside its trading partner and mutually structure their competencies and business processes (Uzzi & Lancaster, 2003).

### Breadth of Information Exchanged

To achieve "breadth of information," firms should share more than the standard, transactional, operational data and be willing to exchange information that informs trading partners of higher-level issues such as changes in marketplace conditions, shifting customer tastes, new product/service attributes, emerging technologies, competitive opportunities (Anand, Manz, & Glick, 1998; Child & Faulkner, 1998; Austin, Lee, & Kopczak, 1997; Fites, 1996).

### Quality of Information Exchanged

In investigating the relationship between knowledge and work performance, Lee and Strong (2004) considered the information quality attributes of relevancy, timeliness, accuracy, and completeness as relevant. O'Brien (2001) looks at the time, content, and form dimensions of information quality. The time dimension refers to timeliness, currency, and frequency; the content dimension refers to accuracy, relevance, completeness, conciseness, performance, and scope; and, finally, the form dimension refers to clarity, media used, order, and manner of presentation. In this study, quality of information is measured in terms of timeliness and completeness of the information exchanged (Malhotra, Gosain, & El Sawy, 2005).

## Absorptive Capacity Outcomes

The absorptive capacity attributes in this study will be associated with two outcomes, operational efficiency and market knowledge creation (Malhotra, Gosain, & El Sawy, 2005). Miles and Snow (2007) found that in the first period of the development of supply chains, their primary focus was on making operations more efficient. A showcase illustration of achieving operational efficiency involves Wal-Mart's deployment of RFID that has led to lower shipment processing costs and lead times, resulting in smaller cycle inventory and safety stocks (Kumar, 2007).

Miles and Snow (2007) also noted that in the second period of their evolution, the focus has shifted from operational efficiency to the achievement of effectiveness as trading partners shared ideas and expertise on how to manage the entire chain and in the more recent time, how to ensure supply chain performance at the industry level. This concept is closely related to "market knowledge creation," the second system deployment outcome explored in this study. Chow, Choy, and Lee (2007) analyzed the applications and technologies adopted when developing the knowledge management system (KMS) in build-to-order supply chains (BOSC). They found that initial KMS applications being developed have focused on single knowledge problem to allow participating firms achieve operational excellence and indicated the need for creating KMS systems supporting knowledge coordination across corporate boundaries to reach BOSC integration. More recent work on supply chain management and KMS appear to be responding to this need for more intensive knowledge creation explorations. In solving a knowledge-based SCM problem involving an IT-driven normative model of vendor-managed inventory system in a two-echelon supply chain comprising of $m$ vendors and $n$ buyers (i.e., two-echelon Multiple Vendor Multiple Buyers Supply Chain (TMVMBSC), Nachiappan, Gunasekaran, and Jawahar (2007) propose using genetic algorithm to discover the optimal transaction quantities for each buyer in the value chain. Raisinghani and Meade (2005) developed a decision model to assist higher management in responding to a dynamic and competitive environment by determining which construct of KM is most important based on an organization's performance criteria, dimensions of agility, and supply-chain drivers.

## Proposed Moderator Variables

### IT Infrastructure Integration Capability

IT infrastructure integration is defined as the degree to which a focal firm has established IT capabilities for the consistent and high-velocity transfer of supply chain-related information within and across its boundaries. This study closely looks at the IT infrastructure integration requirements needed to support the use of RFID within a supply chain context. The formative construct introduced by Rai, Patnayakuni, and Seth (2006) was adopted in this study and used both conceptually and in the instrumentation as well. They define IT infrastructure integration in terms of two subconstructs, data consistency and cross-functional SCM application systems integration. The IT infrastructure needed to support RFID systems should be able to provide real-time information visibility, made possible by collecting data at much lower levels of granularity made possible by RFID. The criticality of data consistency is underlined by this quote: "...having good, usable, uniform data is the foundation for sharing and processing data to achieve a collaborative, tighter, supply chain." (Pagarkar, Natesan, & Prakash, 2005, p. 19). Also, initiatives like Global Data Synchronisation (GDS) will facilitate the achievement of data consistency by synchronizing master data that uniquely describe the product or services being exchanged among trading partners (Patni Americas, Inc., 2008; Holloway, 2006; Pagarkar, Natesan, & Prakash, 2005). The expected "data overload" that will occur with the collection of RFID data only amplifies the need for data consistency. Other data consistency issues include:

1. Missing data attributes: some data attributes needed by retailers, for instance, are not being provided by their suppliers;
2. Dissimilar data characteristics: lack of data similarity may be due to a variety of things—different field sizes, different data formats, etc.; and
3. Varying data nomenclature used: trading partners use different data field names so much so that even if the data entity contains all required attributes, valid data being transmitted from one trading partner might still be rejected by the RFID system of another trading partner—a major problem if specific values for certain fields are expected to trigger downstream transaction processing (Shutzberg, 2004).

### Data Consistency

The extent to which data has been commonly defined and stored in consistent form in databases linked by supply chain business processes is referred to as data consistency (Rai, Patnayakuni, & Seth, 2006). Data consistency is a key requirement in creating a data architecture that defines the structure of the data and the relationships among data entities that is fundamental in establishing interorganizational data sharing (Van Den Hoven, 2004). Simchi-Levi, Kaminsky, and Simchi-Levi (2004) note that recently, many suppliers and retailers observed that despite the lack of variation in customer demand for products, inventory and back-order levels vary, nevertheless, across many supply chains, oddly enough. Sharing consistent data upstream and downstream in the supply chain is one major solution to overcoming the bullwhip effect.

Data from legacy systems of supply chain trading partners need to be accessed to produce useful, integrated data and to be able to transport this data into various datawarehouse structures. Often, data from diverse sources is inconsistent and unusable for the integration purposes required for supply-chain wide initiatives.

## Cross-Functional SCM Application Systems Integration

Cross-functional supply chain management applications systems integration is defined by Malhotra, Gosain, and El Sawy (2005) as the level of real-time communication of a hub firm's functional applications that are linked within an SCM context and their exchanges with enterprise resource planning (ERP) and other related inter-enterprise initiatives like customer relationship management (CRM) applications. At the lowest level, an ERP system is essential in enabling the seamless integration of information flows and business process across functional areas of a focal firm --- this is normally referred to as "ERP I" (Law & Ngai, 2007). ERP functionalities are important control and management mechanisms that are connected with the ERP systems of the firm's trading partner --- referred to as "ERP II". ERP implementations are growing more extensive and interconnected among firms in linked value chains. Karimi, Somers, and Bhattacherjee (2007) found that ERP projects with greater functional, organizational, or geographic scope result in higher positive shareholder returns.

To obtain optimum results, supply chain trading partners have to inevitably approach a collaborative posture in their relationships which would rely heavily on cross-functional interenterprise integration. To facilitate the realization of this goal, the Supply Chain Council (SCC), a not-for-profit corporation, has endorsed the Supply Chain Operations Reference Model (SCOR) as the cross-industry standard for supply chain management (Holloway, 2006).

The SCOR model encompasses the business processes involved in five distinct areas:

1. Demand/supply planning and management
2. Sourcing stocked, make-to-order, and engineer-to-order products
3. Make-to-stock, make-to-order, and engineer-to-order production execution
4. Order, warehouse, transportation and installation management for stocked, make-to-order, and engineer-to-order products; and
5. Return of raw materials (to suppliers) and receipt of returns of finished goods (from customers), including defective products, maintenance, operations, and repair products, and excess products (Holloway, 2006)

This model is used as a reference model by firms in order to address, improve, and communicate supply chain management practices among trading partners. Collaborative supply chain endeavours that involve coordinating inter-firm business processes to achieve supply chain wide integration are facilitated by tools like the SCOR model.

## Supply Chain Process Integration Capability

In this study, supply chain process integration is defined following the construct used by Malhotra, Gosain, and El Sawy (2005): the degree to which a hub firm has integrated the flow of information (Lee, Padmanabhan, & Whang, 1997), physical materials (Stevens, 1990), and financial information (Mabert & Venkatraman, 1998) with its value chain trading partners. This formative construct has three subconstruct components: information flow integration, physical flow integration, and financial flow integration (Mangan, Lalwani, & Butcher, 2008).

Information has the potential to reduce variability in the supply chain, enable suppliers to make better forecasts (i.e., more accurately accounting for effects of promotions and market changes, for instance), enable the coordination of manufacturing and distribution strategies, enable lead time reduction, and enable retailers to service their customers better by providing preferred items and avoiding out of stock situations (Simchi-Levi, Kaminsky, & Simchi-Levi, 2004).

This study uses the construct, information flow integration, to mean the degree to which a firm exchanges operational, tactical, and strategic information with its supply chain trading partners (Malhotra, Gosain, & El Sawy, 2005). The instrument used in this study measures the sharing of production and delivery schedules, performance metrics, demand forecasts, actual sales data, and inventory data, for information flow integration.

Malhotra, Gosain, and El Sawy (2005) define physical flow integration as the level to which the hub firm uses global optimization with its value chain partners to manage the flow and stocking of materials and finished goods. Raw materials, subassemblies, and finished goods constitute downstream physical flows, whereas returned products for defects or repairs make up the upstream physical flows. In this study, physical flow integration is measured in terms of multi-echelon optimization of costs, just-in-time deliveries, joint management of inventory with suppliers and logistics partners, and distribution network configuration for optimal staging of inventory (Malhotra, Gosain, & El Sawy, 2005).

Financial flow integration is defined as the level to which a hub firm and its trading partners exchange financial resources in a manner driven by workflow events (Malhotra, Gosain, & El Sawy, 2005). Value chain participants that do not have well-designed business processes often do not have consistent views of important financial downstream flows such as prices, invoices, credit terms, and upstream financial flows that could include payments and account payables (McCormack & Johnson, 2003).

Accurate financial flows are enabled by event-based workflow systems that trigger electronic payments, for instance, upon delivery of goods. Interacting business processes within a value chain should be reengineered so that participating firms can experience the following benefits:

1. Reduce the costs of billing, payment processing, and dispute handling
2. Shorten the invoicing and receivables cycle time
3. Accelerate the rate of payments
4. Make relevant financial information accessible for high-level decision making
5. Improve customer relationships by gathering information on customer preferences with billing and invoicing transactions; and
6. Positively influence revenue growth by improving cash flow availability for various reasons like production ramp-up with spikes in customer demand or develop new products/services for innovation (Greenfield, Patel, & Fenner, 2001)

In this study, the financial flow integration items measure the automatic triggering of both accounts receivables and accounts payables (Malhotra, Gosain, & El Sawy, 2005).

## HYPOTHESES TO BE TESTED

This study purports to test the following hypotheses:

**H1:** *The positive relationship between each of the absorptive capacity attributes and operational efficiency will be moderated by IT infrastructure integration --- i.e., the higher the level of IT infrastructure integration, the greater the positive relationship between each of the absorptive capacity attributes and operational efficiency.*

**H2:** *The positive relationship between each of the absorptive capacity attributes and market knowledge creation will be moderated by IT infrastructure integration --- i.e., the higher the level of IT infrastructure integration, the greater the positive relationship between each of the absorptive capacity attributes and market knowledge creation.*

**H3:** *The positive relationship between each of the absorptive capacity attributes and operational efficiency will be moderated by supply chain process integration --- i.e., the higher the level of supply chain process integration, the greater the positive relationship between each of the absorptive capacity attributes and operational efficiency.*

**H4:** *The positive relationship between each of the absorptive capacity attributes and market knowledge creation will be moderated by supply chain process integration --- i.e., the higher the level of supply chain process integration, the greater the positive relationship between each of the absorptive capacity attributes and market knowledge creation.*

## RESEARCH METHODOLOGY

Data for this pilot research study was collected using a survey questionnaire administered online to members of the Council of Supply Chain Management Professionals (CSCMP). The specific items used for IT infrastructure integration (i.e., data consistency and cross-functional application integration) and supply chain process integration were borrowed from Rai, Patnayakuni, and Seth (2006), while the items for the nine absorptive capacity attributes were drawn from Malhotra, Gosain, and El Sawy (2005).

Since the organizations have not yet implemented RFID, the survey respondent was asked to indicate their perceptions of the importance of the nine absorptive capacity attributes in order to gain operational efficiency and market knowledge creation, using multiple items per construct. The same approach was used in anticipating their perceptions of the use of the RFID system in achieving data consistency, cross-functional application integration, and supply chain process integration. Seven-point Likert scales were used with minimum-maximum anchoring points appropriate to the construct being measured. The computer program SPSS version 15 was used in conducting a series of simple regression data analyses and their associated moderated regression analysis runs.

## Data Measurement Properties

The internal consistency of the items constituting each construct was assessed using Cronbach's alpha and the results are in conformance with Nunnally's (1978) guidelines of getting values of .70 or above. Generally speaking, the items have internal consistency with values beyond the .70 threshold recommended. The nine absorptive capacity attributes showed the following reliability results:

1. Joint decision making (Cronbach alpha=.973)
2. Business process modularity (Cronbach alpha=.964)
3. Standard electronic business interfaces (Cronbach alpha= .916)
4. Organizational memory systems (Cronbach alpha=.977)
5. Interpretation systems (Cronbach alpha=.953)
6. Breadth of information exchanged (Cronbach alpha=.961)
7. Quality of information exchanged (Cronbach alpha=.974)
8. Privileged information exchanged (Cronbach alpha=.944); and
9. Coordination information exchanged (Cronbach alpha=.906)

The reliability results for the two RFID system deployment outcomes, operational efficiency (Cronbach alpha=.968) and market knowledge creation (Cronbach alpha=.975), are also very satisfactory. The following are the reliability results for IT infrastructure integration which consists of data consistency (Cronbach alpha=.944) and cross-functional application integration (Cronbach alpha=.923), and supply chain process integration, which consists of financial flow integration

(Cronbach alpha=.880), physical flow integration (Cronbach alpha=.945), and information flow integration (Cronbach alpha=.952).

To establish convergent and divergent validity, the item-to-total correlations of the constructs were examined and, in general, the specific items have a stronger correlation with the construct than with other items (Rai, Patnayakuni, & Seth, 2006).

## Sample Profile Description

The convenience sample consists of a total of 104 firms from the CSCMP membership that responded to a certain part of the survey questionnaire—these were the firms that constitute the convenience sample of organizations that are knowledgeable about RFID or may be implementing RFID in the future. About 51.06 percent of the firms had 1,000 or less employees and 32.62 percent had more than 1,000 employees. The following profile shows the membership of the firms in different industry sectors: service (78.57 percent) and manufacturing (21.43 percent). A total of 98 firms identified their firm by nature of industry and number of employees out of the 104 firms; there were missing values for six firms for this descriptive data.

## Moderated Regression Procedure

Moderated regression analysis tests whether the relationship between two variables changes depending on the value of another variable (i.e., interaction effect) (Aguinis, 2004). The moderator variable explains changes in the nature of independent variable to the dependent variable effect, and provides information concerning the conditions under which an effect or relationship is likely to be stronger.

Regression analysis was conducted to test the hypotheses presented in this study. The moderated regression procedure requires testing first order effects, which in this study, will be referred to as "model 1." A model 1 simple regression tests the direct effects of a predictor variable on a dependent variable. As the independent variable, each of the absorptive capacity attributes was regressed against each of the dependent variables, operational efficiency and market knowledge creation. The variance in the dependent variable on account of the independent variable is noted using the $R^2$ value. Then, the regression procedure testing second order effects is conducted, which will be referred to as "model 2" in this study. A model 2 regression duplicates the model 1 regression equation and adds the product term which includes the hypothesized moderator variable.

It is important to determine how large the change in $R^2$ should be in order to qualify as "practically significant" or one that should merit serious attention (Aguinis, 2004). After conducting a Monte Carlo simulation, Evans (1985) stipulated that "...a rough rule would be to take 1% variance explained as the criterion as to whether or not a significant interaction exists in the model...." (p. 320). Evans found that in conducting the simulation, when the population scores included a moderating effect, results based on samples consistently demonstrated an $R^2$ change that was 1 percent or higher. On the other hand, when the population scores did not include a moderating effect, the change in $R^2$ was usually smaller than 1 percent. In conclusion, empirical and simulation results appear to indicate that a statistically significant $R^2$ change of about 1 percent to 2 percent demonstrates an effect size worthy of consideration. The results in this study include significant $R^2$ change values within the range with a maximum value of 4.7 percent and a minimum value of .8 percent, which indicate considerable significant moderating effects of IT infrastructure integration and supply chain process integration.

SPSS was used to run the regression equations and model 1 was specified for block 1 and model 2 was specified for block 2. The resulting $R^2$ values need to be noted for models 1 and 2. If the $R^2$ value is greater for model 2 than for model 1, then, the moderator variable included in the product term is demonstrating a moderating effect.

## FINDINGS

In predicting operational efficiency, supply chain process integration appears to the more powerful moderator variable compared to IT infrastructure integration in accounting for the variance in the increase of operational efficiency over and above the variance explained by the a number of absorptive capacity attributes and supply chain process integration as separate independent variables in the model 1 equations.

As Table 2 shows, with supply chain process integration as the moderator variable, the percentage differences in variance increase for operational efficiency were both considerable and significant for the following absorptive capacity attributes:

1. Business process modularity (4.7 percent, p<.000)
2. Use of standard electronic business interfaces (4.0 percent, p<.000)
3. Breadth of information exchanged (3.2 percent, p<.000)
4. Use of memory systems for interorganizational activities (2.5 percent, p<.01)
5. Joint decision making (2.2 percent, p<.01)
6. Use of interpretation systems for interorganizational information (2.0 percent, p<.01)
7. Exchange of coordination information (1.7 percent, p<.01); and
8. Exchange of privileged information (1.6 percent, p<.01)

In predicting market knowledge creation, supply chain process integration appears to the more powerful moderator variable compared to IT infrastructure integration in accounting for the variance in the increase of market knowledge creation over and above the variance explained

*Table 1. Moderated regression for operational efficiency with it infrastructure integration as moderator (N=104)*

| Independent Variables: AC Attributes 3 | | | | | |
|---|---|---|---|---|---|
| Dependent Variable: Operational Efficiency | | | | | |
| Moderator Variable: ITIntegrateCat1 (Nominal variable for the mean of data consistency and cross-functional process integration --- IT infrastructure integration) | | | | | |
| **AC Attributes** | **Model 1: R2 Without Product Term** | **Model 2: R2 With Product Term** | **% Variance Explained by Moderator with Product Term** | **F Value of Model 2 (degrees of freedom)** | **Significance of F Change** |
| Coordination Information | .771 | .795 | 2.4% | 11.568 (1,100) | P<.01 |
| Business Process Modularity | .726 | .748 | 2.2% | 8.767 (1,100) | P<.01 |
| Privileged Information | .770 | .788 | 1.8% | 8.315 (1,100) | P<.01 |
| Standard Electronic Business Interfaces | .704 | .721 | 1.7% | 6.318 (1,100) | P<.05 |
| Breath of Information | .763 | .768 | .5% | 2.174 (1,100) | Not signif. |
| Organizational Memory Systems | .752 | .757 | .5% | 2.123 (1,100) | Not signif. |
| Interpretation Systems | .742 | .746 | .4% | 1.593 (1,100) | Not signif. |
| Joint Decision Making | .740 | .743 | .3% | 1.463 (1,100) | Not signif. |
| Quality of Information | .837 | .837 | 0% | .221 (1,100) | Not signif. |

*Table 2. Moderated regression for operational efficiency with supply chain process integration as moderator (N=104)*

| Independent Variables: AC Attributes 3 | | | | | |
|---|---|---|---|---|---|
| Dependent Variable: Operational Efficiency | | | | | |
| Moderator Variable: SCMIntegrateCat1 (Nominal variable for the mean of physical, information, and financial flow integration) | | | | | |
| **AC Attributes** | **Model 1: R2 Without Product Term** | **Model 2: R2 With Product Term** | **% Variance Explained by Moderator with Product Term** | **F Value of Model 2 (degrees of freedom)** | **Significance of F Change** |
| Business Process Modularity | .701 | .748 | 4.7% | 18.701 (1,100) | P<.000 |
| Standard Electronic Business Interfaces | .680 | .720 | 4% | 14.188 (1,100) | P<.000 |
| Breadth of Information | .735 | .767 | 3.2% | 13.719 (1,100) | P<.000 |
| Organizational Memory Systems | .722 | .747 | 2.5% | 10.061 (1,100) | P<.01 |
| Joint Decision Making | .709 | .731 | 2.2% | 7.867 (1,100) | P<.01 |
| Interpretation Systems | .721 | .741 | 2% | 7.720 (1,100) | P<.01 |
| Coordination Information | .774 | .791 | 1.7% | 7.787 (1,100) | P<.01 |
| Privileged Information | .762 | .778 | 1.6% | 6.824 (1,100) | P<.01 |
| Quality of Information | .816 | .818 | .2% | .630 (1,100) | Not signif. |

by the a number of absorptive capacity attributes and supply chain process integration as separate independent variables in the model 1 equations.

As Table 4 shows, with supply chain process integration as the moderator variable, the percentage differences in variance increase for market knowledge creation were both considerable and significant for the following absorptive capacity attributes:

1. Business process modularity (4.6 percent, p<.000)
2. Breadth of information exchanged (3.8 percent, p<.000)
3. Use of standard electronic business interfaces (3.2 percent, p<.01)
4. Joint decision making (2.7 percent, p<.01)
5. Use of memory systems for interorganizational activities (2.6 percent, p<.01)
6. Coordination information exchanged (2.3 percent, p<.01)
7. Use of interpretation systems for interorganizational information (2.1 percent, p<.01); and
8. Exchange of privileged information (2 percent, p<.01)

Quality of information was not effectively moderated by both variables.

## IT Infrastructure Integration Capability as Moderator Variable with Operational Efficiency as a System Outcome

More substantial results are shown here in descending order of importance based on the percent $R^2$ change resulting from the introduction of a product term, ITIntegrateCat1, in the multiple

*Table 3. Moderated regression for market knowledge creation with it infrastructure integration as moderator (N=104)*

| Independent Variables: AC Attributes 3 | | | | | |
|---|---|---|---|---|---|
| Dependent Variable: Market Knowledge Creation | | | | | |
| Moderator Variable: ITIntegrateCat1 (Nominal variable for the mean of data consistency and cross-functional process integration --- IT infrastructure integration) | | | | | |
| **AC Attributes** | **Model 1: R2 Without Product Term** | **Model 2: R2 With Product Term** | **% Variance Explained by Moderator with Product Term** | **F Value of Model 2 (degrees of freedom)** | **Significance of F Change** |
| Coordination Information | .675 | .722 | 4.7% | 16.782 (1,100) | P<.000 |
| Privileged Information | .723 | .758 | 3.5% | 14.644 (1,100) | P<.000 |
| Business Process Modularity | .704 | .734 | 3% | 11.236 (1,100) | P<.01 |
| Standard Electronic Business Interfaces | .686 | .707 | 2.1% | 7.051 (1,100) | P<.01 |
| Breadth of Information | .719 | .732 | 1.3% | 5.101 (1,100) | P<.05 |
| Joint Decision Making | .685 | .697 | 1.2% | 3.998 (1,100) | P<.05 |
| Organizational Memory Systems | .703 | .713 | 1% | 3.563 (1,100) | P<.10 |
| Interpretation Systems | .726 | .734 | .8% | 3.052 (1,100) | P<.10 |
| Quality of Information | .780 | .781 | .1% | .578 (1,100) | Not Signif. |

regression equation. This is the nominal variable that represents the mean of data consistency and cross-functional process integration, the two components of IT infrastructure integration. Tables 1and 2 show the results of running two regression models: model 1 showing the relationships between the predictor variables and operational efficiency, without the product term and model 2, the regression results with the inclusion of the product term.

Table 1 shows the results with operational efficiency as the dependent variable and IT infrastructure integration capability as a moderator variable. IT infrastructure integration significantly moderates the relationship between the following predictor variables and operational efficiency in descending order of importance: coordination information exchanged; business process modularity; privileged information exchanged; and standard electronic business interfaces. The table column labelled "% Variance Explained by Moderator with Product Term" indicates the contribution of the product term --- which is the product of the moderator variable, in this case, IT infrastructure integration and the specific predictor variable. And so, for instance, in the case of coordination information exchanged, the product term would be the product of coordination information exchanged and IT infrastructure integration (i.e., Coord3XIntegrate1). The next column label shows "F Value of Model 2 (degrees of freedom)," which means that the F value of model 2 which includes the product term is shown along with the degrees of freedom for that regression model. The significance of the F change from model 1 to model 2 is indicated by the last column.

The relationships between the predictor variables and operational efficiency as moderated by IT infrastructure integration should be interpreted accordingly. Let's take the case of coordination

*Table 4. Moderated regression for market knowledge creation with supply chain process integration as moderator (N=104)*

| Independent Variables: AC Attributes 3 | | | | | |
|---|---|---|---|---|---|
| Dependent Variable: Market Knowledge Creation | | | | | |
| Moderator Variable: SCMIntegrateCat1 (Nominal variable for the mean of physical, information, and financial flow integration) | | | | | |
| **AC Attributes** | **Model 1: R2 Without Product Term** | **Model 2: R2 With Product Term** | **% Variance Explained by Moderator with Product Term** | **F Value of Model 2 (degrees of freedom)** | **Significance of F Change** |
| Business Process Modularity | .715 | .761 | 4.6% | 19.151 (1,100) | P<.000 |
| Breadth of Information | .719 | .757 | 3.8% | 15.682 (1,100) | P<.000 |
| Standard Electronic Business Interfaces | .707 | .739 | 3.2% | 12.211 (1,100) | P<.01 |
| Joint Decision Making | .693 | .720 | 2.7% | 9.596 (1, 100) | P<.01 |
| Memory Systems | .709 | .735 | 2.6% | 9.757 (1,100) | P<.01 |
| Coordination Information | | | 2.3% | P<.01 | |
| Interpretation Systems | .737 | .758 | 2.1% | 8.678 (1,100) | P<.01 |
| Privileged Information | .755 | .775 | 2% | 8.799 (1,100) | P<.01 |
| Quality of Information | .785 | .788 | .3% | 1.470 (1,100) | Not signif. |

information exchanged, the predictor variable whose relationship with operational efficiency is significantly moderated to the greatest extent by IT infrastructure integration.

About 77.1 percent of the variance in operational efficiency is explained by the exchange of coordination information and IT infrastructure integration as indicated by model 1 in Table 1. Model 2 is, then, introduced by including the product term (i.e., Coord3XIntegrate1) which represents the interaction between coordination information exchanged and IT infrastructure integration. As shown on Table 1, the addition of the product term resulted in an $R^2$ change of .024, F (1, 100) = 11.568, p<.01. This result supports the presence of a moderating effect. In other words, the moderating effect of IT infrastructure integration explains 2.4 percent of the variance in the increase of operational efficiency over and above the variance explained by the exchange of coordination information and IT infrastructure integration as separate independent variables.

The relationships between the remaining predictor variables and operational efficiency should be conducted accordingly as well: business process modularity; privileged information exchanged; and standard electronic business interfaces.

## Supply Chain Process Integration as Moderator Variable with Operational Efficiency as a System Outcome

Table 2 shows the results with operational efficiency as the dependent variable and supply chain process integration capability as the moderator variable. Supply chain process integration significantly moderates the relationship between the following predictor variables and operational efficiency in descending order of importance:

1. Business process modularity
2. Standard electronic business interfaces
3. Breadth of information exchanged
4. Organizational memory systems
5. Joint decision making; interpretation systems
6. Coordination information exchanged; and
7. Privileged information exchanged

About 70.4 percent of the variance in operational efficiency is explained by the use of business process modularity and IT infrastructure integration as indicated by model 1 in Table 2. Model 2 is, then, introduced by including the product term (i.e., IOSBusMod3XIntegrate2) which represents the interaction between the use of business process modularity and IT infrastructure integration. As shown on Table 2, the addition of the product term resulted in an $R^2$ change of .047, F(1, 100) = 18.701, p<.000. This result supports the presence of a moderating effect. In other words, the moderating effect of IT infrastructure integration explains 4.7 percent of the variance in the increase of operational efficiency over and above the variance explained by the use of business process modularity and IT infrastructure integration as separate independent variables.

The relationships between the remaining predictor variables and operational efficiency as by supply chain process integration should be conducted accordingly as well: standard electronic business interfaces; breadth of information exchanged; organizational memory systems; joint decision making; interpretation systems; coordination information exchanged; and privileged information exchanged.

## IT Integration Capability as Moderator Variable with Market Knowledge Creation as a System Outcome

Table 3 shows the results with market knowledge creation as the dependent variable and IT infrastructure integration as the moderator variable. IT infrastructure integration significantly moderates the relationship between the following predictor variables and market knowledge in descending order of importance:

1. Coordination information exchanged
2. Privileged information exchanged
3. Business process modularity
4. Standard electronic business interfaces
5. Breadth of information exchanged
6. Joint decision making
7. Organizational memory systems; and
8. Interpretation systems

About 67.5 percent of the variance in market knowledge creation is explained by the exchange of coordination information and IT infrastructure integration as indicated by model 1 in Table 3. Model 2 is, then, introduced by including the product term (i.e., Coord3XIntegrate1) which represents the interaction between coordination information exchanged and IT infrastructure integration. As shown on Table 3, the addition of the product term resulted in an $R^2$ change of .047, F(1, 100) = 16.782, p<.000. This result supports the presence of a moderating effect. In other words, the moderating effect of IT infrastructure integration explains 4.7 percent of the variance in the increase of market knowledge creation over and above the variance explained by the exchange of coordination information and IT infrastructure integration as separate independent variables.

The relationships between the remaining predictor variables and market knowledge creation as by IT infrastructure integration should be conducted accordingly as well:

1. Privileged information exchanged
2. Business process modularity
3. Standard electronic business interfaces
4. Breadth of information exchanged
5. Joint decision making
6. Organizational memory systems; and
7. Interpretation systems

## Supply Chain Process Integration as Moderator Variable with Market Knowledge Creation as a System Outcome

Table 4 shows the results with market knowledge creation as the dependent variable and supply chain process integration as the moderator variable. Supply chain process integration significantly moderates the relationship between the following predictor variables and market knowledge in descending order of importance:

1. Business process modularity
2. Breadth of information exchanged
3. Standard electronic business interfaces
4. Joint decision making
5. Organizational memory systems
6. Coordination information exchanged
7. Interpretation systems; and
8. Privileged information exchanged

About 71.5 percent of the variance in market knowledge creation is explained by the use of business process modularity and supply chain process integration as indicated by model 1 in Table 4. Model 2 is, then, introduced by including the product term (i.e., IOBusMod3XIntegrate2) which represents the interaction between the use of business process modularity and supply chain process integration. As shown on Table 4, the addition of the product term resulted in an $R^2$ change of .046, $F(1, 100) = 19.151$, $p<.000$. This result supports the presence of a moderating effect. In other words, the moderating effect of supply chain process integration explains 4.6 percent of the variance in the increase of market knowledge creation over and above the variance explained by the use of business process modularity and supply chain process integration as separate independent variables.

The relationships between the remaining predictor variables and market knowledge creation as by supply chain process integration should be conducted accordingly as well:

1. Breadth of information exchanged
2. Standard electronic business interfaces
3. Joint decision making
4. Organizational memory systems
5. Coordination information exchanged
6. Interpretation systems; and
7. Privileged information exchanged

## DISCUSSION OF FINDINGS

All four proposed hypotheses were partially supported in this study. Both variables, IT infrastructure integration and supply chain process integration, moderate the relationships between three predictor variables, business process modularity, standard electronic business interfaces, and breadth of information exchange and the two dependent variables examined in this study, operational efficiency and market knowledge creation to a considerable extent. Between the two moderator variables, however, supply chain process integration tempered a greater degree of the variance between these three predictor variables and the dependent variables compared to IT infrastructure integration.

IT infrastructure integration, though, accounted for a greater proportion of the variance in the relationship between coordination information and privileged information exchanged and the two dependent variables. Supply chain process integration, on the other hand, accounted for a greater proportion of the variance in the relationship between use of organizational memory systems and joint decision making and the two dependent variables. Both moderator variables did not significantly regulate the relationships between quality of information exchanged and use of interpretation systems and the two dependent variables to any considerable extent.

These findings affirm the importance of both the IT infrastructure integration and supply chain process integration elements that undergird RFID system implementation. Managers also need to be aware current developments directly relate to these elements of the infrastructure environment as their firms seek either or both operational efficiency and market knowledge creation. In one of the later meetings of GS1, a nonprofit organization that coordinates the development of standards for technologies like barcodes and RFID, a key issue was "data synchronization" or the ability to share accurate data throughout the supply chain which would require data consistency (Roberti, 2008). Planning for the design or reengineering of business processes to achieve cross-functional process integration and supply chain process integration is also a very involved activity as well and one taken seriously by leading-edge firms. Anthony Bigornia, IBM Consulting Services, indicated that firms that will drive the most learning from their RFID implementations are those that address business process issues directly (Wasserman, 2005). ChainLink Research, a research and consulting firm focusing on supply chains, found that more than 40 percent of the manufacturing firms participating in their study are implementing RFID to pursue process improvement goals specifically in the following areas:

1. Manufacturing and plant-floor operations
2. Outbound shipping
3. Distribution and logistics
4. Invoice and dispute resolution
5. Service and support
6. Supply chain and custody tracking (which includes e-pedigrees, or electronic documents used to trace a product's manufacturing and distribution history)
7. Recall and product expiration; and
8. Asset and capital-equipment tracking (Bacheldor, 2006)

About five levels of business process integration have been identified for RFID systems:

1. Level 1 (goal setting assessment)
2. Level 2 (slap and ship)
3. Level 3 (application integration)
4. Level 4 (business process improvement); and
5. Level 5 (collaborative business intelligence) (Blossom, 2005)

Cross-functional application integration and supply chain process integration are both clear issues in levels 3 and 4. Application integration must be planned for carefully because of the need to use preprogrammed adapters or plug-ins that will need to deliver RFID data to the following:

1. Enterprise resource planning (ERP)
2. Customer relationship management (CRM)
3. Supply chain management (SCM); and
4. Other files, databases, and systems (Blossom, 2005)

Managers should also be aware of new technologies such as the use of smart RFID networks that would allow the use of event-driven RFID business applications and react to real-time information and assist the movement to level 4 or the improvement of business processes (Bhargava, 2007). Complex RFID services could be delivered using real-time events such as pushing alerts to a retail store manager when inventory for a particular product is aging or has reached "stale" status.

Considering the scale and scope of RFID systems within a supply chain context, managers need to anticipate the processing of the escalating numbers of events created by RFID technologies that will eventually lead to the generation of increasingly complex business rules that will govern the routing and analysis of signals included in the event stream (Blossom, 2005).

## CONCLUSION

This study's findings reiterate the significance of absorptive capacity attributes in achieving either operational efficiency or market knowledge creation. The relationship between these sets of variables, though, is mitigated either IT infrastructure integration or supply chain process integration. These two moderators, however, were able to mollify the relationships between the two dependent variables and only three absorptive capacity variables, business process modularity, standard electronic business interfaces, and breadth of information exchanged.

Supply chain managers need to be proactive about enabling both IT infrastructure integration and supply chain process integration within their current IT environments to pave the way for a smooth transition to RFID deployment. One way to do so is to support a broad IT application deployment within the firm's supply chains. Whitaker, Mithas, and Krishnan (2007) found a positive association between the range of IT application deployment and RFID adoption. Their study also confirmed the need to have a strong IT infrastructure in place so as not to inflate future expected levels of financial investments in RFID, which may only delay its diffusion within the firm. The results of Bendoly, Citurs, and Konsynski (2007) study involving 234 firms also point in the same direction. They, in turn, found that managers' perceptions of and commitment to RFID deployment are influenced by the explicit (operational and technological) and tacit (knowledge) infrastructural elements of their systems. The tacit component of the infrastructure was measured using cross-functional knowledge and procedural flexibility and the explicit component was measured using effective information processing standards.

## LIMITATIONS AND FUTURE RESEARCH DIRECTIONS

The study has a number of limitations. First of all, study participants that were knowledgeable about RFID or may be implementing RFID in the future were asked to report their perceptions of the absorptive capacity attributes, system deployment outcomes, and the two moderating variables, IT infrastructure integration and supply chain process integration. A future study should capture these perceptions from firms that have actually implemented RFID systems. Second, the data was gleaned from a convenience sample of 104 organizations. A random sample is also needed in order to arrive at representative implications and generalizations.

There are number of future research opportunities that can use this study's findings as a launching pad. The importance of supply chain process integration as a moderator variable between the independent and dependent variables used in this study can be tested by comparing the experiences of firms that have implemented supplier-facing versus customer-facing RFID business applications in their respective supply chains. It would also be interesting to compare the importance of the two moderator variables, IT infrastructure integration and supply chain process integration in retailer-dominated versus manufacturer-dominated supply chains. The importance of the moderator variables could also vary depending on the point-of-view of the respondent: the hub firm (i.e., firm that has initiated and set up the supply chain network) may perceive the importance of these variables differently from its trading partners whom it has enlisted in its network. The ideal outcome, of course, would be for the perceptions of both the hub firm and its trading partners to be congruent. That way, both parties would be willing and accountable for deploying the necessary resources for the absorptive capacity variables and for IT infrastructure integration and supply chain process integration they deem imperative for achieving either operational efficiency or market knowledge creation or both. Should there be differences in opinion, however, then, the issue of how to divide the costs of supporting the supply chain equitably across the supply chain arises.

## REFERENCES

Ackerman, M. S. (1996). Definitional and contextual issues in organizational and group memories. *Information Technology & People*, *9*(1), 10–24. doi:10.1108/09593849610111553

Aguinis, H. (2004). *Regression Analysis for Categorical Moderators*. New York: Guilford Press.

Anand, V., Manz, C. C., & Glick, W. H. (1998). An organizational memory approach to information management. *Academy of Management Review*, *23*(4), 796–809. doi:10.2307/259063

Angeles, R. (2005). RFID technologies: supply-chain applications and implementation issues. *Information Systems Management*, *22*(1), 51–65. doi:10.1201/1078/44912.22.1.20051201/85739.7

Austin, T. A., Lee, H. L., & Kopczak, L. (1997). *Unlocking Hidden Value in the Personal Computer Supply Chain*. Chicago, IL: Andersen Consulting.

Bacheldor, B. (2006). Process improvement drives manufacturers' RFID implementations. *RFID Journal*. Retrieved July 1, 2008, from http://www.rfidjournal.com/article/articleview/2903/1/1/

Bendoly, E., Citurs, A., & Konsynski, B. (2007). Internal infrastructural impacts on rfid perceptions and commitment: knowledge, operational procedures, and information processing standards. *Decision Sciences*, *38*(3), 423–449. doi:10.1111/j.1540-5915.2007.00165.x

Bhargava, H. (2007). Building smart RFID networks. *RFID Journal*. Retrieved July 7, 2008, from http://www.rfidjournal.com/article/articleview/3387/1/82/

Blossom, P. (2005). Levels of RFID maturity, part 2. *RFID Journal*. Retrieved July 5, 2008, from http://www.rfidjournal.com/article/articleview/1347/1/82/

Buckley, P. J., & Casson, M. (1988). A theory of cooperation in international business. In Casson, M. (Ed.), *The Economics of Business Culture*. Oxford, UK: Oxford University Press.

Chen, C. J. (2004). The effects of knowledge attribute, alliance characteristics, and absorptive capacity on knowledge transfer performance. *R & D Management*, *34*(3), 311–321. doi:10.1111/j.1467-9310.2004.00341.x

Child, J., & Faulkner, D. (1998). *Strategies of Cooperation: Managing Alliances, Networks, and Joint Ventures*. New York, NY: Oxford University Press.

Chow, H. K. H., Choy, K. L., & Lee, W. B. (2007). Knowledge management approach in build-to-order supply chains. *Industrial Management & Data Systems*, *107*(6), 882–919. doi:10.1108/02635570710758770

Cohen, W. M., & Levinthal, D. A. (1990). Absorptive capacity: a new perspective on learning and innovation. *Administrative Science Quarterly*, *35*(1), 128–152. doi:10.2307/2393553

Collins, J. (2004). Soap maker cleans up with RFID. *RFID Journal*. Retrieved July 21, 2008, from http://www.rfidjournal.com/article/articleview/1101

Collis, D. J. (1994). Research note: how valuable are organizational capabilities. *Strategic Management Journal*, *15*, 143–152. doi:10.1002/smj.4250150910

Cooper, B. L., Watson, H. J., Wixom, B. H., & Goodhue, D. L. (2000). Data warehousing supports corporate strategy at first American corporation. *Management Information Systems Quarterly*, *24*(4), 547–567. doi:10.2307/3250947

Croasdell, D. T. (2001). IT's role in organizational memory and learning. *Information Systems Management*, (Winter): 8–11.

De Kok, A. G., Van Donselaar, K. H., & Van Woensel, T. (2008). A break-even analysis of RFID technology for inventory sensitive to shrinkage. *International Journal of Production Economics*, *112*, 521–531. doi:10.1016/j.ijpe.2007.05.005

Dos Santos, B. L., & Smith, L. S. (2008). RFID in the supply chain: panacea or pandora's box? *Communications of the ACM*, *51*(10), 127–131. doi:10.1145/1400181.1400209

Eisenhardt, K. M., & Martin, J. A. (2000). Dynamic capabilities: what are they? *Strategic Management Journal*, *21*(10/11), 1105–1121. doi:10.1002/1097-0266(200010/11)21:10/11<1105::AID-SMJ133>3.0.CO;2-E

Evans, M. G. (1985). A monte carlo study of the effects of correlated method variance in moderated multiple regression analysis. *Organizational Behavior and Human Decision Processes*, *36*, 302–323. doi:10.1016/0749-5978(85)90002-0

Fites, D. V. (1996). Make your dealers your partners. *Harvard Business Review*, *74*(2), 84–97.

Galunic, D. C., & Rodan, S. (1998). Resource recombinations in the firm: knowledge structures and the potential for schumpeterian innovation. *Strategic Management Journal*, *19*, 1193–1201. doi:10.1002/(SICI)1097-0266(1998120)19:12<1193::AID-SMJ5>3.0.CO;2-F

Godon, D., Visich, J. K., & Li, S. (2007). An exploratory study of RFID implementation benefits and challenges in the supply chain. In *Proceedings of the 38th Annual Meeting of the Decision Sciences Institute* (pp. 5261-5266).

Greenfield, A., Patel, J., & Fenner, J. (2001). Online invoicing for business-to-business users. *Information Week, November, 863*, 80-82.

Hage, J., & Aiken, M. (1967). Program change and organizational properties: a comparative analysis. *American Journal of Sociology*, *72*, 503–519. doi:10.1086/224380

Hall, R. (1992). The strategic analysis of intangible resources. *Strategic Management Journal*, *13*, 135–144. doi:10.1002/smj.4250130205

Holloway, S. (2006). *Potential of RFID in the Supply Chain*. Chicago: Solidsoft Ltd.

Huber, G. P. (1991). Organizational learning: the contributing processes and literature. *Organization Science*, *2*, 88–115. doi:10.1287/orsc.2.1.88

Huber, G. P. (1996). Organizational learning: a guide for executives in technology-critical organizations. *International Journal of Technology Management*, *11*(7), 821–832.

Jansen, J. J. P., Van Den Bosch, F. A. J., & Volberda, H. W. (2005). Managing potential and realized absorptive capacity: how do organizational antecedents matter? *Academy of Management Journal*, *48*(6), 999–1015.

Karimi, J., Somers, T.M., & Bhattacherjee, A. (2007). The impact of erp implementation on business process outcomes: a factor-based study. *Journal of Management Information Systems, Summer, 24*(1), 101-134.

Kim, L. (1997a). The dynamics of samsung's technological learning in semiconductors. *California Management Review*, *39*(3), 86–100.

Kim, L. (1997b). *From imitation to innovation: The dynamics of Korea's technological learning*. Cambridge, MA: Harvard Business School Press.

Knudsen, T., & Madsen, T. K. (2002). Export strategy: a dynamic capabilities perspective. *Scandinavian Journal of Management*, *18*, 475–501. doi:10.1016/S0956-5221(01)00019-7

Kumar, S. (2007). Connective technology as a strategic tool for building effective supply chain. *International Journal of Manufacturing Technology and Management, 10*(1), 41–56. doi:10.1504/IJMTM.2007.011400

Lane, J. P., & Lubatkin, M. (1998). Relative absorptive capacity and interorganizational learning. *Strategic Management, 19*, 461–477. doi:10.1002/(SICI)1097-0266(199805)19:5<461::AID-SMJ953>3.0.CO;2-L

Larson, A. (1991). Partner networks leveraging external ties to improve entrepreneurial performance. *Journal of Business Venturing, 6*, 173–188. doi:10.1016/0883-9026(91)90008-2

Law, C. C. H., & Ngai, E. W. T. (2007). ERP systems adoption: An exploratory study of the organizational factors and impacts of erp success. *Information & Management, 44*, 418–432. doi:10.1016/j.im.2007.03.004

Lee, H. L., Padmanabhan, V., & Whang, S. (1997). Information distortion in supply chain: the bullwhip effect. *Management Science, 43*(4), 546–558. doi:10.1287/mnsc.43.4.546

Lee, Y. M., Cheng, F., & Leung, Y. T. (2009). A quantitative view on how RFID can improve inventory management in a supply chain. *International Journal of Logistics Research and Applications, 12*(1), 23–43. doi:10.1080/13675560802141788

Lee, Y. W., & Strong, D. M. (2004). Knowing-why about data processing and data quality. *Journal of Management Information Systems, 20*(3), 13–39.

Lindsay, P., & Norman, D. (1977). *Human information processing*. Orlando, FL: Academic Press.

Mabert, V. A., & Venkatraman, M. A. (1998). Special research focus on supply chain linkages: challenges for design and management in the 21st century. *Decision Sciences, 29*(3), 537–550. doi:10.1111/j.1540-5915.1998.tb01353.x

Malhotra, A., Gosain, S., & El Sawy, O. A. (2005). Absorptive capacity configurations in supply chains: gearing for partner-enabled market knowledge creation. *MIS Quarterly, March, 29*(1), 145-187.

Mangan, J., Lalwani, C., & Butcher, T. (2008). *Global Logistics and Supply Chain Management*. Hoboken, NJ: John Wiley & Sons.

McCormack, K. P., & Johnson, W. C. (2003). *Supply Chain Networks and Business Process Orientation*. Boca Raton, FL: St. Lucie Press.

Miles, R. E., & Snow, C. C. (2007). Organization theory and supply chain management: An evolving research perspective. *Journal of Operations Management, 25*(2), 459–463. doi:10.1016/j.jom.2006.05.002

Moon, K. L., & Ngai, E. W. T. (2008). The adoption of RFID in fashion retailing; a business value-added framework. *Industrial Management & Data Systems, 108*(5), 596–612. doi:10.1108/02635570810876732

Moyaux, T., & Chaib-draa, B. (2007). Information sharing as a coordination mechanism for reducing the bullwhip effect in supply chain. *IEEE Transaction on Systems, Man, and Cybernetics, May, 37*(3), 396-409.

Nachiappan, S. P., Gunasekaran, A., & Jawahar, N. N. (2007). Knowledge management system for operating parameters in two-echelon vmi supply chains. *International Journal of Production Research, 45*(11), 2479-2505. Nunnally, J.C. (1978). *Psychometric Theory*. New York, NY: McGraw-Hill.

O'Brien, J. A. (2001). *Management Information Systems: Managing Information Technology in the Internetworked Enterprise*. New York, NY: Mc-Graw Hill.

O'Connor, M. C. (2006). LEGO puts the RFID pieces together. *RFID Journal, February 12*. Retrieved August 1, 2008, from http://www.rfidjournal.com/article/articleview/2145

Padmanabhan, B., & Tuzhilin, A. (2003). On the use of optimization for data mining: theoretical interactions and ecrm opportunities. *Management Science, 49*(10), 1327–1343. doi:10.1287/mnsc.49.10.1327.17310

Pagarkar, M., Natesan, M., & Prakash, B. (2005). *RFID in Integrated Order Management Systems*. Chennai, India: Tata Consultancy Services.

Patni Americas, Inc. (2008). *Thought Paper: Global Data Synchronization: A Foundation Block for Realizing RFID Potential.* Patni Americas, Inc., Cincinnati, Ohio. Retrieved July 24, 2008, from http://www.patni.com/resource-center/collateral/RFID/tp_RFID_Global-Data-Synchronization.html

Perrow, C. (1967). A framework for the comparative analysis of organizations. *American Sociological Review, 32*, 194–208. doi:10.2307/2091811

Pralahad, C. K., & Hamel, G. (1990). The core competence of the corporation. *Harvard Business Review*, (May-June): 79–91.

Rai, A., Patnayakuni, R., & Seth, N. (2006). Firm performance impacts of digitally enabled supply chain integration capabilities. *Management Information Systems Quarterly, 30*(2), 225–246.

Raisinghani, M. S., & Meade, L. L. (2005). Strategic decisions in supply-chain intelligence using knowledge management: an analytic-network-process framework. *Supply Chain Management, 10*(2), 114–121. doi:10.1108/13598540510589188

Rekik, Y., Sahin, E., & Dallery, Y. (2009). Inventory inaccuracy in retail stores due to theft: An analysis of the benefits of RFID. *International Journal of Production Economics, 118*, 189–198. doi:10.1016/j.ijpe.2008.08.048

Reyes, P. M., Frazier, G., Prater, E., & Cannon, A. (2007). RFID: the state of the union between promise and practice. *International Journal of Integrated Supply Management, 3*(2), 192–206. doi:10.1504/IJISM.2007.011976

Roberti, M. (2008). Laying the foundation for RFID. *RFID Journal*. Retrieved August 10, 2008, from http://www.rfidjournal.com/article/articleview/3524/1/435/

Rupnick R., Kukar, M., & Krisper, M. (2007). Integrating data mining and decision support through data mining based decision support system. *Journal of Computer Information Systems, Spring, 47*(3), 89-104.

Shutzberg, L. (2004). *Radio Frequency Identification (RFID) in the Consumer Goods Supply Chain: Mandated Compliance or Remarkable Innovation?* Norcross, GA: Rock-Tenn Company.

Simchi-Levi, D., Kaminsky, P., & Simchi-Levi, E. (2004). *Managing the Supply Chain: The Definitive Guide for the Business Professional*. New York: McGraw-Hill.

Spangler, W. E., Gal-Or, M., & May, J. H. (2003). Using data mining to profile tv viewers. *Communications of the ACM, 46*(2), 66–72. doi:10.1145/953460.953461

Stein, E. W. (1995). Organizational memory: review of concepts and recommendations for management. *International Journal of Information Management, 15*(2), 17–32. doi:10.1016/0268-4012(94)00003-C

Stevens, G. C. (1990). Successful supply chain management. *Management Decision, 28*(8), 25–30. doi:10.1108/00251749010140790

Szmerekovsky, J. G., & Zhang, J. (2008). Coordination and adoption of item-level RFID with vendor managed inventory. *International Journal of Production Economics, 114*, 388–398. doi:10.1016/j.ijpe.2008.03.002

Szulanski, G. (1996). Exploring internal stickiness: impediments to the transfer of best practice within the firm. *Strategic Management Journal*, *17*, 27–43.

Teece, D. J., Pisano, G., & Shuen, A. (1997). Dynamic capabilities and strategic management. *Strategic Management Journal*, *18*(7), 509–533. doi:10.1002/(SICI)1097-0266(199708)18:7<509::AID-SMJ882>3.0.CO;2-Z

Uzzi, B., & Lancaster, R. (2003). Relational embeddedness and learning: the case of bank loan managers and their clients. *Management Science, April, 49*(4), 383-399.

Van Den Hoven, J. (2004). Data architecture standards for the effective enterprise. *Information Systems Management, Summer, 21*(3), 61-64.

Vijayaraman, B., & Osyk, B. (2006). An empirical study of RFID implementation in the warehousing industry. *The International Journal of Logistics Management*, *17*(1), 6–20. doi:10.1108/09574090610663400

Walsh, J. P., & Ungson, G. R. (1991). Organizational memory. *Academy of Management Review*, *16*(1), 57–91. doi:10.2307/258607

Wasserman, E. (2005). Accenture outlines secrets of RFID success. *RFID Journal*. Retrieved August 5, 2008, from http://www.rfidjournal.com/article/articleview/1506/1/1/

Wasserman, E. (2007). Beaver street fisheries automates RFID tagging. *RFID Journal*. Retrieved August 15, 2008, from http://www.rfidjournal.com/article/articleview/3060

Whitaker, J., Mithas, S., & Krishnan, M. (2007). A field study of RFID deployment and return expectations. *Production and Operations Management*, *16*(5), 599–612.

Withey, M., Daft, R. L., & Cooper, W. H. (1983). Measures of perrow's work unit technology: an empirical assessment and a new scale. *Academy of Management Journal*, *26*, 45–63. doi:10.2307/256134

Zahra, S. A., & George, G. (2002). Absorptive capacity: a review, reconceptualization, and extension. *Academy of Management Review*, *27*(2), 185–203. doi:10.2307/4134351

*This work was previously published in International Journal of Information Systems and Supply Chain Management, Volume 3, Issue 2, edited by John Wang, pp. 25-51, copyright 2010 by IGI Publishing (an imprint of IGI Global)*

# Chapter 4
# A Comparison of Information Technology Usage across Supply Chains:
## A Comparison of the U.S. Beef Industry and the U.S. Food Industry

**George. N. Kenyon**
*Associate Professor, Lamar University, USA*

**Brian D. Neureuther**
*State University of New York at Plattsburgh, USA*

## ABSTRACT

*Historically, the growth of the beef industry has been hampered by various entities, i.e., breeders, cow-calf producers, stockers, backgrounders, processors, etc..., within the beef industry's supply chain. The primary obstacles to growth are the large numbers of participants in the upstream side of the supply chain and the lack of coordination between them. Over the last decade significant advances have been made in information and communication technologies, and many new companies have been founded to promote these technical advances. This research looks at both the upstream and downstream participants to determine the degree to which information technologies are currently being utilized and the degree that these new technologies have driven performance improvements in the beef industry's supply chain. Through surveys, the authors find that the beef industry does not use information technologies to their benefit and that the US beef supply chain is not yet strategically poised to enable the use of these technologies.*

## INTRODUCTION AND BACKGROUND

In a study of the U.S. beef industry and the use of information technology (IT) to enable the industry's supply chain, Neureuther and Kenyon (2008) found that the beef industry is not using IT to any significant advantage except in the area of information collection. They further found that IT could enhance supply chain performance and integration, but the supply chain is not yet strategically poised to do so. They attributed this to several reasons:

DOI: 10.4018/978-1-4666-0918-1.ch004

1. Beef in the U.S. is thought of as a commodity product
2. The U.S. beef industry lacks a common vision and industry goals.
3. The mentality of downstream partners in the supply chain constrains incentives and information movement to upper levels in the supply chain.
4. In addition, auction houses have developed into information clearinghouses for much of the information in the supply chain and few are using e-markets or e-commerce tools.
5. Mistrust of internet usage dominates the industry where there is a feeling that small producers may be eliminated by adopting IT.
6. Current supply chain entities partners perceive little or no benefit to moving their processes online with respect to cost savings in procurement.

Based on these findings, they made several recommendations for the beef industry in the area of IT use in the supply chain and with respect to infrastructure changes that will need to occur in order to enable the use of IT. A synopsis of the recommendation is found in Table 1.

Given these conditions in the upstream portion of the Beef Industry's supply chain, it would make sense then, to look at the US beef industry and compare its readiness for the use of IT to that of other food industries. To this avail, this research will examine the use of IT in the food manufacturing industry, using the 2003-2007 Industry Week/MPI Census of US Manufacturers. Specifically, the research will examine the usage of using information technologies in food manufacturing supply chains by examining the impact of information technologies and supply chain performance measures. In the analysis, profitability, costs of goods sold, labor costs, material costs, and overhead costs will be analyzed with each of the technology platforms of enterprise resource planning (ERP and ERPII), demand planning and forecasting (DPF), electronic data interchange (EDI), online purchasing (OP) and online selling (OS). The results will then be compared to the US beef industry.

## OBJECTIVES

Even though a solid foundation of supply chain research exists (Chandra & Kumar, 2000; Levy & Grewal, 2000; Mentzer, Dewit, Keebler, Min,

*Table 1. Beef industry IT recommendations (Neureuther & Kenyon, 2008)*

| **IT Usage** |
|---|
| Increase the use of e-markets and e-commerce for auction houses in order to create visibility and reach, reduce transaction costs, and facilitate asset swap to achieve better utilization of key assets |
| Create tangible rewards for adherence to standards, such as contracts that require a higher price per pound for beef that meets an agreed upon level of supply and/or an agreed upon grade specification |
| The use of electronic data interchange (EDI) technologies (or even internet XML applications) to link animal record keeping information throughout the supply chain – from cow/calf producer to retailer – by individual animal |
| Begin enabling IT at the downstream partners and then pull usage to the upstream participants. |
| **Infrastructure** |
| Coordination mechanisms need to be matched to the market structure in order to improve value |
| Better communication of consumer demands needs to occur throughout the supply chain |
| Better education of the typical rancher of supply chain management benefits, especially in the areas of coordination, vertical integration, and IT usage is a must |

Nix, Smith, & Zacharia, 2001; Lambert, Cooper, & Pagh, 1998; Langley & Holcomb, 1992; Min & Mentzer, 2000; Chandrashekar & Schary, 1999; Cooper, Lambert, & Pagh, 1997; Croxton, Garcia-Dastugue, Lambert, & Rodgers, 2001) there is inconsistent evidence that any of the supply chain management research can be effectively integrated into industry practice or provide sustainable performance improvements (Moberg, Speh, & Freese, 2003). Since it is estimated that poor coordination between the supply chain participants in the U.S. food industry is wasting $30 billion annually (Fisher, 1997), it becomes clear that an analysis of the supply chains in the food industry is of interest. It then becomes important to analyze the degree to which this industry is contributing to the waste. Salin (2000) conducted a study of the beef supply chains and the use of information technology in which three primary issues affecting the success of beef supply chains were discussed. She found that 1) technology and scale, 2) auctions, alliances, and competitiveness and 3) traditional relationships were significant issues in supply chain performance. The purpose of Salin's (2000) study was to use interviews from the beef sector to describe the information technologies used in this industry.

As an extension of Salin's (2000) work the goal of this research is to seek whether or not sustainable process improvements by supply chain integration has been realized in the US food industry. More specifically, the objective of the research is to assess the impact of internet technologies on the food industry's supply chain. In order to meet this objective, the following propositions are proposed:

1. The US food industry is not using information technology to any significant advantage (advantage can be cost reduction of information gathering or transactional).
2. Information technology enhances supply chain integration by use of the internet and other information technologies.

It is paramount that a good understanding of the US food manufacturer's chain is achieved in order to examine these propositions. To this avail, the research will give an in depth overview of the US food manufacture's supply chain by examining change drivers in the supply chain and current coordination strategies and information technology usage in the supply chain. Then, a survey of supply chain participants in the industry will be discussed and analyzed with respect to the above objectives and propositions, with conclusions and findings following.

## LITERATURE REVIEW

Nitschke and O'Keefe (1997) suggest that consumer value is created not by individual firms but rather by business enterprises. They further suggest that the competitiveness of an enterprise is dependent upon both the competitiveness of the individual firms and the nature of the linkages between the firms in the supply chain. A key observation made by Nitschke and O'Keefe (1997) was that coordination mechanisms need to be matched appropriately to the market structure is order to improve value. As an example, auction systems traditionally used in commodity markets would not be appropriate for differentiated products and segmented market strategies. Coordination mechanisms are practices/tools that are intended to enable better communication of information, facilitate continuous improvement, speed up response time, improve product quality, and create closer relationships between customers and suppliers.

One method used to vertically integrate the food industry's supply chain has been the increasing use of contracts. Mighell and Jones (1963) have categorized contracts into three basic groups: market-specification contracts, production-management contracts, and resource-providing contracts. Market-specification contracts are those where the buyer provides a market for a seller's output, while the producer retains control over the production process. Production-management contracts are those where the buyer provides a market and can specify and/or monitor production practices or input usage. Romeo (2006) writes about several restaurant chains that are using contracts at the customer-retailer interface to buy beef at certain prices to hedge against price risks. Resource-providing contracts are defined as those where the buyer provides a market outlet, supervises production practices and supplies key inputs. They assume the greatest proportion of the risk, but may retain ownership of the product.

Narayanan and Raman (2004) have identified some limitations to contract-based solutions for supply chain improvements. First, outcome variables are highly correlated with the unobservable actions of other participants and are often difficult to isolate. Second, designing contracts to overcome actions that impact on the principal's welfare can be difficult when individuals within the supply chain are "risk-averse". Finally, attempts to mitigate actions that impact the individual's welfare with contract-based solutions can exacerbate the problems for the other party.

Vertical partnerships are defined as an arrangement between a buyer and seller to facilitate a mutually satisfying exchange over time, which leaves the operation and control of the two businesses substantially independent (Hughes, 1994). Fearne (1998) states that there are four key aspects to vertical partnerships: partnerships are entered into freely, partnerships must offer mutual benefits, these benefits occur overtime, and partners remain substantially independent. He also describes five benefits to vertical partnerships: improved market access, improved communications, higher profit margins, greater discipline, and higher barriers to entry.

Horizontal partnerships, such as producer co-operatives, have existed for centuries and generally have three key motives (Fearne, 1998), 1) meeting the volume requirements for major customers and increasing bargaining strength in their dealings with major buyers and sellers, 2) accelerating the pace and reducing the cost of penetrating new markets and 3) sharing the costs associated with developing new products and adopting new technologies.

The vast majority of the food supply chain entities do not participate in any of these mechanisms. Even with the development of these coordination mechanisms, problems still exist. Narayanan and Raman (2004) noted that resistance from key stakeholders could often be explained by their refusal to change old ways of doing things or by their inability to perceive the benefits of a proposed change. Lawrence (2006) pleads that the U.S. beef supply chain needs confidence in their products and needs every entity within the supply chain to have strong relationships. In addition, Euclides (2004) notes that the quality of the final product and the need to lower environmental risk are significant aspects of a holistic approach to the beef supply chain in Brazil. Given the failure of these mechanisms to improve the overall coordination of upstream participants in the beef industry's supply chain, how can better information flows drive future improvements?

## INFORMATION TECHNOLOGY AND THE US FOOD INDUSTRY

Narayanan and Raman (2004) stated that changes in supply chain operations often included applications of information technology (IT). Peypoch (1998) discussed how the leveraging of cost-cutting benefits typically available with IT through the streamlining of activities and the utilization of this technology to create electronic communities (networks) drives closer relationships among competitors, suppliers, buyers, and sellers. Through the creation of electronic communities, the traditional barriers that exist between supply chain participants could be reduced. By using a multi-tiered structure that provides for different vertical and horizontal interactions, electronic communities can increase an organization's exposure, thus allowing a more orderly and efficient trading process. Peypoch (1998) further claimed that the potential results would lower the cost structure across an entire industry such that all participants would benefit.

Wise and Morrison (2000) state that future exchanges between supply chain participants (either product or information) would need to occur such that buyers and suppliers were able to form tighter relationships while still maintaining the reach and efficiency of internet commerce. In order for future e-commerce models to succeed in the long run, rewards must flow to both sellers and buyers. While the physical transfer of goods is the goal of all business transactions, the information that defines the transaction can be separated and exchanged electronically. It is this information that is frequently more valuable to companies than the underlying goods themselves (Wise & Morrison, 2000; Lewis & Talalayevsky, 1997; Lewis & Talalayevsky, 2004).

More recently, Choi, Tsai and Jones (2009) examined how a regional supermarket store chain in the retail food industry develops its enterprise network infrastructure to outperform its larger competitors. In this study, critical success factors for firms to build an effective enterprise network infrastructure are highlighted. It is suggested that IT planning in the retail food industry needs to be firmly tied to critical business goals. In addition, Mohtadi (2008) reviews the determinants of information sharing between retailers and their suppliers, in the food industry supply chain. He examines the behavior of food retailers in their adoption of information sharing with suppliers. He finds that retail firms with larger number of suppliers are more inclined to share, rather than to withhold, information. Further, García-Arca and Prado-Prado (2008) in analyzing two Spanish companies in the food sector discuss a methodology for the implementation of information systems; with emphasis on the logistics field.

Yao, Palmer, and Dresner (2007) studied electronically-enabled supply chains (ESCs) using data from a survey of 183 managers in the food industry. They found that top management support and external influences are both important determinants of ESC use in the food industry, that perceived benefits to customers, perceived benefits to suppliers, and perceived internally focused benefits, are all found to influence positively ESC use and Third that distributors are more likely to perceive greater customer benefits from ESC use than are manufacturers and retailers. Finally, Frasquet, Cervera, and Gil (2008) studied the influence of customer orientation on the application of IT to the supply chain and of these on the development of channel relationships based on trust and commitment. A questionnaire-based personal survey was conducted among manufacturers. The found that that customer orientation affects the application of IT to logistics, and IT has a positive impact on manufacturer and supplier commitment to the relationship.

## DATA COLLECTION

Using the data from the Census of Manufacturers Surveys of years 2003 through 2007, as conducted by the Manufacturers Performance Institute, the usage of information technology systems by US Food Manufacturers, NAICS 311, is analyzed. The IW/MPI Census of Manufacturers was conducted using an online questionnaire and a hard-copy questionnaire that was twice mailed from MPI to approximately 12,000 plant leaders (NAICS

*Table 2. Breakout of NAICS industry group 311 – food manufacturers and processors*

| 4 Digit NAICS Codes | 2003 | 2004 | 2005 | 2006 | 2007 | Total |
|---|---|---|---|---|---|---|
| **3111 - Animal Food Mfg** | 1/2.5% | 2/5.7% | 1/6.3% | 3/10.0% | 0 | 7/5.0% |
| **3112 - Grain & Oilseed Milling** | 1/2.5% | 3/8.6% | 0 | 1/3.3% | 2/10.0% | 7/5.0% |
| **3113 - Sugar & Confectionery Product Mfg** | 3/7.5% | 2/5.7% | 1/6.3% | 0 | 1/5.0% | 7/5.0% |
| **3114 - Fruit & Vegetable Preserving & Specialty Food Mfg** | 2/5.0% | 4/11.4% | 1/6.3% | 3/10.0% | 3/15.0% | 13/9.2% |
| **3115 - Dairy Product Mfg** | 7/17.5% | 0 | 3/18.8% | 5/16.7% | 2/10.0% | 17/12.1% |
| **3116 - Animal Slaughtering & Processing** | 2/5.0% | 5/14.3% | 4/25.0% | 4/13.3% | 3/15.0% | 18/12.8% |
| **3117 - Seafood Product Preparation & Packaging** | 0 | 0 | 0 | 2/6.7% | 0 | 2/1.4% |
| **3118 - Bakeries & Tortilla Mfg** | 4/10.0% | 1/2.9% | 3/18.8% | 4/13.3% | 0 | 12/8.5% |
| **3119 - Other Food Mfg** | 7/17.5% | 6/17.1% | 0 | 6/20.0% | 8/40.0% | 27/19.1% |
| **Code not disclosed by respondents** | 13/32.5% | 12/34.3% | 0 | 2/6.7% | 5.0% | 31/22.0% |
| **Total** | 27 | 23 | 13 | 28 | 19 | 110 |

Source: MPI Census of US Manufacturers Survey

31 through 33 levels). Responses are received by MPI, and then entered into a database, edited, and cleansed to ensure answers were plausible, where necessary. All respondent answers to the survey are anonymous. As an incentive, respondents were offered one-time access to a database of this year's findings and entered into a random drawing for monetary awards. The responses to this survey for the years 2003 through 2007 totaled 3,550. A breakout of the respondents from the NAICS Industry Group 311 used in this analysis is shown in Table 2.

Though many of the respondents are invited to participate every year, the responses are anonymous. Therefore, there is no way of tracking a given respondent across several years. Bardhan, Whitaker, and Mithas (2006) analyzed the Census of Manufacturers Surveys response rates and founded the survey to be representative of U.S Census data at the 3 digit NAICS code level.

The IT Systems used in the US food industry and which were examined for this study include Enterprise Resource Planning systems (ERP and ERP II), Demand Management and Forecasting systems, Electronic Data Interchange systems, Online Purchasing systems, and Online Selling systems. The percentages of users of these systems within the 311 Industry respondents are shown in Table 3.

*Table 3. Breakout of IT system usage by NAICS industry group 311 respondents*

| Information Technology Applications | 2003 | 2004 | 2005 | 2006 | 2007 |
|---|---|---|---|---|---|
| **ERP (Enterprise Resource Planning)** | 22.5% | 17.1% | 31.3% | 26.7% | 40.0% |
| **ERP II (Enterprise Resource Planning)** | 7.5% | 8.6% | 0.0% | 3.3% | 5.0% |
| **DPF (Demand-Management/Forecasting)** | 22.5% | 40.0% | 25.0% | 33.3% | 40.0% |
| **EDI (Electronic Data Interchange)** | 45.0% | 48.6% | 50.0% | 30.0% | 50.0% |
| **OP (Online Purchasing)** | 30.0% | 31.4% | 31.3% | 33.3% | 35.0% |
| **OS (Online Selling)** | 20.0% | 14.3% | 25.0% | 23.3% | 10.0% |

Source: MPI Census of US Manufacturers Survey

In comparing the US food industry manufacturers to the US beef industry, it is recognized that upstream organizations within the beef industry supply chain (i.e., Cow-Calf Producers, Stockers, and Order Buyers) are too small to afford these types of IT systems. It is proposed that if any of these systems prove to be beneficial to downstream organizations, that their usage could be extended upstream, improving communication, coordination, and synchronization of activities; thus improving the efficiency of the overall supply chain.

A brief description of the nature of the companies (processes types, product volume and mix structure, size as measured by the number of employees, and age of the plants) found in the downstream food industry group (NAICS 311) are examined in Table 4, Table 5, Table 6, and Table 7.

## ANALYSIS OF DATA

Using the database information, an analysis of variance was conducted using the Proc Mixed procedure in SAS. The dependent variables of profitability, cost of goods sold, labor costs, material costs, and overhead costs were analyzed with respect to the IT systems identified earlier. These variables are key performance measures

*Table 4. Breakout of processing types within the NAICS Industry Group 311 respondents*

| Process Types | 2003 | 2004 | 2005 | 2006 | 2007 |
|---|---|---|---|---|---|
| **Discrete** | 7/17.5% | 3/8.82% | 2/13.33% | 4/13.33% | 2/10% |
| **Continuous** | 24/60% | 26/76.47% | 9/60% | 15/50% | 14/70% |
| **Mixed** | 9/22.5% | 5/14.7% | 4/26.67% | 11/36.67% | 4/20% |

Source: MPI Census of US Manufacturers Survey

*Table 5. Breakout of product volume and mix within the NAICS Industry Group 311 respondents*

| Product Volume and Mix | 2003 | 2004 | 2005 | 2006 | 2007 |
|---|---|---|---|---|---|
| **High Volume/ High Mix** | 0 | 8/23.53% | 3/20% | 9/31.03% | 11/55% |
| **High Volume/ Low Mix** | 0 | 13/38.24% | 5/33.33% | 7/24.14% | 6/30% |
| **Low Volume/ High Mix** | 0 | 11/32.35% | 7/46.67% | 10/34.48% | 1/5% |
| **Low Volume/ Low Mix** | 0 | 2/5.88% | 0 | 3/10.34 | 2/10% |

Source: MPI Census of US Manufacturers Survey

*Table 6. Breakout of plant size within the NAICS Industry Group 311 respondents*

| Plant Size | 2003 | 2004 | 2005 | 2006 | 2007 |
|---|---|---|---|---|---|
| **<100 Employees** | 22/55% | 15/42.86% | 6/37.5% | 11/36.67% | 5/25% |
| **100-249 Employees** | 13/32.5% | 11/31.43% | 5/31.25% | 11/36.67% | 5/25% |
| **250-499 Employees** | 3/7.50% | 4/11.43% | 3/18.75% | 4/13.33% | 6/30% |
| **500-1000 Employees** | 2/5% | 4/11.43% | 1/6.25% | 2/6.67% | 2/10% |
| **>1000 Employees** | 0 | 1/2.86% | 1/6.25% | 2/6.67% | 2/10% |

Source: MPI Census of US Manufacturers Survey

*Table 7. Breakout of plant age within the NAICS Industry Group 311 respondents*

| Plant Size | 2003 | 2004 | 2005 | 2006 | 2007 |
|---|---|---|---|---|---|
| **<5 Years** | 1/2.5% | 2/5.71% | 0 | 0 | 0 |
| **5-10 Years** | 4/10% | 6/17.14% | 2/12.5% | 2/6.67% | 1/5% |
| **11-20 Years** | 4/10% | 5/14.29% | 2/12.5% | 6/20% | 5/25% |
| **>20 Years** | 31/77.5% | 22/62.86% | 12/75% | 22/73.33% | 14/70% |

Source: MPI Census of US Manufacturers Survey

used in analyzing supply chains (Kenyon & Neureuther, 2008). The variable, *year*, was treated as an independent variable to account for any environmental factors that may be associated with a specific year. The 4-digit NAICS code variable was modeled as a random variable to account for differences between industry groups within the food manufacturing industry. Each of the associated IT systems were treated as binary independent variables, with 1 indicating the IT system type is used by the organization and zero indicating it was not. An analysis on each of the dependent variables; profitability, cost of goods sold, labor costs, material costs, overhead costs, and degree of customer and supplier integration with dependent variables of year, NAICS code, and binary variables for each of the IT systems is conducted next.

*Table 8. Profitability analysis: the affect of year*

| | ERP | ERP II | DPF | EDI | OP | OS | |
|---|---|---|---|---|---|---|---|
| **Int.** | <.0001 | 0.0680 | <.0001 | <.0001 | <.0001 | <.0001 | |
| **2003** | 0.2274 | 0.2854 | 0.1686 | 0.2347 | 0.2414 | 0.2499 | |
| **2004** | **0.0957** | 0.1612 | **0.0661** | **0.0964** | 0.1205 | 0.1071 | |
| **2005** | 0.6893 | 0.7264 | 0.6560 | 0.7345 | 0.6932 | 0.7673 | |
| **2006** | 0.1931 | 0.2014 | 0.1574 | 0.1813 | 0.2192 | 0.1755 | |
| **2007** | . | . | . | . | . | . | |

| | OP, & OS | DPF, OP, & OS | ERP, OP, & OS | ERP_II, OP, & OS | EDI, OP, & OS | ERP, DPF, OP, & OS | ERP, EDI, DPF, OP, & OS | ERP, EDI, DPF, OP, & OS | ERP, EDI, DPF, OP, & OS |
|---|---|---|---|---|---|---|---|---|---|
| **Int.** | <.0001 | 0.0209 | 0.0069 | 0.0927 | <.0001 | 0.0428 | 0.0083 | 0.2210 | 0.0524 |
| **2003** | 0.2204 | 0.1383 | 0.2659 | 0.2443 | 0.2292 | 0.1802 | 0.2573 | 0.2435 | 0.1038 |
| **2004** | 0.1295 | 0.0749 | 0.1718 | 0.1827 | 0.1342 | 0.1210 | 0.1660 | 0.1996 | **0.0668** |
| **2005** | 0.6508 | 0.7603 | 0.6668 | 0.6405 | 0.6262 | 0.8087 | 0.5935 | 0.6139 | 0.6506 |
| **2006** | 0.1618 | 0.1950 | 0.2254 | 0.1644 | 0.1869 | 0.2233 | 0.2070 | 0.1900 | 0.1479 |
| **2007** | . | . | . | . | . | . | . | . | . |

*Table 9. Profitability analysis: the affect of IT systems*

| Independent variables | Estimate | Std Error | DF | T Value | Pr > \|t\| |
|---|---|---|---|---|---|
| ERP | -0.04408 | 0.1008 | 73.8 | -0.44 | 0.6633 |
| ERP II | -0.1964 | 0.2208 | 74 | -0.89 | 0.3766 |
| **DPF** | -0.2190 | 0.08525 | 73.7 | -2.57 | **0.0122** |
| EDI | -0.04701 | 0.08271 | 70.1 | -0.57 | 0.5716 |
| OP | -0.08234 | 0.08793 | 73.2 | -0.94 | 0.3521 |
| OS | 0.09916 | 0.1077 | 71.3 | 0.92 | 0.3605 |
| OP<br>OS<br>OP*OS | -0.2265<br>0.1246<br>0.1106 | 0.2172<br>0.1488<br>0.2422 | 70.5<br>71.7<br>71.5 | -1.04<br>0.84<br>0.46 | 0.3006<br>0.4052<br>0.6494 |
| DPF<br>OP<br>OS<br>DPF*OP<br>DPF*OS<br>OP*OS | -0.2278<br>-0.4060<br>0.1403<br>0.2610<br>-0.1507<br>0.2021 | 0.2504<br>0.3167<br>0.2480<br>0.2284<br>0.3282<br>0.2687 | 68.5<br>67.9<br>65.7<br>68.3<br>67.9<br>69 | -0.91<br>-1.28<br>0.57<br>1.14<br>-0.46<br>0.75 | 0.3663<br>0.2042<br>0.5736<br>0.2571<br>0.6477<br>0.4546 |
| ERP<br>OP<br>OS<br>ERP*OP<br>ERP*OS<br>OP*OS | 0.07428<br>-0.1728<br>0.1732<br>-0.06699<br>-0.05670<br>0.1050 | 0.3174<br>0.3455<br>0.3453<br>0.2597<br>0.4016<br>0.2616 | 66.4<br>68.8<br>68.8<br>68.2<br>67.7<br>69 | 0.23<br>-0.50<br>0.50<br>-0.26<br>-0.14<br>0.40 | 0.8158<br>0.6186<br>0.6175<br>0.7972<br>0.8881<br>0.6893 |
| ERP_II<br>OP<br>OS<br>ERP_II*OP<br>ERP_II *OS<br>OP*OS | -0.1155<br>-0.1541<br>0.1111<br>-0.07462<br>0<br>0.1225 | 0.2824<br>0.5156<br>0.1549<br>0.4676<br>.<br>0.2483 | 70<br>69.8<br>69.9<br>69.9<br>.<br>69.6 | -0.41<br>-0.30<br>0.72<br>-0.16<br>.<br>0.49 | 0.6838<br>0.7659<br>0.4758<br>0.8737<br>.<br>0.6232 |
| EDI<br>OP<br>OS<br>EDI*OP<br>EDI*OS<br>OP*OS | 0.1087<br>-0.1395<br>0.1383<br>-0.1517<br>-0.01482<br>0.084336 | 0.2475<br>0.3542<br>0.1944<br>0.2319<br>0.3272<br>0.3205 | 66.1<br>68.7<br>69<br>68.8<br>65.5<br>67.7 | 0.44<br>-0.39<br>0.71<br>-0.65<br>-0.05<br>0.26 | 0.6620<br>0.6949<br>0.4794<br>0.5153<br>0.9640<br>0.7932 |
| ERP<br>DPF<br>OP<br>OS<br>ERP*DPF<br>ERP*OP<br>DPF*OP<br>ERP*OS<br>DPF*OS<br>OP*OS | 0.1570<br>-0.2186<br>-0.2723<br>0.1433<br>-0.05608<br>-0.1682<br>0.2919<br>0.008183<br>-0.1467<br>0.1699 | 0.3384<br>0.3054<br>0.3937<br>0.3732<br>0.2165<br>0.2735<br>0.2395<br>0.4051<br>0.3419<br>0.2824 | 62.6<br>63.9<br>64.2<br>64.7<br>64.8<br>64.1<br>64.3<br>64.3<br>63.7<br>65 | 0.46<br>-0.72<br>-0.69<br>0.38<br>-0.26<br>-0.62<br>1.22<br>0.02<br>-0.43<br>0.60 | 0.6443<br>0.4769<br>0.4916<br>0.7022<br>0.7965<br>0.5394<br>0.2275<br>0.9839<br>0.6692<br>0.5495 |
| ERP<br>EDI<br>OP<br>OS<br>ERP*EDI<br>ERP*OP<br>EDI*OP<br>ERP*OS<br>EDI*OS<br>OP*OS | 0.004063<br>0.2026<br>-0.02497<br>0.06975<br>-0.09972<br>-0.1087<br>-0.1472<br>0.09852<br>-0.02136<br>0.04420 | 0.3525<br>0.3614<br>0.4779<br>0.3854<br>0.2433<br>0.2754<br>0.2485<br>0.4493<br>0.3635<br>0.3469 | 64.3<br>63.5<br>64.9<br>64.9<br>61.9<br>63.8<br>65<br>64.7<br>65<br>65 | 0.01<br>0.56<br>-0.05<br>0.18<br>-0.41<br>-0.39<br>-0.59<br>0.22<br>-0.06<br>0.13 | 0.9908<br>0.5771<br>0.9585<br>0.8569<br>0.6833<br>0.6944<br>0.5559<br>0.8271<br>0.9533<br>0.8990 |

*continued on following page*

*Table 9. Continued*

| | | | | | |
|---|---|---|---|---|---|
| ERP_II | -0.1648 | 0.4036 | 65.2 | -0.41 | 0.6843 |
| EDI | -0.03890 | 0.6360 | 65.4 | -0.06 | 0.9514 |
| OP | 0.07648 | 0.5887 | 65 | 0.10 | 0.9175 |
| OS | 0.1243 | 0.2020 | 66 | 0.62 | 0.5403 |
| ERP_II*EDI | 0.1470 | 0.5887 | 65.2 | 0.25 | 0.8036 |
| ERP_II*OP | -0.2027 | 0.5821 | 65.5 | -0.35 | 0.7289 |
| EDI*OP | -0.1687 | 0.2523 | 66 | -0.67 | 0.5061 |
| ERP_II*OS | 0 | . | . | . | . |
| EDI*OS | -0.00320 | 0.3436 | 64.8 | -0.01 | 0.9926 |
| OP*OS | 0.08325 | 0.3352 | 65.7 | 0.25 | 0.8047 |
| ERP | 0.07175 | 0.3720 | 58.6 | 0.19 | 0.8477 |
| EDI | 0.07970 | 0.4071 | 60 | 0.20 | 0.8454 |
| DPF | -0.2664 | 0.3157 | 58.8 | -0.84 | 0.4022 |
| OP | 0.09587 | 0.4797 | 59.8 | 0.20 | 0.8423 |
| OS | -0.1024 | 0.4026 | 59.7 | -0.25 | 0.8001 |
| ERP*EDI | -0.08229 | 0.2572 | 58.8 | -0.32 | 0.7502 |
| ERP*DPF | -0.07622 | 0.2469 | 60 | -0.31 | 0.7586 |
| EDI*DPF | 0.1830 | 0.2181 | 59.8 | 0.84 | 0.4049 |
| ERP*OP | -0.3578 | 0.3025 | 58.8 | -1.18 | 0.2417 |
| EDI*OP | -0.4888 | 0.2762 | 58.1 | -1.77 | **0.0821** |
| DPF*OP | 0.5253 | 0.2775 | 57.3 | 1.89 | **0.0634** |
| ERP*OS | 0.3271 | 0.4593 | 59.9 | 0.71 | 0.4791 |
| EDI*OS | 0.3217 | 0.3718 | 59.3 | 0.87 | 0.3904 |
| DPF*OS | -0.3917 | 0.3707 | 58.9 | -1.06 | 0.2949 |
| OP*OS | 0.01421 | 0.3452 | 60 | 0.04 | 0.9673 |
| EDI | 0.2286 | 0.1554 | 69.4 | 1.47 | 0.1459 |
| **DPF** | -0.4073 | 0.1518 | 69.8 | -2.68 | **0.0091** |
| OP | -0.1112 | 0.1505 | 68.1 | -0.74 | 0.4627 |
| EDI*OP | -0.2747 | 0.1845 | 69.9 | -1.49 | 0.1410 |
| DPF*OP | 0.2883 | 0.1922 | 69.1 | 1.50 | 0.1381 |

## PROFITABILITY

Plant profitability is measured by the ratio of gross profits divided by net sales (Siegel & Shim, 1991; Williams, Haka, Bettner, & Carcello, 2008). It is expected from literature that information technologies can facilitate greater efficiencies within the production and distribution processes, and when coupled with collaborative forecasting and planning tools can potentially promote higher productivity and more sells: thus, increasing revenues and decreases expenses. Based upon these assertions the following null hypotheses are proposed:

H1: Information Technology system will not improve plant profitability.

$H1_a$: Enterprise Resource Planning systems will increase plant profitability.

$H1_b$: Enterprise Resource Planning II systems will increase plant profitability.

$H1_c$: Electronic Data Interchange systems will increase plant profitability.

$H1_d$: Demand Planning and Forecasting systems will increase plant profitability.

$H1_e$: Online Purchasing systems will increase plant profitability.

$H1_f$: Online Selling systems will increase plant profitability.

In Table 8, the respective p-values for the effect of the year variable on each of the IT systems analyzed are shown. At the 0.05 level of significance, only the year 2004 was found to be marginally significant with respect to the models using ERP, DPF, and EDI.

*Table 10. Cost of goods sold analysis: the affect of year*

| | ERP | ERP II | DPF | EDI | OP | OS | OP, OS | |
|---|---|---|---|---|---|---|---|---|
| **Int.** | <.0001 | <.0001 | <.0001 | <.0001 | <.0001 | <.0001 | <.0001 | |
| **2003** | 0.9020 | 0.9354 | 0.9956 | 0.9563 | 0.9219 | 0.9520 | 0.9269 | |
| **2004** | 0.5576 | 0.5151 | 0.3867 | 0.4654 | 0.5565 | 0.4894 | 0.5673 | |
| **2005** | 0.9636 | 0.9010 | 0.8884 | 0.8947 | 0.8617 | 0.8966 | 0.8563 | |
| **2006** | 0.7484 | 0.7764 | 0.7955 | 0.8265 | 0.6778 | 0.7457 | 0.7172 | |
| **2007** | . | . | . | . | . | . | . | |
| | **DPF, OP, & OS** | **ERP, OP, & OS** | **ERP_II, OP, & OS** | **EDI, OP, & OS** | **ERP, DPF, OP, & OS** | **ERP, EDI, OP, & OS** | **ERP_II, EDI, OP, & OS** | **ERP, EDI, DPF, OP, & OS** |
| **Int.** | <.0001 | <.0001 | <.0001 | <.0001 | <.0001 | <.0001 | <.0001 | <.0001 |
| **2003** | 0.8652 | 0.8678 | 0.9445 | 0.9101 | 0.8810 | 0.8688 | 0.9236 | 0.3622 |
| **2004** | 0.4148 | 0.6532 | 0.5742 | 0.5852 | 0.5933 | 0.6584 | 0.5722 | 0.2454 |
| **2005** | 0.9642 | 0.9365 | 0.8629 | 0.8702 | 0.9890 | 0.9359 | 0.8728 | 0.9747 |
| **2006** | 0.5906 | 0.6803 | 0.7164 | 0.6888 | 0.6710 | 0.6812 | 0.6883 | 0.8701 |
| **2007** | . | . | . | . | . | . | . | . |

Due to excessive skewness and kurtosis in the Profitability metric, a $\text{Log}_{10}$ transformation was used. With respect to potential profitability impacts due the usage of a given IT system or group of systems, only OS systems and DPF coupled with OP systems increased the plants profitability, and none of these systems were found to be significant at the 0.05 level, except for DPF. It is generally accepted that good planning makes for good results, and that the foundations of good planning is accurate and reliable forecasting. Thus, this finding tends to be opposite of what would be expected. There are two possibilities leading to this result; 1) the volatility of demands for products and services is low, making the expense of owning and operating an advanced forecasting /demand management system greater than the revenues associated with the increases accuracy in the planning, or 2) management has not effectively integrated the forecast results in their planning processes.

When the independent variables were analyzed in combinations, along with their possible interactions, similar results were found. In all of the models, none of the main affect variables were found to be significant, and the only interactive affects of significant were between EDI and OP, and DPF and OP; see the ERP|EDI|DPF|OP|OS model. When the non-significant variables and interactions were removed from the ERP|EDI|DPF|OP|OS model, the only affect found to be significant was DPF, and it again negatively affected profitability. Though there is evidence that DPF significantly affect plant profitability, the effect is to decrease profitability. Based upon these finding, we must fail to reject the null hypothesis. The full statistical output can be found in Table 9.

## COST OF GOODS SOLD

In the MPI survey, the cost of goods sold (COGS) variable is reported as a percentage of revenues.

*Table 11. Cost of goods sold analysis: the affect of IT systems*

| Independent variables | Estimate | Std Error | DF | T Value | Pr > \|t\| |
|---|---|---|---|---|---|
| ERP | -0.03383 | 0.08187 | 78.7 | -0.41 | 0.6806 |
| ERP II | 0.02797 | 0.1861 | 78.9 | 0.15 | 0.8809 |
| **DPF** | 0.1559 | 0.07082 | 78.6 | 2.20 | **0.0307** |
| EDI | 0.02496 | 0.06777 | 78.2 | 0.37 | 0.7137 |
| **OP** | 0.1385 | 0.07071 | 78.1 | 1.96 | **0.0538** |
| OS | 0.03759 | 0.09045 | 76.1 | 0.42 | 0.6789 |
| OP<br>OS<br>OP*OS | 0.1666<br>-0.03042<br>-0.01969 | 0.1819<br>0.1231<br>0.2018 | 75.4<br>76.5<br>76.4 | 0.92<br>-0.25<br>-0.10 | 0.3627<br>0.8055<br>0.9225 |
| DPF<br>OP<br>OS<br>DPF*OP<br>DPF*OS<br>OP*OS | 0.1412<br>0.3835<br>-0.09771<br>-0.2746<br>0.1921<br>-0.1144 | 0.2103<br>0.2633<br>0.2080<br>0.1870<br>0.2719<br>0.2226 | 73.1<br>72.4<br>70<br>72.5<br>71.2<br>73.8 | 0.67<br>1.46<br>-0.47<br>-1.47<br>0.71<br>-0.51 | 0.5039<br>0.1497<br>0.6399<br>0.1464<br>0.4821<br>0.6087 |
| ERP<br>OP<br>OS<br>ERP*OP<br>ERP*OS<br>OP*OS | 0.2149<br>0.03495<br>0.2446<br>0.09571<br>-0.3537<br>0.03850 | 0.2617<br>0.2833<br>0.2847<br>0.2104<br>0.3288<br>0.2151 | 71.5<br>74<br>74<br>73.6<br>73.3<br>74 | 0.82<br>0.12<br>0.86<br>0.45<br>-1.08<br>0.18 | 0.4143<br>0.9022<br>0.3930<br>0.6505<br>0.2856<br>0.8584 |
| ERP_II<br>OP<br>OS<br>ERP_II*OP<br>ERP_II *OS<br>OP*OS | -0.06000<br>0.07903<br>-0.03791<br>0.08707<br>0<br>-0.01115 | 0.2352<br>0.4315<br>0.1279<br>0.3908<br>.<br>0.2068 | 75<br>74.7<br>74.8<br>75<br>.<br>74.5 | -0.26<br>0.18<br>-0.30<br>0.22<br>.<br>-0.05 | 0.7994<br>0.8552<br>0.7678<br>0.8243<br>.<br>0.9572 |
| EDI<br>OP<br>OS<br>EDI*OP<br>EDI*OS<br>OP*OS | -0.1434<br>0.2238<br>-0.07583<br>0.03140<br>0.1221<br>-0.09098 | 0.2064<br>0.2934<br>0.1594<br>0.1898<br>0.2698<br>0.2677 | 68.6<br>72<br>73.9<br>72.9<br>65.8<br>70.9 | -0.69<br>0.76<br>-0.48<br>0.17<br>0.45<br>-0.34 | 0.4894<br>0.4481<br>0.6356<br>0.8690<br>0.6522<br>0.7349 |
| ERP<br>DPF<br>OP<br>OS<br>ERP*DPF<br>ERP*OP<br>DPF*OP<br>ERP*OS<br>DPF*OS<br>OP*OS | 0.1380<br>0.03413<br>0.2194<br>0.1911<br>0.1071<br>0.1775<br>-0.3197<br>-0.4408<br>0.2786<br>-0.06376 | 0.2751<br>0.2508<br>0.3172<br>0.3064<br>0.1720<br>0.2197<br>0.1951<br>0.3303<br>0.2801<br>0.2291 | 67.6<br>69.2<br>69.9<br>69.7<br>69<br>69.7<br>69<br>69.3<br>68.2<br>70 | 0.50<br>0.14<br>0.69<br>0.62<br>0.62<br>0.81<br>-1.64<br>-1.33<br>0.99<br>-0.28 | 0.6175<br>0.8922<br>0.4915<br>0.5348<br>0.5356<br>0.4218<br>0.1059<br>0.1864<br>0.3235<br>0.7816 |
| ERP<br>EDI<br>OP<br>OS<br>ERP*EDI<br>ERP*OP<br>EDI*OP<br>ERP*OS<br>EDI*OS<br>OP*OS | 0.3274<br>-0.2426<br>0.07252<br>0.2773<br>0.006467<br>0.09460<br>0.08743<br>-0.4836<br>0.1842<br>-0.03802 | 0.2874<br>0.2981<br>0.3916<br>0.3147<br>0.1957<br>0.2242<br>0.2014<br>0.3644<br>0.2960<br>0.2851 | 70<br>70<br>70<br>70<br>70<br>70<br>70<br>70<br>70<br>70 | 1.14<br>-0.81<br>0.19<br>0.88<br>0.03<br>0.42<br>0.43<br>-1.33<br>0.62<br>-0.13 | 0.2585<br>0.4185<br>0.8536<br>0.3813<br>0.9737<br>0.6743<br>0.6656<br>0.1888<br>0.5357<br>0.8943 |

*continued on following page*

*Table 11. Continued*

| | | | | | |
|---|---|---|---|---|---|
| ERP_II | -0.03065 | 0.3374 | 70 | -0.09 | 0.9279 |
| EDI | -0.06192 | 0.5334 | 70.4 | -0.12 | 0.9079 |
| OP | 0.06137 | 0.6156 | 70.6 | 0.10 | 0.9209 |
| OS | -0.07921 | 0.1653 | 70.9 | -0.48 | 0.6333 |
| ERP_II*EDI | -0.08228 | 0.4939 | 70.1 | -0.17 | 0.8682 |
| ERP_II*OP | 0.1432 | 0.4873 | 70.8 | 0.29 | 0.7697 |
| EDI*OP | 0.05124 | 0.2072 | 70.9 | 0.25 | 0.8053 |
| ERP_II*OS | 0 | . | . | . | . |
| EDI*OS | 0.1035 | 0.2842 | 66.9 | 0.36 | 0.7170 |
| OP*OS | -0.06821 | 0.2806 | 69.8 | -0.24 | 0.8086 |
| ERP | 0.2706 | 0.3001 | 65 | 0.90 | 0.3707 |
| EDI | -0.1562 | 0.3324 | 65 | -0.47 | 0.6400 |
| DPF | 0.02328 | 0.2600 | 65 | 0.09 | 0.9289 |
| OP | 0.06416 | 0.3921 | 65 | 0.16 | 0.8705 |
| OS | 0.3097 | 0.3304 | 65 | 0.94 | 0.3520 |
| ERP*EDI | -0.04410 | 0.2078 | 65 | -0.21 | 0.8326 |
| ERP*DPF | 0.1378 | 0.1977 | 65 | 0.70 | 0.4882 |
| EDI*DPF | -0.08646 | 0.1789 | 65 | -0.48 | 0.6304 |
| ERP*OP | 0.2879 | 0.2470 | 65 | 1.17 | 0.2480 |
| **EDI*OP** | 0.3706 | 0.2188 | 65 | 1.69 | **0.0951** |
| **DPF*OP** | -0.4957 | 0.2226 | 65 | -2.23 | **0.0294** |
| **ERP*OS** | -0.7007 | 0.3739 | 65 | -1.87 | **0.0654** |
| EDI*OS | -0.09088 | 0.2987 | 65 | -0.30 | 0.7619 |
| DPF*OS | 0.4655 | 0.3000 | 65 | 1.55 | 0.1256 |
| OP*OS | -0.02326 | 0.2835 | 65 | -0.08 | 0.9349 |
| ERP | 0.2321 | 0.2538 | 71.2 | 0.91 | 0.3634 |
| EDI | -0.2009 | 0.1323 | 70.4 | -1.52 | 0.1334 |
| **DPF** | 0.3229 | 0.1286 | 71.8 | 2.51 | **0.0143** |
| OP | 0.1757 | 0.1305 | 70.3 | 1.35 | 0.1824 |
| OS | 0.2927 | 0.2487 | 72 | 1.18 | 0.2432 |
| EDI*OP | 0.2320 | 0.1611 | 71.8 | 1.44 | 0.1542 |
| DPF*OP | -0.2567 | 0.1620 | 71 | -1.59 | 0.1153 |
| ERP*OS | -0.3469 | 0.2682 | 71.7 | -1.29 | 0.2001 |

Again, due to excessive skewness and kurtosis, the COGS variable was transformed using a $Log_{10}$ transformation. The hypotheses testing the performance metric of cost of goods sold are as follow:

H2: Information Technology system will not improve the plant's cost of goods sold.

$H2_a$: Enterprise Resource Planning systems will increase the plant's cost of goods sold.

$H2_b$: Enterprise Resource Planning II systems will increase the plant's cost of goods sold.

$H2_c$: Electronic Data Interchange systems will increase the plant's cost of goods sold.

$H2_d$: Demand Planning and Forecasting systems will increase the plant's cost of goods sold.

$H2_e$: Online Purchasing systems will increase the plant's cost of goods sold.

$H2_f$: Online Selling systems will increase the plant's cost of goods sold.

The resulting p-values for the affect of each year can be seen in Table 10. In the single variable models, both ERP and OS systems tended to lower the plant's cost of goods sold, while only DPF and OP systems were found to be significant at the 0.05 level. When the independent variables are analyzed in combinations, along with their possible interactions, similar results were found. None of the main affect variables were found to be significant in any of the models, and the only interactive affects found to be significant were between EDI and OP, and DPF and OP; See the ERP|EDI|DPF|OP|OS model. When the non-significant variables were removed from the ERP|EDI|DPF|OP|OS model, only the main affect

*Table 12. Labor costs analysis: the affect of year*

| | ERP | ERP II | DPF | EDI | OP | OS | OP, OS |
|---|---|---|---|---|---|---|---|
| Int. | <.0001 | <.0001 | <.0001 | <.0001 | <.0001 | <.0001 | <.0001 |
| 2003 | 0.3862 | 0.4494 | 0.4316 | 0.5494 | 0.4495 | 0.4316 | 0.4675 |
| 2004 | 0.7165 | 0.8363 | 0.7997 | 0.9325 | 0.7859 | 0.8330 | 0.8344 |
| 2005 | 0.6836 | 0.6333 | 0.6379 | 0.5230 | 0.5886 | 0.6596 | 0.5253 |
| 2006 | 0.4151 | 0.4535 | 0.4476 | 0.6859 | 0.4387 | 0.4080 | 0.4618 |
| 2007 | . | . | . | . | . | . | . |

| | DPF, OP, & OS | ERP, OP, & OS | ERP_II, OP, & OS | EDI, OP, & OS | ERP, DPF, OP, & OS | ERP, EDI, OP, & OS | ERP_II, EDI, OP, & OS | ERP, EDI, DPF, OP, & OS |
|---|---|---|---|---|---|---|---|---|
| Int. | <.0001 | <.0001 | <.0001 | <.0001 | <.0001 | <.0001 | 0.0020 | <.0001 |
| 2003 | 0.4235 | 0.2762 | 0.5132 | 0.5416 | 0.2601 | 0.3446 | 0.5865 | 0.3386 |
| 2004 | 0.7785 | 0.6144 | 0.8043 | 0.9645 | 0.6053 | 0.7918 | 0.9049 | 0.8520 |
| 2005 | 0.5058 | 0.5071 | 0.5289 | 0.5756 | 0.4895 | 0.5503 | 0.5704 | 0.5023 |
| 2006 | 0.4583 | 0.2608 | 0.4567 | 0.4991 | 0.2960 | 0.3326 | 0.5033 | 0.4420 |
| 2007 | . | . | . | . | . | . | . | . |

variable DPF was still found to be significant; and it effect is to increase the plant's cost of goods sold. From these results there is evidence that both DPF and OP significantly affect the plant's cost of goods sold. There is also evidence that ERP, EDI, and OS systems when used with other IT systems can significantly affect the plant's cost of goods sold. Unfortunately, the singular, and net interactive affects in all cases was to increase the plant's cost of goods sold. Thus, we again fail to reject the null hypothesis. The full statistical output can be found in Table 11.

## LABOR COSTS

In the MPI survey, labor costs were reported as a percentage of the cost of goods sold. Due to excessive skewness and kurtosis, the cost of labor variable was transformed using a $Log_{10}$ transformation. The hypotheses testing the performance metric of labor costs are as follow:

H3: Information Technology system will not improve the plant's labor costs.

$H3_a$: Enterprise Resource Planning systems will increase the plant's labor costs.

$H3_b$: Enterprise Resource Planning II systems will increase the plant's labor costs.

$H3_c$: Electronic Data Interchange systems will increase the plant's labor costs.

$H3_d$: Demand Planning and Forecasting systems will increase the plant's labor costs.

$H3_e$: Online Purchasing systems will increase the plant's labor costs.

$H3_f$: Online Selling systems will increase the plant's labor costs.

The p-value of each year in the study can be seen in Table 12, where it can be seen that none of the years were significant at the .05 level.

In the single variable models, only ERP, ERP II, and DPF system showed evidence of lowering the plant's labor costs; and with the exception of EDI, none of these IT systems were found to be

*Table 13. Labor costs analysis: the affect of IT systems*

| Independent variables | Estimate | Std Error | DF | T Value | Pr > \|t\| |
|---|---|---|---|---|---|
| ERP | -0.04177 | 0.07445 | 82.8 | -0.56 | 0.5763 |
| ERP II | -0.03467 | 0.1822 | 83.4 | -0.19 | 0.8496 |
| DPF | -0.01728 | 0.06716 | 82.5 | -0.26 | 0.7976 |
| **EDI** | 0.1164 | 0.06444 | 84.9 | 1.81 | **0.0743** |
| OP | 0.08786 | 0.06660 | 83.5 | 1.32 | 0.1907 |
| OS | 0.086667 | 0.08145 | 83.3 | 1.06 | 0.2904 |
| OP<br>OS<br>OP*OS | 0.2056<br>0.1179<br>-0.1708 | 0.1623<br>0.1133<br>0.1821 | 79.8<br>82.8<br>81.2 | 1.27<br>1.04<br>-0.94 | 0.2090<br>0.3012<br>0.3513 |
| DPF<br>OP<br>OS<br>DPF*OP<br>DPF*OS<br>OP*OS | -0.08895<br>0.1664<br>0.09276<br>0.07357<br>0.006739<br>-0.1581 | 0.1907<br>0.2482<br>0.1821<br>0.1821<br>0.2464<br>0.2141 | 77.5<br>76.9<br>77<br>79.3<br>77.5<br>78.7 | -0.47<br>0.67<br>0.51<br>0.40<br>0.03<br>-0.74 | 0.6423<br>0.5044<br>0.6119<br>0.6873<br>0.9783<br>0.4625 |
| ERP<br>OP<br>OS<br>ERP*OP<br>ERP*OS<br>OP*OS | -0.2789<br>0.2353<br>-0.09793<br>-0.01486<br>0.2949<br>-0.1930 | 0.1910<br>0.2159<br>0.1981<br>0.1723<br>0.2248<br>0.1838 | 78.2<br>78.7<br>78.1<br>77.8<br>77.8<br>78.1 | -1.46<br>1.09<br>-0.49<br>-0.09<br>1.31<br>-1.05 | 0.1482<br>0.2792<br>0.6225<br>0.9315<br>0.1933<br>0.2971 |
| ERP_II<br>OP<br>OS<br>ERP_II*OP<br>ERP_II *OS<br>OP*OS | -0.1835<br>-0.2251<br>0.09270<br>0.4253<br>0<br>-0.1394 | 0.2324<br>0.4219<br>0.1176<br>0.3848<br>.<br>0.1859 | 81.1<br>80.4<br>81.7<br>79.5<br>.<br>80 | -0.79<br>-0.53<br>0.79<br>1.11<br>.<br>-0.75 | 0.4322<br>0.5951<br>0.4326<br>0.2724<br>.<br>0.4554 |
| EDI<br>OP<br>OS<br>EDI*OP<br>EDI*OS<br>OP*OS | -0.05233<br>0.3894<br>0.01328<br>-0.1572<br>0.3003<br>-0.3253 | 0.1809<br>0.2567<br>0.1455<br>0.1737<br>0.2392<br>0.2336 | 78.8<br>77.9<br>80.4<br>80<br>80.1<br>78.2 | -0.29<br>1.52<br>0.09<br>-0.91<br>1.26<br>-1.39 | 0.7732<br>0.1334<br>0.9275<br>0.3681<br>0.2129<br>0.1677 |
| ERP<br>DPF<br>OP<br>OS<br>ERP*DPF<br>ERP*OP<br>DPF*OP<br>ERP*OS<br>DPF*OS<br>OP*OS | -0.2809<br>-0.00076<br>0.1726<br>-0.07173<br>0.009815<br>-0.04244<br>0.08906<br>0.3215<br>-0.1036<br>-0.1439 | 0.2234<br>0.2354<br>0.2782<br>0.2263<br>0.1647<br>0.1959<br>0.1918<br>0.2438<br>0.2615<br>0.2167 | 75.2<br>74.4<br>73.5<br>72.4<br>72.4<br>74.8<br>75.5<br>74.7<br>73.7<br>74.5 | -1.26<br>-0.00<br>0.62<br>-0.32<br>0.06<br>-0.22<br>0.46<br>1.32<br>-0.40<br>-0.66 | 0.2125<br>0.9974<br>0.5369<br>0.7521<br>0.9526<br>0.8291<br>0.6438<br>0.1913<br>0.6931<br>0.5088 |
| ERP<br>EDI<br>OP<br>OS<br>ERP*EDI<br>ERP*OP<br>EDI*OP<br>ERP*OS<br>EDI*OS<br>OP*OS | -0.2592<br>0.04328<br>0.3953<br>-0.1522<br>0.005100<br>-0.06188<br>-0.1521<br>0.2886<br>0.1951<br>-0.2918 | 0.2190<br>0.2458<br>0.3080<br>0.2093<br>0.1650<br>0.1785<br>0.1788<br>0.2377<br>0.2602<br>0.2437 | 75.8<br>74<br>72.7<br>74.4<br>75.4<br>73.7<br>75.6<br>74.6<br>75.3<br>72.9 | -1.18<br>0.18<br>1.28<br>-0.73<br>0.03<br>-0.35<br>-0.85<br>1.21<br>0.75<br>-1.20 | 0.2402<br>0.8607<br>0.2035<br>0.4695<br>0.9754<br>0.7298<br>0.3975<br>0.2284<br>0.4557<br>0.2350 |

*continued on following page*

*Table 13. Continued*

| | | | | | |
|---|---|---|---|---|---|
| ERP_II | -0.1147 | 0.3225 | 74.9 | -0.36 | 0.7231 |
| EDI | -0.01117 | 0.5035 | 75.5 | -0.02 | 0.9824 |
| OP | -0.05457 | 0.5744 | 75.8 | -0.10 | 0.9246 |
| OS | -0.00205 | 0.1499 | 77.7 | -0.01 | 0.9891 |
| ERP_II*EDI | -0.04404 | 0.4708 | 75.5 | -0.09 | 0.9257 |
| ERP_II*OP | 0.4203 | 0.4664 | 75.9 | 0.90 | 0.3704 |
| EDI*OP | -0.1354 | 0.1854 | 76.5 | -0.73 | 0.4673 |
| ERP_II*OS | 0 | . | . | . | . |
| EDI*OS | 0.2851 | 0.2486 | 76.7 | 1.15 | 0.2550 |
| OP*OS | -0.2882 | 0.2432 | 75.7 | -1.18 | 0.2398 |
| ERP | -0.2554 | 0.2471 | 71.5 | -1.03 | 0.3048 |
| EDI | 0.08521 | 0.2861 | 70.7 | 0.30 | 0.7667 |
| DPF | -0.01871 | 0.2464 | 69.6 | -0.08 | 0.9397 |
| OP | 0.3459 | 0.3418 | 68.4 | 1.01 | 0.3151 |
| OS | -0.1209 | 0.2352 | 67.6 | -0.51 | 0.6088 |
| ERP*EDI | 0.005548 | 0.1797 | 71 | 0.03 | 0.9755 |
| ERP*DPF | 0.03975 | 0.1790 | 68.7 | 0.22 | 0.8250 |
| EDI*DPF | -0.05808 | 0.1624 | 68.3 | -0.36 | 0.7218 |
| ERP*OP | -0.1367 | 0.2025 | 70.1 | -0.67 | 0.5020 |
| EDI*OP | -0.1847 | 0.1929 | 69.3 | -0.96 | 0.3417 |
| DPF*OP | 0.1644 | 0.2011 | 68.9 | 0.82 | 0.4164 |
| ERP*OS | 0.3389 | 0.2565 | 70.2 | 1.32 | 0.1907 |
| EDI*OS | 0.2490 | 0.2735 | 70.2 | 0.91 | 0.3656 |
| DPF*OS | -0.1958 | 0.2678 | 68.6 | -0.73 | 0.4671 |
| OP*OS | -0.2381 | 0.2622 | 68.8 | -0.91 | 0.3672 |
| ERP | -0.08213 | 0.1381 | 80.3 | -0.61 | 0.5427 |
| EDI | 0.1826 | 0.1299 | 79.1 | 1.41 | 0.1638 |
| DPF | -0.05115 | 0.09648 | 80.4 | -0.53 | 0.5975 |
| OP | 0.04613 | 0.1381 | 80.3 | 0.33 | 0.7392 |
| EDI*DPF | -0.06328 | 0.1506 | 77.7 | -0.42 | 0.6758 |
| ERP*OP | 0.03711 | 0.1601 | 79.8 | 0.23 | 0.8173 |

significant. When the independent variables are analyzed in combinations along with their possible interactions similar results were found. By automating the transmission of data to suppliers the work load on purchasing managers should be significantly reduced; allowing them to spend more time on other activities, and/or reducing staff. From these results, we fail to reject the null hypothesis. The full statistical output can be found in Table 13.

## MATERIAL COSTS

In the MPI survey, the cost of materials was reported as a percentage of the cost of goods sold.

The hypotheses testing the performance metric of labor costs are as follow:

H4: Information Technology system will not improve the plant's material costs.

$H4_a$: Enterprise Resource Planning systems will increase the plant's material costs.

$H4_b$: Enterprise Resource Planning II systems will increase the plant's material costs.

$H4_c$: Electronic Data Interchange systems will increase the plant's material costs.

$H4_d$: Demand Planning and Forecasting systems will increase the plant's material costs.

$H4_e$: Online Purchasing systems will increase the plant's material costs.

$H4_f$: Online Selling systems will increase the plant's material costs.

As can be seen in Table 14 of Appendix A, the intercept was strongly significant at the 0.05 level in all models, and the year 2006 was found to be significant in one model and marginally

*Table 14. Materials cost analysis: the affect of year*

| | ERP | ERP II | DPF | EDI | OP | OS | OP, OS |
|---|---|---|---|---|---|---|---|
| Int. | <.0001 | <.0001 | <.0001 | <.0001 | <.0001 | <.0001 | <.0001 |
| 2003 | 0.2861 | 0.2875 | 0.3043 | 0.3681 | 0.3033 | 0.3014 | 0.2939 |
| 2004 | 0.2890 | 0.2687 | 0.3076 | 0.4413 | 0.3053 | 0.3164 | 0.3144 |
| 2005 | 0.7264 | 0.7827 | 0.7543 | 0.8384 | 0.7582 | 0.7369 | 0.6731 |
| 2006 | **0.0636** | **0.0705** | **0.0674** | 0.1175 | **0.0663** | **0.0596** | **0.0569** |
| 2007 | . | . | . | . | . | . | . |

| | DPF, OP, & OS | ERP, OP, & OS | ERP_II, OP, & OS | EDI, OP, & OS | ERP, DPF, OP, & OS | ERP, EDI, OP, & OS | ERP_II, EDI, OP, & OS | ERP, EDI, DPF, OP, & OS | ERP, & OP | ERP, EDI, DPF, & OP |
|---|---|---|---|---|---|---|---|---|---|---|
| Int. | <.0001 | <.0001 | <.0001 | <.0001 | <.0001 | <.0001 | 0.0066 | <.0001 | <.0001 | <.0001 |
| 2003 | 0.3919 | 0.2927 | 0.3073 | 0.3746 | 0.3064 | 0.4065 | 0.3558 | 0.6586 | 0.3537 | 0.6269 |
| 2004 | 0.3615 | 0.4275 | 0.2418 | 0.4310 | 0.4134 | 0.6763 | 0.4000 | 0.9693 | 0.4367 | 0.8896 |
| 2005 | 0.5166 | 0.7509 | 0.7064 | 0.6176 | 0.6450 | 0.7251 | 0.6850 | 0.7730 | 0.7637 | 0.8480 |
| 2006 | **0.0830** | **0.0428** | **0.0590** | **0.0857** | **0.0621** | **0.0663** | 0.1019 | 0.2317 | **0.0649** | 0.2071 |
| 2007 | . | . | . | . | . | . | . | . | . | . |

significant most of the other models. In each year the estimate effect tended to decrease material costs, with the largest decreased effect in 2006.

In the single effect models, with the exception of ERP systems, all of the IT systems tended to lower the plants profitability, and none of the effects were found to be significant. When the independent variables are analyzed in combinations, along with their interactions, several models showed one or more IT systems to be significant or marginally significant. In the DPF|OP|OS model, the DPF system was marginally significant at the 0.05 level, and tended to lower material costs. In both the ERP|OP|OS and ERP|EDI|OP|OS models, the ERP|OP interaction was found to be marginally significant and significant respectively. In the case of the ERP|OP|OS model's interaction, the net effect was a tendency to increase material costs, while the net effect of the interaction in the ERP|EDI|OP|OS model was to lower material costs. With the ERP|EDI|DPF|OP|OS model, both the ERP|OP and the EDI|DPF interactions were found to be significant at the 0.05 level, and while the net effect of the ERP|OP interaction was to increase material costs, the EDI|DPF interaction tended to decrease material costs.

When the non-significant variables and interactions were removed from the ERP|EDI|DPF|OP|OS model, both ERP|OP and the EDI|DPF interactions remained marginally significant and significant respectively, the EDI main variable became significant at the 0.05 level. Then net estimated effect of the ERP|EDI|DPF|OP model was to decrease material costs. Furthermore, in each multi-variable model where one or more variables and/or interactions were significant, the net effect was to decrease materials costs. Based upon these results, we would have to conclude that ERP, DPF, and OP systems have the potential to improve material costs. Due to this conclusion being based upon the interactive affects, there does not excess

*Table 15. Materials cost analysis: the affect of IT systems*

| Independent variables | Estimate | Std Error | DF | T Value | Pr > \|t\| |
|---|---|---|---|---|---|
| **ERP** | 1.2982 | 4.5767 | 81.6 | 0.28 | 0.7774 |
| **ERP II** | -6.9982 | 11.1476 | 81.7 | -0.63 | 0.5319 |
| **DPF** | -0.03197 | 4.1207 | 81.1 | -0.01 | 0.9938 |
| **EDI** | -5.4114 | 3.9962 | 83.6 | -1.35 | 0.1793 |
| **OP** | -0.6819 | 4.1241 | 82 | -0.17 | 0.8691 |
| **OS** | -4.0100 | 5.0122 | 82 | -0.80 | 0.4260 |
| **OP** | 5.3271 | 10.0661 | 78.2 | 0.53 | 0.5982 |
| **OS** | -2.1301 | 7.0580 | 81.8 | -0.30 | 0.7636 |
| **OP*OS** | -5.7381 | 11.3138 | 79.8 | -0.51 | 0.6134 |
| **DPF** | -20.6672 | 11.4593 | 75.3 | -1.80 | **0.0753** |
| **OP** | 1.3727 | 14.8918 | 74.8 | 0.09 | 0.9268 |
| **OS** | -13.3626 | 10.9302 | 74.9 | -1.22 | 0.2253 |
| **DPF*OP** | 11.5069 | 10.9873 | 77.2 | 1.05 | 0.2982 |
| **DPF*OS** | 14.7924 | 14.8063 | 75.5 | 1.00 | 0.3210 |
| **OP*OS** | -8.2236 | 12.9018 | 76.7 | -0.64 | 0.5258 |
| **ERP** | -3.2205 | 11.7611 | 76.4 | -0.27 | 0.7850 |
| **OP** | -10.8010 | 13.3030 | 76.6 | -0.81 | 0.4193 |
| **OS** | 4.3852 | 12.1979 | 76.4 | 0.36 | 0.7202 |
| **ERP*OP** | 19.2928 | 10.5974 | 75.8 | 1.82 | **0.0726** |
| **ERP*OS** | -10.6567 | 13.8286 | 75.9 | -0.77 | 0.4433 |
| **OP*OS** | -3.2791 | 11.3179 | 76.4 | -0.29 | 0.7728 |
| **ERP_II** | 3.1517 | 14.3885 | 79.7 | 0.22 | 0.8272 |
| **OP** | 36.3789 | 26.0736 | 78.5 | 1.40 | 0.1669 |
| **OS** | -1.5934 | 7.2926 | 80.8 | -0.22 | 0.8276 |
| **ERP_II*OP** | -30.8304 | 23.7488 | 77.6 | -1.30 | 0.1981 |
| **ERP_II *OS** | 0 | . | . | . | . |
| **OP*OS** | -6.8113 | 11.4823 | 78.4 | -0.59 | 0.5548 |
| **EDI** | -5.2038 | 11.2229 | 77.1 | -0.46 | 0.6442 |
| **OP** | -6.0425 | 15.9080 | 76.4 | -0.38 | 0.7051 |
| **OS** | 1.6442 | 9.0536 | 79.3 | 0.18 | 0.8564 |
| **EDI*OP** | 14.7491 | 10.8019 | 78.8 | 1.37 | 0.1760 |
| **EDI*OS** | -12.6998 | 14.8697 | 78.6 | -0.85 | 0.3957 |
| **OP*OS** | 1.6894 | 14.4799 | 76.6 | 0.12 | 0.9074 |
| **ERP** | 7.5316 | 13.3927 | 73 | 0.56 | 0.5756 |
| **DPF** | -17.7583 | 14.0816 | 72.3 | -1.26 | 0.2113 |
| **OP** | -9.2544 | 16.6018 | 71.2 | -0.56 | 0.5790 |
| **OS** | -3.8362 | 13.4757 | 70.5 | -0.28 | 0.7767 |
| **ERP*DPF** | -5.6148 | 9.8106 | 70.5 | -0.57 | 0.5689 |
| **ERP*OP** | 18.0601 | 11.7272 | 72.3 | 1.54 | 0.1279 |
| **DPF*OP** | 6.5864 | 11.5104 | 73.4 | 0.57 | 0.5689 |
| **ERP*OS** | -18.0016 | 14.5892 | 72.3 | -1.23 | 0.2212 |
| **DPF*OS** | 20.0076 | 15.6194 | 71.7 | 1.28 | 0.2043 |
| **OP*OS** | -8.1815 | 12.9701 | 72.5 | -0.63 | 0.5302 |
| **ERP** | -7.1907 | 13.3087 | 73.8 | -0.54 | 0.5906 |
| **EDI** | -9.1337 | 14.8702 | 72 | -0.61 | 0.5410 |
| **OP** | -23.0055 | 18.5845 | 70.6 | -1.24 | 0.2199 |
| **OS** | 8.7510 | 12.6746 | 72.4 | 0.69 | 0.4921 |
| **ERP*EDI** | 6.7398 | 10.0172 | 73.4 | 0.67 | 0.5032 |
| **ERP*OP** | 22.1118 | 10.7876 | 71.4 | 2.05 | **0.0441** |
| **EDI*OP** | 12.6900 | 10.8577 | 73.6 | 1.17 | 0.2463 |
| **ERP*OS** | -10.1775 | 14.3896 | 72.2 | -0.71 | 0.4817 |
| **EDI*OS** | -14.7075 | 15.7805 | 72.9 | -0.93 | 0.3544 |
| **OP*OS** | 4.0395 | 14.7108 | 70.9 | 0.27 | 0.7844 |

*continued on following page*

*Table 15. Continued*

| | | | | | |
|---|---|---|---|---|---|
| **ERP_II** | 10.7788 | 19.6156 | 72.4 | 0.55 | 0.5844 |
| **EDI** | 20.5959 | 30.6654 | 73.1 | 0.67 | 0.5039 |
| **OP** | 4.3024 | 35.0181 | 73.6 | 0.12 | 0.9026 |
| **OS** | 1.5946 | 9.1934 | 76.1 | 0.17 | 0.8628 |
| **ERP_II*EDI** | -26.3569 | 28.6864 | 73.3 | -0.92 | 0.3612 |
| **ERP_II*OP** | -12.3279 | 28.4378 | 73.7 | -0.43 | 0.6659 |
| **EDI*OP** | 17.1335 | 11.3242 | 74.6 | 1.51 | 0.1345 |
| **ERP_II*OS** | 0 | . | . | . | . |
| **EDI*OS** | -15.1123 | 15.1864 | 74.3 | -1.00 | 0.3229 |
| **OP*OS** | 3.0767 | 14.8223 | 73.4 | 0.21 | 0.8361 |
| **ERP** | 4.9319 | 14.1374 | 68 | 0.35 | 0.7283 |
| **EDI** | -24.2474 | 16.3181 | 67.8 | -1.49 | 0.1419 |
| **DPF** | -16.1550 | 13.9729 | 66.5 | -1.16 | 0.2517 |
| **OP** | -20.9702 | 19.2684 | 65.1 | -1.09 | 0.2805 |
| **OS** | -1.9146 | 13.2362 | 64.9 | -0.14 | 0.8854 |
| **ERP*EDI** | 8.4878 | 10.2572 | 67.9 | 0.83 | 0.4109 |
| **ERP*DPF** | -14.2506 | 10.1213 | 66 | -1.41 | 0.1638 |
| **EDI*DPF** | 18.5009 | 9.1697 | 65.7 | 2.02 | **0.0477** |
| **ERP*OP** | 26.0717 | 11.4863 | 66.3 | 2.27 | **0.0265** |
| **EDI*OP** | 10.2030 | 10.9260 | 66.3 | 0.93 | 0.3538 |
| **DPF*OP** | 0.9023 | 11.3613 | 65.6 | 0.08 | 0.9369 |
| **ERP*OS** | -19.2928 | 14.5525 | 64.9 | -1.33 | 0.1895 |
| **EDI*OS** | -14.0404 | 15.5170 | 66.4 | -0.90 | 0.3688 |
| **DPF*OS** | 25.1224 | 15.1246 | 65.7 | 1.66 | 0.1015 |
| **OP*OS** | -1.2920 | 14.8086 | 65.5 | -0,09 | 0.9307 |
| **ERP** | -10.6682 | 7.9718 | 77.2 | -1.34 | 0.1847 |
| **EDI** | -21.7784 | 7.6739 | 76 | -2.84 | **0.0058** |
| **DPF** | -3.5862 | 5.7303 | 77.5 | -0.63 | 0.5333 |
| **OP** | -13.2467 | 8.1905 | 77.1 | -1.62 | 0.1099 |
| **EDI*DPF** | 18.4487 | 8.8590 | 75.1 | 2.08 | **0.0407** |
| **ERP*OP** | 18.6798 | 9.4700 | 76.3 | 1.97 | **0.0522** |

strong conclusive evident to support the failing of the null hypothesis. The full statistical output can be found in Table 15.

## OVERHEAD COSTS

In the MPI survey, the overhead costs were reported as a percentage of the cost of goods sold.

The hypotheses testing the performance metric of overhead costs are as follow:

H5: Information Technology system will not improve the plant's overhead costs.

$H5_a$: Enterprise Resource Planning systems will increase the plant's overhead costs.

$H5_b$: Enterprise Resource Planning II systems will increase the plant's overhead costs.

$H5_c$: Electronic Data Interchange systems will increase the plant's overhead costs.

$H5_d$: Demand Planning and Forecasting systems will increase the plant's overhead costs.

$H5_e$: Online Purchasing systems will increase the plant's overhead costs.

$H5_f$: Online Selling systems will increase the plant's overhead costs.

As can be seen in Table 16, the intercept (coincides with year 2007), was strongly significant at the 0.05 level in all the analysis. In none of the models were any of the other years significant.

In the analysis of the single main effect variables, none were found to be significant, and their estimated effects were minor. When the variables were analyzed in combinations, along with their interactions, OP and DPF were found to be mar-

*Table 16. Overhead cost analysis: the affect of year*

| | ERP | ERP II | DPF | EDI | OP | OS | OP, OS |
|---|---|---|---|---|---|---|---|
| Int. | <.0001 | <.0001 | <.0001 | <.0001 | <.0001 | <.0001 | <.0001 |
| 2003 | 0.9892 | 0.9357 | 0.9590 | 0.9027 | 0.9351 | 0.9331 | 0.9975 |
| 2004 | .09279 | 0.9893 | 0.9966 | 0.9248 | 0.9696 | 0.9594 | 0.9856 |
| 2005 | 0.2700 | 0.2991 | 0.2857 | 0.3120 | 0.2876 | 0.2843 | 0.1819 |
| 2006 | 0.2562 | 0.2831 | 0.2695 | 0.3206 | 0.2849 | 0.2605 | 0.2101 |
| 2007 | . | . | . | . | . | . | . |

| | DPF, OP, & OS | ERP, OP, & OS | ERP_ II, OP, & OS | EDI, OP, & OS | ERP, DPF, OP, & OS | ERP, EDI, OP, & OS | ERP_II, EDI, OP, & OS | ERP, EDI, DPF, OP, & OS | ERP, & OP | DPF, & OP | ERP, DPF, & OP | ERP, DPF, OP, & ERP*OP (only) |
|---|---|---|---|---|---|---|---|---|---|---|---|---|
| Int. | <.0001 | <.0001 | <.0001 | <.0001 | <.0001 | <.0001 | 0.0066 | <.0001 | <.0001 | <.0001 | <.0001 | <.0001 |
| 2003 | 0.8340 | 0.9503 | 0.9552 | 0.8721 | 0.9213 | 0.8494 | 0.9020 | 0.7353 | 0.8729 | 0.7332 | 0.7323 | 0.8798 |
| 2004 | 0.8986 | 0.79991 | 0.9316 | 0.8301 | 0.8316 | 0.5971 | 0.8141 | 0.4934 | 0.7801 | 0.8819 | 0.7861 | 0.7741 |
| 2005 | 0.1096 | 0.1940 | 0.1839 | 0.1796 | 0.1411 | 0.2203 | 0.1904 | 0.2000 | 0.2805 | 0.1618 | 0.1755 | 0.2803 |
| 2006 | 0.2580 | 0.2004 | 0.2093 | 0.3324 | 0.2315 | 0.2932 | 0.3489 | 0.4549 | 0.2631 | 0.2520 | 0.2357 | 0.2583 |
| 2007 | . | . | . | . | . | . | . | . | . | . | . | . |

ginally significant at the 0.05 level in two models, and OP was significant in one model. In several models the interactions between ERP and OP, and DPF and OP were found to be significant and marginally significant. In a model of just these variables (ERP|DPF|OP), only OP was found to be significant, and the interaction between DPF and OP was found marginally significant. The net effect of all of the multi-variable models was to increase overhead costs. Reducing the model to only DPF|OP, only the interaction effect was significant, with a net estimated effect of a minor increase in overhead costs. There does not excess strong conclusive evident to support the failing of the null hypothesis. The full statistical output can be found in Table 17.

## DEGREE OF SUPPLIER AND CUSTOMER INTEGRATION

The trend in most industries today is to use IT systems not only to improve the efficiency of internal operations, but to also improve external efficiency through closer integration with customers and suppliers. Table 17 shows the degree to which the US food manufacturers have integrated with their supply chain (self reported) (Table 18).

When supplier and customer integration was analyzed with respect to profitability, cost of goods sold, labor costs, material costs, and overhead costs, only supplier integration was found to be significant at the 0.05 level, and then only when associated with labor costs. The full statistical output can be found in Table 19.

*Table 17. Overhead cost analysis: the affect of IT systems*

| Independent variables | Estimate | Std Error | DF | T Value | Pr > \|t\| |
|---|---|---|---|---|---|
| ERP | -0.03181 | 0.06491 | 81.9 | -0.49 | 0.6254 |
| ERP II | 0.01526 | 0.1586 | 44.7 | 0.10 | 0.9236 |
| DPF | -0.02607 | 0.05842 | 81.5 | -0.45 | 0.6566 |
| EDI | 0.01985 | 0.05715 | 84.2 | 0.35 | 0.7292 |
| OP | -0.02898 | 0.05838 | 82.3 | -0.50 | 0.6209 |
| OS | 0.04740 | 0.07118 | 82.4 | 0.67 | 0.5073 |
| **OP** | -0.2355 | 0.1412 | 78.5 | -1.67 | **0.0993** |
| OS | -0.01588 | 0.09863 | 82.4 | -0.16 | 0.8725 |
| OP*OS | 0.2296 | 0.1585 | 80.3 | 1.45 | 0.1513 |
| **DPF** | 0.2769 | 0.1601 | 75.8 | 1.73 | **0.0877** |
| OP | -0.1420 | 0.2081 | 75.2 | -0.68 | 0.4971 |
| OS | 0.1304 | 0.1528 | 75.3 | 0.85 | 0.3960 |
| DPF*OP | -0.1942 | 0.1532 | 77.9 | -1.27 | 0.2089 |
| DPF*OS | -0.1862 | 0.2068 | 75.9 | -0.90 | 0.3708 |
| OP*OS | 0.2476 | 0.1800 | 77.3 | 1.38 | 0.1726 |
| ERP | 0.1313 | 0.1614 | 76.8 | 0.08 | 042.59 |
| OP | -0.01505 | 0.1855 | 77.2 | -0.08 | 0.9355 |
| OS | -0.02029 | 0.1701 | 76.8 | -0.12 | 0.9054 |
| **ERP*OP** | -0.2690 | 0.1479 | 76.2 | -1.82 | **0.0728** |
| ERP*OS | 0.03213 | 0.1930 | 76.2 | 0.17 | 0.8682 |
| OP*OS | 0.2004 | 0.1579 | 76.8 | 1.27 | 0.2081 |
| ERP_II | -0.1307 | 0.2022 | 80.9 | -0.65 | 0.5199 |
| OP | -0.6543 | 0.3671 | 80 | -1.78 | 0.0785 |
| OS | -0.03752 | 0.1022 | 81.6 | -0.37 | 0.7146 |
| ERP_II*OP | 0.4126 | 0.3350 | 78.8 | 1.23 | 0.2217 |
| ERP_II *OS | 0 | . | . | . | . |
| OP*OS | 0.2578 | 0.1618 | 79.3 | 1.59 | 0.1151 |
| EDI | 0.2207 | 0.1574 | 77.6 | 1.40 | 0.1650 |
| OP | -0.1882 | 0.2233 | 76.7 | -0.84 | 0.4019 |
| OS | 0.02643 | 0.1268 | 79.8 | 0.21 | 0.8354 |
| EDI*OP | -0.1775 | 0.1513 | 79.3 | -1.17 | 0.2443 |
| EDI*OS | -0.06334 | 0.2083 | 79.2 | -0.30 | 0.7619 |
| OP*OS | 0.2419 | 0.2032 | 76.9 | 1.19 | 0.2374 |
| ERP | 0.01686 | 0.1877 | 73.8 | 0.09 | 0.9286 |
| DPF | 0.2260 | 0.1975 | 72.9 | 1.14 | 0.2563 |
| OP | 0.003729 | 0.2331 | 71.8 | 0.02 | 0.9873 |
| OS | 0.07117 | 0.1894 | 70.8 | 0.38 | 0.7082 |
| ERP*DPF | 0.04136 | 0.1379 | 70.8 | 0.30 | 0.7651 |
| ERP*OP | -0.2325 | 0.1645 | 73.2 | -1.41 | 0.1617 |
| DPF*OP | -0.1355 | 0.1612 | 74.2 | -0.84 | 0.4034 |
| ERP*OS | 0.1055 | 0.2046 | 73.1 | 0.52 | 0.6076 |
| DPF*OS | -0.2078 | 0.2192 | 72.2 | -0.95 | 0.3465 |
| OP*OS | 0.2397 | 0.1819 | 73.1 | 1.32 | 0.1917 |
| ERP | 0.1021 | 0.1876 | 74.1 | 0.54 | 0.5877 |
| EDI | 0.2854 | 0.2098 | 72.1 | 1.36 | 0.1780 |
| OP | 0.004327 | 0.2624 | 70.5 | 0.02 | 0.9869 |
| OS | -0.03841 | 0.1788 | 72.6 | -0.21 | 0.8305 |
| ERP*EDI | -0.1008 | 0.1412 | 73.7 | -0.71 | 0.4776 |
| **ERP*OP** | -0.2655 | 0.1522 | 71.5 | -1.74 | **0.0854** |
| EDI*OP | -0.1419 | 0.1530 | 73.9 | -0.93 | 0.3567 |
| ERP*OS | 0.09583 | 0.2030 | 72.4 | 0.47 | 0.6383 |
| EDI*OS | -0.04252 | 0.2225 | 73.2 | -0.19 | 0.8490 |
| OP*OS | 0.2198 | 0.2077 | 70.9 | 1.06 | 0.2935 |

*continued on following page*

*Table 17. Continued*

| | | | | | |
|---|---|---|---|---|---|
| ERP_II | -0.2772 | 0.2779 | 73.1 | -1.00 | 0.3214 |
| EDI | -0.2004 | 0.4341 | 73.9 | -.046 | 0.6457 |
| OP | -0.2789 | 0.4954 | 74.4 | -0.56 | 0.5751 |
| OS | 0.008037 | 0.1296 | 77 | 0.06 | 0.9507 |
| ERP_II*EDI | 0.4225 | 0.4060 | 74 | 1.04 | 0.3014 |
| ERP_II*OP | 0.1223 | 0.4023 | 74.5 | 0.30 | 0.7620 |
| EDI*OP | -0.2143 | 0.1600 | 75.4 | -1.34 | 0.1845 |
| ERP_II*OS | 0 | . | . | . | . |
| EDI*OS | -0.02290 | 0.2146 | 75.4 | -0.11 | 0.9153 |
| OP*OS | 0.2325 | 0.2097 | 74.2 | 1.11 | 0.2713 |
| ERP | -0.00994 | 0.2080 | 69.4 | -0.05 | 0.9620 |
| EDI | 0.3418 | 0.2403 | 68.9 | 1.42 | 0.1593 |
| DPF | 0.1687 | 0.2062 | 67.4 | 0.82 | 0.4161 |
| OP | 0.01890 | 0.2849 | 65.8 | 0.07 | 0.9473 |
| OS | 0.05847 | 0.1958 | 65.4 | 0.30 | 0.7661 |
| ERP*EDI | -0.1174 | 0.1510 | 69.1 | -0.78 | 0.4397 |
| ERP*DPF | 0.1149 | 0.1495 | 66.7 | 0.77 | 0.4449 |
| EDI*DPF | -0.07485 | 0.1355 | 66.3 | -0.55 | 0.5825 |
| **ERP*OP** | -0.2858 | 0.1695 | 67.4 | -1.69 | **0.0964** |
| EDI*OP | -0.1168 | 0.1613 | 67.2 | -0.72 | 0.4713 |
| DPF*OP | -0.07810 | 0.1678 | 66.4 | -0.47 | 0.6432 |
| ERP*OS | 0.1771 | 0.2147 | 67.5 | 0.83 | 0.4123 |
| EDI*OS | -0.02841 | 0.2290 | 67.5 | -0.12 | 0.9016 |
| DPF*OS | -0.2430 | 0.2234 | 66.4 | -1.09 | 0.2808 |
| OP*OS | 0.2674 | 0.2188 | 66.3 | 1.22 | 0.2260 |
| ERP | 0.1529 | 0.1117 | 80.3 | 1.37 | 0.1748 |
| OP | 0.1676 | 0.1150 | 79.7 | 1.46 | 0.1489 |
| **ERP*OP** | -0.2646 | 0.1336 | 79.1 | -1.98 | **0.0512** |
| ERP | 0.1090 | 0.1196 | 80.1 | 0.91 | 0.3649 |
| DPF | 0.1159 | 0.1062 | 80.9 | 1.09 | 0.2785 |
| **OP** | 0.2573 | 0.1261 | 77.5 | 2.04 | **0.0447** |
| ERP*OP | -0.1996 | 0.1409 | 78.9 | -1.42 | 0.1604 |
| **DPF *OP** | -0.2216 | 0.1327 | 80.3 | -1.67 | **0.0988** |
| DPF | 0.1533 | 0.09895 | 81.9 | 1.55 | 0.1252 |
| OP | 0.1400 | 0.09592 | 80.4 | 1.46 | 0.1482 |
| **DPF *OP** | -02752 | 0.1264 | 81.6 | -2.18 | **0.0323** |

## CONCLUSION

Our study has shown that although there is significant usage of information technologies in the downstream portion of the food manufacturing supply chains (similar to the beef industry), these systems are not significantly contributing to improvements in plant performance. The systems that seem to be making the most contributions are enterprise resource planning systems, forecasting/demand management systems, electronic data interchange systems, and online purchasing systems; and these contributions are mixed and inconclusive. There is also indication that extensive supplier integration helps to reduce labor costs.

Theory would indicate that all of these systems should improve plant performance. ERP systems are used to improve the collection and dissemination of information, and improve the coordination of planning and production activities, both inside the plant and across the supply chain. EDI systems typically improve the timeliness to communications between plants. Forecasting and demand management systems can improve the effectiveness of planning by providing increased accuracy in forecasted demands. Online purchasing systems are expected to reduce search costs in the procurement process. Given this, the food industry should be able to better coordinate activities all along their respective supply chains;

*Table 18. Analysis of supplier and customer integration*

| | Effect | Treatment | Estimate | Error | DF | t Value | Pr > \|t\| |
|---|---|---|---|---|---|---|---|
| Profit | Supp_Int | Extensive | 0.0243 | 0.1312 | 68.1 | 0.19 | 0.8536 |
| | Supp_Int | None | -0.1107 | 0.111 | 69.3 | -1.00 | 0.3221 |
| | Supp_Int | Some | 0 | . | . | . | . |
| | Cust_Int | Extensive | 0.02274 | 0.1441 | 69.8 | 0.16 | 0.8750 |
| | Cust_Int | None | -0.01013 | 0.1157 | 69.6 | -0.09 | 0.9304 |
| | Cust_Int | Some | 0 | . | . | . | . |
| COGS | Supp_Int | Extensive | -0.03616 | 0.1026 | 65.4 | -0.35 | 0.7256 |
| | Supp_Int | None | 0.04938 | 0.09234 | 74.8 | 0.53 | 0.5944 |
| | Supp_Int | Some | 0 | . | . | . | . |
| | Cust_Int | Extensive | -0.1042 | 0.1155 | 74.8 | -0.90 | 0.3698 |
| | Cust_Int | None | -0.00969 | 0.09654 | 74.7 | -0.10 | 0.9204 |
| | Cust_Int | Some | 0 | . | . | . | . |
| Labor Costs | Supp_Int | Extensive | 0.237 | 0.09569 | 80.6 | 2.48 | **0.0153** |
| | Supp_Int | None | 0.07954 | 0.08608 | 79 | 0.92 | 0.3583 |
| | Supp_Int | Some | 0 | . | . | . | . |
| | Cust_Int | Extensive | -0.09909 | 0.1068 | 78.4 | -0.93 | 0.3566 |
| | Cust_Int | None | 0.05709 | 0.09082 | 79.8 | 0.63 | 0.5314 |
| | Cust_Int | Some | 0 | . | . | . | . |
| Material Costs | Supp_Int | Extensive | -7.648 | 5.8891 | 79.4 | -1.30 | 0.1978 |
| | Supp_Int | None | -5.3336 | 5.2784 | 77.7 | -1.01 | 0.3154 |
| | Supp_Int | Some | 0 | . | . | . | . |
| | Cust_Int | Extensive | 6.4419 | 6.5434 | 76.7 | 0.98 | 0.3280 |
| | Cust_Int | None | -4.8887 | 5.5779 | 78.6 | -0.88 | 0.3835 |
| | Cust_Int | Some | 0 | . | . | . | . |
| Overhead Costs | Supp_Int | Extensive | -0.02488 | 0.08512 | 80 | -0.29 | 0.7708 |
| | Supp_Int | None | 0.000133 | 0.07646 | 77.8 | 0.00 | 0.9986 |
| | Supp_Int | Some | 0 | . | . | . | . |
| | Cust_Int | Extensive | -0.04587 | 0.09486 | 76.9 | -0.48 | 0.6301 |
| | Cust_Int | None | 0.05645 | 0.08072 | 78.9 | 0.70 | 0.4864 |
| | Cust_Int | Some | 0 | . | . | . | . |

and many areas of the food industry have made significant progress such as in the poultry and pork industries. From our results it is indeterminate as to how much information technologies are adding to these industries performance. One reason for these results may be due to the level of compatibility within the food industries supply chain (Posey & Bari, 2009), thus creating limitations on the supply chain's ability to process shared information within its structure.

Though online technologies are being used by upstream participants in the beef industry's supply chain, they have yet to make any significant impact. Clearly, the use of internet technologies is in its infancy within this industry and most supply chain participants have not placed an adequate

*Table 19. Breakout of NAICS industry group 311 – food manufacturers and processors*

| 4 Digit NAICS Codes | | | 2003 | 2004 | 2005 | 2006 | 2007 |
|---|---|---|---|---|---|---|---|
| 3111 - Animal Food Mfg | Supplier Integration | None | 0 | 1/50% | 1/100% | 1/33% | 0 |
| | | Some | 1/100% | 1/50% | 0 | 2/67% | 0 |
| | | Extensive | 0 | 0 | 0 | 0 | 0 |
| | Customer Integration | None | 0 | 0 | 0 | 0 | 0 |
| | | Some | 1/100% | 0 | 1/100% | 2/67% | 0 |
| | | Extensive | 0 | 2/100% | 0 | 0 | 0 |
| 3112 - Grain & Oilseed Milling | Supplier Integration | None | 0 | 0 | 0 | 0 | 0 |
| | | Some | 1/100% | 3/100% | 0 | 1/100% | 2/100% |
| | | Extensive | 0 | 0 | 0 | 0 | 0 |
| | Customer Integration | None | 0 | 0 | 0 | 1/100% | 0 |
| | | Some | 1/100% | 2/100% | 0 | 0 | 2/100% |
| | | Extensive | 0 | 1/100% | 0 | 0 | 0 |
| 3113 - Sugar & Confectionery Product Mfg | Supplier Integration | None | 0 | 0 | 1/100% | 0 | 0 |
| | | Some | 2/67% | 1/50% | 0 | 0 | 1/100% |
| | | Extensive | 1/33% | 1/50% | 0 | 0 | 0 |
| | Customer Integration | None | 0 | 0 | 0 | 0 | 0 |
| | | Some | 2/67% | 2/100% | 1/100% | 0 | 0 |
| | | Extensive | 1/33% | 0 | 0 | 0 | 1/100% |
| 3114 - Fruit & Vegetable Preserving & Specialty Food Mfg | Supplier Integration | None | 1/50% | 0 | 0 | 2/67% | 0 |
| | | Some | 1/50% | 3/75% | 1/100% | 1/33% | 1/33% |
| | | Extensive | 0 | 1/25% | 0 | 0 | 2/67% |
| | Customer Integration | None | 0 | 1/25% | 0 | 2/67% | 0 |
| | | Some | 2/100% | 2/50% | 1/100% | 1/33% | 3/100% |
| | | Extensive | 0 | 1/25% | 0 | 0 | 0 |
| 3115 - Dairy Product Mfg | Supplier Integration | None | 2/29% | 0 | 1/33% | 0 | 1/50% |
| | | Some | 2/29% | 0 | 2/67% | 4/80% | 0 |
| | | Extensive | 3/42% | 0 | 0 | 1/20% | 1/50% |
| | Customer Integration | None | 2/29% | 0 | 2/67% | 1/20% | 0 |
| | | Some | 3/43% | 0 | 1/33% | 4/80% | 2/100% |
| | | Extensive | 2/28% | 0 | 0 | 0 | 0 |
| 3116 - Animal Slaughtering & Processing | Supplier Integration | None | 0 | 1/25% | 2/50% | 1/25% | 2/100% |
| | | Some | 0 | 2/50% | 1/25% | 2/50% | 0 |
| | | Extensive | 2/100% | 1/25% | 1/25% | 1/25% | 0 |
| | Customer Integration | None | 0 | 2/40% | 2/50% | 2/50% | 0 |
| | | Some | 0 | 3/60% | 2/50% | 2/50% | 2/100% |
| | | Extensive | 2/100% | 0 | 0 | 0 | 0 |

*continued on following page*

*Table 19. Continued*

| | | | | | | | |
|---|---|---|---|---|---|---|---|
| **3117 - Seafood Product Preparation & Packaging** | Supplier Integration | None | 0 | 0 | 0 | 2/100% | 0 |
| | | Some | 0 | 0 | 0 | 0 | 0 |
| | | Extensive | 0 | 0 | 0 | 0 | 0 |
| | Customer Integration | None | 0 | 0 | 0 | 1/50% | 0 |
| | | Some | 0 | 0 | 0 | 1/50% | 0 |
| | | Extensive | 0 | 0 | 0 | 0 | 0 |
| **3118 - Bakeries & Tortilla Mfg** | Supplier Integration | None | 0 | 0 | 1/33% | 1/25% | 0 |
| | | Some | 4/100% | 0 | 2/67% | 3/75% | 0 |
| | | Extensive | 0 | 0 | 0 | 0 | 0 |
| | Customer Integration | None | 0 | 0 | 0 | 0 | 0 |
| | | Some | 4/100% | 0 | 3/100% | 4/100% | 0 |
| | | Extensive | 0 | 0 | 0 | 0 | 0 |
| **3119 - Other Food Mfg** | Supplier Integration | None | 4/67% | 1/17% | 0 | 1/20% | 2/29% |
| | | Some | 2/33% | 3/50% | 0 | 4/80% | 4/57% |
| | | Extensive | 0 | 2/33% | 0 | 0 | 1/14% |
| | Customer Integration | None | 4/66% | 0 | 0 | 2/40% | 0 |
| | | Some | 1/17% | 3/50% | 0 | 3/60% | 6/86% |
| | | Extensive | 1/17% | 3/50% | 0 | 0 | 1/14% |
| **Code not disclosed by respondents** | Supplier Integration | None | 1/7% | 0 | 0 | 0 | 0 |
| | | Some | 8/65% | 8/73% | 3/100% | 2/100% | 1/100% |
| | | Extensive | 4/31% | 3/27% | 0 | 0 | 1/100% |
| | Customer Integration | None | 3/23% | 1/9% | 1/50% | 0 | 0 |
| | | Some | 10/77% | 8/73% | 1/50% | 2/100% | 1/100% |
| | | Extensive | 0 | 2/18% | 0 | 0 | 0 |

Source: MPI Census of US Manufacturers Survey

degree of their effort to this new medium. As of yet, information technologies do not seem to be supplying the aid needed for improvements to occur.

## REFERENCES

Bardhan, I., Whitaker, J., & Mithas, S. (2006). Information technology, production process outsourcing, and manufacturing plant performance. *Journal of Management Information Systems*, *23*(2), 13–40. doi:10.2753/MIS0742-1222230202

Chandra, C., & Kumar, S. (2000). Supply chain management in theory and practice: a passing fad or a fundamental change. *Industrial Management & Data Systems*, *100*(3), 100–113. doi:10.1108/02635570010286168

Chandrashekar, A., & Schary, P. B. (1999). Toward the virtual supply chain: the convergence of IT and organization. *International Journal of Logistics Management*, *10*(2), 27–39. doi:10.1108/09574099910805978

Choi, B., Tsai, N., & Jones, T. (2008). Building enterprise network infrastructure for a supermarket store chain. *Journal of Cases on Information Technology*, 31–46.

Cooper, M. C., Lambert, D. M., & Pagh, J. D. (1997). Supply chain management: more than new name for logistics. *International Journal of Logistics Management*, *8*(1), 1–14. doi:10.1108/09574099710805556

Croxton, K. L., Garcia-Dastugue, S. J., Lambert, D. M., & Rodgers, D. S. (2001). The supply chain management processes. *International Journal of Logistics Management*, *12*(2), 13–36. doi:10.1108/09574090110806271

Euclides, F. K. (2004). Supply chain approach to sustainable beef production from a Brazilian perspective. *Livestock Production Science*, *90*(1), 53–61. doi:10.1016/j.livprodsci.2004.07.006

Fearne, A. (1998). The evolution of partnerships in the meat supply chain: insights from the British beef industry. *Supply Chain Management, Bradford*, *3*(4), 214. doi:10.1108/13598549810244296

Fisher, M. F. (1997). What is the right supply chain for your product? A simple framework can help you figure out the answer. *Harvard Business Review*, 105–116.

Frasquet, M., Cervera, A., & Gil, I. (2008). The impact of IT and customer orientation on building trust and commitment in the supply chain. *International Review of Retail, Distribution and Consumer Research*, *18*(3), 343. doi:10.1080/09593960802114164

García-Arca, J., & Prado-Prado, J. C. (2007). The implementation of new technologies through a participative approach. *Creativity and Innovation Management*, *16*(4), 386. doi:10.1111/j.1467-8691.2007.00450.x

Hughes, D. (1994). *Breaking with tradition: building partnerships and alliances in the European food industry*. Ashford, UK: Wye College Press.

Lambert, D. M., Cooper, M. C., & Pagh, J. D. (1998). Supply chain management: implementation issues and research opportunities. *International Journal of Logistics Management*, *9*(2), 1–19. doi:10.1108/09574099810805807

Langley, C. J., & Holcomb, M. C. (1992). Creating logistics customer value. *Journal of Business Logistics*, *13*(2), 1–27.

Lawrence, C. (2006). Friend or foe. *Farmers Weekly*, *144*(25), 18–19.

Levy, M., & Grewal, D. (2000). Overview of the issues of supply chain management in a networked economy. *Journal of Retailing*, *76*(4), 415–429. doi:10.1016/S0022-4359(00)00043-9

Lewis, I., & Talalayevsky, A. (1997). Logistics and information technology: a coordination perspective. *Journal of Business Logistics*, *18*(1), 141–157.

Lewis, I., & Talalayevsky, A. (2004). Improving the interorganizational supply chain through optimization of information flows. *Journal of Enterprise Information Management*, *17*(3), 229–237. doi:10.1108/17410390410531470

Mentzer, J. T., DeWitt, W., Keebler, J. S., Min, S., Nix, N. W., Smith, C. D., & Zacharia, Z. G. (2001). Defining supply chain management. *Journal of Business Logistics*, *22*(2), 1–25.

Mighell, R. L., & Jones, L. A. (1963). Vertical coordination in agriculture. *USDA-ERS Agricultural Economics Report*, 19.

Min, S., & Mentzer, J. T. (2000). The role of marketing in supply chain management. *International Journal of Physical Distribution & Logistics Management*, *30*(9), 765–787. doi:10.1108/09600030010351462

Moberg, C. R., Seph, T. W., & Freese, T. L. (2003). SCM: making the vision a reality. *Supply Chain Management Review*, *7*(5), 34–39.

Mohtadi, H. (2008). Information sharing in food supply chains. *Canadian Journal of Agricultural Economics*, *56*(2), 163. doi:10.1111/j.1744-7976.2008.00123.x

Narayanan, V. G., & Raman, A. (2004). *Aligning incentives in supply chains*. Boston: Harvard Business School Publishing.

Neureuther, B. D., & Kenyon, G. N. (2008). The impact of information technologies on the US beef industry's supply chain. *International Journal of Information Systems and Supply Chain Management*, *1*(1), 48–65.

Nitschke, T., & O'Keefe, M. (1997). Managing the linkage with primary producers: experiences in the Australian grain industry. *Supply Chain Management, Bradford*, *2*(1), 4. doi:10.1108/13598549710156295

Peypoch, R. (1998). The case for electronic business communities. *Business Horizons*, *4*(6), 17–20. doi:10.1016/S0007-6813(98)90073-8

Posey, C., & Bari, A. (2009). Information sharing and supply chain performance: Understanding complexity, compatibility, and processing. *International Journal of Information Systems and Supply Chain Management*, *2*(3), 67–76.

Romeo, P. (2006). Buyers embrace more sophisticated supply-purchasing procedures. *Nations Restaurant News*, *40*(21), 92–93.

Salin, V. (2000). Information technology and cattle-beef supply chains. *American Journal of Agricultural Economics*, *82*(5), 1105–1111. doi:10.1111/0002-9092.00107

Siegel, J. G., & Shim, J. K. (1991). *Financial management*. Hauppauge, NY: Barron's Business Library.

Williams, J. R., Haka, S. F., Bettner, M. S., & Carcello, J. V. (2008). *Financial accounting* (13th ed.). New York: McGraw-Hill Publishing.

Wise, R., & Morrison, D. (2000). Beyond the exchange: The future of B2B. *Harvard Business Review*, *78*(6), 86–96.

Yao, Y., Palmer, J., & Dresner, M. (2007). An interorganizational perspective on the use of electronically-enabled supply chains. *Decision Support Systems*, *43*(3), 884. doi:10.1016/j.dss.2007.01.002

*This work was previously published in International Journal of Information Systems and Supply Chain Management, Volume 3, Issue 4, edited by John Wang, pp. 42-69, copyright 2010 by IGI Publishing (an imprint of IGI Global)*

# Section 2
# Supply Chain Collaboration

# Chapter 5
# Strategies for E-Procurement: Auto Industry Hubs Re-Examined

**Mickey Howard**
*University of Exeter, UK*

**Richard Vidgen**
*University of New South Wales, Australia*

**Philip Powell**
*Birkbeck, University of London, UK, & University of Groningen, The Netherlands*

## ABSTRACT

*Amid the turmoil of the current economic crisis, the wild expectations for business-to-business electronic marketplaces or 'e-hubs' as transformative modes of exchange for all industries have subsided. However, e-hubs continue to elicit interest in industries such as car production. Yet, there is little research that investigates firms' strategies for e-procurement in the automotive industry and the potential benefits of e-hubs to them. This research re-examines the transition from bespoke electronic data interchange to generic electronic procurement, and conflicting motivations and complex barriers at firm and industry level are revealed. The article develops a framework that examines the benefits and barriers to firms joining e-hubs, applies the framework to the car industry, and proposes an e-procurement matrix that offers alternative strategies. Six cases from vehicle manufacturers and component suppliers demonstrate a shallow industry structure that lacks supplier integration, where a particular concern is the emergence of consortium e-hubs that combine a transactional approach for reducing price, with a collaborative approach that requires sharing knowledge. While this dispels the myth of simplistic e-commerce models, the governance of e-procurement across collaborative supply chains is still uncertain.*

## INTRODUCTION

The car industry is undergoing profound changes as the global financial crisis hits the firms', their suppliers' and their customers' ability to borrow. Some major automotive manufacturers such as GM and Ford are closing production plants while others are moving to short working and periodic shutdowns (Professional Engineering, 2008). There has been rampant press speculation as to the ability of some firms to survive, though government assistance appears to have reduced the immediate threat of bankruptcy. Despite its current problems, the world automotive industry could still

DOI: 10.4018/978-1-4666-0918-1.ch005

produce over 94 million new cars per annum even though there are now buyers for fewer than 60 million (Economist, 2009), and the industry represents a significant proportion of gross domestic product in developed countries, for instance 5% in the UK (Crain, 2008). However, though the current crisis has been sharp and sudden, significant change has been occurring in the car industry for some time. The rise of the Internet and the rapid spread of electronic procurement across world markets have left few industries unchanged. Since its inception in the early 1990s, electronic commerce has been a focus of world markets seeking new solutions to business models and dramatic reductions in transaction costs (Timmers, 1998; Min & Galle, 1999). While initially acclaimed for the re-structuring of old-world economies and enabling inter-organization collaboration, e-commerce has since endured a 'dotcom' style crash and has been criticised for failing to deliver value (Boot & Butler, 2001; Connelly, 2001). This is particularly true of the automotive sector, whose old economy origins, information technology (IT) legacy systems and complex, hierarchical supply chains mean information system (IS) transformation is more difficult than in other industries such as grocery retailing (Smaros et al., 2000; Vanany et al., 2009). Intense competition in high-volume passenger vehicle manufacturing, an emphasis on cost, and the collaborative opportunities offered by electronic supply networks, has sustained interest in business-to-business electronic marketplaces or 'e-hubs' in this sector (Bakos, 1998; Kaplan & Sawhney, 2000; Baldi & Borgman, 2001; Barratt & Rosdahl, 2002; Jun, et al., 2008).

In 2000 the launch of 'Covisint', the biggest and most powerful automotive e-marketplace was heralded as the beginning of a new era in auto industry purchasing and supply chain management. The founder members, Ford, General Motors and DaimlerChrysler, anticipated significant component price reductions and customer responsiveness by combining purchasing economies of scale and Internet technology (Baldi & Borgman, 2001). However, rival vehicle manufacturers (VMs) and component suppliers were already developing their own solutions and were reluctant to subscribe over fears of accepting a subordinate role. As private trade exchanges proliferated, Covisint's vision of offering collaborative procurement, lower transaction costs and the introduction of a universal system standard began to diminish (Kisiel & Whitbread, 2000; Helper & MacDuffie, 2002). While the e-hub has survived the on-going world economic and socio-political upheaval, their number has dwindled to around 10 per cent of the 2000 peak (Kisiel, 2002). The automotive industry needs to become leaner, more efficient and better integrated with its supply chain in order that it can survive the credit crunch and prosper thereafter.

With this in mind, this research investigates the strategies for e-procurement being implemented in the auto industry. It assesses why firms joined e-hubs and the benefits they foresaw and the barriers that emerged. The article first develops a framework that examines the motivation and barriers for firms joining e-hubs, and applies the framework to the car industry. This 'industry transformation framework' enables a cross-case comparison of the motivations and barriers to e-hub adoption between six firms consisting of vehicle manufacturers and component suppliers. The article then offers an e-procurement matrix that suggests alternative strategies for the cases by focusing on supply chain interaction in electronic markets and networks. The findings suggest that despite the growth in electronic business, neither the means for effective e-procurement integration, nor the impacts on industry are well understood.

The article is structured as follows. Section two explains the research method. Section three reviews the emergence of e-procurement and e-hubs in the auto industry, and models the benefits and barriers to integration. Section four conducts the analysis of the cases and applies the framework. Section five develops the e-procurement strategies, and section six presents the conclusions and implications.

## METHOD

Much has been written in the trade and professional press about e-hubs, but there is little academic work that investigates the actual, as opposed to the theoretical, role of e-hubs in the auto industry (Arbin & Essler, 2005). This article begins by presenting a snapshot of auto e-hubs in terms of firm affiliation and the nature of their relationship with other partners across the supply chain. Having established an industry view and demonstrated that this does not accord with existing theoretical propositions, the research explores e-hub benefits and barriers by modelling a transformation framework (Table 2) that is applied to selected case studies. The strategies of each firm are then explored through an e-procurement strategy matrix (Figure 3).

This research builds on work by the 3DayCar programme, launched in 1999, to study the role of customer order fulfilment in the UK auto industry. A key finding of the original study is that 85% of delay in the order pipeline is derived from information systems; order entry, order processing and scheduling, *not vehicle manufacturing* (Holweg & Pil, 2001). The case study presented here is particularly suited to IS research, because it copes with the technically distinctive situation in which there will be more variables of interest than data points (Yin, 1994). It presents a multiple, exploratory case study as a collaborative process of critical enquiry (Eisenhardt, 1989; Yin, 1994; Miles & Huberman, 1994). This approach requires an *'iterative tabulation of evidence'* based on replication (not sampling) logic across cases as a means of building internal validity, and confirming and extending theory (Eisenhardt, 1989). The research adopts the view where early identification of the research question and *a priori* specification of constructs and frameworks can benefit data collection and analysis (Eisenhardt, 1989; Checkland, 1991). The unit of analysis is the automotive industry, and the phenomenon under investigation is the e-hub.

An industry transformation framework provides the foundation for exploring e-hubs and is based on archival research and longitudinal interviews since e-hubs emergence in the auto industry in 1998 (Howard et al., 2001; 2002). Meetings with the founding firms of Covisint in Dearborn, US, and in Europe established that e-hubs represented a radical departure from earlier business practices: they required high levels of collaboration between multiple partners involving complex interactions and information exchange, which presented a significant opportunity for organizational research. Three vehicle manufacturers and three component suppliers were chosen for study on the basis of their roles in e-procurement, and their contribution in terms of understanding benefits and barriers affecting inter-firm relationships using e-hubs.

After the preliminary investigation, structured interviews were conducted with senior procurement managers at Ford Motor Company, Volvo Car Corporation, Volkswagen, Bosch, Delphi, and Johnson Controls, using the framework in Table 2 as the basis for questioning. These firms were chosen for the research because of their active involvement with e-hubs. Approximately 30 interviews and focus groups (Table 1) were conducted with vehicle manufacturers and suppliers, all of which were adopting e-procurement hubs either as part of their ICT development strategy, or as a subscriber. Interviews lasted, on average, for 1.5 hours and were not recorded electronically due to the often sensitive nature of the difficulties over information sharing and exchange with other firms. Where possible, two investigators attended the meeting, one to ask questions and one to write down the response.

A summary of the key points of the meeting was sent back to the individual (in confidence) for verification. Data analysis involved within-case analysis and cross-case pattern searching for themes and issues using divergent techniques, such as tables and matrices, assisted by the triangulation of multiple investigators (Eisenhardt, 1989). Closure was considered reached when improvements to the research became small.

## FROM BESPOKE EDI TO GENERIC E-PROCUREMENT

E-procurement can be defined as the electronic buying, selling and tendering of goods and services (Timmers, 1998), or '*..using Internet technology in the purchasing process*' (De Boer et al., 2002 p26). It is a subset of e-commerce that has existed for more than twenty years in a variety of guises, for instance: Electronic Data Interchange (EDI), Computer-assisted Lifecycle Support (CALS), and Computer-aided Design (CAD) (Webster, 1995; Kuroiwa, 1999; Croom, 2001). E-procurement is concerned with buy-side, B2B electronic markets, focusing on procurement, supply chain management and product development. This research examines the integration of vertical and horizontal, buy-side electronic markets (Baldi & Borgman, 2001), specifically the effect of web-enabled applications on inter-organizational relationships.

Traditional EDI originated from firms wishing to automate the exchange of data internally and with partners, as a secure link between firms offering a reliable means of communicating purchase orders, build schedules and forecasts (Swatman & Swatman, 1992). However, EDI integration faces a number of difficulties in the auto industry: high entry costs, a proliferation of standards, and coercive pressures from powerful vehicle manufacturers (VM). Firms using EDI have found themselves tied into a technology that replicates the hierarchical nature of traditional, adversarial customer-supplier relationships (Lamming, 1993). For instance, the Ford Motor Company in the mid-1980s had a basic objective in developing its EDI network 'Fordnet': to gain competitive advantage by locking its suppliers and customers into its systems, and locking its competitors out (Webster, 1995).

E-commerce differs from EDI as it provides an inter-organizational IS that fosters market-based exchanges between agents in all transaction phases (Bakos, 1997). It overcomes the technological barrier of a bespoke system through web-enabled technology that uses the Internet for low cost, real-time information exchange with multiple partners. 'Bespoke' information technology was typical in 70s and 80s system design where, prior to the common architectures available today, both software and hardware was custom built around firm-specific requirements making connectivity to other systems very difficult. The emergence of Web-EDI in the late 90s, using eXtensible Markup Language (XML) and a PC, offers a low cost solution for suppliers seeking connection to their business partners via the Internet (Vidgen & Goodwin, 2000). E-commerce also offers on-line development using 'virtual spaces' in which manufacturers and suppliers can collaborate on joint projects. Research and development provides the most important link between vehicle assemblers and suppliers since it is one of the clearest manifestations of collaboration on the part of both partners (Lamming, 1993). This forms the basis for ensuring on-line information sharing activities: product development, procurement, and operations planning and scheduling. The potential benefits are not focused solely on transaction cost, but on enabling buyer-supplier partnerships and new product collaboration. This explosion in world-wide connectivity enabled the emergence of the on-line trade exchange, or e-hub.

The term 'hub' is used here to cover portal, Internet trade exchange, B2B electronic marketplace, or cyber-purchasing system (Min & Galle, 1999). The appeal of e-hubs was clear: by bringing together a large number of buyers and sellers and automating transactions, they expanded the choice available to buyers, gave sellers access to new customers, and reduced transaction

*Table 1. Data sources*

| *Organization* | *Type* | *Preliminary interviews* | *Semi-structured interviews* | *Focus groups* |
|---|---|---|---|---|
| Ford Motor Company | Vehicle manufacturer (VM) | 2 | 5 | 1 |
| Volvo Car Corporation | VM | 3 | 9 | 1 |
| Delphi | Tier 1 supplier | 2 | 4 | 1 |
| Volkswagen | VM | 1 | - | - |
| Johnson Controls | Tier 1 supplier | 1 | 2 | - |
| Bosch | Tier 1 supplier | 1 | - | - |
| Covisint | VM consortium e-hub | 1 | - | - |

costs (Bakos, 1998; Kaplan & Sawhney, 2000). E-hubs provided a virtual marketplace, where operators earned revenue by extracting fees for the transactions. Industry stalwarts such as GM and Ford anticipated dual benefits; first from the cost reductions using Internet auctions, and second from the additional revenue arising from the subsequent flotation of Covisint.

Rubin and Fogarty's (2000) illustration of the emergence of the web-enabled organization in the auto industry (Figure 1) represents an early understanding of the e-hub, before the e-commerce downturn in April 2000. They argue that VMs and component suppliers (tier 1 and tier 2) shift from a hierarchical top-down structure (A), towards a centrally aligned hub structure providing both an online marketplace and a one–stop communication point for commercial trading and new product collaboration (B). However, this is an over-simplification of the role of electronic markets that ignores the fundamental conflict of interest between those procurement systems focusing on price, and fostering supply chain collaboration.

This research demonstrates that the impacts of the e-hub on traditional supply chains were more complex than suggested by Figure 1. The barriers to change in the auto industry structure between buyer/supplier and supplier/supplier transactions involved not only technological, but also social and market constraints, particularly ownership and competition.

*Table 2. Industry transformation framework (Adapted: Kwon & Zmud, 1987; Kirveennummi et al., 1998)*

| *Benefits* | |
|---|---|
| Expected | Cost<br>Market share<br>Collaboration & knowledge |
| Realised | - |
| ***Industry barriers*** | ***Indicative factors*** *(Inter-organizational)* |
| Environmental | Market organizations<br>Anti-trust legislation<br>Power |
| ***Firm barriers*** | ***Indicative factors*** *(Intra-organizational)* |
| Structural | Organizational form<br>Decision making<br>Financial resources |
| Cultural | Politics<br>Trust<br>Attitude to risk |
| Managerial | Control<br>Leadership<br>Information management skills |
| User | Resistance<br>Fear of change<br>IT skills |
| Technical | Infrastructure<br>Standards<br>Data quality |

## E-Market Classification: Ownership and Competition

E-markets are generally classified in terms of what they exchange, e.g. products and services, and how they exchange it, for example, either

*Figure 1. Current and web-enabled industrial organization (Rubin and Fogarty, 2000)*

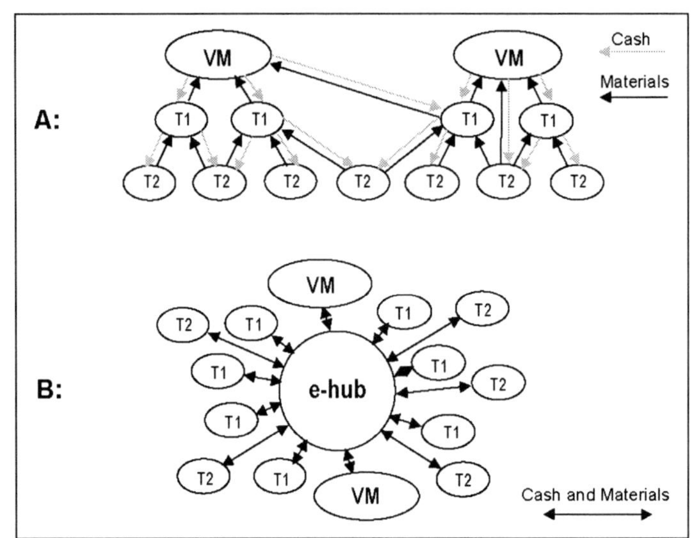


systematic sourcing and negotiated contracts, or spot sourcing and commodity trading (Kaplan & Sawhney, 2000). Systematic sourcing involves negotiated contracts with qualified suppliers, involving long-term contracts and close relationships between buyers and sellers (Sako, 1992; Lamming, 1993). In spot sourcing the buyer's goal is to fulfil an immediate need at the lowest cost, such as commodity trading for steel by NewView Technologies, formerly 'e-Steel'. In their description of the governance structures of e-markets Baldi and Borgman (2001) consider two dimensions to be of particular importance. First, *the role of the owner*: the owner of the market can be an active market participant or an independent third party. Second, the *competitive relation* of the owners: where the firms owning and operating the market can be direct competitors outside this venture. This results in different ownership structures for electronic markets:

*A private trade exchange* is owned and operated by a single firm.

*A third party exchange* is operated by a group of non-competing firms or one that is not considered a trading partner.

*A consortium trade exchange* is where ownership is shared between competing firms.

The Typology of Business Webs (Tapscott et al., 2000) classifies e-markets as differentiated along two primary dimensions: control (self-organizing or hierarchical) and value integration (low or high). High economic control requires a leader that controls the content of the value proposition, the pricing, and the flow of transactions. High value integration requires facilitating the production of product or service offerings by integrating contributions from multiple sources. Five classifications make up the typology: Agora, Aggregation, Distributive Network, Alliance, and Value Chain. However, Tapscott reports that only a handful of firms had made progress towards this approach by the mid-90s, where industrial-era firms in particular are restricted by central, 'hub and spoke' IT architecture, and hampered by changes that require too much of a break from established management culture.

The issue of ownership is critical to this research, where introducing electronic networks to 'old world' industries such as the auto sector potentially provided significant benefits related

*Figure 2. Automotive industry e-hub structure (Howard et al., 2006)*

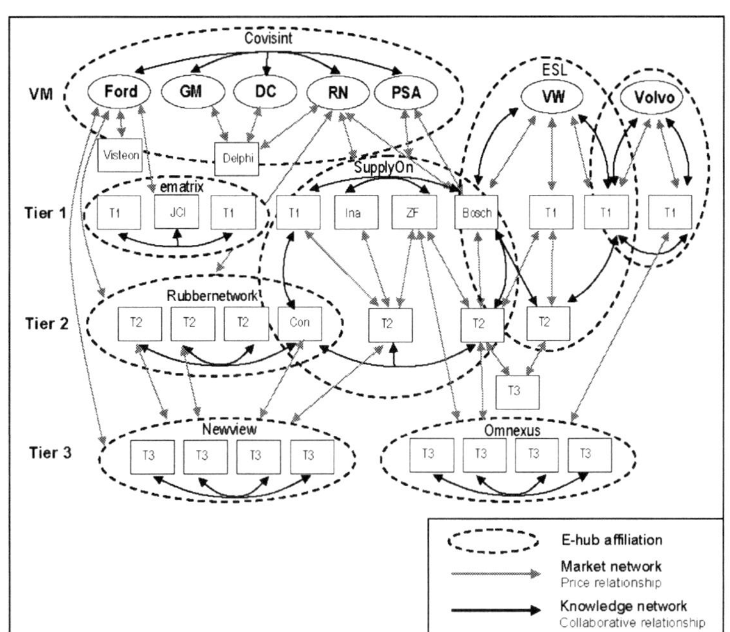


to transaction cost and exchange of data, while raising concerns over the creation and protection of firm knowledge (Nonaka & Takeuchi, 1995). Although suppliers may be willing to participate in systems which co-ordinate and manage the supply chain (Núñez-Muñoz & Montoya-Torres, 2009), extending this towards collaborative new product development initiatives that involve sharing sensitive information, may mean partners become concerned over revealing intellectual property and eroding core competence (Prahalad & Hamel, 1990).

While e-hubs represent the focus or point of exchange for the aggregate supply and demand of goods and services in e-markets (Reynolds, 2000), auto firms experienced difficulties setting up hubs for procurement and product development. Consortium hubs such as Covisint represented a horizontal model of e-procurement that excluded suppliers from participating as an equal partner, because they were not being offered shares in the venture. Suppliers were sceptical of win-win promises because they feared their profit margins would be slashed through Internet auctions and their product knowledge commoditised. Moreover, other exchanges emerged in the industry such as a consortium of tyre manufacturers; 'Rubbernetwork', a consortium of injection moulding and blow moulders; 'Omnexus', and private purchasing hubs run by BMW and Volkswagen which both rejected joining Covisint. While the consortium hubs offered all the functional requirements needed in terms of connectivity and system standardisation, they did not overcome the reluctance of smaller suppliers to participate with more powerful members. Firms faced immense costs and socio-political barriers relating to their transformation into collaborative communities (Nokkentved et al., 2000). Only if mutual, value added benefits existed for all players: customer, dealer, vehicle manufacturer, supplier and logistics firms, would the original vision by Covisint of one industry e-hub evolve as the dominant model.

Electronic commerce has evolved from a maze of bespoke EDI systems, however, to a structure that bears little resemblance to the single industry hub design originally envisaged by Covisint's founders. Unlike the simplistic models in literature

*Figure 3. e-Hub Strategy Matrix*

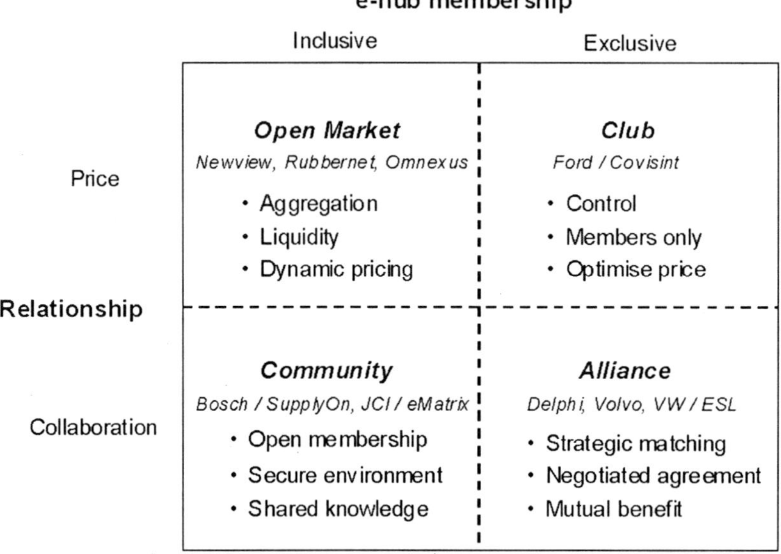


(e.g. Rubin and Fogarty, 2000), the introduction of e-commerce resulted in a mess of overlapping networks. In the race to re-engineer the industry from its old economy origins, e-procurement structures began to appear closer to a loose arrangement of spokes than a hub. Overlapping networks competed for limited membership across the industry, resulting in isolated pockets of collaboration and irregular information flow. Figure 2 presents the structure of resources and collaboration flow across all tiers in the supply chain by classifying e-hubs in terms of relationship and membership. This research demonstrates that, if introduced without sufficient planning in terms of consideration of both price and collaboration, e-commerce and the introduction e-hubs simply *speeds up the mess.*

## E-hub Benefits & Barriers: The Industry Transformation Framework

While the potential benefits of e-hubs have been consistently reported as reducing cost, increasing market share, and increasing firm knowledge through collaborative capability (Bakos, 1998; Kaplan & Sawhney, 2000), there is no clear consensus in the literature over factors which constrain e-hub implementation. This is particularly significant for the vulnerable e-commerce community in the auto industry, faced with rebuilding confidence after the dotcom crash and the departure in 2001 of ardent supporters such as Jac Nasser (CEO of Ford). A starting point for considering such features is the Barriers Information Framework by Kirveennummi et al. (1998). This has four categories, classified as structural, managerial, user and technical that operate at firm level to consider impediments to IS-related change. More generically, Heeks and Davies (2001) categorize the barriers to integration of e-government services as structural, cultural, political, and technical. They argue that barriers exist between organizations as well as within, the chief factors being trust and power.

These concepts are synthesized here to create the Industry transformation framework (Table 2). While IS researchers have identified barriers such as structural, cultural, managerial and user at firm level in the past (Kwon & Zmud, 1987; Kirveennummi et al., 1998), this framework's contribution is by prompting the investigator to record both the expected and realised firm benefits, which in nearly all cases exposes a 'benefit gap'. This is examined over multiple levels of analysis, classified as inter-organizational (industry) and inter-organizational (firm) barriers to IS-related change. A brief review of the barriers that constrain the adoption and use of e-hubs is now considered.

*Industry level, environmental barriers* are classified as market organizations, anti-trust legislation, and power. Market organizations are defined as the structure of supply chain relationships, imposed by both institutional, market or power related hierarchy, and informal inter-business networks. Market organizations are affected by the emergence of new electronic markets driven by sophisticated technology, where sustained competition would be likely to destroy some e-hubs after the rapid growth of electronic markets on the Internet. Huber et al. (2000) estimated that there would be room for only three vertical portals in each industry, which implied considerable consolidation for the auto sector.

Anti-trust legislation in the US, the Competition Commission in the UK, and the Competition Commissioner in Europe inhibited the growth of consortium trade exchanges such as Covisint, which was prevented from trading fully until after a federal enquiry. Limited competition such as monopoly or oligopoly also represents a barrier to e-procurement where the main force compelling sellers to conform to consumer wants and to hold prices near cost is not competition, but countervailing power exercised by strong buyers (Scherer, 1980). The original proliferation of e-hubs may be explained through rival VMs and supplier groups attempting to offset the power of Covisint by forming their own consortia or exchange systems of their own.

Power is the capability of a firm to exert influence on another firm to act in a prescribed manner (Pfeffer & Salancik, 1978; Hart & Saunders, 1997). More powerful firms that control resources that other firms rely upon, can preserve their power while influencing their trading partners to adopt new technology (Webster, 1995). The introduction of e-commerce meant suppliers were wary of vehicle manufacturers introducing new IT, abusing sensitive information, and reaping the benefits from information sharing, such as cost data on the purchase price of parts (Damsgaard & Lyytinen, 1998; Lee & Whang, 1998).

*Organizational level, structural barriers* in firms are defined as 'organizational form', based on firm size, complexity and function, or as informal social networks based on interpersonal communication (Kwon & Zmud, 1987; Ciborra, 1993). Understanding structures requires knowledge of how information and decision processes flow through organizations (Mintzberg, 1979). One of the commonest forms of formal structural barrier in firms is departmental hierarchy – known as chimneys or silos - which can lead to varying degrees of inflexibility, resistance to change and slow reactions (Merton, 1968; Davenport et al., 1992). Labour organizations such as unions, who perceive IT-related change to threaten themselves or their members, can also hinder or delay the process of change.

Cultural barriers in any organization can present a formidable challenge to those attempting to implement change, where a closed atmosphere and the sunk costs of groups and individuals in maintaining the status quo are considerable (Argyris, 1990). 'Information politics' (Davenport et al., 1992) describes the negotiation of information at a departmental or individual level. Hart and Saunders (1997) argue that while the adoption of new technology can reflect existing power arrangements, a critical condition of its successful use over time is determined by the extent of trust of individuals between firms. An important challenge to building trust in a buyer-supplier relationship involves individuals avoiding the use of power to persuade or coerce resistant or hesitant partners to adopt EDI (Helper, 1991; Sako, 1992).

Managerial barriers are defined where the chief executive officer and senior managers are in a unique position to 'disconfirm' assumptions, attitudes or routines in the firm that are no longer working (Schein, 1994). Top management is the most powerful driving force or barrier for change, combining a task-orientated role and a supportive role for groups and teams. They can neither avoid IT, nor delegate the issues it raises to others, because business strategy and IT have become so intertwined that large corporate failures frequently lead to the demise of the CEO (Connelly, 2001). Sophisticated IT applications complicate the change process in a number of ways; it moves the locus of knowledge and hence power in the organization, and changes the time dimension of processes and decisions (Scott Morton, 1991).

User barriers are defined as people's inability to change and not technology, as the prime limiting factor in transforming organizations (Benjamin and Blunt, 1994). Poor user perceptions have led to difficulties in conceptualising change, and disuse of the collaborative system (Ciborra 1993). In order to successfully handle change, it must be perceived as a learning process for individuals, who require creativity and imagination to transfer it into practice (Argyris 1990). Otherwise, fear of failure, loss of power and control may cause user resistance (Kirveennummi et al., 1998; Balakrishnan et al., 2008,). It is common for users to fall back into their traditional information values and ignore the possibilities introduced by a new system, with limited understanding and appreciation for the value of information for program planning, monitoring and evaluation (Heeks & Davies, 1999).

Technical barriers are often reported by firms implementing IS-related change who find the process costly, time consuming and risky. Firm infrastructure often includes layers of 'legacy systems' from an age where technology was associated with control, and built for a different world of IT capability. Where existing processes have been computerised, they can represent an 'electronic concrete' that renders reform practically impossible (Heeks & Davies, 1999). A common technical barrier in the auto industry occurs when partners cannot agree on a common technical specification. The proliferation of incompatible electronic standards in terms of protocol (the language used during transmission) and format (the label layout or visual interface) is widely reported in auto manufacturing (Howard et al., 2001; Helper & McDuffie, 2002). Data quality is defined in terms of reliability and timeliness, but this represents some difficulties in distinguishing between technical and processual problems.

In summary, the introduction of an electronic point of information exchange allowed buyers and suppliers to aggregate the data of their components and products, where consistency of data was essential in order to integrate with external trading partners and develop a synchronised supply chain (Lee & Whang, 1998; Barratt & Oliveira, 2001). Yet adopting e-hubs and sharing information across supply chains represented a significant social as well as technical challenge, involving timeliness, accuracy of information and developing collaborative capability. The next section applies the framework to cases in the automotive industry to assess the development and consequences of e-hubs.

## AUTOMOTIVE CASE ANALYSIS

The case interviews identify the perceived benefits of e-procurement membership together with an analysis of industry and organizational barriers. The six cases and their associated e-hubs are now considered in more depth and the outcomes summarized.

### Ford Motor Company

Ford undertook a major procurement initiative driven by price involving the introduction of a standard application to all purchasing departments. Covisint was perceived as an opportunity to develop the electronic marketplace as a means of utilising the reverse auction on commodity and production materials. The expected benefits include price reduction, minimisation of paper transactions and electronic audits, leading to lower material costs, increased transaction efficiency and greater control over maverick spending. However, despite a joint business/IT team of 300 people, IS-related problems persisted. In adopting e-procurement, Ford felt it must *'institutionalise'* the new system to comply with the needs of the central purchasing commission who approve all spending. This encountered resistance by managers who already had well-established purchasing relationships. In addition, system operators who were used to having bespoke software, had to use only standard e-procurement systems. Thus, Ford is still far from reaching its goal of: *'Moving the buying community away from the transaction [*and*] to give them the tools which will help them in negotiation, strategic thinking and disseminating information from a lower level.'*

Despite success in price reductions internally, a key industry market barrier for Ford remained the reluctance by suppliers to subscribe to Covisint over fears of the effect of reverse auctions on component prices. Ford is renowned for its cost focus, and this affected the willingness of its partners to accept e-procurement. There was increasing concern that the reluctance to join Covisint would harm on-line, collaborative product development initiatives: *'They [*Ford*] may lose a captive market if they are not customer-focused enough.'*

## Volvo Car Corporation

Volvo was one of the first manufacturers in Europe to become seriously committed to slimming its distribution pipeline and reducing inventory costs well over a decade ago. In order to continue its 'Customer Ordered Production' initiative it needed to integrate its programme of sequenced in-line supply (SILS) of parts and component modules further upstream to 2nd tier suppliers. While at one time Volvo believed that visibility throughout the supply chain (promised by Covisint) could be solved through better planning, Volvo subsequently believed that the key is not visibility, but better synchronization of orders in the supply chain. The increasing emphasis on in-sequence component supply places a greater responsibility between Volvo and its suppliers. While 1st tier suppliers have been successful in implementing SILS, this has had a consequential knock-on effect on the requirements for suppliers further upstream. 2nd tier suppliers would like to see the production sequence in advance because they are being asked to hold higher levels of stocks, but Volvo says it will *'never de-centralise the decision making process.'*

Volvo was reluctant to increase visibility across the supply chain *'because simply showing information to everyone isn't the solution [*to building to order*].'* In keeping with its reputation in the industry as an independent manufacturer, proud of its Swedish origins (despite ownership by Ford), Volvo was sceptical of consortium e-hubs and preferred to pursue its own private links with its partners. While it managed to improve component quality, decreased fixed assets and increased efficiency in the car plant, it had difficulty demonstrating lower total cost from SILS, suggesting that many problems had simply been pushed back up the supply chain. In overcoming a mutual technical challenge by developing an open book relationship with its 1st tier, it encountered conflict over increased stock holding and cases of labour disputes at 2nd tier level: *'Our 1st tier is good, but we are having some problems with 2nd tier.'*

### Volkswagen

Volkswagen established its private hub; Electronic Supplier Link (ESL) in 1999 as a fully Internet-enabled e-procurement system for 1st tier suppliers. Its aims were to improve procurement-based processes, where ESL was offered free to selected suppliers. On-line product collaboration using

CAD such as Catia and ProEngineer were also offered, but on a separate system to ESL. VW's approach to supplier relationships and e-hubs was different from the Ford/Covisint model, where: *'The policy of VW is not to earn money [*directly*] from the Internet.'*

VW's expected benefits were to optimise business processes across its supply chain, not just for itself. Its supplier hub saved considerable time in the procurement process and eliminated most unproductive work. While it represented a relatively simple system in comparison with Covisint, it faced a number of barriers during implementation and use. For instance, ESL retained its central purchasing system that, as it was not internet-based, required an additional interface connectionwith its mainframe computer to link it to suppliers, adding cost and time to the process. Also, users found that the high number of system standards required considerable knowledge of procurement, currency and logistics. As an internally based private hub, ESL experienced few problems of market barriers or supplier acceptance, particularly as the system was offered free. However, VW was vulnerable to an increase in individual systems and standards that Covisint, with its 'one-stop communication' philosophy, had attempted to eradicate. VW's approach to e-procurement could be summarised as: *'to optimise business processes with our suppliers [*with*] the whole supply chain, not just VW.'*

The outcomes of the case research with the three vehicle manufacturers is summarised in Table 3. The table shows the expected and realised benefits, and industry/firm level barriers (or 'disbenefits') to adopting e-procurement. Its purpose is to force investigators to look beyond initial impressions and see the evidence through multiple lenses, in order to sharpen research definition and validity (Eisenhardt, 1989). In instances where little or no evidence was found to support the case of either a benefit or barrier, this is recorded as 'no strong evidence' in the table.

## Bosch

Bosch was a major shareholder in the supplier consortium exchange 'SupplyOn', founded in 2000 to host online procurement for 1st and 2nd tier component suppliers. The expected benefits focused on reduction in order-to-delivery lead-time, reduction in manual processes and cost transparency. Bosch realised a number of benefits: fully integrated e-procurement, simultaneous engineering and logistics, enhanced core competency in building electronic supply chains, and also led negotiations of industry e-standards. Bosch is renowned for innovative engineering and was concerned it should develop its on-line engineering and purchasing capability. Like Volkswagen, its motivation for developing an e-hub differs significantly from Ford: *'I hope reduced price will be one of the secondary benefits, but it's not our focus.'*

Introducing an on-line materials purchasing and development system encountered reservations from managers and users. Some managers wanted to wait until other suppliers had implemented systems. However, Bosch was eager to gain first-mover advantage, particularly over system standards. Some users became reluctant after the reality of using an Internet system did not live up to expectations. The lack of a common standard during system trials meant suppliers were not involved until Bosch had achieved full back-end integration of its own mainframes. This followed the strategy of SupplyOn to become an integrated concept across all members, not just a supplier directory or on-line bidding website.

Bosch's goal to develop SupplyOn as a collaborative supplier engineering portal, based on building relationships in the supply chain and not simply price, meant that it took an ambivalent view towards Covisint as a potential competitor: *'There is conflict, but also a complementary approach.'* SupplyOn is positioned as a developer and operator of electronic marketplaces for auto suppliers seeking to *'optimize and accelerate the business*

*Table 3. Vehicle manufacturer e-procurement benefits and barriers*

| *Benefits* | **Ford** - *Covisint* | **Volvo** – *SILS* | **Volkswagen** - *ESL* |
|---|---|---|---|
| *Expected* | • Price reduction<br>• Minimise paper transactions<br>• Electronic audits | • Improve quality<br>• Decrease fixed assets<br>• Increase flexibility / efficiency<br>• Lower total cost | • Improve business process<br>• Supplier development |
| *Realised* | • Lower material costs<br>• Transaction efficiency<br>• Control maverick spending | • Improve quality<br>• Decrease fixed assets<br>• Increase flexibility / efficiency | • Reduced order leadtime<br>• Reduced unproductive waste |
| ***Industry Barriers*** | | | |
| *Environmental* | • Reluctance by suppliers to subscribe: reverse auctions<br>• Anti-trust legislation | • 2nd tier supplier conflict over increased stock levels | • Private hub: may suffer from individual system standards in the future |
| ***Firm Barriers*** | | | |
| *Structural* | • New systems must align with the needs of the central purchasing commission | • Union disputes over labour agreements from outsourcing<br>• 'eVEREST' e-procurement initiative introduced by Ford: concern over limited choice of suppliers and cost of catalogue | • Centrally based purchasing system |
| *Cultural* | • Resistance by departments who already have established purchasing relationships | No strong evidence | No strong evidence |
| *Managerial* | No strong evidence | • Difficultly in justifying system due to lack of a business case | No strong evidence |
| *User* | • Staged implementation means it is difficult to identify the system as a definitive product<br>• Operators used to bespoke system design: difficulties in adapting to standard package | No strong evidence | • Proliferation of individual systems and standards requires considerable knowledge by users: UN standards for currency, EDI standards used for logistics, but not purchasing |
| *Technical* | • 15 - 20 year old, mainframe systems | • Lack of integration between supplier systems during installation | • Initial problems in Internet technology caused by high system use |

*processes between suppliers and their partners... to drive industry growth through efficient communication and transaction development.'*

## DELPHI

Delphi is a tier 1 supplier of systems, modules and components such as wiring harnesses and vehicle suspension assemblies. It led the development of sequenced in-line supply (SILS) at supplier parks owned by Volvo and Vauxhall Motors UK (Opel). While their SILS system for delivery of production parts was based on conventional extranet infrastructure, Delphi explored the potential of e-hubs: *'We can't wait for Covisint to develop this [*part of the business*]*'.

A key challenge for the internal organization was in implementing unproven sequencing technology, with little knowledge transfer from other departments. A further barrier during the programme was political issues arising over the alignment and hierarchy of seven former business units. Delphi overcame these barriers, cutting costs by reducing component inventory at 1st tier level, and established an *'open book relationship with the vehicle manufacturer through mutual technical challenge'*. It also maintained good relations with its competitor Bosch, despite some concerns over rival interests. Delphi scheduled all suppliers through the plant, regardless of business origin. Delphi planned to pursue its implementation of SILS on the web by seeking a partnership-style relationship with vehicle manufacturers, and by developing its approach to organizational change by focusing on *'skill sets, tool sets, and mind sets'*.

## JOHNSON CONTROLS

Johnson Controls International (JCI), supported by software company MatrixOne, set up the web-enabled product development hub 'eMatrix' to eliminate inaccuracy in engineering drawing changes, increase productive design engineering time and act as acentral repository of knowledge between auto engineers across all tiers of the supply chain. Early in its life, the project realised benefits which include more efficient conversion of supplier part numbers and faster downloading of drawings, where drawings previously sent by e-mail could take up to 4 hours because of bandwidth restrictions. As a system set up by the engineering community to improve the interface internally between engineers, and between engineering and purchasing, the eMatrix concept required refinement to eliminate the manual loading of data and overcome problems of incompatible CAD systems (i.e. CATIA, ProEngineer, AutoCAD). Progress was made over communication between suppliers and Covisint by establishing ground rules over a common operating environment: *'JCI selected Covisint and persuaded them to adopt MatrixOne'*. Yet suppliers increasingly questioned why they should join Covisint, viewed as another means for automakers to cut prices, when they can choose specialist collaborative services from private hubs such as eMatrix.

The outcomes of the case analysis with the three component suppliers is summarised in Table 4. Both frameworks serve as a focus for analysis by logically leading the investigator to reflect on the gap between expected and realised benefits, and by considering the level and type of barrier responsible.

## E-PROCUREMENT STRATEGIES

The six cases highlight the different motivations and barriers during the development and deployment of e-hubs within and between firms. Ford's policy was to use the hub as a lever to gain further component and material price reductions globally, despite the reluctance of some suppliers to collaborate with Covisint. Bosch and Johnson Control's motivation for SupplyOn and eMatrix respectively, stemmed from their desire to develop highly innovative products through collaborative online engineering, as well as improving purchasing capability with suppliers. VW and Volvo aimed to optimise business processes across its supply chain, not just for themselves.

These characteristics can be classified in terms of hub membership, and the buyer-supplier relationship. Figure 3 provides a matrix that classifies the four types of strategy encountered in this research. Analysis of the cases leads to two critical dimensions: exchange membership and the buyer-supplier relationship. Relationship corresponds with the Kaplan and Sawhney (2000) dimension of procurement practice (i.e. systematic versus spot sourcing), which is generalized here into relationships based primarily on price and relationships concerned with building collaboration.

The second dimension is membership of the trade exchange. Although this does relate to ownership (Tapscott et al. 2000, Baldi & Borgman 2001) it is more concerned with the structural aspects of the exchange community, whether it is inclusive (anybody can join) or exclusive (members are invited to join). In order for trade exchanges to create value the buyer-supplier relationship must be considered as significant to the transaction process, as is membership of the exchange. Four types of buyer-supplier procurement strategy are

*Table 4. Supplier e-procurement benefits and barriers*

| *Benefits* | **Bosch** – *SupplyOn* | **Delphi** – *SILS* | **Johnson Controls** – *eMatrix* |
|---|---|---|---|
| *Expected* | • Reduced time to market<br>• Less manual processes<br>• Cost transparency | • Elimination of waste<br>• Linear material flow | • Eliminate inaccuracy in engineering drawing changes<br>• Increase engineering time<br>• 'Central repository' |
| *Realised* | • Collaborative engineering & logistics.<br>• Core competence in building electronic supply chains<br>• A leader in e-standards | • Information transparency<br>• Reduced inventory/transport<br>• 'Open book' relationship with VM through a 'mutual technical challenge' | • More efficient conversion of part numbers<br>• Faster downloading of drawings<br>• Able to design in colour |
| ***Industry Barriers*** | | | |
| *Environmental* | • Competitive conflicts of interest with Covisint | • Concern by Delphi that Bosch may be supplying its competitors | • Some questions over to join Covisint or choose specialist services from a private hub |
| ***Firm Barriers*** | | | |
| *Structural* | No evidence | • Hierarchy of 7 business units | • Problems with cross referencing with other partners |
| *Cultural* | • General reservations against e-business | • Political issues over changes to corporate alignment | No evidence |
| *Managerial* | • Some desire to 'wait and see' how other suppliers conduct e-business strategy | • Little knowledge transfer from other departments | No evidence |
| *User* | • User expectations of 'big benefits' not realised over the short term | No evidence | • Different part numbers used for the same parts |
| *Technical* | • Lack of a universal system standard: suppliers involved only after Bosch achieved integration | • Some mis-sequencing during initial operation | • Manual loading of part numbers<br>• Incompatible CAD systems |

identified here to demonstrate the significance of price-based versus collaborative relations in e-hubs.

The *open market* is perhaps the simplest of the four types that emerge from Figure 3. The open market is an agora (Tapscott et al, 2000) that brings together sufficient buyers and sellers to establish prices and create liquidity, moving to a frictionless marketplace that anybody can join.

The *community* is also inclusive but is aimed at fostering collaboration with a key aim of sharing knowledge of limited commercial sensitivity. Such an exchange might include job opportunities, technical advice, and product ratings in addition to procurement.

The *alliance* is concerned with collaboration but is rather more directed. An alliance might be established for motor manufacturers to work together on the co-design of a new vehicle, or for a manufacturer to expose its inventory levels and manufacturing schedule to a tier 1 supplier to even out inventory fluctuations in the supply chain. Clearly, this environment requires trust and security, and as a result membership is by invitation.

The *club* is concerned with control and price optimization for buyers, who are also the owners of the exchange. However, buyers and sellers might collude, giving rise to concerns about the reduction of competition and legal issues of anti-trust behaviour and the establishment of a cartel.

The 'club' of powerful vehicle manufacturers, represented here by the Ford and other Covisint partners, sought to create one buyer in the market.

Concerns over monopolistic practices caused Covisint to be delayed under federal legislation in the US, yet little action was taken to allay supplier fears during the launch in Europe. The restriction of entry to new shareholders and the focus on material cost in the supply chain meant that Covisint reduced supplier goodwill to collaborate, affecting the willingness to share information that is essential for supply responsiveness. Hence the club strategy was problematic in the industry where, unlike the other strategies in Figure 3, it was driven by control and position in the supply chain and value may not be shared fairly between firms. This research argues that Open Market, Community and Alliance type strategies represent legitimate and long-term approaches to building e-procurement between firms, where each can co-exist with the other, unlike the Club, which has cartel-like properties.

## CONCLUSION AND IMPLICATIONS

This article re-examines the perceived wisdom of the transition from supply chain to electronic hub. The research confirms that a consortium-based approach is adopted if price and control are the prime motivation for founding an e-hub, e.g. Covisint (Baldi & Borgman, 2001). If supply chain management, supplier development and product innovation is of greater importance, then a private-based e-hub is adopted, e.g. SupplyOn and ESL. When applied at industry level, this research shows the resulting combination of structures to be sub-optimal.

While technology exists to create electronic marketplaces, acting as a 'one stop' procurement and product development hub, a considerable barrier involves collaboration from a critical mass of industry players for the e-hub to succeed. Thus, the advantages of early co-operation amongst competitors brings mutual benefit, not least in the agreement of IT standards, such as the Button programme in Australia, and the Auto Industry Action Group (AIAG) in North America. The initial proliferation of e-hubs by rival VMs and component suppliers in Europe and North America can be explained by the presence of long-term e-commerce development projects by VW and Bosch, countervailing power exercised by suppliers concerned over the lack of representation within Covisint, and the desire to develop core competence in electronic markets. Overcoming the problems of an overcrowded e-commerce environment requires refocusing the strategic objectives of existing hubs, by applying the tools developed here: the Industry transformation framework (Table 2), and the e-Procurement strategy matrix (Figure 3). Three findings are paramount:

1. The introduction of e-commerce resulted in a mess of overlapping networks in the auto industry. This was a shallow structure, lacking in supplier integration particularly of tiers 2 and 3, and resembles an ad-hoc arrangement of spokes rather than a singular hub design.
2. While suppliers worried over higher than expected levels of complexity in the exchange of information using electronic media, of particular concern was the introduction of an inter-firm system that combined a transactional approach for reducing price, with a collaborative product development approach that required sharing organizational knowledge. This research demonstrates that the monopolistic club strategy, most notably demonstrated by the Covisint hub, to be an unsustainable business model for the automotive industry.
3. The apparent 'strategies' in the industry were not long-term. Manufacturers and suppliers cannot sustain a system characterised by duplicated services, multiple standards and restrictions on membership that result in less than optimal numbers of partners. Having recognised the threat of rival hubs, policy changes by Covisint indicated a

> softening in its approach towards suppliers, encouraging membership by emphasising the capability of its product collaboration software (Kisiel, 2002a). However this was a case of 'too little, to late': considering the subsequent shake-out of e-hubs, the auto industry structure displayed symptoms of a mid-transitory phase in the e-procurement development lifecycle. Indeed, the demise of Covisint as an automotive hub and its rebirth in 2004, under new ownership, as a supplier of collaborative solutions to the health care, finance, and public sector is testament to this.

Many organisations were taken in by simplistic, technology-driven models professing to enhance competitive and co-operative capability. In the journey from bespoke supply chain to electronic hub, this research found multiple barriers to information sharing at firm and industry level, overlapping networks competing for membership, and isolated pockets of collaboration. Hence the auto industry structure was shallow, lacking in supplier integration, and resembled an ad-hoc arrangement of spokes rather than a singular hub design. This transitory phase involved conflict in motivation between price, collaboration and membership exclusivity, which prevented e-hubs from becoming fully integrated and the industry realising full benefit from electronic markets.

As concerns research, this work examines the value of e-markets and the motivations and barriers to relationships that govern e-procurement processes. The exploratory cases question the simplistic portrayal in the literature of one dominant hub aligned in the centre of an industry. In re-examining the transition from bespoke EDI to generic e-procurement, it discovered conflicting motives and complex barriers at firm and industry level.

The technique of mapping e-hub structures in Figure 2 provided a snapshot industry overview, though it omits the complete picture of resource and collaboration flow between all stakeholders. The classification of networks, represented by arrows, require further definition beyond market/price and knowledge/collaboration. While this dispels the myth of simplistic, monolithic e-commerce models in the auto industry, more work is required in developing a method that reflects the underlying complexity of flows of goods, services and knowledge.

The e-Procurement strategy matrix in Figure 3 highlights the difficulties of implementing e-market strategies that attempt to reconcile the mutually exclusive nature of buyer-supplier relationships: price vs. collaboration, and inclusive vs. exclusive hub membership. While Tapscott et al.'s typology of business webs provides a useful grounding it assumes a level of diversity in electronic markets whereas this research shows a consolidation of e-hubs, driven by a revision in the hugely optimistic forecasts in e-commerce revenue. Given that the nature of e-hub benefits is somewhat different from original expectations, particularly in the case of Covisint, further work could extend the case analysis and focus on the motivation and barriers to information sharing by industry stakeholders.

## ACKNOWLEDGMENT

The authors wish to acknowledge support provided by the Engineering Physical Science Research Council in the UK during this research.

## REFERENCES

Arbin, K., & Essler, U. (2005). Covisint in Europe: Analysing the auto B2B e-marketplace. *International Journal of Automotive Technology & Management*, *5*(1).

Argyris, C. (1990). *Overcoming organizational defenses: Facilitating organizational learning.* Boston, Allyn & Bacon.

Bakos, Y. (1998). The emerging role of electronic marketplaces on the Internet. *Communications of the ACM, 41*(8).

Balakrishnan, J., Bowne, F., & Eckstein, A. (2008). A strategic framework for managing failure in JIT supply chains. *International Journal of Information Systems and Supply Chain Management, 1*(4), 20–38.

Baldi, S., & Borgman, H. (2001). Consortium-based B2B e-marketplaces: A case study of the automotive industry. *14th Bled Electronic Commerce Conference,* Slovenia.

Barratt, M., & Oliveira, A. (2001). Exploring the experiences of collaborative planning initiatives. *International Journal of Physical Distribution & Logistics Management, 31*(4).

Barratt, M., & Rosdhal, K. (2002). Exploring business-to-business marketsites. *International Journal of Purchasing and Supply Management,* 8.

Benjamin, R., & Blunt, J. (1994). IS plans in context: The IS planning environment: Critical IT issues in the year 2000. In R. Galliers, D. Leidner, & B. Baker (Eds.), *Strategic Information Management.* Oxford: Butterworth.

Boland, J., & Hirschheim, R. (1987). *Critical issues in information systems research,* Wiley, New York.

Boot, R., & Butler, J. (2001). *Momentum in the automotive industry.* KPMG report.

Checkland, P. (1991). From framework through experience to learning: the essential nature of action research. In H. Nissen, H. Klein, & Hirschheim (Eds.), *Information Systems Research.* Elsevier Science.

Ciborra, C. (1993). *Teams, markets and systems: Business innovation and information technology.* Cambridge University Press.

Connelly, M. (2001). Where Jacques Nasser went wrong. *Automotive News.*

Crain, K. (2008). Global market data book. *Automotive News Europe.* Crain communications.

Croom, S. R. (2001). The dyadic capabilities concept: examining the processes of key supplier involvement in collaborative product development. *European Journal of Purchasing and Supply Management, 6.*

Damsgaard, J., & Lyytinen, K. (1998). Contours of diffusion of electronic data interchange in Finland: Overcoming technological barriers and collaborating to make it happen. *The Journal of Strategic Information Systems,* 7.

Davenport, T., Eccles, R., & Prusak, L. (1992). Information politics. *Sloan Management Review.* Fall.

De Boer, L., Harink, J., & Heijboer, G. (2002). A conceptual model for assessing the impact of electronic procurement. *European Journal of Purchasing & Supply Management, 8.*

Desanti, S. (2000, June). Evolution of electronic B2B marketplaces. *Federal Trade Commission Public Workshop.*

Earl, M., & Feeney, D. (2000). How to be a CEO for the information age. *Sloan Management Review.*

Economist. (2009, 17 January). The big chill (pp. 65-67).

Eisenhardt, K. (1989). Building theories from case study research. *Academy of Management Review, 14*(4).

Galliers, R., Leidner, D., & Baker, B. (1999). *Strategic information management.* Oxford: Butterworth Heinemann.

Harland, C. (1996). Supply chain management: relationships, chains and networks. *British Journal of Management*, 7.

Hart, P., & Saunders, C. (1997, Jan/Feb). Power and trust: Critical factors in the adoption and use of electronic data interchange. *Organization Science*, *8*(1).

Heeks, R., & Davies, A. (1999). Different approaches to information age reform. In Heeks (Ed.), *Reinventing Government in the Information Age – International practice in IT-enabled public sector reform.* Routledge, London.

Helper, S. (1991). How much has really changed between US automakers and their Suppliers? *Sloan Management Review*, *32*(4).

Helper, S., & MacDuffie, J. (2000). *Evolving the auto industry: E-business effects on consumer and supplier relationships.* Fischer Center, UC Berkeley.

Helper, S., & MacDuffie, J. P. (2002). B2B and modes of exchange: Evolutionary and transformative effects. In B. Kogut (Ed.), *The Global Internet Economy.*

Holweg, M., & Pil. F. (2001). Successful build-to-order strategies: Start with the customer. *MIT Sloan Management Review*.

Howard, M., Vidgen, R., & Powell, P. (2006). Automotive e-hubs: exploring motivations and barriers to collaboration and interaction. *The Journal of Strategic Information Systems*, *15*(1), 51–75.

Howard, M., Vidgen, R., Powell, P., & Graves, A. (2001). Planning for IS-related industry transformation: The case of the 3DayCar. *Procs of 9th European Conference on Information Systems*, Slovenia.

Howard, M., Vidgen, R., Powell, P., & Graves, A. (2002). Are hubs the centre of things? E-procurement in the automotive industry. *Procs of 10th European Conference on IS*, Poland.

Huber, N., Ward, H., Goodwin, B., & Simons, M. (2000, July). The e-procurement dilemma. *Computer Weekly.*

Jun, M., Cai, S., & Kim, D. (2008). The strategic implications of e-network integration and transformation paths for synchronizing supply chains. *International Journal of Information Systems and Supply Chain Management*, *1*(4), 39–59.

Kanter, R. (1994, July/August). Collaborative advantage: The art of alliances. *Harvard Business Review*.

Kaplan, S., & Sawhney, M. (2000, May-June). E-hubs: The new B2B marketplaces. *Harvard Business Review*.

Kirvennummi, M., Hirvo, H., & Eriksson, I. (1998). Framework for barriers to IS-related change: development and evaluation of a theoretical model. In De Gross et al., (Eds.), *Proceedings of Joint IFIP Working Conference*, Finland.

Kisiel, R. (2002, April 29). Online trade exchanges: What went wrong? *Automotive News Europe, 2.*

Kisiel, R. (2002a, July 8). Covisint gets one last chance. *Automotive News Europe, 3.*

Kisiel, R., & Whitbread, C. (2000). E-business. *Automotive News Europe* (pp. 37–40).

Kuroiwa, S. (1999). Growing an open business environment from CALS, JIT & supply chain. *Logistics in the Information Age, 4th Conference Proceedings ISL*, Florence, Italy.

Kwon, K., & Zmud, R. (1987). Unifying the fragmented models of information systems implementation. In R.J. Boland,& R.A. Hirschheim (Eds.), *Critical Issues in Information Systems Research.* Chicester: Wiley.

Lamming, R. (1993). *Beyond partnership: Strategies for innovation and lean supply.* Prentice Hall.

Lane, M., & Koronios, A. (2000). Using stakeholder salience theory to facilitate management of stakeholder requirements in business-to-customer web information systems. *13th International Bled Electronic Commerce Conference*, Slovenia.

Lee, H., & Whang, S. (1998). Information sharing in the supply chain. *Research Paper Series*, working paper, Stanford University.

Li, F., & Williams, H. (1999). Interfirm collaboration through interfirm networks. *Information Systems Journal*, 9.

Merton, R. (1968). *Social theory and social structure*. Free Press, NY.

Miles, M., & Huberman, M. (1994). *Qualitative data analysis: An expanded sourcebook.* Sage: London.

Min, H., & Galle, W. (1999). Electronic commerce usage in business-to-business purchasing. *International Journal of Operations & Production Management*, *19*(9).

Mintzberg, H. (1979). *The structure of organisations*. New York: Prentice Hall.

Nonaka, I., & Takeuchi, H. (1995). *The knowledge-creating company: How japanese companies create the dynamics of innovation*. Oxford University Press.

Núñez-Muñoz, M., & Montoya-Torres, J. (2009). Analyzing the impact of coordinated decisions within a three-echelon supply chain. *International Journal of Information Systems and Supply Chain Management*, *2*(2), 1–15.

Pfeffer, J., & Salancik, G. (1978). *The external control of organizations: A resource dependent perspective.* Harper and Row.

Prahalad, C. K., & Hamel, G. (1990). The core competence of the corporation. *Harvard Business Review*, *68*(3).

(2008, November 26)... *Professional Engineering*, *21*(21), 5.

Reynolds, J. (2000). Supply chain, distribution and fulfilment. *International Journal of Retail and Distribution Management*, *28*(10).

Rubin, S., & Fogarty, T. (2000). *Automotive B2B: The creation of B2B enterprises should add value for GM and Ford shareholders*. Warburg Dillon Read, Global Equity Research.

Sako, M. (1992). *Prices, quality and trust: Interfirm relations in Britain and Japan*. Cambridge University Press.

Schein, E. H. (1994). The role of the CEO in the management of change: the case of information technology. In R. Galliers, D. Leidner, & B. Baker (Eds.), *Strategic information management.* Oxford: Butterworth Heinemann.

Scherer, F. (1980). *Industrial market structure and economic performance*. Rand McNally.

Scott Morton, M. S. (1991). *The corporation of the 1990s.* Oxford University Press, NY.

Smaros, J., Holmstrom, J., & Kamarainen, V. (2000). New service opportunities in the e-grocery business. *International Journal of Logistics Management*, *11*(1).

Swatman, P., M., & Swatman, P., A. (1992). EDI system integration: A definition and literature survey. *The Information Society*, 8.

Tapscott, D., Ticoll, D., & Lowy, A. (2000). *Digital capital: Harnessing the power of business webs.* Nicholas Brealey.

Threlkel, M., & Kavan, B. (1999). From traditional EDI to Internet-based EDI: managerial considerations. *Journal of Information Technology*, 14.

Timmers, P. (1998). Business models for electronic markets. *Electronic Markets*, *8*(2).

Vanany, I., Zailani, S., & Pujawan, N. (2009). Supply chain risk management: Literature review and future research. *International Journal of Information Systems and Supply Chain Management*, *2*(1).

Vidgen, R., & Goodwin, S. (2000). XML: What is it good for? *Computing and Control Engineering Journal*, *11*(3), 119–124.

Webster, J. (1995). Networks of collaboration or conflict? Electronic data interchange and power in the supply chain. *The Journal of Strategic Information Systems*, *4*(1).

Yin, R. (1994). *Case Study Research, design and methods*. Sage: Thousand Oaks, CA.

*This work was previously published in International Journal of Information Systems and Supply Chain Management, Volume 3, Issue 1, edited by John Wang, pp. 21-42, copyright 2010 by IGI Publishing (an imprint of IGI Global)*

# Chapter 6
# Industry-Wide Supply Chain Information Integration:
## The Lack of Management and Disjoint Economic Responsibility

**Stefan Henningsson**
*Centre for Applied Information and Communication Technology, Copenhagen Business School, Denmark*

**Jonas Hedman**
*Centre for Applied Information and Communication Technology, Copenhagen Business School, Denmark*

## ABSTRACT

*Experiences from enterprise-wide integration initiatives during more than four decades indicate that industry-wide information integration could render substantial benefits. Two ways in which industry-wide integration differs from enterprise-wide integration are that there is no common management level and the economic units in the integration are the constituent units, not the industry. Management involvement has been emphasized as perhaps the most critical success factor for enterprise-wide information integration. The common economic unit enables increased costs in one part of the organization to lower the total cost in the company as a whole. In this article the authors address which consequence these two differences have for the development of information integration in four industry-wide supply chains. The authors find the existing methods for enterprise-wide information integration, such as BPR, virtually impossible to apply on industry-wide information integration and that the disjoint economic responsibility is a hampering aspect in reaching potential benefits of industry-wide information integration.*

DOI: 10.4018/978-1-4666-0918-1.ch006

## INTRODUCTION

If management commitment is the most critical success factor, what then if there is no management level? After companies have become integrated in their internal information flows there is also a need to address information integration across organizational borders in industries. During the last decades, organizational boundaries have been blurred and technological innovations have introduced the alternative of integrating business with partners higher up or further down in the value chain. This type of integration is in many aspects different from integration that takes place inside an organization, not at least because of the absence of a higher managerial force. Industries with their cross-organizational spanning supply chains have no orchestrating function, something that could make existing integration and business process reengineering methodologies inadequate. In addition, experience from intra-organizational information integration tells that frequently what in the end becomes an efficiency gain for the organization as a whole may impose an increased burden on the single business unit – the factor of asynchronous efficiency gains (c.f. Hedman & Kalling, 2003). Asynchronous gains are possible to enforce since the organization act as one single economic unit, but this is not the case for industries which have disjoint economic responsibility.

Much is written on information integration in intra-organizational value chains and how to make it work efficiently (e.g. Alsene, 1999; Karuppan & Karuppan, 2008; Konsynski, 1993). Similarly much research has been directed towards two-part integration with inter-organizational systems (IOS) and electronic data-interchange (EDI) in Business-to-Business (B2B) and Business-to-Government (B2G) relationships (e.g. Henriksen, 2006; Krcmar, Bjørn-Andersen & O'Callaghan, 1995; Lim & Palvia Prashant, 2001; Masetti & Zmud, 1996; White, Daniel, & Mohdzain, 2005). Less is, however, known about the industry-wide value chain with focus on the whole chain from initial producers to end consumers (Browne, Sockett & Wortmann, 1995; Konsynski, 1993). The industry-wide context introduces issues of inter-organizational collaboration among several actors, an increased multitude of IS, organizational cultures and organizational objectives. Although the technical challenges of information integration may be similar regardless of intra- or inter-organizational context, the organizational and managerial challenges in industry-wide information integration needs special attention as they presents a different integration context (Neureuther & Kenyon, 2008).

This article answers to an explicit call for research by Pant, Sethi and Bhandari (2003) in which they state that "*there is a need to revisit intra-organizational change management theories to ascertain if they will as effectively apply to inter-organizational or supply chain-wide change.*" We do so by addressing the question which consequences the lack of management level to orchestrate the process and the disjoint economic responsibility has on development of industry-wide information integration. Our general purpose is thus to provide new knowledge on information integration in industry-wide supply chains. Our contributions to this purpose are several. We both theoretically and empirically explore two contextual circumstances to information integration in industries: the lack of common management unit and the factor of asynchronous gains. In this process we also describe and explain the existing information integration, and the lack of integration, in one particular industry – the Swedish agri-food industry. Finally, we also contribute with ideas of processes and factors that would lead to increased integration in industry-wide supply chains.

This article is structured as follows. In the next we will present previous research that can be useful for explaining information integration in industry-wide supply chains. Thereafter we will introduce our structured case study approach and the four cases in the agri-food industry that create the empirical foundation for the article. We will then discuss which aspects of our cases that is covered by existing theory and which are not. Finally we address how the barriers for information integration in industry supply chains can be overcome and draw general conclusions from our research.

## INFORMATION INTEGRATION

The development in information technology (IT), in combination with extended pressure of globalization, environmental consideration and transformation of organizational structures, is said to transform organizational boundaries, blurring the frontiers to customers and suppliers (Browne & Jiangang, 1999; Markus, 2000). As the technological platform continues to develop, the foundation for extending the corporations business process becomes more and more relevant. Manufacturing companies can no longer be seen as individual systems, but rather as participators in an extended value chain (Browne & Jiangang, 1999). Optimizing this value chain is one major challenge in order to achieve business success (Jähn, Zimmermann, Fischer & Käschel, 2006).

### Industry Wide Information Integration

To describe what organizations are doing when they engage in initiatives to align and cooperate with their peers further up or down in the supply chain through advanced technology one could use terms as information systems (IS) integration (Alaranta & Henningsson, 2008), IT integration (Wijnhoven, Spil, Stegwee, & Fa, 2006), or inter-organizational systems (Henriksen, 2006), but what creates the benefits for the companies is not that two or more IT-systems are interrelated. What creates advantages is that information is present where and when it is needed. Therefore we use the term 'information integration' (Evgeniou, 2002) in this article.

Since the 1970's integration in various contexts and shapes has been a recurrent topic on the IT professionals' agenda. Starting with the Total Systems concept (Blumenthal, 1969) followed by the Management Information Systems (e.g. Adelberg, 1975; Kashyap, 1972; Lidd, 1979), companies have engaged in a continuing quest for free flowing information that would enable more efficient and completely new ways of doing business. At the end of the 1990's many companies, enthusiastically encourage by consultants as well as academics, approached ERP-based, enterprise-wide monolith systems based on the promise of one single, homogenous and standardized information system that should consist the organizational spine. Although most companies never saw this enterprise-wide IS coming true (Hanseth & Braa, 2001), the efforts of the late 1990's revealed what integrated information flows could do for the business of company. At its best, integrated information that flows in the global IS provides every part of the organization with unlimited information access that enables numerous benefits, including organizational flexibility, increased productivity, integration of business processes, improved quality and standardized quality of output (Hedman & Kalling, 2003).

Although the status of the internal information integration is far from a state when researchers and IS professionals can go home a declare victory, today internal information flows in most companies is not a matter of existent or non-existent, but of quality, reliability and resource utilization. However, for integration researchers and IS professionals new integration challenges are lurking in the dark. It is necessary to draw the attention also towards integration of information flows within complete industries.

Integration of intra-organizational information flows has showed that information integration of business activities can improve both efficiency and effectiveness of business activities (Shang & Seddon, 2002), but also lead to greater dependencies and decrease organizational flexibility

(White et al., 2005). However, it has been acknowledged that dependent business activities exist also across organizational borders. Several studies address information integration between business activities of two different organizations (e.g. Alaranta &Henningsson, 2008; Henriksen, 2006; Saeed, Malhotra & Grover, 2005). In this article we take the study of inter-organizational information integration one step further. Studying two-part information integration has doubtlessly led to important knowledge contribution, but only recognizing the dependency of two parts is like only study integration of two activities of intra-organizational integration and not recognizing the advantages, disadvantages, and difficulties of integrating the whole organization. Much of the benefits created by intra-organizational integration are directly dependent on not only two activities becoming integrated, but on the orchestrating and harmonizing of all related activities. The same should logically be true for integration of industries, but the decision making business units have different requirements on the information-integration of the supply chain (Pant et al., 2003). Much of the benefits can only be derived if all activities are integrated, meaning that the integrated information flows should also be studied on an industry level.

## Management of Information Integration

To address the consequence of the lack of common management function it is necessary to understand the role of management in information integration. An IS finds its raison d´être in how it contributes to the organization to reach its objectives (Alter, 1999). It is the managerial task to ensure this is the outcome. IS management is thus about making decisions and taking action that leads IS issues in a certain direction. To be able to make decisions on direction, it must be known which alternatives are available and which decisions have to be made. IS management can thus be divided into two parts: 1) knowledge of alternatives and basic structural choices of IS and 2) how the choices affect the organization.

Existing explanations as well as methodologies, methods and approaches for intra-organizational information integration can be argued inappropriate for addressing information integration of industries. These approaches generally assume the existence of a managerial level balancing needs and available means, technological and business requirements, and devote sufficient time and resources for succeeding with the initiative. If one lessons was learned from the initiatives of the late 1990's is was that management support and involvement is essential for this type of coordinating initiatives (Finney & Corbett, 2007; Lam, 2005). For industries, there is no managerial function that can support coordination. This follows the ideas of Segars and Grover's (1999) how discusses the differences between top-down and bottom-up processes for IS deployment. A common managerial level represents a hierarchical management mechanism (Brunsson & Adler, 1989). By the position in the organization individuals are given the possibility to orchestrate and use a common resource base as it best serves the entire organization. Alternatives to hierarchical management includes management through markets (Brunsson & Adler, 1989) where the individual's or organization's bargain power determines which decisions and actions that are taken and management through standards in which a all actors that identifies themselves with a certain group performs a task according to a standardized template (Brunsson & Jacobsson, 2000).

The field of IS governance (and IT governance) is closely related to IS management, even though some authors choose to make a distinction. Weill (2004) describes the difference as being that IS governance is about systematically determining who makes each type of decision (a decision right), who has input to a decision (an input right) and how these people (or groups) are held accountable for their role. IS management is about what specific decisions are made. However, not all authors make such a clear cut distinction between IS management and IS governance. For example, in introducing the IT governance track for Hawaiian International Conference on System Sciences in 2005, Van Grembergen defined IT governance as "the organizational capacity exercised by the board, executive management and IT management to control the formulation and implementation of IT strategy and in this way ensuring the fusion of business and IT" (Van Grembergen, 2005, p. 1). This definition approaches IS management as defined in this text as it talks about what management has to do to relate IT decisions to organizational goals, and also how to implement the IT strategy. Further, when discussing IS management in terms of how to manage or govern IS to a desired outcome, the very typical questions of IS governance, such as division of labor, responsibility and organization of the task, are tightly intertwined with the actual tasks carried out. Compared to the managerial discussion as being part of the IS capability domain, the similarity is striking. Both traditions have arrived at the same understanding to focus on basic, structural options and the contingency factors determine which options are appropriate during specific settings.

IS integration management can be seen as a subtask of IS management that is specifically related to integration of various IS. Not much is written about IS integration management, but the task is similar to what was concluded above about IS management in general. Drawing on general IS management research (Clemons & Row, 1991; Gottschalk, 2000; Kalling, 2003; Mata, Fuerst & Barney, 1995; Meyer, 2004; Pyburn, 1983; Weill & Broadbent, 1998) and research on IS and IT governance (Brown & Grant, 2005; Meyer, 2004; Van Grembergen, 2005; Webb, Pollard & Ridley, 2006; Willcocks, Feeny & Olson, 2006), IS integration management is refined into two tasks: a) the basic, structural options of IS integration, and b) the options effects on the business of the organization. As top management involvement has been found essential, the absence of a unit for housing these two activities may have far reaching consequences for industry-wide information integration.

## Asynchronous Gains

A feature of large scale IS projects that to some degree is related to the above described absence of management mechanism is the factor of asynchronous gains. "Asynchronous gains" means that when implementing a large scale IS in an organization some parts of the organization and the people affected by the implementation may be affected negatively while the organization as a whole is beneficial (Hedman & Kalling, 2003). For example, in implementing a new enterprise-wide system with one objective to have only one point of data entry order reregistering may be more bulky and time consuming. The sales organization may also become less flexible and adaptable. If the implemented IS' scope is within one single economic unit the new system can be enforced, but what if the potential benefits requires standardization and integration among several different economic units of whom only a few may experience benefits?

The resource dependency perspective (Pfeffer & Salancik, 2003) explains that organizations attempt to alter their dependence relationships by minimizing their own dependence or by increasing the dependence of other organizations on them. Within this perspective, organizations are viewed as coalitions alerting their structure and patterns of behavior to acquire and maintain needed external resources (Pfeffer & Salancik, 2003). Information systems can be designed explicitly to secure required resources (Tiliquist, King & Woo 2002). As long as information integration takes place within a single organizational unit resource dependency is not an issue, since the driving force in the mechanism is that the resource is out of the control scope.

Research has found a general positive relationship between inter-organizational supply chain integration and organizational performance (Lummus, Vokurka & Krumwiede, 2008). The results of Lummus et al. (2008) confirms that companies that are part of well integrated supply chains do better on various performance measures than others, but this does not mean that every company in a value chain benefit from the integration. Generally there is more evidence that backwards integration from customer to suppliers are beneficial than forward integration from suppliers to customers. Backward integration is associated with efficiency gains in for example reduced inventory, faster time to market, and more reliable output (Hedman & Kalling, 2003). Forward integration is associated with decreased demand uncertainties, development of market specific strategies, quality assurance and lock in effects (Childerhouse & Towill, 2006; Christopher & Towill, 2002). The fact that has been hard to prove positive financial impact on forward integration should make companies further down the supply chain generally more reluctant to information integration than companies near the end customer.

## METHODOLOGY

The study presented in this article is a structured case study mainly inspired by Carroll and Swatman (2000). The object of study is information integration in the food chain industry, and the focus is on factors describing, explaining and prescribing the existence or the absence of information integration in an industry. The goal of our study is to induct concepts and theories from empirical observations (Carroll & Swatman, 2000; Eisenhardt, 1989). The structured case study approach suggested by Carroll and Swatman (2000) and applied by, for instance, Grimsley and Meehan (2007), includes guidelines for the process of developing knowledge and theory based on empirical data. It does not prescribe specific data collection techniques or ways of analyzing the data, but outlines a framework for how to develop knowledge and theory. The main steps are in their approach to develop an initial conceptual framework, to collect and analyze empirical evidence, and to reflect on the result in order to induce knowledge. In that regard it is similar to the approach suggest by Eisenhardt's (1989) approach to develop theories from case studies. However, there are several differences between the approaches suggested by Carroll and Swatman (2000) and Eisenhardt (1989). The first is that Carroll and Swatman approach has a strong interpretative legacy whereas Eisenhardt belongs to a more positivistic tradition. Another difference is the view on theory building. Eisenhardt (1989) describes this as a fairly straight forward linear process with iterations (i.e. follow the eight steps and theory will emerge at the end), whereas Carroll and Swatman view the process of theory building as process based on iterations between conceptual framework, data collection and analysis, and reflection where new theories, data, conceptual framework may be introduced until saturation is reached. In the reminder of the section we will describe how we have adopted the structured case approach.

### Study Context, Phases and Methods

As stated previously the core topic for this study is information integration from an industry perspective. Information integration, as such, has been explored in previous research (see section two). However, it has mainly been studied as intra-organizational integration or inter-organizational integration with an emphasis on technical issues. Instead we decided to broaden the scope of integration by studying an industry from a managerial perspective. This complements existing research and provides an opportunity of knowledge contributions.

A second issue in relation to the study context is: Why the food chain? There were several reasons for this the agri-food industry. The first reason is the importance of the industry. The industry is fundamental to our society both in terms of output of food products and consist a substantial part of the world economy. Second, the food industry is relatively stable in terms of shape actors in the supply chains. Third, there are obvious reasons for integration in the industry as it at first glance faces several classical problems normally addressed by integration such as traceability needs and synchronization of processes. Forth, the research was carried out in relation to the EU-funded research project ITAIDE (www.itaide.org) which has focus on governance (customs, control and traceability issues) of international trade without adding additional administrative burden to the industry actors. One of four industries in focus of the ITAIDE project is the agri-food industry which in addition to accentuate the importance of the agri-food industry by making it focal point of a € 7 M research project also opened up the empirical ground for this study. Thus, the industry selection also contained an element of pragmatism.

## First Phase (Conceptual Framework, Data Collection and Analysis

The first phase of the study involved the development of an initial framework used for both empirical data collection and data analysis. It was based on an integration of two streams of research: The first was the business model concept, which integrates both internal and external inputs of an organization (Hedman & Kalling, 2003). The business model concept includes five interrelated components: 1) customers, 2) offering, 3) business processes and how these are organized, 4) resources and 5) suppliers. The interrelationships between the components refer to changes in one of the components effects the others. In addition to the business model concept we applied Konsynski's (1993) framework for understating the external control of the firm. Business drivers for intra-organizational integration includes higher ability for organizational learning, better ability to respond to market change, and in the end more efficient management due to new or smoother information flows (Konsynski, 1993). The business model concept and external control of the firm was used to guide the empirical data collection, analyze the data, and provided us with a structure for summarizing individual case. These two sets of theories were used as tools for capturing the information integration context and not as theoretical domains to which this article aim to make contributions. Hence, the omitance in section 2 that presents the articles theoretical foundation. The contextual theories were complemented with questions conserving the actual information integration based on the review in section 2.

In total nine companies were investigated including three farmers, four food processors, one corporate function of one grocery chain, and one large grocery store. The nine investigated companies:

- Askliden AB is a milk producer, with 250 milk cows.
- Bramstorp Gård AB produces sugar beets and peas.
- Coop Norden is the corporate function of Scandinavia's largest grocery chain.
- Danisco Sugar's facility at Örtofta refines sugar beets into raw sugar – has a virtual monopoly in Scandinavia.
- Findus AB is specialized in frozen food, such as vegetables (illustrated by peas), meat and fish.
- ICA Tuna is a local grocery store and belongs to the ICA group.
- Skånemejerier, a cooperative owned by milk farmers, is a leading actor among dairy products.
- Swedish Meats is the leading slaughter house in Sweden and also a cooperative.
- Tygelsjö Mölla is a pig farmer, who delivers 4500 piglets to Swedish Meats.

The research questions were organized in three parts based on theory. The first part was loosely centered upon customers, products, business processes, work activities, organizational structure, and suppliers in order to get a background to each company. The second part addressed the external control of the firm. The final and third part of the empirical investigation questions aimed at determining the existing IS, information systems integration, and information flows.

The main data collection method used was interviewing. In total 13 semi-structured interviews were made. Each respondent was asked to describe his or hers firm according to customer base, offering, business processes and organizational structure, resources, suppliers, external control, information systems, and integration. In addition public available documents, such as annual reports and web pages were used. The individual respondents were owners of the firms (framers), chief information officers, or financial officers.

## Second Phase (Analysis and Reflection)

Based on the initial framework, data collection and analysis, nine rich case stories of about 2000 word each were written. These were used as input for first round of analysis and reflection. It has to be noted there was no clear demarcation between the phases. The first round of analysis and reflection was mainly done in relation to the first theoretical framework (business models and extern control of the firm) and written up the nine case stories. A contribution of the first reflection phase was in the concept of lack of management level, which was consistent with Konsynski's framework, but based on a reversed logic. In addition, to the concept of lack of management level another observation emerged, namely the differences between the agricultural products (milk, pigs, peas, and sugar). It seemed that each product had its own unique properties influencing the degree and type of integration. So, following the structured case approach we re-analyzed the data and presented it as four cases of information integration. These four cases are presented in the fourth section below. When re-analyzing, the cases, we developed several competing conceptual frameworks focusing on different aspects of the cases. The conceptual frameworks functions as a way of coding the empirical cases. The conceptual frameworks express the researchers understanding and interpretation of the data and the research themes (Carroll and Swatman, 2000). For example one conceptual framework was based on the degree of product sensitivity and a competing could be based on what type of data that was integrated between firms (see Figure 1 for an example of one conceptual framework). In addition, to re-analyzing and re-presenting the cases we also introduced a new theoretical perspective to stimulate analysis, namely alliance theories (e.g. Bronder & Pritzl, 1992; Kale, Dyer & Singh, 2002). This follows the suggestion made by Carroll and Swatman (2000) who argue that researchers should not stop at a confirmative point, but should strive for critical reflection. In order stimulate this process a second round of reflection to further explore the absence of an overall management level. Alliance theories provided us with another set of lenses in which yet another observation emerged. This was dissimilarities in time when efficiency gains and cost arises along the food industry.

This led us to our final step and last reflection, so far. Based on the first two phases of reflection two themes had emerged the absence of managerial level and the factor of asynchronous gains. For this we introduced a third set of theoretical lenses in order to explain these two concepts. The first is managerial theories and models and the other is resource dependency theory, both revisited in section two. The third analysis and reflection phase also involved the creation of different and competing conceptual frameworks.

*Figure 1. Outline of investigated organizations. End consumers not included in the study*

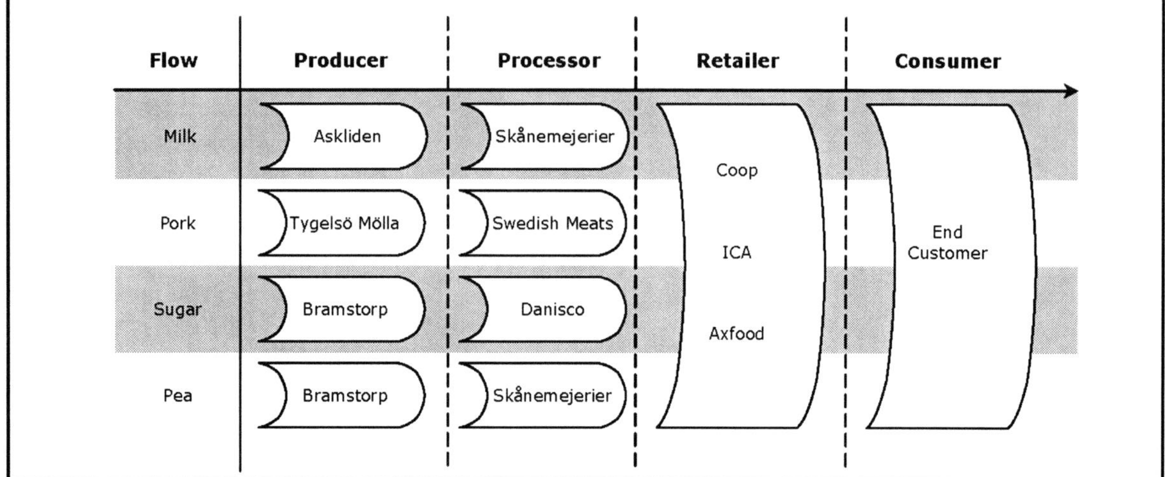


## FOUR INTEGRATION CASES

The four cases below have in common that the industry-wide supply chains all were initiated in the south of Sweden. Some of the physical goods ended up locally, but others were exported worldwide to the final consumer. The four cases all present different stories of the industry-wide information integration.

### The Swedish Food Industry

The number of end consumers is just above ten million consumers, whereof one million is Norwegians, who cross the border to Sweden to shop for food and alcohol. The Swedish market is dominated by three large retail chains, ICA, Coop, and Axfood, with a total market share of 72%. ICA and Axfood are privately owned, whereas Coop is a cooperative owned by the consumers. During the last few years two new low price retail chains (Lidl and Netto) have entered the market.

There are several major food producers and processors in the southern part of Sweden – Skåne, such as Procordia Food AB, Findus Sverige AB, Skånemejerier, and Pågen AB. Skåne is the most important agricultural area of Sweden with some 8700 farmers. The main food products from Skåne are different types of crops, dairy products, sugar beets, and meat.

In addition to the companies directly involved in the extended value chain, there are several other important actors in the food industry. These actors are not actively involved in the production, but have a potential to influence the end customers and their preferences, the products produced by the farmers or a general influence in the whole chain. Other actors are: KRAV (certifier of organically produced food), European Union (EU) and its Common Agricultural Policy (CAP), National Food Administration, Consumers in Sweden, and Agricultural Universities.

In the following sections, we present the empirical data from the nine cases along the four product flows. Figure 1 presents an outline of the companies' positions in the chains. The focus is on various types of IS integration among the actors. The specific information flow and ERP systems of the four products (milk, meat, sugar and peas) from the farmer, through the producer and to the retailer, will also be described.

## Milk Flow

Milk production at Askliden AB is supported by milk robots and automatic feeding machines, without almost any human interaction. The data collected by the milk robots (amount and quality etc) is linked trough an advanced IS to the Swedish Diary Association (SDA), who makes analysis of data and provides feedback, e.g. what to feed each individual cow. SDA and Askliden also diffuse data and information to other milk producers, such as quality of milk. In addition, Askliden uses a number of other IS to support their business. For instance, PC-Stall Journal to manage all their live stock and Genvägen which is used to pick out the best bull for each cow.

The entire production is sold to Skånemejerier. The price is based on quality (fat and taste). When the milk arrives from the farmers (about 900 dairy farmers) to the dairy it is stored in special transition halls where controls are made before the milk is pumped into storage silos. Thereafter it is processed. The milk is cooled down and the cream is separated from the milk. Both products are being pasteurized and mixed together again to meet specific percentages of fat. The milk is also homogenized before packing. To support these activities Skånemejerier use Movex (a large ERP-system from Lawson) to handle logistics, purchasing, resource management, financial assets, maintenance, supply chain management and data warehousing. A system called EDI/Link-XLM is used to manage the electronic information flow (order, invoicing, and payment) to and from farmers and customers. It is fully integrated with Movex. In the beginning of 2005 the EU passed a law that requires the possibility to track the origins of food products. To comply with this law, Skånemejerier implemented a system that could be used for extracting up-to-date packing data. When managing customers, Skånemejerier uses a CRM.

In an experimental project Skånemejerier implemented a system where the customer could use a code printed on one of their products, type the code into a internet site and then see a picture of the farm where the original milk was produced. The project was abandoned since it was not considered meeting a consumer demand at this stage and led to increased production costs in that the milk from different farmers could not be mixed during the processing. Since the milking of cows is automated and at an initial stage there is data linking all milk to a specific cow on a system, it would be possible to create a full integration connecting a single package of milk to a specific cow. However, currently the milk is blended on the farm and later among farms at the dairy. Additional costs are associated with not blending at these places in the process. The packaged milk is sold to the retail chains in southern part of Sweden. In total Skånemejerier have 1 million end costumers in Skåne.

The retailers collect a substantial amount of data through their sales terminal and customer loyalty card, but hardly not data is passed on to Skånemejerier and no data further than that. The retailers employ an automated inventory control system which communicates the supply need to Skånemejerier. End users have no automated information integration with any other actor than the retail chains. The data that is passed on is of transactional kind reporting on amounts sold and needed. No informational data on for example customers purchase pattern and behavior is passed on from the retailers. This description of information integration including retailers is also true for the three forthcoming flows and will not be repeated in the text.

## Pork Flow

Tygelsjö Mölla is specialized on pig breeding and has one costumer, namely Swedish Meats. The farmer makes a yearly agreement on production quotes. The current production quote is 4500 Pig-gham pigs which are delivered on regularly basis. The quality of the pig is based on percentage of fat in the meat. Low percentage of fat increases the value, because in makes slaughter easier. However, low fat percentage affects the taste in a negative way. In order to benchmark the individual farmer Swedish Meats provide the farmers with PIGWIN. Tygelsjö Mölla use PIGWIN to compare their own productivity with other breeders. They also use a web portal supplied by Swedish Meats with information such as the quality of the animals they have delivered, and how much Swedish Meats are willing to pay for these. In addition Tygelsjö Mölla informs Swedish Meats about how many animals they will deliver to the slaughter.

Swedish Meats is one of the largest slaughter houses in Sweden. It is also owned by the breeders. The information flow starts with the communication between the farmer and Swedish Meats. The farmer notifies Swedish Meats via the Internet, SMS or telephone, on how many and what kind of animal that he/she wants to deliver. Swedish Meats uses several different systems to collect data about the animals, for example their weight, age and origin. All of the information from these systems is sent to the ERP system. Swedish Meats uses approximately 4-5 systems when interacting with the farmers for handling payment, butchering notifications and so on. They also use a CRM system when collecting the information from the farmers which is used to keep track of all of the 17 000 breeders. Swedish Meats has decreased their client list from over 10 000 customers when almost every super market was their customer, down to a customer basis consisting of 3 grocery chains (ICA, COOP and Dagab) and 100 industrial customers. Even though the system handles the whole process from the farmer to delivery, no detailed information is passed on to the customer. It is possible to have a continuous information flow from the origin to the end customer, if requested.

## Pork Flow

Tygelsjö Mölla is specialized on pig breeding and has one costumer, namely Swedish Meats. The farmer makes a yearly agreement on production quotes. The current production quote is 4500 Pigham pigs which are delivered on regularly basis. The quality of the pig is based on percentage of fat

in the meat. Low percentage of fat increases the value, because in makes slaughter easier. However, low fat percentage affects the taste in a negative way. In order to benchmark the individual farmer Swedish Meats provide the farmers with PIGWIN. Tygelsjö Mölla use PIGWIN to compare their own productivity with other breeders. They also use a web portal supplied by Swedish Meats with information such as the quality of the animals they have delivered, and how much Swedish Meats are willing to pay for these. In addition Tygelsjö Mölla informs Swedish Meats about how many animals they will deliver to the slaughter.

Swedish Meats is one of the largest slaughter houses in Sweden. It is also owned by the breeders. The information flow starts with the communication between the farmer and Swedish Meats. The farmer notifies Swedish Meats via the Internet, SMS or telephone, on how many and what kind of animal that he/she wants to deliver. Swedish Meats uses several different systems to collect data about the animals, for example their weight, age and origin. All of the information from these systems is sent to the ERP system. Swedish Meats uses approximately 4-5 systems when interacting with the farmers for handling payment, butchering notifications and so on. They also use a CRM system when collecting the information from the farmers which is used to keep track of all of the 17 000 breeders. Swedish Meats has decreased their client list from over 10 000 customers when almost every super market was their customer, down to a customer basis consisting of 3 grocery chains (ICA, COOP and Dagab) and 100 industrial customers. Even though the system handles the whole process from the farmer to delivery, no detailed information is passed on to the customer. It is possible to have a continuous information flow from the origin to the end customer, if requested.

## Sugar Flow

To grow and harvest sugar beets, Bramstorp Gård uses a web portal (www.sockerbetor.nu) supplied by Danisco Sugar. All information exchange, for example invoices and dates for seed distribution, between the farmer and Danisco is done through the portal. The information flow is basically one-way, from Danisco to the farmer. Danisco, as the leading sugar producer in the region, uses SAP's ERP-system R/3 to cover the IT needs of the entire organization, internally as well as externally. As we are focusing our study on the information flow concerning sugar beets, we will not discuss the company's internal systems (for maintenance for example), but instead concentrate on the part of the ERP-system that handles external information exchange. The SAP modules used in the sugar beet information flow are: Agri, Sales & Distribution, and Logistics. The Sales & Distribution module (SD) is used to handle the information exchange between Danisco and their customers, while the Logistics module aids the transportation of the processed product (feed and sugar) stored at Örtofta. Agri is used to control the delivery of beets from farmers by creating delivery plans. The module is connected to the web portal www.sockerbetor.nu. Danisco aims to guide the farmer on how to best cultivate sugar beets by providing information, for example appropriate PH levels, protecting against erosion, balanced fertilization and numerous hints and tips on how to protect and salvage parasite infected crops and soil. After the sugar beets have been harvested and transported to Danisco's processing plants the sugar is extracted from the beet and mixed, thus breaking the information chain.

## Pea Flow

When we investigated Bramstorp, the pea farmer, we found that the cultivation of peas is not controlled by the farmer. The production of peas is a very controlled and regulated process which is dominated by Findus with a market share of 60%. The process has an 18-month time horizon, i.e. the foundation that is laid in March should produce a harvest in August the following year. To support this Findus has developed a concept called LISA (Low Input Sustainable Agriculture) which aims to structure the process and minimize the weaknesses. The base in LISA is the selection of fields for growing peas by analyzing the soil in different fields, picking the most suitable fields and monitoring the development of the crops while looking for signs of harmful organisms. The subsequent harvest and processing of the peas is also a highly controlled and automated process. It is Findus who controls the information gathering, and they more or less tells the farmer what and where to grow the peas.

Findus uses ERP-systems from both SAP (financials and administration) and Movex (logistics and production). They supply the farmers with information about which fields are suitable for pea cultivation, when to plant seeds, how much and what kind of fertilizing. This information is extracted from Findus's data-bases which are based on soil samples from the farmers' fields. This means that in many cases Findus knows more about a field than the farmer that owns it. In addition Findus even harvest the peas with their own machines. In the production at Findus's plant data about peas is gathered, such as quality and origin, so that Findus can provide feedback for the farmer. In the future even the end consumer can take part of this information, i.e. know from which field the peas have been grown. Today the information flow is broken when the peas are packaged for consumers. There is no integration between the Findus production system and the packaging system. Findus also collects a lot of information from the market and competitors, but this is not an issue for the pea information flow.

## FINDINGS AND DISCUSSION

The above described industry-wide supply chains shows that the fundamental idea behind this article of a lack of management and a disjoint economic responsibility are present in many processes that shapes the information integration. This is hardly surprising, but noteworthy since it's the postulates that the article relies upon. What is more interesting and not so self-evident is the possibility to discern consequences of these two particularities. In this section we make a cross-case comparison of the four flows and address which general lessons can be learnt regarding industry-wide information integration. We combine our findings with an integrated discussion on what these findings actually imply, and in accordance with the structured cases study approach we relate findings to theories that were not considered in the initial framework but can assist in further the understanding on the impact of our findings.

### Findings of Industry Wide Information Integration

The four flows studied were all part of the agri-food industry and as such have a general outline of different actors in common. All four flows start with a farmer, continue with food processors, grocery chains and finalize with end customers. The flows are to some degree simplified in order to emphasize the chain character of the industries and better illustrate findings.

With exception from the pea-farmers, the farmers do not receive any information from the other actors in the industry chains. The pea-case is special since the farmer only works as an executer of the instructions by Felix. Felix is in complete control of the production process.

The food processors - Felix, Findus, Swedish Meats, and Danisco – are the ones dominating the industry chains and dictates conditions for the farmers. They receive information required for their production (which to a large extent is due to legislation and regulation demands) from the farmers, but generally only receive transactional data from the retailers.

The grocery chains seem to collect substantial amounts of data that to a large extent is unused. It could be used by farmers and processors to achieve the benefits of forward integration and learn more about how their products are perceived and purchased by the end customer, but the retailers only pass on transaction data.

End customers do only receive limited information from the previous steps in the industry chain. In best case they get information about from which country the products come from and by brand name can identify the processor. For example, regarding the pork flow it would have been possible to trace information about where a pig is raised. Even such items as the diet of the pig and health condition during the life would be possible to transfer to the end customer.

### Consequences of Managerial Absence

A common theme in all four cases above is that the supply chains are dominated by one major actor. In the milk flow it was the dairy company, Skånemejerier, who had an almost monopolistic position in the southern part of Sweden. In the pig flow, Swedish meet is the dominating actor which holds true for also the rest of Sweden. Findus dominates the pea flow and Danisco the sugar flow. The theme has been noticed being typical for the agri-food industry (Salin, 1998). At the beginning of the flows there are a large number of farmers and suppliers, in the middle a small number of processors and in the end a large number of consumers. The industry-wide supply chain has the shape of a sandglass.

With respect to the information integration in our four cases, the few actors in the middle of the supply chain seems to have dominating roles. We will in the following refer to these dominant actors as supply chain captains. To the extent information integration has taken place, it has mainly been information integration that answers directly to their need.

In the milk flow it was proven that information integration was possible from the individual farmer to the consumer. However, Skånemejerier who is the captain of the milk flow did not see benefits enough to support the technique. From the consumer there could have been the improvements of traceability. Governments and control agencies responsible to track down the source of origin once there is an outbreak of any disease could also have benefited from information integration. For the individual farmer quality assurance would have been improved. Also, once there is an outbreak most often all farmers in a region or in a group of suppliers are affected until the origin has been identified. Apparently there are many actors that would have benefited from information integration, such integration that has proven possible, but since it's not directly affecting the supply chain captain it is not implemented.

Another consequence of the lack of common managerial level is the seemingly inappropriateness of existing methodologies for supply chain improvements. Business process reengineering (BPR) is used by most companies that want to improve their efficiency in their activities (Attaran, 2004). One study of Singaporean companies shows that about 87% of firms surveyed were either engaged in BPR projects, or indicating their intention to take up BPR projects in the next few years (Ranganathan & Dhaliwal, 2001). BPR was extensively used in the 1990's to primarily refocus organization on their customers' demands. In 2001 Hammer and Champy revived the topic in their book, "Reengineering the Corporation,". They argue that the next wave of BPR will be between a company and its customers, suppliers and partners. Cisco, Dell, and Intel are put forward as good examples. With our view of information integration in the food industry in mind, BPR for this type of reengineering becomes troublesome. BPR is strongly related to a top-down flow where the improvements are related to a holistic view. Top management addresses organizational problems for which solutions are redesigned. Industry-wide information integration is, unless completely dominated by value chain captains, by its nature to be seen as bottom-up processes.

Segars and Grover (1999) addresses differences between top-down and bottom-up processes for IS development. They find that the two processes are compatible with different process characteristics in order to be successful (Table 1). A top-down planning process is, roughly speaking, effective when associated with a planning process that is comprehensive, consistent, formal, have a high number of contributing members, and a control focus rather than a creativity focus. Bottom-up processes, Segars and Grover finds effective when associated with a creativity focus, informality, inconsistency, and have a narrow group of contributing members as well as a narrow scope, as opposed to a comprehensive scope. Without going to deep into the details of each characteristic BPR roughly corresponds to the top-down associated characteristics which may be argued explaining part of its success. However, for industry-wide integration which is bottom-up driven it is according to this analysis not appropriate. As far as the authors have been able to identify no methodology for information integration that incorporates the characteristics of a bottom-up process exists, which may hamper the development.

*Table 1. Association of IS planning characteristics for process efficiency (Based on Segars and Grover, 1999)*

| *Characteristic\ Flow* | **Top-down process** | **Bottom-up process** |
|---|---|---|
| *Scope* | Comprehensive | Incomprehensive |
| *Formality* | Formal | Informal |
| *Focus* | Control | Creativity |
| *Participation* | High | Low |
| *Consistency* | Consistent | Inconsistent |

## Consequences of Disjoint Economic Responsibility

One of the key characteristics of large scale IS implementations is that input and gains are not homogenously spread over the implementing organization. When implementing an enterprise-wide IS this means that the organization as a whole may make substantial efficiency gains but for the individual units or even employees there might be an increased workload that gives nothing in return – the factor of asynchronous efficiency gains. Thus, comparing information integration in internal supply with information in an industry-wide supply chain there is besides the lack of common management level described above, logically also the issue of asynchronous savings by integration efforts.

In the light of the disjoint economic responsibility once again we see the role of the value chain captains. The captains are using their dominant positions to enforce information integration that lays in their direct interest. The costs have to be taken by the smaller actor in the supply chain, while savings mainly are made within the realm of the captain. The four cases confirm the hypothesis expressed earlier in this article that backward integration is more prominent than forward integration. The question to ask is thus whether this is due to benefits of forward integration are fewer or just harder to proof in numbers.

Information is also used to ensure the availability of a critical resource. By making the farmers using LISA for their production, Findus soon knows much more about the soil, appropriate fertilizers, and harvest time than the individual farmer. Thus, a small advantage has been turned into a complete dependency. However, when the key resource, as the pea, is outside the organizational boarder of the chain captain an initiative such as the one by Felix is tempting. By securing one-way information integration Findus also secure a resource on which they are depending on, in that the farmer cannot switch customer or crop. Switching crop without knowing the exact status and history of the soil is very costly for the farmer since effective production is highly dependent on a fine tuned infusion of fertilization. A consequence of the disjoint economic responsibility is thus that the resource dependency mechanism becomes a factor for information integration - or rather a factor working against information integration.

In our four cases information integration has taken place either because the chain captains have enforced integration that answers directly to their needs or due to legislation and regulation. When governments put new demands on the industry, such as traceability of origin of meat, it answers as one common economic unit for its survival. But except from these rare occasions the organizations in the industry-wide supply chain that we studied did not recognize dependability of the success of another. Therefore the asynchronous cost and savings in information integration yet becomes an unsolved issue. One could only imagine what would trigger an increased collaboration and at least an alliance perspective on the chain, which would lead to increased possibilities for information integration.

## CONCLUSION

In this article we have addressed industry-wide information integration, as opposed to intra-organizational integration and two-part inter-organizational integration. We argue that the industry level is an additional viewpoint from which information integration needs to be studied since it differs from intra-organizational and two part inter-organizational integration in that no managerial function is available, which has been found essential for integration projects, and that the industry present an integration context of disjoint economic responsibility. We investigated the consequences of these two particularities in four agrifood industries: milk, sugar, pea, and pork. In these four cases we found the following consequences:

- Information integration was dominated by supply chain captains, dominant companies in the middle to the supply chain that controlled the integration agenda. Information integration that would have been desirable by other actors was hardly ever implemented.
- To the extent other actors information integration needs were met it was due to legislation or other governmental demand.
- Existing integration methodologies seems to be inappropriate for industry-wide information integration in that they requires a top-down flow which is contradictory to the chain that lacks managerial level.
- Asynchronous costs and savings due to perceived or real difference in backward and forward integration worked hampering on the industry-wide integration. As long as the complete chain is within one economic unit this is not an issue, but as companies exists at different places their benefits are not equal and if possible the unbeneficial contest integration.
- One way information integration was used to ensure availability of key resources and as lock on less powerful actors in the supply chain.

Our study shows that currently there are substantial amounts of unused information available in the companies' IS that if regarding the industry as a whole could lead to efficiency and effectiveness improvements. However, these improvements are hampered by the issues discussed above. What would then lead to the full potential of industry-wide information integration being unleashed? Hierarchy is the absent mechanism in industry-wide integration. The market mechanism is the force behind chain captain emergence and enforces integration that suits their needs. What is left to ensure the needs of the other actors in the chain is the standardization mechanism. Research on standardization as a mean for industry-wide integration is thus desirable.

For future research it is clear that theory building on information integration is still surprisingly sparse given the practical importance of the matter. We see that the concept of integration needs development in two directions. First, the concept itself is sparsely explored. What means such things degree of integration? Can units be more or less integrated? Is it possible to develop an integration measurement? These and many similar questions remain unsolved. In addition, the importance of the integration context needs more attention. We have here addressed information integration on an industry level. However, much still remains unexplained regarding this context's relation to information integration. The same is true for other contexts, such as the context of information integration being a part of a mergers or acquisitions, or in nonprofit organizations where the economic rationality plays a minor part. We recommend that future research both seeks to develop the concept of integration and understand how it is affected by different contexts.

A number of developments related to standardization as a means of information integration is currently taking place. One example is the work of the Supply-Chain Council, which is developing the Supply-Chain Operations Reference (SCOR) model. SCOR "lays out a top-level supply chain process in five key steps: plan, source, make, deliver, and return." (Davenport, 2005). The ERP-company SAP has begun to include SCOR flows and metrics in its supply chain software packages. Another approach is the The MIT Process Handbook Project (Malone, Crowston & Herman, 2003). The project involved collecting examples of how organizations perform similar processes. The process on-line repository includes knowledge about over 5000 business processes and activities as well as tools to edit and view this knowledge repository (http://ccs.mit.edu/ph/). The

Handbook has been used by, for example, Dow Corning Corporation in a major SAP implementation and project supply chain management project (Phios, 1999).

We have already mentioned that governmental intervention can trigger changes as governments' implement new laws or regulation. Governments can also intervene in other ways, for example by sponsoring the development of industry-wide integration solutions in order to increase the competitiveness of the national industry. In the EU we see the commission takes an active role in projects like Traceback and ITAIDE, to create solutions for traceability and electronic customs. By taking the costs centrally, the asynchronous dilemma can partly be overcome.

The increased activity of university-business partnerships in so called Mode-2 research (Gibbons et al., 1994; Nowotny, Scott & Gibbons, 2001) is also a way forward to creating solutions that are beneficial for industries rather than the specific company. For the companies, increased competition may lead to a refocus of industry-wide value chains into alliances simple because no actor can afford to be in a non-optimized supply chain. In Sweden, it has long been rumored that German retail-giants will enter the Scandinavian market in full scale and thus put additional pressure on the fairly isolated and protected market. We hope that we will have the ability to follow these changes and see if it will impact the industry-wide information integration in the agri-food industry.

## ACKNOWLEDGMENT

This work was in part supported by the DREAMS project via a grant from the Danish Agency of Science and Technology (grant number 2106-04-0007) and by the ITAIDE project, funded by the European Union under the 6th Framework Information Society Technology (IST) programme.

## REFERENCES

Adelberg, A. H. (1975). Management information systems and their implications. *Management Accounting*, *53*(9), 328–328.

Alaranta, M., & Henningsson, S. (2008). An approach to analyzing and planning post-merger IS integration: Insights from two field studies. *Information Systems Frontiers*, *10*(3), 307–319. doi:10.1007/s10796-008-9079-2

Alsene, E. (1999). The computer integration of the enterprise. *IEEE Transactions on Engineering Management*, *46*(1), 26–35. doi:10.1109/17.740033

Alter, S. (1999). A general, yet useful theory of Information Systems. *Communications of the AIS*, *1*(13).

Attaran, M. (2004). Exploring the relationship between information technology and business process reengineering. *Information & Management*, *41*(5), 585–596. doi:10.1016/S0378-7206(03)00098-3

Blumenthal, S. (1969). *Management Information Systems: A Framework for Planning and Development.* Englewood Cliffs, NJ: Prentice-Hall.

Bronder, C., & Pritzl, R. (1992). Developing strategic alliances: A successful framework for co-operation. *European Management Journal*, *10*(4), 412–420. doi:10.1016/0263-2373(92)90005-O

Brown, A. E., & Grant, G. G. (2005). Framing the frameworks: a review of IT governance research. *Communication of the AIS*, *15*, 696–712.

Browne, J., & Jiangang, Z. (1999). Extended and virtual enterprises - similarities and differences. *International Journal of Agile Management Systems*, *1*(1), 30–36. doi:10.1108/14654659910266691

Browne, J., Sockett, P. J., & Wortmann, J. C. (1995). Future manufacturing systems - Towards the extended enterprise. *Computers in Industry*, *25*, 235–254. doi:10.1016/0166-3615(94)00035-O

Brunsson, N., & Adler, N. (1989). *The Organization of Hypocrisy: Talk, Decisions, and Actions in Organizations*. New York: Wiley.

Brunsson, N., & Jacobsson, B. (2000). *A World of Standards*. Oxford: Oxford University Press.

Carroll, J. M., & Swatman, P. A. (2000). Structured-case: A methodological framework for building theory in information systems research. *European Journal of Information Systems*, *9*(4), 235–242. doi:10.1057/palgrave/ejis/3000374

Childerhouse, P., & Towill, D. R. (2006). Enabling seamless market-orientated supply chains. *International Journal of Logistics Systems and Management*, *2*(4).

Christopher, M. G., & Towill, D. R. (2002). An Integrated Model for the Design of Agile Supply Chains. *International Journal of Physical Distribution & Logistics*, *31*(4), 262–264.

Clemons, E. K., & Row, M. C. (1991). Sustaining IT Advantage: The Role of Structural Differences. *MIS Quarterly*, *15*(3), 275–293. doi:10.2307/249639

Davenport, T. H. (2005). The coming commoditization of processes. *Harvard Business Review*(June).

Eisenhardt, K. M. (1989). Building Theories from Case Study Research. *Academy of Management Review*, *14*(4), 532–550. doi:10.2307/258557

Evgeniou, T. (2002). Information Integration and Information Strategies for Adaptive Enterprises. *European Management Journal*, *20*(5), 486–494. doi:10.1016/S0263-2373(02)00092-0

Finney, S., & Corbett, M. (2007). ERP implementation: a compilation and analysis of critical success factors. *Business Process Management Journal*, *13*(3), 329–347. doi:10.1108/14637150710752272

Gibbons, M., Limoges, C., Nowotny, H., Schwartzman, S., Scott, P., & Trow, M. (1994). *The New Production of Knowledge: the Dynamics of Science and Research in Contemporary Societies*. London: Sage.

Gottschalk, P. (2000). Studies of key issues in IS management around the world. *International Journal of Information Management*, *20*(3), 169–180. doi:10.1016/S0268-4012(00)00003-7

Grimsley, M., & Meehan, A. (2007). e-Government information systems: Evaluation-led design for public value and client trust. *European Journal of Information Systems*, *16*(2), 134–148. doi:10.1057/palgrave.ejis.3000674

Hanseth, O., & Braa, K. (2001). Hunting for the treasure at the end of the rainbow. Standardizing corporate IT infrastructure. *Journal of Collaborative Computing*, *10*(3-4), 261–292. doi:10.1023/A:1012637309336

Hedman, J., & Kalling, T. (2003). The business model concept: theoretical underpinnings and empirical illustrations. *European Journal of Information Systems*, *12*, 49–59. doi:10.1057/palgrave.ejis.3000446

Henriksen, H. Z. (2006). Motivators for IOS adoption in Denmark. *Journal of Electronic Commerce in Organizations*, *4*(2), 25–39.

Jähn, H., Zimmermann, M., Fischer, M., & Käschel, J. (2006). Performance evaluation as an influence factor for the determination of profit shares of competence cells in non-hierarchical regional networks. *Robotics and Computer-integrated Manufacturing*, (22): 526–535. doi:10.1016/j.rcim.2005.11.011

Kale, P., Dyer, J. H., & Singh, H. (2002). Alliance capability, stock market response, and long-term alliance success: The role of the alliance function. *Strategic Management Journal*, (23): 747–767. doi:10.1002/smj.248

Kalling, T. (2003). ERP systems and the strategic management processes that lead to competitive advantage. *Information Resources Management Journal*, *16*(4), 46–67.

Karuppan, C. M., & Karuppan, M. (2008). Resilience of super users' mental models of enterprise-wide systems. *European Journal of Information Systems*, *17*(1), 29–46. doi:10.1057/palgrave.ejis.3000728

Kashyap, R. N. (1972). Management information systems for corporate planning and control. *Long Range Planning*, *5*(2), 25–31. doi:10.1016/0024-6301(72)90042-8

Konsynski, B. R. (1993). Strategic control in the extended enterprise. *IBM Systems Journal*, *30*(1), 111–142.

Krcmar, H., Bjørn-Andersen, N., & O'Callaghan, R. (Eds.). (1995). *EDI in Europe: How it Works in Practice*. Chichester: John Wiley & Sons.

Lam, W. (2005). Investigating success factors in enterprise application integration: a case-driven analysis. *European Journal of Information Systems*, *14*(2), 175–187. doi:10.1057/palgrave.ejis.3000530

Lidd, S. J. (1979). The Pressing Need for Management Information Systems. *Management Focus*, *26*(5), 44–44.

Lim, D., & Palvia Prashant, C. (2001). EDI in strategic supply chain: Impact on customer service. *International Journal of Information Management*, *21*(3), 193–211. doi:10.1016/S0268-4012(01)00010-X

Lummus, R. R., Vokurka, R. J., & Krumwiede, D. (2008). Supply chain integration and organizational success. *S.A.M. Advanced Management Journal*, *73*(1), 56–62.

Malone, T. W., Crowston, K., & Herman, G. A. (2003). *Organizing Business Knowledge: The MIT Process Handbook*: MIT Press.

Markus, M. L. (2000). Paradigm shifts - E-Business and business/systems integration. *Communication of the AIS*, *4*(10), 1–45.

Masetti, B., & Zmud, R. (1996). Measuring the extent of EDI usage in complex organizations: Strategies and illustrative examples. *MIS Quarterly*, *20*(3), 331–345. doi:10.2307/249659

Mata, F. J., Fuerst, W. L., & Barney, J. B. (1995). Information technology and sustained competitive advantage: A resource-based analysis. *MIS Quarterly*, *19*(4), 487–506. doi:10.2307/249630

Meyer, N. D. (2004). Systemic IS governance: An introduction. *Information Systems Management*, *21*(4), 23–34. doi:10.1201/1078/44705.21.4.20040901/84184.3

Neureuther, B. D., & Kenyon, G. N. (2008). The impact of information technologies on the US beef industry's supply chain. *International Journal of Information Systems and Supply Chain Management*, *1*(1), 48–65.

Nowotny, H., Scott, P., & Gibbons, M. (2001). *Re-thinking Science. Knowledge and the Public in the Age of Uncertainty*. Oxford: Polity Press.

Pant, S., Sethi, R., & Bhandari, M. (2003). Making sense of the e-supply chain landscape: An implementation framework. *International Journal of Information Management*, *23*(3), 201–221.

Pfeffer, J., & Salancik, G. R. (2003). *The External Control of Organizations: A Resource Dependence* Stanford: Stanford University Press.

Phios. (1999). *New Tools for Managing Business Processes*. Cambridge, MA: Phios Corporation.

Pyburn, P. J. (1983). Linking the MIS plan with corporate strategy: an exploratory study. *MIS Quarterly*, *7*(2), 1–14. doi:10.2307/248909

Ranganathan, C., & Dhaliwal, J. S. (2001). A survey of business process reengineering practices in Singapore. *Information & Management*, *39*(2), 125–134. doi:10.1016/S0378-7206(01)00087-8

Saeed, K. A., Malhotra, M. K., & Grover, V. (2005). Examining the impact of interorganizational systems on process efficiency and sourcing leverage in buyer-supplier dyads. *Decision Sciences*, *36*(3), 365–396. doi:10.1111/j.1540-5414.2005.00077.x

Salin, V. (1998). Information technology in Agri-Food supply chains. *International Food and Agribusiness Management Review*, *1*(3), 329–334. doi:10.1016/S1096-7508(99)80003-2

Segars, A. H., & Grover, V. (1999). Profiles of Strategic Information Systems Planning. *Information Systems Research*, *10*(3), 199–233. doi:10.1287/isre.10.3.199

Shang, S., & Seddon, P. B. (2002). Assessing and managing the benefits of enterprise systems: the business manager's perspective. *Information Systems Journal*, *12*(4), 271–299. doi:10.1046/j.1365-2575.2002.00132.x

Tiliquist, J., King, J. L., & Woo, C. (2002). Analyzing IT and organizational dependency. *MIS Quarterly*, *26*(2), 91–118. doi:10.2307/4132322

Van Grembergen, W. (2005). Introduction to the Minitrack "IT Governance and its Mechanisms". *System Sciences, 2005. HICSS '05. Proceedings of the 38th Annual Hawaii International Conference on* (pp. 235-235).

Webb, P., Pollard, C., & Ridley, G. (2006). Attempting to define IT governance: Wisdom or folly? *System Sciences, 2006. HICSS '06. Proceedings of the 39th Annual Hawaii International Conference on, 8*, 194-199.

Weill, P. (2004). Don't just lead govern: How top-performing firms govern IT. *MIS Quarterly Executive*, *3*(1), 1–17.

Weill, P., & Broadbent, M. (1998). *Leveraging the New Infrastructure* Boston: Harvard Business School Press.

White, A., Daniel, E. M., & Mohdzain, M. (2005). The role of emergent information technologies and systems in enabling supply chain agility. *International Journal of Information Management*, *25*(5), 396–410. doi:10.1016/j.ijinfomgt.2005.06.009

Wijnhoven, F., Spil, T., Stegwee, R., & Fa, R. T. A. (2006). Post-merger IT integration strategies: An IT alignment perspective. *The Journal of Strategic Information Systems*, *15*(1), 5–28. doi:10.1016/j.jsis.2005.07.002

Willcocks, L., Feeny, D., & Olson, N. (2006). Implementing core IS capabilities: Feeny-Willcocks IT governance and management framework revisited. *European Management Journal*, *24*(1), 28–37. doi:10.1016/j.emj.2005.12.005

*This work was previously published in International Journal of Information Systems and Supply Chain Management, Volume 3, Issue 1, edited by John Wang, pp. 1-20, copyright 2010 by IGI Publishing (an imprint of IGI Global)*

# Chapter 7
# The Strategic Role of Human Collaboration in Supply Chain Management

**Kenneth Saban**
*Duquesne University, USA*

**John Mawhinney**
*Duquesne University, USA*

## ABSTRACT

*Supply chain performance is often equated with acquiring the best technology or process. However, current studies suggest that supply chain performance also requires human collaboration. To change conventional thinking, this paper proposes a holistic approach to supply chain management (SCM), clarifies the forces that facilitate human collaboration, and identifies the steps management can take to create a more collaborative network.*

## INTRODUCTION

Advances in Information Technology (IT) coupled with the creation of the World Wide Web (WWW) and Internet have not only generated a more efficient means for manufacturers to interact with customers, but also for customers to access competitive products. This latter change has, in many ways, shifted the power from the seller to the buyer. As a result, manufacturers are transforming their supply chains to provide more value laden products and services. Supply chains represent the material and information interchanges stretching from acquisition of raw materials to delivery of finished products, and consists of vendors, service providers, and customers (Chopra and Meindl, 2001; McAdam & McCormick, 2001; Council of Supply Chain Management Professionals, 2005).

To help with this transformation, many manufacturers have adopted the latest best in class strategies like Customer Relationship Manage-

DOI: 10.4018/978-1-4666-0918-1.ch007

ment, Supplier Relationship Management, and Collaborative Planning Forecasting and Replenishment. While 48 percent of U.S. businesses have implemented these approaches and their associated tools, only nine percent are considering future updates, with the remaining 91 percent not sure how to proceed (Forrester, 2005). This finding suggests that adopting the latest supply chain program does not guarantee success. Success also requires the support of the people responsible for implementing such programs.

This paper will review the challenges associated with transforming supply chains, and discuss the factors that facilitate human collaboration as well as the managerial implications to achieve supply chain collaboration.

## SUPPLY CHAIN CHALLENGES

There are a number of challenges with transforming today's supply chain. Foremost is the trap of becoming more enamored with the technology than its implementation (Mills, Schmitz, and Frizelle, 2004). As one executive was quoted as saying "too often, (CIOs) hear about a new technology and think 'we have to have one of those' without stopping to think about whether or not this is true" (Gain, 2005). Information systems and technology are critical enablers of many best in class SCM practices. However, selecting enablers without a clear vision of business goals and understanding of the role that people play makes any decision a risky proposition.

Bowersox, Closs and Drayer (2005) contend that lasting supply chain performance can only be achieved when an organization develops an integrated approach to supply chain management. A good example is IBM. After deciding to transform the company into an adaptive organization that could profitably respond to customer needs, Sam Palmisano, CEO, developed a multi-dimensional supply chain transformation program shaped by five insights:

1. Cultural transformation access when leaders walk the talk.
2. CEO backing and trust are keys to sustained cross-unit integration.
3. Customer focus must permeate end-to-end supply chain processes.
4. Employees must be measured and rewarded for end-to-end efforts.
5. Technology deployment must be backed by sound IT governance.

IBM's strategic transformation netted: a savings of about $27 million per day; 38.6% more time for IBM's sales force to spend time with customers; helped reduce IBM services businesses' costs by more than $3 billion between 2002-2004; and improved the company's ability to sell integrated solutions made-up of IBM hardware, software and services (Gartner, 2005). This case underscores the importance of dealing with human behavior, the right technology and business processes as integrated and dependent factors.

Another challenge is that vendors and suppliers may not have the resources and expertise to deal with advanced supply chain management initiatives. This is especially true with small and medium enterprises (SMEs) who can constitute a large percentage of a supplier base (Chan, 2004; Emiliani & Stec, 2004). Having limited resources and expertise, SMEs tend to focus on internal operational issues - unless forced to comply (Morrell & Ezingeard, 2001; Estrin, Foreman, & Garcia, 2003; Harney, 2005). When pressured to comply, many SMEs simply walk away. Ohno (1988) found that only nine percent of SMEs saw reverse auctions as an opportunity. This behavior can be linked to a mindset which is "sell as much as possible, make payroll, and keep the lights on" (Davidow & Malone, 1992). Toyota dealt with these shortcomings by providing technical support and on-site training which increased profits for both the company and its suppliers.

A third challenge is not to overlook the strategic role that people play in implementing an advanced supply chain strategy (National Research Council, 2000; Handfield & Nichols, 2002; Russell & Hoag, 2004; Maku & Collins, 2005). There are specific costs with underestimating social and organizational issues. Ernst & Whinney (1987) found that companies that did not place a strong emphasis on training performed lower than companies who made employee development a priority. Michigan State University (Bowersox, Droge, Rogers, & Wardlow, 1989), Pennsylvania State University (Novack, Langly, & Rinehart, 1995), and Fawcett and Magnan (2005) also found that human resource development plays a major role in supply chain performance. After surveying 358 executives from leading manufacturers and service providers, Gowen and Tallon (2003) concluded that proper human resource management enhances the "value added" of supply chains by providing more effective resources in terms of better trained employees and enthusiastic employees and managers, which can equate to a more efficient and effective supply chain and one that is hard to emulate.

Social and organizational issues are especially important when dealing with virtual supply chain's which is defined as "groups of people who work interdependently with shared purpose across space, time, and organizational boundaries using technology to communicate and collaborate" (Lipnack & Stamps, 1997; Kirkman, Rosen, Gibson, Tesluk, & McPherson, 2002). Sabre, Inc., the inventor of electronic commerce for the travel industry, overcame this problem by building trust among their partners distributed around the globe (Kirkman et al., 2002). The company processes over 400 million travel bookings annually and is used by more than 60,000 travel agents in 114 countries. Building trust in virtual partnerships is often very difficult to achieve for two reasons. First, due to the unique nature of virtual enterprises, managers cannot rely on traditional trust-building methods based on social interaction, face-to-face meetings, and direct observations of fellow team member commitment. Second, while managers can generally dictate how employees interact and work in their own enterprise, directing transactions outside company boundaries is quite challenging – to say the least (Welty & Becerra-Fernandez, 2001).

This discussion has highlighted three human challenges with transforming today's supply chain; management being more enamored with technology than its implementation, the lack of expertise among vendors and suppliers to implement today's advanced supply chain management methods, and not overlooking the strategic role that people play. Each challenge underscores the various ways people (professionals and management) impact supply chain collaboration. It is therefore important to develop a better understanding of human collaboration

## HUMAN COLLABORATION

This discussion will define human collaboration, introduce relevant organizational theory that supports human collaboration, and discuss the particular factors that facilitate human collaboration.

One way to think about human collaboration is to consider a large philharmonic orchestra. If each orchestra member were to use a different sheet of music and play independently, the performance would be chaotic. However, when you employ the same sheet of music and add a conductor to organize individual play, real harmony is achieved (McClellan, 2003). Through harmonization, individuals/teams/organizations direct their respective talents in such a fashion that they achieve an outcome that can go beyond their own limited vision of what is possible (Gray, 1989). The same can be said about supply chains

"What is the best way to get all the members in a supply chain to collaborate?" To some, supply chain collaboration is achieved when information technology systems are in place and operational (Barrat, 2004; Sanders and Premus, 2005). While this approach works well for the transfer of explicit information, it does little to facilitate the transfer of tacit information which is critical to complex problem solving (Dyer, 2000; Handfield & Nichols, 2002; Davis & Spekman, 2004; Russell & Hoag, 2004; Cohen & Roussel, 2005). These topics will be discussed in greater detail under "communications". A more suitable approach is to employ a strategy that aligns people, technologies and processes in such a fashion as to be able to address the many nuances of each supply chain partnership (Bal & Teo, 2000, 2001, 2001a). We call this the "Human-Technology-Process Linkage" (Figure 1) whereby human collaboration facilitates technology adoption which in turn facilitates the supply chain processes in place.

This linkage is based upon contingency theory - which from a supply chain point of view – suggests that managers recognize change in the marketplaces and then reconfigure their supply chains to meet and/or exceed those challenges. For example, increased competition, shortened technology cycles, and heightened customer demands requires companies to collaborate more up and/or down the supply chain. This *contingent* response toward collaboration shifts the competitive focus from firms competing against each other to supply chains competing against supply chains.

For the purposes of this paper, we modified Fawcett, Magnan and McCarter's (2008) contingency framework (See Figure 2). This framework contains four components. The first component represents the forces that drive change. For example, changing customer demands, increased competition, shortened product life-cycles, etc. The second component represents the strategic management initiatives like building network collaboration. We used a gear because it to illustrate how human collaboration engages the advanced supply chain technology and processes in place. This graphic also illustrates the fact that human collaboration is contingent upon a number of factors. The third component, "Resisting Forces", represents the forces that drive people to resist communicating with others, adopting new technologies/processes, working toward common goals and strategies, etc. Managers must deal with these resisting forces by employing the right collaboration strategy. When successful, people communicate well; adopt the technology/ processes in place, and work towards the achievement of common goals and strategies. The forth component represents the desired collaborative performance outcomes such as improved innovation, abbreviated new product launches, and better problem solving. By addressing the "Resisting Forces", components one and two can generate the level of collaboration required to meet desired performance levels.

"What factors facilitate human collaboration?" Our review of the literature generated a large number of factors that were linked to supply chain

*Figure 1. Human-technology-process linkage©*

*Figure 2. Contingency framework to understanding supply chain collaboration*

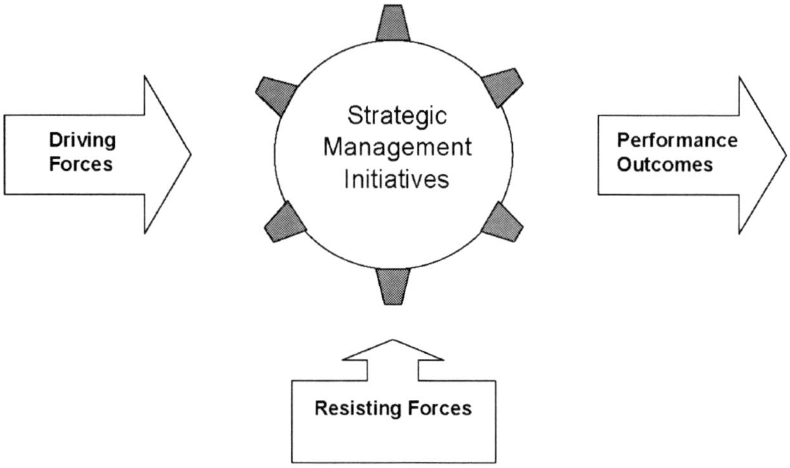

Driving Forces

Strategic Management Initiatives

Performance Outcomes

Resisting Forces

collaboration. To validate these factors, we developed a focus group of 10 regional supply chain executives (representing Fortune 500 corporations as well as consulting organizations). The attendees were asked to rate the relevance of each factor based on their day-to-day business experience. This event resulted in the identification of six factors that facilitate human collaboration.

## TRUST

Grenier and Metes, 1995; Kumar, 2000; Chopra and Meindl, 2001; Jonsson and Zineldin, 2003; Ryssel, Ritter and Gemunden, 2004 agree that building trust among supply chain partners is a critical starting point. However, trust is often the most difficult collaboration factor to achieve. The reason is managers often assume personal relationships will fall into place due to the implementation of today's collaborative technology. However, this is far from the truth. Building trusting relationships where each party has confidence in the other members' capabilities and actions requires more than the right technology and processes, it also requires building personal relationships. As one executive noted: "Supply chain management is one of the most emotional experiences I've ever witnessed." There have been so many methodologies that have developed over the years, people blaming other people for their problems, based on some incident that may or may not have occurred sometime in the past. Once you get everyone together into the same room, you begin to realize the number of false perceptions that exist. People are still very reluctant to let someone else make decisions within their area. It becomes especially tricky when you show people how 'sub-optimizing' their functional area can 'optimize' the entire supply chain." (Handfield & Nichols, 2002).

Trust also requires linking the belief systems of individuals and teams such that each is interested in the other partner's welfare and would not take any action without considering its impact on the other partner (Fukuyama, 1996). Consider this example: a supplier has been brought in by a manufacturer to provide the aluminum wings for an airplane. However, late in the design process, it is discovered that a composite wing will better suite the customer. In addition, the supplier does not have the ability to make composite wings, and has sunk considerable costs into the project. In a trust-based relationship, the supplier has to trust that the manufacturer will treat them fairly whether they decide to stay with the aluminum wing or go with the composite wing. The manufacturer must also trust that each supplier will optimize the enterprise, even when it reduces or eliminates its own role (Goranson, 1999).

## BELIEFS

Collaboration requires that partners from different organizations not only trust each other, but also believe that the whole is as important as the individual pieces – especially as it relates to meeting the needs of others. A good example is the Kariya Number 1 plant of Aisin Seiki - a major Japanese supplier of brake fluid-proportioning valves to Toyota. Within hours after the plant burned to the ground, Aisin engineers met with their counterparts at Toyota and their other tier one suppliers and passed along their blueprints for valves to any supplier that requested them and distributed whatever undamaged tools, raw materials, and work in process that could be saved. To everyone's surprise, Kyoritsu Sangyo – a small tier two supplier of welding electrodes – delivered 1,000 production-quality values within 85 hours after the accident (Evans & Wolf, 2005).

To be able to work in this fashion, individual/team/organizational beliefs must be aligned. Beliefs are the lenses through which workers see their purpose and responsibilities, set expectations for success and reward, and generally make sense of their place in this environment. Belief system alignment is especially critical when forging virtual supply chains (Grenier & Metes, 1995):

- *Beliefs about working with networked information* - selecting electronic information over paper information, being more comfortable working in distant locations over traditional office settings, proactively seeking information over reacting to information sent you, and sharing information over hoarding information.
- *Beliefs about work processes* - virtual processes are simultaneous, people need to think more about solving the problem now not working on it later. When they plan their work, people need to think about getting the right mix of people together over where do these people work.
- *Beliefs about teaming* - people need to trust other stakeholders they may have not worked with before. You are finished only when everyone's work is done and recognition and rewards are distributed for team performance over individual performance.
- *Beliefs about communications* - just because a meeting ends, communications do not end, even if I cannot see someone I can still communicate with them 24/7.
- *Beliefs about learning* - you learn as much from success as you do failure. Learning is not a one-time experience, rather a series of personal experiences from which to draw upon.

Once belief systems are aligned, people are more inclined to share information about one's self as well as process knowledge and domain specific knowledge; for example developing a working knowledge of given protocols to communicate electronically with other team members (Cleland, Bidinda, & Chung, 1995). Based on these observations we posit that beliefs and knowledge contribute to individuals, teams and organizations ability to collaborate.

## COMMUNICATIONS

While the essence of collaboration is the interaction between two parties, it is communication that links individuals and companies. Jonsson and Zineldin (2003) call communications the "factory" of human society, as it is people and not accounting systems, computer terminals or trading agreements that can communicate effectively with each other. While technology plays an increasingly important role in the transfer of information between organizations, it is still an enabler and not the driver of success (Cohen & Roussel, 2005).

Therefore, it is important to understand the types of information that are exchanged in any relationship. Most researchers divide information into two types. Explicit information i.e., facts and figures is easily codified and can be transmitted across computer networks. Tacit information i.e., know-how involves knowledge that is complex and difficult to codify. To facilitate both exchanges, Toyota formed its supplier association (*Kyohokia*) which focused on: 1) information exchange between suppliers and Toyota, 2) mutual training and development of suppliers, and 3) establishing socializing events. This has allowed Toyota to develop superior knowledge transfer capabilities which in turn allows all supply chain members to understand each others goals and coordinate their efforts to achieve common goals while maintaining a satisfactory working relationship (Dyer, 2000).

## CULTURE

By aligning trust, beliefs, knowledge and communications, organizations are able to establish a culture which supports decision-making and work. Schein (1985) defines organizational culture as "a pattern of basic assumptions – invented, discovered, or developed by a given group as it learns to cope with its problems of external adaptation and internal integration – that has worked well enough to be considered valid and, therefore, to be taught to new members as the correct way to perceive, think, and feel in relation to those problems." This definition suggests that culture resides both inside the company and with external partners. While problems with internal integration typically revolve around power/status, leadership and standards, external integration issues normally entail external environments [obtaining a shared understanding of key actors in the environment], mission [developing consensus on who are we, why are we here, and what are our strengths and weaknesses] and correction [developing consensus on how external issues should be solved] (Schultz, 1994). As noted earlier, IBM recognized that if they were going to regain their market dominance, the company had to change their culture that was out-of-sync with its customers (Forrester, 2005). IBM took a number of corrective actions:

- The company appointed some 400 top supply chain executives as "evangelists" to "walk the talk".
- The company communicated its supply chain strategy anchored around four imperatives [Drive focus-flexibility-quality-cost competitiveness, roll-out IBM's core strategic processes across the globe, extend supply chain principles to IBM's labor-based businesses, provide industry leading solutions integration and delivery capability].
- The company secured bottom-line support by changing success metrics.
- The company stayed the course no matter if the strategy changed.

Because each supply chain partnership has unique qualities, the possibility exists that management may be confronted with a wide-range of assumptions pertaining to the value of collaboration. By educating front-line managers as to the nuances of creating a collaborative culture, the company will increase the probability that individuals and teams will be more supportive of the networks mission and goals.

## REWARD SYSTEMS AND METRICS

To maintain a collaborative culture, it is critical that the proper reward systems and metrics be established as they reinforce certain types of behavior. It is therefore important that manufacturers pay considerable attention to the reward systems that are in place for both their employees as well as the employees of their suppliers (Harrington & Harrington, 1995). The reason for doing so is that these programs not only reward individuals and teams on how they improve corporate performance, but also recognize individuals and teams that deliver high customer value (Forrester, 2005). Therefore, when designing a supply chain reward system, manufacturers need to: a) focus on what is valued, b) demonstrate a clear connection between behavior, the rewards, and metrics, and c) recognize that changes in behavior patterns – such as the adoption of new supply chain technology and processes – may require different types of incentives.

The second requirement is to establish the right metrics. Cohen and Roussell (2005) argue that any effective metrics program must include:

1. Internally focused metrics that address such items as the cost of goods sold, labor rates, and asset turns.
2. Financial and non-financial metrics that address such items as production quality and flexibility, and forecast accuracy.
3. Customer metrics that address such items as on-time delivery, order-fulfillment, and fill rates.
4. Innovation and collaborative metrics that address such items as bringing innovative products to market in record time, developing break-through technology, and solving complex business issues.

Once the metrics are defined, it is important to periodically review one's reward systems and metrics to insure they support the current mission and objectives of the supply chain network.

## Synergy

To maintain ongoing collaboration, a number of synergistic activities must take place (Harrington, Hefner, & Cox, 1995). For example, making sure that the flow of information between and among partners is maintained. While team interactions within an office are easy to arrange and manage, interactions among virtual teams can be more complex. Therefore, it is important that team building exercises be included in virtual situations coupled with formal exercises like creating charters, mission statements, goals, and operating norms (Kirkman et al., 2002). Another activity is to encourage team members to accept and utilize the dynamics of the team, thus eliminating the traditional view that "I'm right and you're wrong." The Toyota Group's *kyson kyoei* (co-existence and co-prosperity) philosophy underscores the view that what is good for the extended enterprise is good for me, and what is good for me is good for the extended enterprise (Dyer, 2000). The next activity involves internalization. That is, team members must learn to tolerate ambiguity, modify their own views and belief systems, and be receptive to new ways to make group decisions (Goldman, Nagel, & Preiss, 1995). The last activity requires that the team implement their plans and strategies from which they will learn, grow and become more synergistic.

This discussion suggests that trust, beliefs, communications, culture, reward systems/ metrics, and synergy encourage people to collaborate and therefore become the elements or teeth of the human collaboration gear (see Figure 3). Once the elements are identified, each drives harmonious behavior which leads to great supply chain collaboration and improved performance.

*Figure 3. Human collaboration gear©*

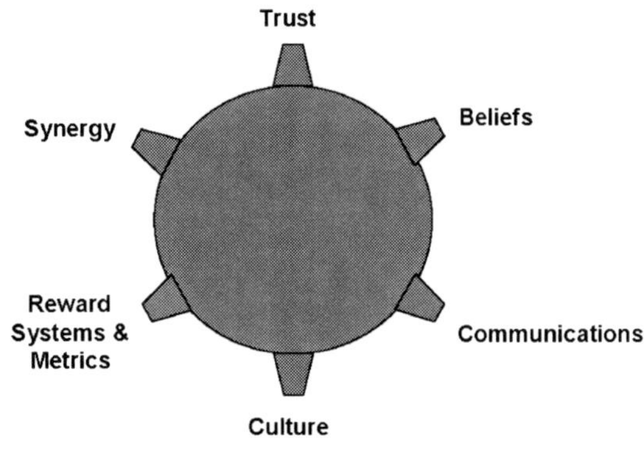


## MANAGERIAL IMPLICATIONS

While technology can be tested to ensure performance to specifications and processes can be mapped to confirm the expected results, human collaboration is more difficult to develop and maintain. There are four steps management can take to transform their conventional supply chains into more collaborative networks.

### First Step

The starting point is to recognize that to be competitive: a) the company will require an ongoing stream of innovations, b) the ability to create those innovations may reside outside corporate boundaries, and c) to access those capabilities, the company will have to become more collaborative with external partners. A study conducted by IBM (2006) of some 750 global CEOs shows that only about half of the operations were capable of collaboration within their own boundaries - let alone with external partners. Recognizing the criticality of collaboration is a starting point on which to build a collaborative organization. After conducting 51 in-depth interviews across various supply chain positions, Fawcett, Magnan and McCarter (2008) concluded that managers must recognize the importance of collaboration if they are going to be able to compete in the 21st century. They also reported that performance improvement hinged on a company's ability to improve collaboration with its supply chain partners.

Therefore, the first step is for the leadership of the organization to make a specific decision to focus on collaboration and supply chain integration as a strategy for competitive success. This will often require a culture that is able to change through the introduction of new processes and technology. With this commitment in place the following three steps will provide a roadmap to success.

### Second Step

After recognizing the importance of supply chain collaboration, supply chain managers must then alter their views about achieving it. That is, management needs to realize that all supply chain collaborations are not equal because the products and services purchased can vary significantly in terms of value and criticality. For example, the relationship with a supplier of jet engines will be viewed differently than one with a supplier of cleaning products. As jet engines are more complex to make, have a higher value, and more critical to the success of operating an airplane than cleaning products, a higher degree of collaboration will be required with the jet engine manufacturer. It is important to employ a framework to understand and be able to deal with the different types of collaborative arrangements. After studying numerous supply leaders, Cohen, and Rousel (2005) concluded that each recognize that every supply chain relationship required different attention. To which, they suggest that collaborative arrangements can be grouped into one of four categories:

1. Transactional collaborations occur when low risk and low value purchases are made. These partnerships require little collaboration as the purchases are normally straight or modified rebuys.

2. Cooperative collaborations demand a higher level of information exchange because the products or services are more critical to the manufacturer. As the level of information increases so to does the level of collaboration.
3. Coordinated collaborations require that manufacturers and suppliers work much closer due to the higher criticality of the products. Due to the functionality of the products, more collaboration is required between the manufacturer and the supplier.
4. Synchronized collaborations occur when manufacturers and suppliers invent and/or co-engineer a new product. As these tasks are very complex and timely, extensive collaboration is required between the partners.

By employing this framework, management can determine the level of collaboration required with each supply chain partner and then devise a plan to achieve the desired collaboration through the proper alignment of people, processes and technology (Fung, Fung, & Wind, 2007). This is an important step as it reduces the many problems associated the "one size fits all" approach to supply chain management (Stonebraker & Afifi, 2004).

Given the differentiation of suppliers based on complexity, risk, and cost, the foundation is set to develop the specific goals and plans expected from a collaborative approach. It is important to recognize that all suppliers identified in target categories can be included in a collaborative initiative at one time. Therefore, it is important to develop a collaborative plan.

## Third Step

The next step is to develop a collaborative plan that aligns the network's mission, goals, collaborative strategies, and structure. Mission and goals not only drive investments, but also social networks. Therefore, before a collaborative network can be created, the leadership must insure that its mission and goals are sound. Garrett Engine Boosting Systems Inc, a supplier to the automotive industry, set-out to improve both the quality of their products and reduce costs through improved supplier collaboration and integration. Garrett's leadership team established a goal to synchronize planning with their suppliers to meet the rapid changes in the automotive industry. This was achieved by aligning people with collaborative supply chain technology, processes, and metrics. By providing the technology and process to allow people to better communicate, develop stronger relationships, and work together creatively, Garrett achieved a reduction in change notice cycle time of 50 percent and the quality of finished product improved 600 percent (Rockstroh, 2002). This case underscores the importance of incorporating the human collaboration in one's mission and goals.

Collaborative strategies should not only deal with technology and process implementation, but also human integration. The employees hired by the company – as well as its partners – need to have a strong degree of trust, an open belief system, good communication skills, and work in a collaborative environment (Andraski, 1998; Rutherford, 2000; Collins, 2001; Nelson, Moody, & Stegner, 2001; Burt, Dobler & Starling, 2003; Gowen & Tallon, 2003; Koulikoff-Souviron & Vida, 2005).

After interviewing supply chain executives from six medium/large size firms in the information/electronic industry, Hsu (2005) came to the same conclusion. That is, supply chain collaboration is contingent upon a number of such as mission and goals, leadership, reward systems, culture, trust, and beliefs.

An appropriate recommendation then would be to assess how well each critical partner that is targeted for collaborative integration in your supply chain is contributing to the overall mission of the network. Those not providing the level of support and performance required to meet the network goals would be candidates for collaboration review. This first requires establishing minimum requirements for the six collaboration factors described earlier. With this baseline in place an assessment of the current status of the non-conforming critical partner's collaboration factors can be conducted. By then benchmarking the collaborative practices of supply chain leaders in the deficient areas, plans for change and improvement can be established. This gap analysis will not only demonstrate the importance of human collaboration, but also pinpoint areas of strength and weakness.

With the collaborative goals and strategies in place including areas of weakness to be improved, the next step is to establish specific metrics to monitor improvement and goal achievement.

## Fourth Step

Given the proper people and/or partners are "on the bus," it is also important to develop the right controls to ensure that the company's mission and goals can be met. Controls normally take two forms: reward systems and metrics. Reward systems represent the formal and informal mechanisms by which employee performance is defined, evaluated, and rewarded. Reward systems are important as they affect attitudes, behaviors and motivation which influence technology and process adoption that eventually impacts the execution of Key Performance Indicators (KPIs). Metrics which are internally, financial/nonfinancial, customer, and innovation/collaboration focused provide a way to monitor progress toward a particular outcome or set of goals (Manetti, 2001). Benchmarking supply chain leaders is a good place to start to insure that you have the right metrics in place (Leenders, Johnson, Flynn, & Fearon, 2006). Therefore, it is incumbent that reward systems and metrics be aligned.

After surveying supply chain professionals across 11 different industries, Lockamy and McCormack (2004) found that delivery process measures – on inter-relationships and supply chain processes - have a significant impact on supply chain performance. These measurements are best used to reward and recognize the process participants.

In short, to make your supply chain more collaborative, management must not only employ the right technology and processes, but also manage the human resources distribute across the network. This is achieved by embracing the strategic role of human collaboration, altering conventional thinking about the role of people, developing plans geared to leveraging one's human resources, and establishing controls that facilitate and not deter supply chain collaboration.

## SUMMARY

Supply chain performance is often equated with acquiring the best technology or process. However, current studies suggest that supply chain performance also requires human collaboration. To change conventional thinking, this paper has: a) proposed a holistic approach to supply chain management – one that links human, technology and processes and b) identified six factors (trust, beliefs, communications, culture, reward systems/ metrics, and synergy) that facilitate human collaboration, and c) identified the four steps management can take to create a more collaborative network.

While the literature on supply chain management is quite robust, less attention has been given to the impact of human collaboration. This void has left many CEOs questioning how they are going to make their supply chain networks more collaborative given the extreme pressure for innovative products and services. We believe that the holistic approach outlined herein, will allow manufacturers to better leverage their technology and process investments by producing the type of collaboration necessary to meet if not exceed customer expectations.

As a next step, the authors plan on conducting a comprehensive field study to validate the impact of the six collaborative factors, and by doing so validate not only validate the factors that drive human collaboration, but also the strategic role of human collaboration in supply chain management.

## REFERENCES


Andraski, J. C. (1998). Leadership and the realization of supply chain collaboration. *Journal of Business Logistics*, *19*(2), 9.

Bal, J., & Teo, P. K. (2000). Implementing virtual team working: Part 1 - a literature review of best practice. *Logistics Information Management*, *13*(6), 346. doi:10.1108/09576050010355644

Bal, J., & Teo, P. K. (2001). Implementing virtual team working: Part 2 - a literature review. *Logistics Information Management*, *14*(3), 208. doi:10.1108/09576050110390248

Bal, J., & Teo, P. K. (2001a). Implementing virtual team working: Part 3 – a methodology for introducing virtual team working. *Logistics Information Management*, *14*(4), 276. doi:10.1108/EUM0000000005722

Barrat, M. (2004). Understanding the meaning of collaboration in the supply chain. *Supply Chain Management*, *9*(1), 30. doi:10.1108/13598540410517566

Bowersox, D. J., Closs, D.J., & Drayer, R.W. (2005, January 1). The digital transformation: Technology and beyond. *Supply Chain Management Review*.

Bowersox, D. J., Daugherty, P. J., Droge, C. L., Rogers, D. S., & Wardlow, D. L. (1989). Leading-edge logistics: Competitive positioning for the 90s. *Council of Logistics Management*, Oak Brook, IL.

Burt, D. N., Dobler, D. W., & Starling, S. L. (2003). *World class supply management: The key to supply chain management* (7th ed.) Boston: McGraw-Hill Irwin.

Chan, H. (2004, February). The supply-chain squeeze. *Optimize*.

Chopra, S., & Meindl, P. (2001). *Supply chain management: Strategy, planning and operation*. Upper Saddle River: Prentice Hall.

Cleland, D. I., Bidanda, B., & Chung, C. A. (1995). Human issues in technology integration – Part 1. *Industrial Management (Des Plaines)*, *37*(4), 22.

Cohen, S., & Roussel, J. (2005). *Strategic supply chain management*. New York: McGraw-Hill.

Collins, J. (2001). *Good to great*. New York: Harper Business.

Council of Supply Chain Management Professionals. (2005). *Glossary of supply chain and logistics terms and glossary*. http://www.cscmp.org, (February 2005).

Davidow, W. H., & Malone, M. S. (1992). *The virtual corporation*. HarperBusiness.

Davis, E. W., & Spekman, R. E. (2004). *The extended enterprise*. Upper Saddle River: Prentice Hall.

Dyer, J. H. (2000). *Collaborative advantage*. Oxford University Press.

Emiliani, M. L., & Stec, D. J. (2004). Aerospace parts suppliers' reaction to online reverse auctions. *Supply Chain Management*, *9*(2), 139. doi:10.1108/13598540410527042

Ernst & Whinney (1987). *Corporate profitability & logistics*. Oak Brook: Council of Logistics Management.

Estrin, L., Foreman, J. T., & Garcia, S. (2003). *Overcoming barriers to technology adoption in small manufacturing enterprises (SMEs).* White paper Carnegie Mellon University, Software Engineering Institute (June).

Evans, P., & Wolf, B. (2005, July-August). Collaboration rules. *Harvard Business Review, 83*(7/8), 96.

Fawcett, S. E., & Magnan, G. M. (2005). *Achieving world-class supply chain alignment: Benefits, barriers, and bridges.* White paper Center for Advanced Purchasing Studies, *http://.*www.capsresearch.org.

Fawcett, S. E., Magnan, G. M., & McCarter, M. W. (2008). A three-stage implementation model for supply chain collaboration. *Journal of Business Logistics, 29*(1), 93.

Forrester (2005, September 19). *APAC study chain apps spending outlook.* http://www.forrester.com.

Fukuyama, F. (1996). *Trust: The social virtues and the creation of prosperity.* New York: Free Press.

Fung, V. K., Fung, W. K., & Wind, Y. (2007). *Competing in a flat world.* Upper Saddle River: Wharton School Publishing.

Gain, S. (2005, September 19). *Perfect projects.* ITP Technology. http://.www.itp.net.

Gartner (2005, March 24). *IBM transforms its supply chain to drive growth.* http://www.gartner.com.

Goldman, S. L., Nagel, R. N., & Preiss, K. (1995). *Agile competitors and virtual organizations.* New York: Van Nostrand Reinhold.

Goranson, H. T. (1999). *The agile virtual enterprise.* Westport: Quorum Books.

Gowen, C. R., & Tallon, W. J. (2003). Enhancing supply chain practices through human resource management. *Journal of Management Development, 22*(1-2), 32.

Grenier, R., & Metes, G. (1995). *Going virtual.* Upper Saddle River: Prentice Hall.

Grey, B. (1989). *Negotiations: Arenas for reconstructing meaning.* Unpublished working paper, Pennsylvania State University, Center for Research in Conflict and Negotiation, University Park, PA.

Hanfield, R. B., & Nichols, E. L., Jr. (2002). *Supply chain redesign.* Upper Saddle River: Prentice Hall.

Harney, J. (2005). Enterprise content management for SMBs. *AIIM E-Doc Magazine, 19*(3), 59.

Harrington, H. J., & Harrington, J. S. (1995). *Total quality management.* New York: McGraw-Hill, Inc.

Harrington, H. J., Hefner, M. B., & Cox, C. K. (1995). Environmental change plans: Best practices for improvement planning and implementation. *Harrington and Harrington's Total Improvement Management.* New York: McGraw-Hill, Inc.

Hsu, L. (2005). SCM system effects on performance for interaction between suppliers and buyers. *Industrial Management (Des Plaines), 105*(7), 857.

IBM (2006). *Expanding the innovation horizon.* Somers: IBM Global Services.

Jonsson, P., & Zineldin, M. (2003). Achieving high satisfaction in supplier-dealer working relationships. *Supply Chain Management, 8*(3-4), 224.

Kirkman, B.L., Rosen, B., Gibson, C.B., & Tesluk, P.E., & McPherson, Simon O. (2002). Five challenges to virtual team success: Lessons from Sabre, Inc. *The Academy of Management Executive, 16*(3), 67.

Koulikoff-Souviron, M., & Pascal, V. (2005). A dose of collaboration. *European Business Forum, 22*, 59.

Kumar, N. (2000). The power of trust in manufacturer-retailer relationships. *Harvard Business Review's Managing The Value Chain.* Boston: HBS Publishing.

Leenders, M. R., Fraser, J. P., Flynn, A. E., & Fearson, H. E. (2006). *Purchasing and supply management.* Boston: McGraw-Hill Irwin.

Lipnack, J., & Stamps, J. (1997). *Virtual teams: reaching across space, time, and organizations with technology.* New York: John Wiley and Sons.

Lochamy, A. III, & McCormack, K. (2004). Linking SCOR planning practices to supply chain performance. *International Journal of Operations & Production Management, 24*(11/12), 1192.

Maku, T. C., & Collins, T. R. (2005). The impact of human interaction on supply chain management practices. *Performance Improvement, 44*(7), 26. doi:10.1002/pfi.4140440708

Manetti, J. (2001). How technology is transforming manufacturing. *Production and Inventory Management Journal, 1st Quarter.*

McAdam, R., & McCormack, D. (2001). Integrating business processes for global alignment and supply chain management. *Business Process Management, 7*(2), 113. doi:10.1108/14637150110389696

McClellan, M. (2003). *Collaborative manufacturing.* New York: St. Lucie Press.

Mills, J., Schmitz, J., & Frivol, G. (2004). A strategic view of supply networks. *International Journal of Operations & Production Management, 24*(9/10), 1012. doi:10.1108/01443570410558058

Morrell, M., & Ezinheard, J. (2001). Revisiting adoption factors of inter-organizational information systems in SMEs. *Logistics Information Management, 15*(1-2), 46.

National Research Council. (2000). *Surviving supply chain integration.* Washington: National Academy Press.

Nelson, D., Moody, P. E., & Stegner, J. (2001). *The purchasing machine*. New York: The Free Press.

Novack, R.A., Langly, Jr., C.J., & Rinehart, L.M. (1995). *Creating logistics value: themes for the future.* Oak Brook: Council on Logistics Management.

Ohno, T. (1988). *Toyota production system.* Portland: Productivity Press.

Rockstroh, J. (2002). Achieving quality ROI across the supply chain. *Quality, 41*(6), 54.

Rogers, S. (2004). *Supply chain management: Six elements of superior design.* http://www.manufacturing.net.

Russell, D. M., & Hoag, A. M. (2004). People and information technology in the supply chain: Social and organizational influences on adoption. *International Journal of Physical Distribution & Logistics Management, 34*(1-2), 102. doi:10.1108/09600030410526914

Rutherford, T. D. (2000). Re-embedding, Japanese investment and the restructuring buyer-supplier relations in the Canadian automotive components industry during the 1990s. *Regional Studies, 34*(8), 739. doi:10.1080/00343400050192838

Ryssel, R., Ritter, T., & Gemunden, H. G. (2004). The impact of information technology deployment on trust, commitment and value creation in business relationships. *Journal of Business and Industrial Marketing, 19*(3), 197. doi:10.1108/08858620410531333

Sanders, N. R., & Premus, R. (2005). Modeling the relationship between firm IT capability, collaboration, and performance. *Journal of Business Logistics, 26*(1), 1.

Schein, E. H. (1985). *Organizational culture and leadership*. San Francisco: Jossey-Baas Publishers.

Schultz, M. (1994). *On studying organizational culture*. New York: Walter de Gruyter.

Simatupang, T. M., & Sridharan, R. (2004). Benchmarking supply chain collaboration: An empirical study. *Benchmarking*, *11*(5), 484. doi:10.1108/14635770410557717

Stonebraker, P. W., & Afifi, R. (2004). Toward a contingency theory of supply chains. *Management Decision*, *42*(9), 1131. doi:10.1108/00251740410565163

Welty, B., & Becerra-Frenandez, I. (2001). Managing trust and commitment in collaborative supply chain relationships. *Communications of the ACM*, *44*(6), 67. doi:10.1145/376134.376170

*This work was previously published in International Journal of Information Systems and Supply Chain Management, Volume 3, Issue 1, edited by John Wang, pp. 43-57, copyright 2010 by IGI Publishing (an imprint of IGI Global)*

# Chapter 8
# Supply Chain Dispute Resolution:
## A Delphi Study

**Frank Wolf**
*Nova Southeastern University, USA*

**Lee Pickler**
*Baldwin Wallace College, USA*

## ABSTRACT

*This paper examines how supply chain conflicts across domestic and international jurisdictions arise and become resolved given that conventional conflict resolution tribunals cannot effectively settle fast enough to serve the needs of supply chain partners. Observations from the field should guide practitioners, and in combination with information technology, may lead to best practice rules in dispute resolution. For this study, the Delphi Method was selected, in which a panel of 14 experts participated in three rounds of successive surveys over a one-year period. Survey data was collected by mail as well as via telephone conversations and interviews, while under the Delphi method, the content of the second questionnaire was derived from the responses of the first questionnaire. All participants were supply chain experts in the United States from eight different industrial sectors, and none of the participants interacted with one another. End results show that supply chain's relationships are very private trade arrangements and that disputes arise, predictably, from common performance criteria such as quality, timely delivery and payment issues.*

## INTRODUCTION

For over a decade supply chain management has emerged as the concept for squeezing greater efficiencies out of a delivery system for products and services. The evolving body of supply chain management knowledge is akin to the biological model of business, in which one seeks to survive and prosper through cooperation. The goal of this study is to add to that body of knowledge by discovering dispute resolution practices in a cross section of industrial US sectors, and to encourage future study in dispute prevention. In particular, we aim to understand the degree of central control over disputes along supply chain tiers, the nature of disputes, and supply chain design changes in

DOI: 10.4018/978-1-4666-0918-1.ch008

times of economic contraction. Certain terms used in the context of supply chain studies are listed in Appendix II at the end of the paper.

Frequently mentioned in the literature are the mechanics of achieving coordination, and most discussions focus on the Internet, enterprise software and the integration of disparate systems among supply chain partners. A void in the discussion, however, is how to deal with conflict that is bound to arise across global supply chains relative to cultural differences, jurisdiction, and economic stress etc., without destroying the trust among supply chain partners. Supply chains can and do fail. An example is reported by Lunsford (2007) in a case for the Boeing 787 Dream Liner aircraft. Management's main concern at the time was the making of large aircraft sections out of carbon fiber plastics. What they did not count on were shortages of titanium bolts to keep the carbon fiber plastic sections together. The fasteners capacity shrank due to consolidations in the fastener industry. Inasmuch as Boeing had contingency plans, this is actually a good example of supply chain management. Dispute resolution processes themselves can fail also, which was the case in an international arbitration between Turkish and US business partners, when a judge disallowed the panel's finding because of an undisclosed conflict of interest by the arbitration panel's chairman, Tait (2006).

## LITERATURE REVIEW

Given the recent phenomenon of Supply Chain Management as a field of study, a considerable body of literature has accumulated on the subject. However, very little has been written about supply chain disputes, and even less about methods of supply chain dispute resolution. On that basis the literature review is focused mainly on the period after the year 2000. Dispute settlements can occur formally in a court of competent jurisdiction, or less formally in another forum such as the American Arbitration Association, AAA (2001), or informally through negotiation. There may be several reasons for this gap in the literature, one being the novelty of the subject, another, is conflict experience is not what supply chain managers wish to publicize. In the current literature the terms "conflict" and "risk" are often used interchangeably. One such reference is Chopra et al. (2004), in which he discusses risks in terms of disruptions, delays, systems, forecasts, intellectual property, procurement, receivables, inventory, and capacity risks. For each of these, there is a comprehensive discussion of remedies and risk avoidance without touching on the subject of dispute resolution conventions. Given the link between risk and dispute, risk avoidance is of course preferable to dispute resolution. The drivers for supply chain disruptions, for example, are natural disasters, labor strikes, supplier bankruptcies, war and terrorism, and a single source dependency. A mitigating strategy for disruptions in supply chains consists of carrying larger inventories and developing supplier redundancies. For each of the other risk categories, mitigating strategies are offered by the authors. Lee (2008) addresses one risk that Chopra (2004) does not address, namely the risk of having too many supply chain partners, which issue arose among this Delphi Study's respondents. The author approaches the subject by suggesting a model to compute the optimum number of supply chain partners by a mean-value approach with deterministic failure cost. Denning (2008) approaches the subject of shortening supply chains from the practical side, observing that bottlenecks in ports, roads, and rail networks have increased cost greatly.

Li, Du, and Wong (2005) define supply chains in terms of closer integration among partners through planning, design, development and services through information integration. The higher the integration, the lower is the risk of a bullwhip effect. Access control conflicts occur when two people work on a shared project; one has access to data while the other has not. Access control policies, by which conflict can be resolved, include terms of access, interference, data- and semantic integrities, accountability and auditing. The authors define schema conflict as differing access control policies within and between organizations. Propagation conflict is when organization "A" allows access of its data bases to company "B", when that data base contains data about company "C". Close supply chain integration can also be achieved by a revenue contract among the partners, described by Giannoccaro (2004). Such contracts are appropriate in supply chains with centralized management and come in two parts; one is the purchase price of the object paid by the retailer to the wholesaler for example, and the other part is a percentage of sales volume the retailer generates, which is paid to the wholesaler. Such a contractual mechanism forces the parties to coordinate well and thereby deflects from potential conflicts. Runge (2008) achieves supply chain harmony through a conflict competent team of managers who will have to undergo training in negotiations and sensitivities. Given that training experience, the incidence of conflicts may be reduced.

Vanany et al. (2009) present a comprehensive literature review on supply chain risk to guide future research directions. They find that the higher the risk, the higher the probability of all manner of conflict. Since supply risk is not the same across all industrial sectors, this work is a good industry-specific literature reference. Here the supply chain risk literature is also categorized by type of relationship, such as dyadic relations, chains and networks.

Patterson (1996) offers a theoretical paper on inter-organizational conflict among otherwise cooperative partners. The author is focused on managing the process by breaking conflict down into several levels with feedback. These levels are latent, perceived, or manifest conflicts. Latent conflict is the condition where the parties only dimly perceive a conflict. In a perceived conflict the parties understand a conflicts' existence, while manifest means open conflict under which condition the parties try to block each other. Possible outcomes are: clear winners or relationship change. He views such conflicts as both unavoidable and not dysfunctional. Dysfunctional power imbalances such as interdependences and incompatible goals are the type of conflicts that wind up in litigation. A theoretical paper by Blackhurst (2006) solves the problem of dispute detection prior to its occurrence by a Petri Net method. There is clearly a great management value in potential conflict detection early on.

Patel (2006) does not address conflict resolution methods within supply chains, but defines the attributes of good contract management by reducing the likelihood of conflict from occurring in the first place. He addresses best practices composed of enforcing standard processes, storing contract language online, making terms and conditions retrievable to all participants, monitoring clauses, and compliance with Sarbanes-Oxley (SOX) legislation. These practices apply to good supply chain management in general, as without them a supply chain relationship is not sustainable. Distinctly related to good contract management, is the work of Gulati (2007), who advocates that consideration for contract termination be given early in the process. The authors stress the need to define an exit process early in the relationship, which may actually enhance the partnership in the long term. A good disengagement plan invariably will have a well defined trigger point- or event. When and if such an event occurs, such as insolvency or inability to meet a milestone, the rights of the partners in terms of assets and products and process flow must be defined in a service level agreement. Difficulties can arise in jointly owned operations, or multiple product lines from jointly owned operations. Rights always come with obligations relative to employees and customers which must be considered and spelled out. The termination clause may even cover buy-out options, first rights of refusal and certainly has to lay out a disengagement process. The process should be managed by a team of managers who were not involved in the formation of the alliance. The process must have an end-date.

Wang (2008) analyzes international sub-contracting in the context of the relationships, ranging from arms-length to fully integrated multinational enterprises. The relationships are categorized by (1) lose sub-contracts,(2) unequal economic partners in which either party (a) or party (b) can have the upper hand, and (3) that of mutual dependence. Situations, in which there is a mutual dependence, the author finds, are usually due to specialized investments and shared technology. The likelihood of opportunistic behavior under the latter condition is minimal, as the stakes for failure are great. Practical supply chain failures are demonstrated by Sanchanta (2007) and Lundsford (2007), both describing supply chain issues at The Boeing Company in which relatively small and inexpensive items became the bottleneck and contingency plans had to be put into action. It also shows the importance of having contingency plans in the first place.

Speckman (2004) zeroes in on expanded enterprise thinking to identify various risks in supply chains. The author observes that in times past, corporate purchasing was an adversarial process in which the outcome of low purchase price implied high value. The new way is for all members in a supply chain to consider themselves party to an adaptive network delivering value through organizational learning, with a joint focus on the customer. Understanding risk then leads to two priorities, one avoiding conflict through early detection, and to have settlement procedures in place for disputes that do occur. Speckman (2004) then identify six dimensions of risk. These are:

1. Inbound supply, dealing with traditional purchase department issues
2. Information flow risks, to address the incompatibility of information systems in supply chain partners, trust in sharing confidential data such as strategic plans and sales forecasts.
3. Money flows, to address timely payments and pricing of end product
4. Security disputes, which arise when customers and suppliers access ERP systems that also become open to hackers.
5. Building and growing a trusting relationship is an art-form and in need of continuous nurturing. Otherwise opportunistic behavior and self-interest will begin to dominate
6. The reputation of any one supply chain partner affects all other; for example, one scandalous partner may reflect poorly on all others

Balakrishnan (2008) categorized supply chain dispute causes into two dimensions, namely location and unpredictability, for situations where JIT manufacturing technologies are applied. Gulati (2007) continues where Speckman (2004) leaves off, namely by considering a potential supply chain breakup and the need to memorialize that possibility in the beginning of the relationship in the form of exit strategies. This is hard to do because it is like laying out a divorce process at the time of a marriage proposal. As a practical matter, one's responsibility to clients require that one knows what exactly triggers failure in the supply chain, how to separate assets, inventories and intellectual properties in jointly owned operations with multiple product lines, and above all, to stipulate the need for open communication in order to preserve the reputation of the former supply chain partners. The separation of jointly owned properties becomes an issue only after an ordinary supply chain relationship has become a full partnership and the integration between two firms is now called a strategic alliance. Under the former arrangement, service level agreements will cover the separation process. Under the latter, a more complex partnership agreement will have to cover the separation of jointly owned properties, such as inventories and intellectual properties.

Panayides (2005) developed a model that actually measures the positive relationship of supply chain cooperative behavior on quality of service, what he calls relationship orientation. Panayides (2005) does not discuss the possibility of disputes, however. Anonymous (2004) addresses ADR (alternative dispute resolution) directly. The article suggests the incorporation of mediation and arbitration into supply chain contracts to avoid conventional litigation costs, time delays, and dispute management across international jurisdictions. The American Arbitration Association's supply chain offerings are described in reference AAA (2001). Here conflicts can be submitted online to a trained case worker ushering the process through resolution in either mediation- or arbitration format. The process is relatively inexpensive and above all private. Arbitration has become the preferred forum to settle disputes, as it is discreet and usually quicker and less expensive than litigation. But the process is not always perfect because panel members may have undisclosed conflict of interest (Tait, 2006).

Vassiliadis (2009) focuses on Information Technology as a way to ameliorate supply chain risk and disputes, in that information sharing is a vital component of any supply chain. He maintains a futuristic view in which grid computing enables data sharing and processing across international borders. The term 'grid computing' is defined as a heterogeneous computational resource, transparent and on-demand, in other words, a commodity like electricity, hence, the word 'grid'.

Carter-Steele (2009) also focuses on IT standards to produce a relatively conflict-free and well-functioning supply chain performance. The standards deal with (a) the service delivery process, such as accounting and reporting, (b) relationship processes, such as supplier management, (c) resolution process, such as incident- and problem management, (d) control process, such as configuration, and (e) release process. It is noteworthy that (b) and (c) suggest a standard for dispute resolution in IT Services. In fact, a global standard for Information Technology Services is evolving under the ISO umbrella. This standard, ISO/IEC 20000 (2009) infers a quality standard to IT, just as ISO 9000 infers quality standards for other industrial sectors by certifying a quality process, which then assumes that a quality process will produce a quality product, or service in this case.

Carter-Steel (2008) and Brown (2000) are in agreement with respect to the need of constraining opportunistic supply chain behavior through a governance mechanism for jointly owned asset classes. Brown's (2000) research is based on the hotel industry and he investigates situations in joint ownership, investments in specific assets, and norms in relational exchange and combinations thereof. The argument for restraining opportunism is that the short term and unilateral gain of one party comes at the expense of long term gain of both parties. While the trends in supply chain law suits are not known, Dillard (2006) found that contract litigation was second in frequency after employment.

## METHODOLOGY

The general supply chain literature is concerned with relationship management, the famous bull whip effect, logistic networks, inventory and outsourcing, all activities that aim to increase productivity. Within all these some disputes have been observed, and those relate to delays, capacity constraints, collaboration and good contract management, IT services and chain length. The reduction of supply chain conflict by prevention or fair resolution methods is enormously important because it has an effect on many parties. Hence, this investigation into conflict resolution was undertaken. The Delphi method was deemed appropriate because it's about building consensus on the future by successively consulting a panel of experts. A Delphi definition is also provided in the Glossary of terms at the end of this paper. The panel of experts was composed of senior supply chain managers in the United States, with international members. Of the 14 panel members, industrial sectors represented were super markets(1), commercial aviation(1), materials(1), consumer electronics(2), government(1), the judiciary(1), defense(1), aerospace(1) and agriculture(2), consumer products (3). Data collection started in late 2007 and ended in late 2008. Once data collection had begun, no new panel members were solicited.

Panel experts were selected on the basis of current supply chain managerial responsibility, both line managers and policy makers. At first, a mail solicitation was sent to one hundred executive members of a logistics association whose titles reflected supply chain responsibility. That effort produced only one panel member. Secondly, through personal contacts, a multi-divisional defense contractor firm participated via lengthy conference calls and a federal judge participated from the dispute resolution perspective as well as that of government contracting. A former business associate, who is now engaged in venture capital investing, was a policy level participant. Thirdly, the remaining panel members were front line supply chain managers and recent MBA graduates in South Florida. Field level panel members were at the manager level, policy level members at the vice presidential level, and one Federal bankruptcy judge participated with respect to the process of settlements. All members are employed with public corporations. Panel members' expertise is inferred by the fact that all serve in active managerial supply chain activities, either at the field level or that of policy making. This study is not intended to be industry-specific; hence we have a cross-section of industries represented on the panel. An industry-specific study into supply chain disputes is certainly an area to expand this research on.

Three rounds of questionnaires were presented to individual panel members over a one year period. Round I was intended to understand supply chain descriptive parameters such as number of chains under management and their respective tiers and dispute characteristics. Round II questions came out of the Round I discussions, focusing on disputes arising from compliance issues. Similarly, Round III questions evolved from those of Round II, and zeroed in on supply chain disputes in a deteriorating global economy and supply chain redesign challenges as a response to economic issues.

What follows are the questions for each Round, tabulated responses and discussion. The questions are in italics for ease of reading.

## SUMMARY OF FINDINGS, ROUND I (COMPOSED OF 5 QUESTIONS)

*Question 1*

From your company's management perspective, about how many supply chains are monitored or managed? A domestic supply chain is composed of only home country-based operations, while an international one has at least one or more non-home country suppliers.

*Answer 1*

The Number of Supply Chains under Management, as detailed in Table 1

It appears that the largest companies also have the most number of supply chains under active management. For example, for those respondents with more than 50 supply chains, one quarter were international in nature. In the US government sector, supply chain management is rather different, in that the GSA manages the contract and the agencies draw supplies from a central contract without concern about the chain management. In industry, as opposed to the government sector, technology and risk factors determine style and form of management.

*Table 1. Number of supply chains under management*

| Managed Supply Chains | Percent of Total in Study | Percent of Total International |
|---|---|---|
| 1 – 10 | 32 Percent | Approx.25 percent of 32 percent |
| 11 - 50 | 28 Percent | Approx.10 - 25 percent of 28 percent |
| > 51 | 40 Percent | Approx 25 percent of 40 percent |

*Question 2*

How deeply into the levels, or tiers of contractors, sub-contractors etc., do you involve yourself to resolve conflict?

*Answer 2*

All panel members responded with "it depends" as follows:

A. When a dispute in any tier becomes a constraint for the entire chain limiting overall performance, top level managers will get involved in any level.
B. When the dispute involves very large and very important supply relationships measured by significant revenue and decades of working together, top managers will get involved at that level.
C. When the dispute involves an inadvertent event, such as the discovery that intellectual property rights have accidentally been violated due to component purchases or mergers between companies. In such cases, managers will work it out and figure out a way to preserve the relationship, shake hands, and move on.
D. When the dispute resolution process between parties in the chain is a judicial one, like arbitration for example, management will not get involved.

More than two thirds of the respondents expressed the feeling, that less involvement in disputes more than one tier away from the core, is preferable to more involvement.

*Question 3*

Do your supply chain agreements/understandings, stipulate conditions of termination, as in a prenuptial divorce agreement? Have terminations happened before and how often, and under what conditions?

A. Do your supply chain agreements or contracts call for mandatory arbitration, and what code of arbitration would apply in those cases?
B. Do your supply chain agreements or contracts call for voluntary mediation to precede either arbitration or litigation?
C. In case of litigating international supply chain disputes, is the term "court of competent jurisdiction" defined?

*Answer 3*

Large firms and government agencies publish terms and conditions on their websites. These cover causes such as force majeure, default conditions and courts of competent jurisdiction, which vary from company to company. Chain partnerships are not self-regulatory organizations operating under an agreed upon forum, at least not yet. Law departments usually handle such issues and the State law of the supply chain leader is presumed to be applicable. New York state law is preferred because, and those may specify the forum or terminal jurisdiction. Terminations of supply chain relationships do occur in about ten percent of supply chain members. With respect to item (a), calls for arbitration are not uncommon. Rather than specifying the arbitration code or forum, the contract may call for mutually acceptable arbitrators in a panel. In case of three-person panels, each party often names its own arbitrator, and the two so named in turn agree on a third panel member. This process is sometimes called "baseball" arbitration .With respect to item (b), voluntary mediation options may appear in a contract or service level agreement (SLA), however, panel members did not recall ever having seen this put into practice. With respect to item (c), when disputes reach the point at which the term "court of competent jurisdiction" becomes an issue, the relationship is presumed to have failed. In US domestic cases, as mentioned before, the State of the supply chain leaders' headquarter is selected. In international disputes, the parties tend to agree on US law, or New York law, as defined in Appendix II. As mentioned earlier, Gulati (2007) describes a supply chain disengagement process to prevent expensive haggling if a chain fails.

*Question 4*

Supply Chain disputes can arise for many reasons and at many levels in the chain. The following eight are examples with respect to possible supply chain risk. Please feel free to ad your own dimension of dispute risk and then rank them, 1 being the highest risk.

*Answer 4*

D. The Consensus of all respondents with respect to supply chain dispute sources, in order of importance are shown in Table 2

One can infer a relationship between Speckman's (2004) supply chain risk factors with the dispute causes from this study as follows:

## SUMMARY OF FINDINGS, ROUND II (COMPOSED OF TWO QUESTIONS)

1. Has your firm experienced supply chain disputes arising from the compliance needs under the Sarbanes-Oxley legislation (SOX), such as accounting for goods in transit?
2. The United Nations has drafted a model laws and conciliation rules to harmonize international trade, known under the name UNCITRAL. In your experience, have these terms and conditions and arbitration codes ever been applied or considered? (Answers to questions 1 and 2 are shown in Table 3).

| Speckman (2004) Ranking | Delphi Study Ranking |
|---|---|
| Information flow risk (#2 of 6) | Information Flow (#6 of 8) |
| Security dispute risk (#4 of 6) | Data security (#5 of 8) |
| Inbound supply risk (#1 of 6) | Quality and timeliness (#1 of 8) |

It is recognized that SOX has changed supply chain relationships and added work and cost; but in terms of actual supply chain disputes, the parties will often find a way to work problems outside the accounting environment. Respondents to whom Sarbanes-Oxley was a significant supply chain issue, that is those who said "yes" to question 1, offered two interpretations. One was that SOX had indeed changed supply chain relationships in that accountants and compliance officers are dictating financial and other terms, as a result of which compliance is assured. On the other side, in well-established supply chain relationships that are characterized by a high level of trust and familiarity, undocumented arrangements are sealed by a hand-shake in order to keep the supply chain relationship going. The constitutionality of Sarbanes-Oxley is yet to be settled, as the US Supreme Court has agreed to hear such a case (Wall Street Journal, 2009).

With respect to the second question, the United Nation's UNCITRAL rules for the purpose of harmonizing international trade are not widely

*Table 2. Supply chain risks in order of importance*

| Order of Importance | Supply Chain Dispute Risk |
|---|---|
| 1 | Quality and timeliness of inbound supplies |
| 2 | Timely payments and pricing |
| 3 | Labor disputes |
| 4 | Opportunistic behavior to the detriment of chain partners |
| 5 | Security of data and trade secrets, intellectual property |
| 6 | Information flow, incompatible software and data access |
| 7 | Public scandals that negatively reflect on the whole chain |
| 8 | Economic nationalism |

known. UNCITRAL (1976) rules were accepted by the US, the UK and Japan in 2005, but not by the EU. The European industry prefers to practice self-regulation by using their own terms and conditions in supply chain management. Here the "center of main interest" has to be settled among the parties. In practice that "center of main interest" often turns out to be the United States, because of the fact that under US bankruptcy law reorganization is at least possible. Since 2005 only about 160 cases involving UNCITRAL have come up, and fewer than 25 percent of those involved reorganization.

## SUMMARY OF FINDINGS, ROUND III (ONE FINAL QUESTION)

Many current supply chains appear to have been designed and established prior to the year 2000 under assumptions of energy costs which no longer hold. Consequently, supply chain management is in the process of going back to the drawing board for the purpose of shortening the chain to reduce costs and risk.

*Table 3. Sarbanes-Oxley and UNCITRAL effects on dispute resolution*

| SOX Question 1 57% Yes 43%No |
|---|
| UNCITRAL Question 2 33% Yes 67%No |

A. Is your company currently re-designing any or all supply chains for the above reasons?
B. If the answer to (A) is yes, do you expect new conflict issues to arise
   a. We expect no new conflict issues to arise
   b. We expect conflict issues to arise related to relationship changes
   c. We expect conflict issues to arise related to task changes.
   d. We expect conflict issues to arise related to both relationship-and task changes.
C. Explanations/comment (Results detailed in Table 4).

Supply chain re-design appears to be a current preoccupation for a variety of reasons, one reason being that stable fuel prices we enjoyed ten years ago are not part of the landscape any longer. From discussion we found that those respondents expecting no conflict, fall into a category of supply chains that are very well established, have worked well over five years and usually involve products in the very high technology rubric, such as aerospace and defense. In addition, those supply chains have already gone through various stages of pruning because of their maturity. High technology products and parts are very valuable and therefore air transport is the means of getting them to their

*Table 4. Supply chain re-design*

| A) Respondents redesigning supply chains 90% |
|---|
| B) Of those respondents redesigning supply chains |
| a. Expecting no conflict 22% |
| b. Expecting relationship and task conflict 67% |
| c. Expecting relationship related conflict only 11% |

final destination, if for no other reason than the time value of money outweighing the additional cost of air transport. If conflicts do arise, they will come from poorly administered contractual relationships, rather than from the supply chain structure itself.

Respondents expecting conflicts to arise out of supply chain redesign efforts either plan to (a) shorten the supply chain, or (b) reduce the tiers of a chain to specifically no more than five. However, difficulties appear not to be overwhelming. For example, a manufactured part may go through casting, welding, machining and plating; each task being made by a different supplier. Under redesign, the casting organization, for example, simply takes responsibility of the whole process, while the others continue to do the work without being part of the chain. If one party loses business in the redesign process, the pain is often minimal as it's not the biggest piece of business for either party. OEMs are often happy to do the distribution for off-the-shelf products themselves.

Supply Chain costs are always subject to review, especially now as the labor cost advantage of China is getting smaller. High air transport cost was offset by a low labor rate. Now that gap has narrowed as air transport is being replaced by surface means. That, in turn, stretches the time in transit and increases the risk of on-time delivery, which was ranked the #1 risk in question 1, Round I. The time value of money of large quantities in transit is also a consideration. It's a given that nobody keeps sufficient inventory because of costs and currency fluctuations, which can and does, lead to payment disputes, Dvorak (2009). Finally, the limits of logistics economies of scale are observable in port congestion, adding another supply chain failure point.

Utility companies do not usually appear in the supply chain literature, but are in fact part and parcel of the supply chain business model. In cases of large economic developments, in which an anchor enterprise and its JIT-based vendors settle in a new geographic area, utility companies have to guarantee water, sewer and energy needs for a long time to come. Utility companies then commit to investments, payments for which services are often assured by bonds or large down payments. Disputes from that source are usually settled after court hearings.

## CONCLUSION

This Delphi study concentrated on discovering supply chain dispute resolution practices from a panel of US supply chain executives in several industries, such as defense, materials, government, the judiciary, food marketing, commercial aviation, aerospace industries and electronics. Three rounds of surveys were sent, the first concentrating on supply chain demographics such as number of tiers and nature of disputes that arise; the second round focused on disputes related to compliance issues, and the third round on disputes coming from second generation supply chain re-design as a consequence of global economic issues.

Based on survey instruments, conference calls and in some cases personal interviews with panel member, we found that of the supply chains we studied, three-fourth are predominantly US domestic, and cover about four tiers on average. Disputes between tiers tend to be left for resolution among the parties themselves, unless such a conflict constrains all partners in the chain. As soon as a conflict becomes a legal issue, the relationship is often considered damaged. However, when it does occur, New York law is generally agreed upon as this is most familiar to the parties. The top three dispute categories are quality and timeliness, payment and pricing, plus labor issues, followed by culture, language for international cases. There was unanimous agreement among the panel members that government has no useful role to play in supply chain disputes. The quasi judicial processes for resolving disputes start with a scheduling conference in which the issues are aired and the process is laid out, followed by a neutral person managing the process through discovery, and ultimately a settlement conference.

Feedback from the panels led this inquiry into considering Sarbanes - Oxley (SOX) compliance for US supply chain disputes. The finding is that SOX adds new costs to the management of the process. In practice, the supply chain relationship itself is far more important than SOX compliance. This leads to the conclusion that the parties work around SOX without violating it. The United Nations model law for harmonizing trade, UNCITRAL, UNCITRAL (1976), was unknown to most panel members in the USA.

Supply chains are subject to redesign, and the reasons for this are many. For example: inventory optimization has led to low buffers in supply chains, endangering smooth flow and timely supply. In addition, many formerly low labor cost nations are losing the cost advantage as their wage rates increase. Such cost increases are compensated by switching from air shipment to surface mode, and that in turn can lead to delays and port congestion. Two thirds of the panel members expect relationship- and task conflict to arise from these issues.

The current literature does not generally include utilities in the supply chain discussion. However, in large scale infrastructure projects, utilities have to make significant long-term investments that require financial guarantees, should the development fail. Disputes arising from those issues, which could wind up in court, can resemble the nature of conventional supply chains conflict.

Future researchers on supply chain disputes may wish to concentrate on specific industries. Being able to define best practices in one industrial sector may lead to the transfer of such practices to another industry. Of course, mechanisms for avoiding conflict in the first place are equally important. The study of conflict in supply chains is especially relevant in times of economic stress for several reasons: The drive to shorten supply chains and getting closer to the end- market will require rearrangement of current supply chain relationships, and that can ultimately lead to disputes. Finally, the lean- and green movement in supply chains may require products with longer lives, requiring fewer resources with less material to recycle. Changes lead to turmoil and turmoil leads to disputes, as well as opportunities (Sodhi, 2009).

Preferable to conflict management is conflict avoidance, and such a strategy involves instituting good communication among key personnel. Avoidance builds trust and strengthens the supply chain, and also promotes longevity of the relationship. Specifically helpful in this regard are shared information technology in the form of enterprise resource software and databases. Such data bases should measure the supply chain performance with respect to inventory levels, correct billing and on-time deliveries, and cause of disputes when they do occur. Service Level Agreements (SLA) should be easily accessible through online information technology and subject to frequent review with respect to compliance.

All supply chains, from Henry Ford's onward, share common attributes. In this report a distinguished panel of experts contributed to conflict settlement and avoidance, insights that may apply in many industries. Looking towards the future, supply chain managers could institute shared experiments in conflict avoidance to make the relationship more sustainable and ultimately benefit the end consumer.

## REFERENCES

American Arbitration Association. (2001). E-commerce dispute risk management services offered by AAA. *Dispute Resolution Journal*, *56*(3), 5.

Anonymous. (2004, September). *Using ADR in Supply Chain Disputes*. Paper presented at the Purchasing b2b, PMAC-AGM conference, Halifax, Nova Scotia.

Balakrishnan, B., & Bowen, F. (2008). A strategic framework for managing failure in JIT supply chains. *Journal of Information Systems and Supply Chain Management*, *1*(4), 20–38.

Blackhurst, J., Wu, T., & Graighead, C. (2006). A systematic approach to supply chain conflict detection with a hierarchical Petri Net extension. *Omega. International Journal of Management Science*, *36*(5), 680.

Brown, J., Dev, C., & Lee, D. (2000). Managing marketing channel opportunism: The efficacy of alternative governance mechanisms. *Journal of Marketing*, *64*(2), 51–66. doi:10.1509/jmkg.64.2.51.17995

Carter-Steel, A. (2009). IT service departments struggle to adopt a service-oriented philosophy. *International Journal of Information Systems in the Service Sector*, *1*(2), 69–77.

Chase, R., Jacobs, R., & Aquilano, J. (2004). *Operations Management*. New York: McGraw Hill Irwin.

Chopra, S., & ManMohan, S. (2004, fall). Managing risk to avoid supply chain breakdown. *MIT Sloan Management Review*, *3*, 53-60.

Denning, L. (2008, December 3). Ships ahoy: New world's supply chain. *The Wall Street Journal*.

Dillard, S. (2006, November 25). Litigation nation. *The Wall Street Journal*, A9.

Dvorak, P. (2009, May 18). Ups and downs whipsaw supply chains. *The Wall Street Journal*, A1.

Giannoccaro, L., & Pontrandolfo, P. (2004, February). Supply chain coordination by revenue sharing contracts. *International Journal of Production Economics*, *89*(2), 131. doi:10.1016/S0925-5273(03)00047-1

Gulati, R., Sytch, M., & Mehrotra, P. (2007, March 3). Preparing for the exit. *The Wall Street Journal*, R11. *ISO/IEC- 20000*. (2009). Retrieved from www.isoiec20000certification.com

Lee, T. (2008). Supply chain risk management. *International Journal of Information and Decision Sciences*, *1*(1), 98–111. doi:10.1504/IJIDS.2008.020050

Li, E., Du, T., & Wong, J. (2005). Access control in collaborative commerce. *Decision Support Systems*, 1–11. Retrieved from www.elsevier.com/locate/dsw.

Lundsford, J., & Glader, P. (2007, June 9). Boeing's nuts-and-bolts problem. *The Wall Street Journal*, A8.

Panayides, P., & So, M. (2005). The impact of integrated logistics relationships on third party quality performance. *Maritime Economics & Logistics*, *7*(1), 36–48. doi:10.1057/palgrave.mel.9100123

Patel, V. (2006). Contract management: The new competitive edge. *Supply Chain Management Review*, *10*(3), 42–44.

Patterson, J. (1996). An investigation of conflict resolution between buyers and suppliers in strategic long-term cooperative relationships. In *Proceedings of the ISM 81st Annual International Conference*, Chicago.

Runde, C., & Flannegan, T. (2008). *Building Conflict Competent Teams* (p. 238). San Francisco, CA: Jossey-Bass.

Sanchanta, M. (2007, February 2). Boeing suppliers face production hitches. *Financial Times (North American Edition)*, 22.

Smith, R. (2008). Reality of supply chain models. *APICS-Extra online, APICS*. Retrieved May 28, 2008, from www.apics.com

Sodhi, M., & Tang, C. (2009). Rethinking links in the global supply chain. *Financial Times (North American Edition)*, 11.

Speckman, R., & Davis, E. (2004, May). Risky business: expanding the discussion on risk and the extended enterprise. *International Journal of Physical Distribution & Logistics Management*, *34*(5), 414–429. doi:10.1108/09600030410545454

Tait, N., & Peel, M. (2006, August 8). Tool for dispute resolution loses its certainty. *Financial Times (North American Edition)*, 14.

UNCITRAL. (1976). *United Nations Commission on International Trade Law*. Retrieved from http://www.united.nations.org

UNCITRAL. (1994). *Model Law on International Commercial Arbitration*. Retrieved from http://www.uncitral.org

Vanany, I., & Zailani, S. (2009). Supply chain risk management: A literature review and future research. *International Journal of Information Systems and Supply Chain Management*, *2*(1), 16–33.

Vassiliadis, W. (2009). The grid as a virtual enterprise enabler. *International Journal of Information Systems in the Service Sector*, *1*(1), 78–92.

Journal, W. S. (2009, May 19). Sarbox and the constitution. *Review and Outlook*, A16.

Wang, Y. (2008, March). Strategic management of international subcontracting. *International Journal of Information Systems and Supply Chain Management*, *1*(3), 21–32.

## APPENDIX I: DELPHI QUESTIONS

Round One Questions

1. From your company's management perspective, about how many supply chains are monitored or managed? A domestic supply chain is composed of only home country-based operations, while an international one has at least one or more non-home country suppliers.
2. How deeply into the levels, or tiers of contractors, sub-contractors etc., do you involve yourself to resolve conflict?
3. Do your supply chain agreements/understandings, stipulate conditions of termination, as in a prenuptial divorce agreement? Have terminations happened before and how often, and under what conditions?
   a. Do your supply chain agreements or contracts call for mandatory arbitration, and what code of arbitration would apply in those cases?
   b. Do you supply chain agreements or contracts call for voluntary mediation to precede either arbitration or litigation?
   c. In case of litigating international supply chain disputes, is the term "court of competent jurisdiction" defined?
4. Supply Chain disputes can arise for many reasons and at many levels in the chain. The following eight are examples with respect to possible supply chain risk. Please feel free to ad your own dimension of dispute risk and then rank them, 1 being the highest risk.

| | |
|---|---|
| 1 | Quality and timeliness of inbound supplies |
| 2 | Timely payments and pricing |
| 3 | Labor disputes |
| 4 | Opportunistic behavior to the detriment of chain partners |
| 5 | Security of data and trade secrets, intellectual property |
| 6 | Information flow, incompatible software and data access |
| 7 | Public scandals that negatively reflect on the whole chain |
| 8 | Economic nationalism |
| | |

5. What would be the nature of an ideal Supply Chain ADR (alternative dispute resolution) environment sometime in the future? For example, could a hypothetical Supply Chain settlement method involve governments, is there a risk reduction role for government to play? Have you had an experience in which a government involved itself, directly or indirectly, in settling a Supply Chain dispute? Please comment:

Round Two Questions:

1. Has your firm experienced supply chain disputes arising from the compliance needs under the Sarbanes-Oxley legislation (SOX), such as accounting for goods in transit?
2. The United Nations has drafted a model laws and conciliation rules to harmonize international trade, known under the name UNCITRAL. In your experience, have these terms and conditions and arbitration codes ever been applied or considered?

Round Three Question:

Many current supply chains appear to have been designed and established prior to the year 2000 under assumptions of energy costs, which do no longer hold. Consequently, supply chain management is in the process of going back to the drawing board for the purpose of shortening the chain to reduce costs and risk.

A. Is your company currently re-designing any or all supply chains for the above reasons?
B. If the answer to (A) is yes, do you expect new conflict issues to arise
    a. We expect no new conflict issues to arise
    b. We expect conflict issues to arise related to relationship changes
    c. We expect conflict issues to arise related to task changes.
    d. We expect conflict issues to arise related to both relationship-and task changes.
C. Explanations/comment

## APPENDIX II: GLOSSARY OF TERMS

**Conflict:** When the long term strategies and goals of the supply chain partners are not being achieved and there is little hope for achieving them without outside intervention.

**Risk:** Probability of variance from expected outcome.

**A.D.R:** Alternative dispute resolution covers mediation and arbitration, for which there are established codes and procedures in many jurisdictions.

**The Delphi Technique:** The technique is defined as a group of experts responding to questionnaires under the guidance of a moderator who compiles results and formulates new questionnaire and submits it to the group. Implied is a learning process leading to consensus. Chase (2004).Delphi applications are creative explorations that produce information for decision making, as in forecasting for example. The technique is based on a structured process for collecting and distilling knowledge from a panel of experts by means of questionnaires with feedback. Here a panel of experts, without interacting with each other, are given a number of questions relative to their expertise, which consolidated answers lead to new round of questions, ultimately leading to a consensus of views on a subject matter that is otherwise hard to predict. Having its origin ancient Greece, oracles were temples to go to for good counsel, the Oracle of Delphi being the prime example. The high priestess could infer the state of society from the questions, and use that information for giving advice. More recently, the Delphi method became a forecasting technique perfected by The Rand Corporation's Olaf Helmer.

**Sarbanes-Oxley:** The Sarbanes-Oxley Act of 2002 (Pub.L. 107-204, 116 Stat. 745, enacted July 30, 2002), also known as the Public Company Accounting Reform and Investor Protection Act of 2002 and commonly called Sarbanes-Oxley, Sarbox or SOX, is a United States federal law enacted as a reaction to a number of accounting scandals.

**UNCITRAL Arbitration Rules:** Provides a comprehensive set of procedural rules upon which parties may agree for the conduct of arbitral proceedings arising out of their commercial relationship that are widely used in ad hoc arbitrations as well as administered arbitrations. The Rules cover all aspects of the arbitral process, providing a model arbitration clause, setting out procedural rules regarding the appointment of arbitration and the conduct of arbitral proceedings and establishing rules in relation to the form, effect and interpretation of the award. UNCITRAL(1976).

**New York Law:** Laws constitute a system of rules that are enforced by institutions, which together, affect society, economics, and the relationship between people and companies. The term "New York Law" refers to the laws of the State of New York, as these affect contracts, properties, trusts and torts, constitutional and administrative considerations. New York City being a global commercial hub, it follows that a familiarity with the laws of New York exists between supply chain parties, and that those parties are likely to select it as a forum for dispute resolution.

**Opportunistic Behavior:** The term is defined as lack of candor in supply chain transactions, governed by self-interest and guile.

*This work was previously published in International Journal of Information Systems and Supply Chain Management, Volume 3, Issue 1, edited by John Wang, pp. 50-65, copyright 2010 by IGI Publishing (an imprint of IGI Global)*

# Chapter 9
# Fair Distribution of Efficiency Gains in Supply Networks from a Cooperative Game Theory Point of View

**Stephan Zelewski**
*University of Duisburg-Essen, Germany*

**Malte L Peters**
*Center for Education and Science of the Federal Finance Administration, Germany*

## ABSTRACT

*In this paper, the authors address the distribution of efficiency gains among partially autonomous supply network actors in a manner they will accept as fair and as an incentive to cooperation. The problem is economically significant because it requires substantiating efficiency gains in an understandable manner. Moreover, supply networks suffer from a conflict potential because the partially autonomous actors seek to maximize their own shares of the efficiency gain. The method applied appropriates a model from cooperative game theory involving the τ-value. The special nature of the τ-value ensures that it seems rational to the actors to cooperate in the supply network. The proposed method for the distribution problem offers a fair distribution of efficiency gains in the supply network and ensures that the distribution results can be communicated easily.*

DOI: 10.4018/978-1-4666-0918-1.ch009

## INTRODUCTION

Over the past several years, a considerable amount of research has been carried out in the area of supply chain management (e.g., Croson & Donohue, 2006; Krol et al., 2005; Wu et al., 2007; Zokaei & Simons, 2006). One principal aim of this research is to find ways of realizing efficiency gains by coordinating the activities of all actors in a supply chain or, more precisely, in a supply network. Since this paper's findings are not restricted to supply chains, but also apply to supply networks, the latter term will be used throughout.

Several effects must be considered as sources of efficiency gains. The most prominent effect is the so called bullwhip effect (Lee et al., 1997; see also Croson & Donohue, 2006; Krol et al., 2005; McCullen & Towill, 2002; Metters, 1997). The bullwhip effect describes especially how companies build inventory buffers based on the demand of their customers: the further the company is from the final customer the greater the "safety stock" is in times of rising demand. The cost of capital invested in oversized inventory buffers in the stocks causes inefficiency and thus efficiency gains can be realized by avoiding or reducing the bullwhip effect. Evidence of the practical relevance of the bullwhip effect to supply chain management is provided by studies of its financial consequences (McCullen & Towill, 2002; Metters, 1997). Based on available estimates of the cost of the bullwhip effect, companies should be able to increase their profits—depending on the source—by 8.4 to 20.1% (McCullen & Towill, 2002) or by 10 to 30% (Metters, 1997) by avoiding it.

When efficiency gains are realized in supply networks, a distribution problem arises. The cooperating actors know that they are realizing the efficiency gains by mutually coordinating their activities. Moreover, each actor is interested in maximizing his own gain at the expense of the other actors in the supply network. Thus, supply networks suffer from a built-in conflict between cooperation and defection. The problem lies in distributing efficiency gains among partially autonomous actors in a manner that the actors will accept as fair and advantageous to cooperation. If, on the other hand, it would be advantageous for at least one of the actors to leave the grand coalition, the supply network would collapse. With this scenario in mind, a stability requirement can be posited for the solutions of efficiency gain distribution problems. These problem solutions are regarded as desirable only, if they ensure that all actors in a supply network are willing to cooperate with each other. The fulfillment of this stability requirement is often circumscribed by the actors' acceptance of the distribution of the efficiency gains as fair.

In economic literature, significant research efforts have been devoted to developing concepts of fairness (e.g., Fehr & Schmidt, 1999; Pazner, 1977; Varian, 1976). Unfortunately, to the knowledge of the authors, these concepts have not yet been adapted to solve the distribution problem outlined above. Furthermore, cooperative game theory offers a number of solution concepts for distribution problems worthy of closer consideration. These are especially the Shapley value (Shapley, 1953; see Derks & Tijs, 2000) and the Nucleolus (Schmeidler, 1969; see also Meertens & Potters, 2006). Yet these solution concepts suffer from a serious drawback. The fairness of the distribution results is hard to justify, since the 'logic' of these approaches is difficult to communicate. The central issue of this paper is to present a solution concept from cooperative game theory for the problem of distributing efficiency gains in supply networks that aims at distributing efficiency gains in a fair manner. From a management point of view, this concern is motivated by the fact that the (game) theoretic solution concepts have to withstand acceptance problems in management practice, since they are not intuitively comprehensible. This is why there is a considerable management demand for solutions of the efficiency gain distribution problem that are both theoretically sound and involve an easily, if not intuitively comprehensible operationalization of fairness. The issue of an intuitively and easily comprehensible fairness of efficiency gain distributions is tackled by means of cooperative game theory. In contrast to the literature, no "completed" solution concepts are taken as starting points and analyzed with respect to their theoretical characteristics. Furthermore, no abstract mathematical axioms are taken as bases for showing then that a certain game theoretic concept is implicated. Instead, a novel and primarily economic, argumentation is presented. This argumentation seeks to derive a known solution concept – the so-called τ-value – from economically plausible assumptions by means of which the space of possible solution concepts is restricted until the τ-value results as the solution concept for the fair distribution of efficiency gains.

The main results of this paper concern three aspects which serve to advance the findings in the pertinent literature. Firstly, the τ-value solution concept is derived in a novel, non mathematical, economically motivated way. Secondly, this results in a new type of the τ-value formula as yet unknown in game theoretic literature. Thirdly, this novel method of representing the τ-value facilitates demonstrating that it ensures the fair distribution of efficiency gains in a manner that is theoretically substantiated and easily, if not intuitively, comprehensible.

## THE τ-VALUE SOLUTION CONCEPT FOR THE FAIR DISTRIBUTION OF EFFICIENCY GAINS IN SUPPLY NETWORKS

### The Generic Distribution Problem for Supply Networks

The generic distribution problem lies in fairly allocating a positive efficiency gain realized through supply network collaboration among its actors. It involves the class of all instances which, on the one hand, have the basic structure of the generic distribution problem and in which, on the other hand, each parameter describing the problem is substantiated by numerical values.

Since efficiency gains already realized are taken for granted, the generic distribution problem relates to existing supply networks exclusively. The only question of interest in this context is how to distribute the efficiency gains among the actors in a supply network in such a fashion that all actors will accept the distribution as fair and have no reason to leave the existing supply network. This question is touched upon in the first section of this paper as the stability requirement for existing supply networks. Such questions as the emergence of supply networks or the granting of incentives to increase realizable efficiency gains are not thematized. While of no less interest from management point of view, they require analytical approaches other than cooperative game theory. Coalition formation games from non-cooperative game theory, for example, could be utilized to analyze the emergence of supply networks. Moreover, models from principal agent theory are preferred for the analysis of incentives for increasing the realizable efficiency gains. Both of these last mentioned analytical approaches have to be considered in detail in other publications.

In conventional cooperative game theory, the starting point for the generic distribution problem is the efficiency gain G to be distributed with $G \in R_+$ where $R_+$ is the set of all positive real numbers. The method of calculating efficiency gain G is immaterial, since values are considered as given in game theory. From a management point of view, however, this is quite unsatisfactory, since solving the real problem of fair distribution of efficiency gains takes priority over the solution of a formally specified distribution problem (formal problem). Conceptualizing the real problem requires notions about how to obtain the information assumed to be known in the formal problem. While it is assumed that efficiency gains refer to those of one business year, other periods can be analyzed as required without changing the solution concept proposed below.

There are two ways to determine quantitatively the efficiency gain G. Firstly, it is possible to fall back on rough estimations of the efficiency gains based on expert opinions. These experts could be managers of the considered supply network. Furthermore, third parties like consultants specialized on analyzing supply networks may be considered. Secondly, it is possible to estimate the efficiency gain G to be distributed by comparative modelling. This analytic determination of the efficiency gain G rests upon information about the supply network for a given number of past periods provided by controlling departments.

Comparative modelling relies upon two different approaches which are applied jointly. For each supply network actor, a partial model assists determining the amount of profit that would have been made were optimal adaptations based on past period information (partial profit) carried out. Subsequently overall profit is calculated as the sum of the partial profits of all supply network actors. In the second approach, a total model of the supply network is constructed. This model contains all actors of the supply network including all mutual material and information flows. Sophisticated models have been developed for this holistic modelling approach (e.g., Zhang, 2006; especially both with game models of supply networks: Mahdavi et al., 2008; Xiao et al., 2009). The total model is utilized to determine the total profit the supply network actors would have made when all actors cooperate. Such cooperation includes the optimal coordination of the actors' partial plannings in such a manner that the total profit of all actors is maximized. In the event of isolated optimization, this total profit of cooperative optimization is then compared to the overall profit. If the total profit (cooperative optimization) is greater than the overall profit (isolated optimization), then the difference is the efficiency gain G.

From a management point of view, the second procedure for determining probable efficiency gains is preferred to the first one, both since the procedure involves formal models and since the underlying information is easily comprehensible and open to critical discussion. Rough estimations, on the other hand, are usually difficult to discuss. In addition, it is often difficult to question the assessment of experts.

If the efficiency gain of a supply network is estimated by expert opinion or calculated by comparative modelling, the following four requirements are considered to be important for the game theoretic modelling of the generic distribution problem: (1) It should be possible to explicate the different scopes for alternative distribution results that emerge from different assumptions regarding the rationality of the actors. From a management point of view, it is interesting to assess how rationality assumptions, which, prima facie, appear plausible affect the scopes for alternative distribution results which are each consistent with the concept of rational actors by restricting the solution space. (2) Distribution results determined by game theoretic solution concepts must be justified by good reasons in order to facilitate acceptance of the distribution result. (3) The solution of the distribution problem can be communicated easily in the supply network. (4) The existence and the uniqueness of the solution of the distribution problem should be ensured.

The generic distribution problem of allocating an efficiency gain G with $G \in R_+$ among N actors of a supply network (where $N \in N_+$ and $N \geq 2$) can be modelled in a first approach with basic elements of cooperative game theory. But this is only correct as long as it is abstracted from the guideline that good reasons can be formulated for accepting a distribution of the efficiency gain G among the N actors of a supply network as fair. Thus, only a reduced form of the generic distribution problem defined above is modelled, since the fairness condition is temporarily neglected. At first, from a management point of view, this reduced generic distribution problem is seen as accentuating the relevance of the τ-value solution concept.

From the viewpoint of cooperative game theory the reduced generic distribution problem for supply networks is to determine a distribution function v to allocate a share $v_n$ of the efficiency gain G to each actor $A_n$ with $n \in \{1,...,N\}$ in the grand coalition $C_0 = \{A_1,...,A_N\}$ of all actors in the supply network. If, in addition, it is assumed that the distribution function must not assign a negative amount as share of the efficiency gain G, then the distribution function v with $R_{\geq 0}$ as the set of all non-negative real numbers can be specified as follows: $v : A \rightarrow R_{\geq 0}$ with $A_n \mapsto v(A_n) = v_n$ for each actor $A_n$. The solution of the reduced distribution problem is every N-tupel $\underline{v}$ with $\underline{v} = (v_1,...,v_N)$ which assigns a share $v_n$ of the efficiency gain G to each actor $A_n$ in the set of actors A. This N-tupel is the solution point $L_v$ in the N-dimensional non-negative real number space $R^N_{\geq 0}$. The solution point $L_v$ is represented as a column vector $\vec{v}$, whose transposed representation denoted by a superscript letter ($^T$) is: $\vec{v} = (v_1,...,v_N)^T$.

The standard approach of cooperative game theory for solving distribution problems like that described above consists of two steps. In the first step, a characteristic function is developed. This function refers to all possible coalitions which could be formed of the actors in the considered supply network. Moreover, "degenerate" coalitions formed by one actor are also feasible. Therefore, a coalition $C_m$ is a non empty subset of the set $A = \{A_1,...,A_N\}$ of all actors in a supply network: $\varnothing \subset C_m \subseteq A$.

The following holds for the characteristic function c with $\wp$ as power set operator: $c : \wp(A) \rightarrow \mathbb{R}_{\geq 0} R_{\geq 0}$ with $C_m \mapsto c(C_m)$ for each coalition $C_m$ with $\varnothing \mapsto c(\varnothing) = 0$. This characteristic function assigns the amount $c(C_m)$ the respective coalition $C_m$ can claim with good reasons. In the case of a grand coalition $C_0 = A$ this is the overall efficiency gain G: $c(C_0) = G$. For all other coalitions $C_m$ with $\varnothing \subset C_m \subset A$ these are the amounts $c(C_m)$ these coalitions could realize on their own outside the grand coalition $C_0$ and therefore in competition to the rest of the coalition $C_0 \setminus C_m$.

In the second step, the shape of the distribution function v is determined by calculating the distribution function values $v(A_n) = v_n$ for each actor $A_n$ in the set $A = \{A_1,...,A_N\}$ of all actors in the supply network. Only two information sources are considered to calculate these values. On the one hand, these are the amounts each feasible coalition can claim due to the characteristic function c from the first step. On the other hand, the applied solution concept specifies how the distribution function values $v(A_n) = v_n$ are calculated based on the values $c(C_m)$ of the characteristic function c for all feasible coalitions $C_m$ with $m = 0,1,...,2^N-2$. When all distribution function values $v(A_n) = v_n$ are determined, there is a N-tupel $\underline{v} = (v_1,...,v_N)$ as a solution $\underline{v}$ for the respective regarded instance of the generic distribution problem in its reduced form.

From a management point of view, this standard approach of cooperative game theory is unsatisfactory. The main weakness of this standard approach lies in the characteristic function c which is assumed to be known in conventional game theoretical analyses. This information premise can be viewed as quite distant from reality, since in actual managerial practice it is often not known for each feasible coalition $C_m$ which value $c(C_m)$ is reasonably appropriate for the respective coalition.

It is conceivable, however, to calculate the values $c(C_m)$ of the characteristic function c by applying calculation procedures a priori. But due to the great number $2^N$-1 of feasible coalitions exploding exponentially with the number N of actors in a supply network it must be assumed, at least for great supply networks, that the calculation of all values $c(C_m)$ of the characteristic function c will fail because of the demanding calculation effort. If, for example, the values $c(C_m)$ are not known from the start, a supply network with N = 12 actors requires to accomplish up to $2^{12}$-1 = 4,095 calculation procedures. From a management point of view, a game theoretic solution concept—one which makes it possible to calculate the values $v_n$ of the distribution function v for all actors $A_n$ in a supply network as a solution for the generic distribution problem without having complete knowledge about the characteristic function c —is of great interest. Instead, such a non state-of-the-art solution concept should refer to as few as possible coalitions to calculate the values $v_n$ of the distribution function v for all actors $A_n$ in a supply network. This requirement of minimal knowlegde is added to the four already mentioned requirements for the reconstruction of the real problem of distributing efficiency gains in supply networks.

## The τ-value Solution Concept

### Motivation for a New Perspective on the τ-Value Solution Concept

Initially, the τ-value was proposed as a solution concept for cooperative games by Tijs (1981). Subsequently, it was further developed, especially by Tijs and Driessen (Driessen, 1985; Driessen & Tijs, 1982; Driessen & Tijs, 1983; Driessen & Tijs, 1985; Tijs, 1987; Tijs & Driessen, 1983; Tijs & Driessen, 1986; see also for the τ-value Curiel, 1997). Todate, only a few research efforts have attempted applying the τ-value solution concept in the areas of economics and management. The corresponding papers, however, are focused on the distribution of fixed costs or overhead costs (e.g., Tijs & Driessen, 1986). To the knowledge of the authors, until now relatively little research has been carried out applying the τ-value to the solution of the distribution problem of efficiency gains in supply networks.

In standard game theory, it is common to introduce a new solution concept by specifying its applicability conditions and its calculation formulas or its calculation algorithms. If the applicability conditions of a solution concept are available in an axiomatized form, it is often argued that the proposed solution concept is the only one that is capable of fulfilling a set of formally specified axiomatic requirements. In this case the solution concept can be justified by fulfilling axioms which are assumed to be "given" or "reasonable". Critical reflection regarding the justification of such axioms is seldomly found. As a result, it is rather difficult to detect the connection between some axioms and the respective real problem. Sometimes it even appears that some axioms are introduced to enforce the application of a certain solution concept which fulfills the axioms.

This paper, avoids such a standard way of proceeding, since it suffers from a formalistic approach in regarding a rigorous solution concept primarily with respect to its formal applicability and calculation. From a management point of view, it lacks orientation towards the relevant real problem. Instead, it would be desirable to justify a game theoretical solution concept in the following manner: Starting from the real problem of distributing efficiency gains among the actors in a supply network, the solution concept should be developed in an easily understandable manner so that it is derivable from plausible requirements oriented towards the real problem. Moreover, it should be possible to have good reasons to accept the resulting solution as fair distribution. In the following, the authors try to develop a justification program for game theoretic solution concepts oriented towards real problems exemplified by the τ-value solution concept. A major concern of this paper is to reconstruct the well known τ-value solution concept in a new way with regard to management concerns.

## Calculation of the τ-Value

The initial point for distributing the efficiency gain G with $G \in R_+$ among the N actors of a supply network with $N \in N_+$ and $N \geq 2$ is the reduced generic distribution problem. The basic idea of the reconstruction of the τ-value solution concept is to restrict the solution space $R^N_{\geq 0}$ for the generic distribution problem by successively adding five requirements which stem from the real problem of the fair distribution of supply network efficiency gains. The following chain of reasoning yields the τ-value as a "senseful" solution to the generic distribution problem, which is acceptable as fair.

The first requirement for a solution concept is the condition of individual rationality. This condition assumes that every actor in the supply network acts rationally in the sense of the conventional concept of perfect rationality. This means that each actor maximizes her or his individual utility. Furthermore, it is assumed that no so called envy effects are present, i.e., each actor evaluates her or his individual utility without considering the shares of the other actors in the supply network. Moreover, the information processing capacity of the actors is not restricted so that the actors are capable of calculating their individual utility maximums. The condition of individual rationality causes a restriction of the solution space $R^N_{\geq 0}$, since it would not be rational for an actor $A_n$ to participate in the supply network within the coalition $C_0$, if this coalition yields a smaller utility in comparison to leaving the coalition and realizing the amount $c(\{A_n\})$ outside the supply network. Thus, the condition of individual rationality can be formulated with the characteristic function c and the feasible solution point $L_v$ within the solution space $R^N_{\geq 0}$ as follows:

$$\forall L_v \in R^N_{\geq 0}:$$

$$\forall L_v \in \mathbb{R}^N_{\geq 0}: \quad L_v = \begin{pmatrix} v_1 \\ \ldots \\ v_N \end{pmatrix} \geq \begin{pmatrix} c(\{A_1\}) \\ \ldots \\ c(\{A_N\}) \end{pmatrix} \tag{1}$$

The second requirement is the efficiency condition. This condition postulates that the efficiency gain G is distributed exactly among the coalition $C_0 = \{A_1,...,A_N\}$ of all actors, since it would be irrational to distribute less than the efficiency gain G and it would necessarily entail a loss of Pareto optimality (just as it would be impossible to distribute more than the efficiency gain G). Thus, the following equation will hold true with $L_v$ as a feasible solution point in the solution space $R^N_{\geq 0}$:

$$\forall L_v \in R^N_{\geq 0}:$$

$$\forall L_v \in \mathbb{R}^N_{\geq 0}: \quad L_v = \begin{pmatrix} v_1 \\ \ldots \\ v_N \end{pmatrix} \rightarrow \sum_{n=1}^{N} v_n = G \tag{2}$$

Furthermore, the efficiency condition implies a further restriction of the solution space $R^N_{\geq 0}$, since all solutions of the distribution problem, which fulfill the requirement of individual rationality as well as the efficiency condition, are solution points $L_v$ on a hyperplane H in the N-dimensional solution space $R^N_{\geq 0}$. This hyperplane H is the set of all solutions $\underline{v} = (v_1,...,v_N)$ of the distribution problem, which fulfill the restriction $\sum_{n=1}^{N} v_n = G$.

The third requirement is the rationality condition for maximum allocable shares of the gain. This condition has the character of a condition of collective rationality, since it mirrors the rational consideration of all N-1 actors of the so called marginal coalition $MC_n$ with $MC_n = C_0 \backslash \{A_n\} =$

*Exhibit 1.*

$$\forall \varnothing \subset AC_{n.q} \subset A: \quad \{A_n\} \subset AC_{n.q} \rightarrow c\left(\{A_n\} \middle| AC_{n.q}\right) = c\left(AC_{n.q}\right) - \sum_{m \in \left(IN_{n.q} \setminus \{n\}\right)} v_{m.\max} \qquad (4)$$

$\{A_1,...,A_{n-1,}A_{n+1},...,A_N\}$ to grant actor $A_n$ at most the share $v_{n.max}$ of the efficiency gain G, so that the efficiency gain G would decrease, if actor $A_n$ would leave the grand coalition $C_0 = \{A_1,...,A_N\}$. This rationality condition satisfies the following:

$$\forall n = 1,..., N \quad \forall v_n \in R_{\geq 0}:$$
$$v_n \leq v_{n.\max} \wedge v_{n.\max} = c(C_0) - c(MC_n) = G - c(MC_n) \qquad (3)$$

In the solution space $R^N_{\geq 0}$ the point in which the maximum allocable share $v_{n.max}$ of the efficiency gain G is assigned to each actor $A_n$, is called the upper bound UB or ideal point for the distribution of efficiency gains G.

The fourth requirement for the solution concept is a rationality condition for minimum allocable shares of the efficiency gain. This condition has the character of a condition of collective rationality as well, since the condition reflects the rational consideration of all N-1 actors of the marginal coalition $MC_n$ with $MC_n = C_0 \setminus \{A_n\}$, to grant actor $A_n$ at least the share $v_{n.min}$ of the efficiency gain G, so that she or he could believably threaten to found at least one so called outsider coalition $AC_{n.q}$. An outsider coalition is a coalition $AC_{n.q}$ of former actors of the supply network, who leave the grand coalition $C_0 = \{A_1,...,A_N\}$, at least hypothetically, and has at least the actor $A_n$ as „leader". Since the same actor $A_n$ can lead several outsider coalitions, the second index q is used to differentiate all outsider coalitions led by the same actor $A_n$. Furthermore, an outsider coalition can never contain all actors of the grand coalition $C_0$, since no non-empty residual coalition $RC_{n.q} = C_0 \setminus AC_{n.q}$ would exist, whose actors could generate the efficiency gain G to be distributed.

It is important for the solution concept of the τ-value which outsider coalitions $AC_{n.q}$ enable an actor $A_n$ to threaten in a believable manner. In this paper, it is assumed that the characteristic function is partially known due to the amounts $c(AC_{n.q})$ for each outsider coalition led by an actor $A_n$. The actor $A_n$ offers all other actors of the outsider coalition $AC_{n.q}$ an optimal incentive to defect. This incentive consists of so called side payments and it ensures that the utility of each other actor out of the considered outsider coalition $AC_{n.q}$ is the same as his or her maximum utility in the grand coalition $C_0$. In this case the actors in an outsider coalition have no incentives to remain in the grand coalition $C_0$. The operationalization of the side payments takes place in the following way with the amount $c(\{A_n\}|AC_{n.q})$ realizable by actor $A_n$ in the outsider coalition $AC_{n.q}$ and with the index set $IN_{n.q}$ of all indices of actors in this outsider coalition: (see exhibit 1).

The amounts $c(\{A_n\}|AC_{n.q})$ utilized by actor $A_n$ in threatening to found an outsider coalition may be negative. This can be explained by two reasons. Firstly, the sum $\sum_{m \in \left(IN_{n.q} \setminus \{n\}\right)} v_{m.\max}$ of the side payments can be greater than the amount $c(AC_{n.q})$ realized by the outsider coalition $AC_{n.q}$. In this case the leading actor $A_n$ has to withdraw the partial amount

$$\sum_{m \in \left(IN_{n.q} \setminus \{n\}\right)} v_{m.\max} - c\left(AC_{n.q}\right)$$

from savings or even incur debt. Secondly, if actor $A_n$ is the sole actor in the outsider coalition $AC_{n.q}$

*Exhibit 2.*

$$c_{n.1} = c\left(\left\{A_n\right\} \mid AC_{n.q}\right) = c\left(\left\{A_n\right\}\right) \quad for \quad AC_{n.q} = \left\{A_n\right\} \tag{5}$$

$$c_{n.2} = \max\left\{ c\left(\left\{A_n\right\} \mid AC_{n.q}\right) = c\left(AC_{n.q}\right) - \sum_{m \in \left(IN_{n.q} \setminus \{n\}\right)} v_{m.\max} \;\middle|\; \begin{matrix} \ldots \\ \varnothing \subset AC_{n.q} \subset A \wedge \left\{A_n\right\} \subset AC_{n.q} \end{matrix} \right\}$$

and thus the above mentioned side payments are not required, then the amount c({$A_n$}) may be negative as well. Actor $A_n$, for example, may be not competitive in the market without collaborating in the supply network. In both cases with c({$A_n$}|$AC_{n.q}$)<0 a threat would not be believable. Thus, both cases are excluded from the rationality condition for minimum allocable shares of the efficiency gain. Summa summarum the complete rationality condition for minimum shares of the efficiency gain to be allocated is as follows:

$$\forall n = 1,\ldots,N \quad \forall v_n \in R_{\geq 0}:$$
$$v_n \geq v_{n.\min} \wedge v_{n.\min} = \max\left\{ c_{n.1}; c_{n.2}; 0\right\}$$

with (see exhibit 2)

The lower bound LB for the distribution of the efficiency gain G is that point in the solution space $R^N_{\geq 0}$, in which the minimum allocable share $v_{n.min}$ of the efficiency gain G is assigned to each actor $A_n$ with n = 1,...,N. The lower bound LB is often called the threat point (Kuhn et al., 1996); in the τ-value literature especially, it is called minimal right vector (Bilbao et al., 2002; Tijs, 1981).

The following integrity condition is introduced for the relation of the lower bound LB to the upper bound UB for the shares of the efficiency gain G to be distributed as well as for the hyperplane H for the compliance with the efficiency condition to avoid particular complications, which lie outside the scope of this paper:

$$\forall L_v, LB, UB \in R^N_{\geq 0} \ \forall G \in R_+:$$

$$\forall L_v, LB, UB \in \mathbb{R}^N \ \forall G \in \mathbb{R}_+:$$

$$\left( L_v = \begin{pmatrix} v_1 \\ \ldots \\ v_N \end{pmatrix} \wedge LB = \begin{pmatrix} v_{1.\min} \\ \ldots \\ v_{N.\min} \end{pmatrix} \wedge UB = \begin{pmatrix} v_{1.\max} \\ \ldots \\ v_{N.\max} \end{pmatrix} \right)$$
$$\rightarrow \left( \sum_{n=1}^{N} v_{n.\min} \leq \sum_{n=1}^{N} v_n = G \leq \sum_{n=1}^{N} v_{n.\max} \wedge LB \leq UB \right) \tag{6}$$

The fifth requirement for the solution concept is that the concept must fulfill an operational fairness criterion: the greater the bargaining power of an actor $A_n$, the greater her or his share $v_n$ of the efficiency gain G. The bargaining power of an actor $A_n$ is affected by two opposed effects: on the one hand the bargaining power of an actor $A_n$ is measured by the contribution the actor would make, if she or he would take part in the marginal coalition $MC_n$ with $MC_n = C_0 \setminus \{A_n\}$ and thus would make this marginal coalition a grand coalition $C_0$. This positive network effect is the maximum allocable share $v_{n.max}$ of the efficiency gain in the third requirement mentioned above. On the other hand, the bargaining power of an actor $A_n$ is measured by her or his threat potential build up of the believable threat to found at least one outsider coalition $AC_{n.q}$. This negative network effect has been specified as minimum allocable share $v_{n.min}$ of the efficiency gain in the fourth requirement above.

The good reasons for accepting the distribution of the efficiency gain G among the actors $A_n$ in a supply network, and therefore a feasible solution $L_v$ of the generic distribution problem as fair, can be specified as follows: it is regarded as fair to grant actor $A_n$ a share $v_n$ of the efficiency gain G which is positively correlated with the actor's contribution to build the grand coalition (positive network effect) and with the threat potential to prevent the grand coalition from coming into existence (negative network effect).

The aforementioned characterization of a fair distribution of the efficiency gain G is mainly qualitative. Thus, this characterization offers some scope of interpretation regarding the numerical determination of the shares $v_n$ for all actors $A_n$ in a supply network. A quantification of the fairness criterion in the form of a calculation rule for the $\tau$-value is desirable. This calculation rule should be as easy as possible to understand to gain acceptance in management practice. The calculation rule employed is the following *new type* of $\tau$-value formula:

$$\forall n = 1, ..., N :$$

$$v_{n.\tau} = \begin{cases} \alpha \bullet \dfrac{v_{n.\max}}{\sum_{n=1}^{N} v_{n.\max}} \bullet G + \beta \bullet \dfrac{v_{n.\min}}{\sum_{n=1}^{N} v_{n.\min}} \bullet G; & if \sum_{n=1}^{N} v_{n.\max} \neq \sum_{n=1}^{N} v_{n.\min} \\ v_{n.\max} = v_{n.\min} & ; \quad if \sum_{n=1}^{N} v_{n.\max} = \sum_{n=1}^{N} v_{n.\min} \end{cases} \qquad (7a)$$

with:

$$\alpha = \frac{G - \sum_{n=1}^{N} v_{n.\min}}{\sum_{n=1}^{N} v_{n.\max} - \sum_{n-1}^{N} v_{n.\min}} \bullet \frac{\sum_{n=1}^{N} v_{n.\max}}{G} \qquad (7b)$$

$$\beta = \frac{\sum_{n=1}^{N} v_{n.\max} - G}{\sum_{n=1}^{N} v_{n.\max} - \sum_{n=1}^{N} v_{n.\min}} \bullet \frac{\sum_{n=1}^{N} v_{n.\min}}{G}$$

The *new type* of $\tau$-value formula is in its form in equation (7a,b) unknown in the literature. It has been developed to work out the economic interpretation of the $\tau$-value in a more concise way. For interested readers, the derivation of equation (7a,b) can be found in the Appendix. The special characteristics of the calculation rule for $\tau$-value according to equation (7a,b) are discussed in the following.

Firstly, this calculation rule is characterized by capturing the bargaining power of the actor by two summands. The first summand reflects the bargaining power of actor $A_n$ due to her or his contribution $v_{n.max}$ to the coalition (positive network effect) by means of the share $v_{n.max}$ of the efficiency gain. The second summand represents the bargaining power of actor $A_n$ due to her or his threat potential $v_{n.min}$ to dissolve the grand coalition (negative network effect) by means of the share $v_{n.min}$ of the efficiency gain.

Secondly, the coalition contribution and the threat potential of actor $A_n$ are not measured absolutely, but are relativized with respect to the sums of the coalition contributions and the threat potentials respectively. This is a normalization of the coalition contribution and of the threat potential of actor $A_n$ regarding the upper bound UB and the lower bound LB, respectively.

Thirdly, the coalition contribution and the threat potential of actor $A_n$ are weighted with $\alpha$ and $\beta$. This weighting allows representation of the $\tau$-value in a more compact but equivalent form (cf. the appendix for the proof) as the linear combination of the upper bound (ideal point) and the lower bound (threat point) lying on the hyperplane H within the solution space and satisfying the efficiency condition (see figure 1):

$$\forall n = 1, ..., N : \; v_{n.\tau} = \gamma \bullet v_{n.\max} + \left(1 - \gamma\right) \bullet v_{n.\min} \qquad (8a)$$

with ("∨" is the logical OR operator):

*Figure 1. The τ-value*

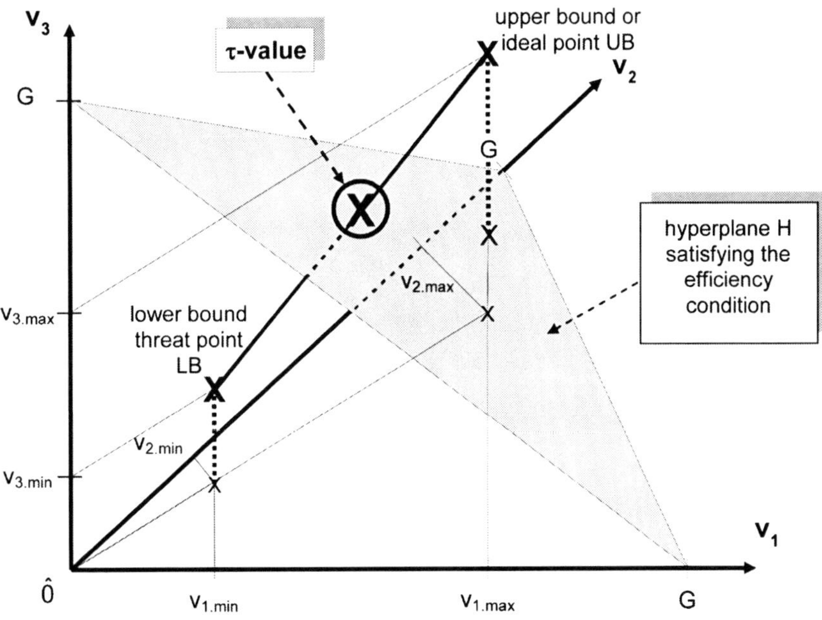


*Exhibit 3.*

$$\sum_{k=1}^{N} v_k^* = v_n^* + \sum_{m=1,m\neq n}^{N} v_m^* = v_{n.\tau} + \varepsilon + \sum_{m=1,m\neq n}^{N} v_m^*$$

$$\gamma = \begin{cases} \dfrac{G - \sum_{n=1}^{N} v_{n.\min}}{\sum_{n=1}^{N} v_{n.\max} - \sum_{n=1}^{N} v_{n.\min}} & ; \ if \sum_{n=1}^{N} v_{n.\max} \neq \sum_{n=1}^{N} v_{n.\min} \\ 0 \vee 1 & ; \ if \sum_{n=1}^{N} v_{n.\max} = \sum_{n=1}^{N} v_{n.\min} \end{cases} \tag{8b}$$

The τ-value according to equation (8a), which is the most widespread representation in literature (e.g. Branzei et al., 2005; Curiel, 1997; Tijs, 1987), represents a compromise solution for the distribution problem. It is characterized by two properties. Firstly, the compromise solution is Pareto efficient in terms of the above mentioned efficiency condition (2). Thus, the τ-value can be understood as an efficient compromise (Driessen, 1987). The Pareto efficiency of the τ-value results from an elementary consideration: If attempts were made to make at least one actor $A_n$ better off ($v_{n.\tau}^* > v_{n.\tau}$) in comparison to the share $v_{n.\tau}$ of the efficiency gain that the actor receives according to the τ-value $v_\tau$, by means of an alternative solution v* for the efficiency gain distribution problem without making any other actor $A_m$ with $m \neq n$ worse off ($v_{m.\tau}^* \geq v_{n.\tau}$), then it should hold for all assigned shares of the efficiency gain with $\varepsilon > 0$: (see Exhibit 3). Because of $v_{m.\tau}^* \geq v_{n.\tau}$ it follows:

$$\sum_{k=1}^{N} v_k^* \geq v_{n.\tau} + \varepsilon + \sum_{m=1,m\neq n}^{N} v_{m.\tau} \,.$$

Since by definition

$$v_{n.\tau} + \sum_{m=1,m\neq n}^{N} v_{m.\tau} = \sum_{k=1}^{N} v_{k.\tau} \text{ holds,}$$

it follows $\sum_{k=1}^{N} v_k^* \geq \sum_{n=1}^{N} v_{n.\tau} + \varepsilon$ and – because of $\varepsilon > 0$ – $\sum_{k=1}^{N} v_k^* > \sum_{n=1}^{N} v_{n.\tau}$. From the efficiency condition (2) for the τ-value $\sum_{n=1}^{N} v_{n.\tau} = G$ follows immediately: $\sum_{k=1}^{N} v_k^* > G$. This contradicts the prerequisite that only the efficiency gain G can be distributed among the actors $A_n$ of the supply network. It is therefore *impossible* to realize an alternative solution v* for the efficiency gain distribution problem that would make at least one actor $A_n$ better off without making any other actor $A_m$ worse off. Thus, the τ-value is Pareto efficient (q.e.d). Secondly, the compromise solution represents the simplest compromise between the threat point and the ideal point. It is impossible to construct a simpler connection between these two points in the solution space $R^N_{\geq 0}$ than the direct rectilinear distance determined by equation (8a).

The characterization of the τ-value as a compromise solution for the distribution problem yields a further good reason to accept the distribution of the efficiency gains as fair, since intuitive preconceptions about what is accepted as fair contain the normative connotation that fair distributions should be based on a compromise between the interests of the involved actors. In the τ-value solution concept these interests are operationally specified with the aid of the threat point and the ideal point. This is illustrated in Figure 1 below.

For each instance of the generic distribution problem which fulfills the integrity condition (6) exactly one solution exists that satisfies the above mentioned requirements of individual and collective rationality as well as the requirements of the efficiency and fairness of the efficiency gain distribution. This solution is the τ-value which can be determined by one of the equivalent calculation rules (7a and b) versus (8a and b).

## NUMERICAL EXAMPLE FOR THE SUPPLY NETWORK DISTRIBUTION GAME

### Goal and Scope of the Numerical Example

In the following, a numerical example is presented which shows how the τ-value can be applied concretely in management practice to solve the problem of a fair distribution of efficiency gains in supply networks. For illustrative purposes, a simply structured hypothetical example is considered. It is restricted to the number of 5 actors. The numerical values are chosen so that the required calculations remain relatively easy. The calculations can also be carried out by computer using spreadsheet software such as Microsoft Excel®. Furthermore, the following example should illustrate what information is required in management practice in order to apply the τ-value for the calculation of efficiency gain distributions. Finally, the fictive example has the virtue of being usable as some kind of benchmark for the comparison of game theoretic solution concepts for distribution problems, since, in principle, it uses the same data as used in Curiel (1997) and Fromen (2004) to calculate the Shapley value and the Nucleolus as alternative game theoretic solution concepts for distribution problems. In this regard, the following benchmark example is in keeping with the pertinent literature. It should be noted, however, that the example is not concerned with the gathering of information. In management practice, it could be particularly problematic to obtain all values of the characteristic function for possible coalitions. This information gathering problem is not specific to the τ-value, but concerns all solution concepts of cooperative game theory. This is the reason why the information gathering problem is not approached closer in this paper.

*Table 1.*

| Values of the characteristic function c for all coalitions $C_m$ | | | | | |
|---|---|---|---|---|---|
| $C_m$ | $c(C_m)$ | $C_m$ | $c(C_m)$ | $C_m$ | $c(C_m)$ |
| $C_0=\{A_1,A_2,A_3,A_4,A_5\}$ | 100,000 | | | | |
| $C_1=\{A_1\}$ | 0 | $C_{11}=\{A_2,A_4\}$ | 25,000 | $C_{21}=\{A_1,A_4,A_5\}$ | 55,000 |
| $C_2=\{A_2\}$ | 0 | $C_{12}=\{A_2,A_5\}$ | 30,000 | $C_{22}=\{A_2,A_3,A_4\}$ | 50,000 |
| $C_3=\{A_3\}$ | 0 | $C_{13}=\{A_3,A_4\}$ | 30,000 | $C_{23}=\{A_2,A_3,A_5\}$ | 55,000 |
| $C_4=\{A_4\}$ | 5,000 | $C_{14}=\{A_3,A_5\}$ | 35,000 | $C_{24}=\{A_2,A_4,A_5\}$ | 65,000 |
| $C_5=\{A_5\}$ | 10,000 | $C_{15}=\{A_4,A_5\}$ | 45,000 | $C_{25}=\{A_3,A_4,A_5\}$ | 70,000 |
| $C_6=\{A_1,A_2\}$ | 0 | $C_{16}=\{A_1,A_2,A_3\}$ | 25,000 | $C_{26}=\{A_1,A_2,A_3,A_4\}$ | 60,000 |
| $C_7=\{A_1,A_3\}$ | 5,000 | $C_{17}=\{A_1,A_2,A_4\}$ | 35,000 | $C_{27}=\{A_1,A_2,A_3,A_5\}$ | 65,000 |
| $C_8=\{A_1,A_4\}$ | 15,000 | $C_{18}=\{A_1,A_2,A_5\}$ | 40,000 | $C_{28}=\{A_1,A_2,A_4,A_5\}$ | 75,000 |
| $C_9=\{A_1,A_5\}$ | 20,000 | $C_{19}=\{A_1,A_3,A_4\}$ | 40,000 | $C_{29}=\{A_1,A_3,A_4,A_5\}$ | 80,000 |
| $C_{10}=\{A_2,A_3\}$ | 15,000 | $C_{20}=\{A_1,A_3,A_5\}$ | 45,000 | $C_{30}=\{A_2,A_3,A_4,A_5\}$ | 90,000 |

## Description of the Numerical Example

In the numerical example, a supply network with 5 actors $A_1,\ldots,A_5$ is considered. In the last business year, the actors in the supply network have jointly realized an efficiency gain G in the amount of $100,000. This efficiency gain should be distributed among the supply network actors in a manner that these actors accept as fair. First of all, to ensure the comparability with other game theoretical solution concepts like the Shapley value and the Nucleolus it is assumed that the values of the characteristic function for the supply network distribution game are known completely. Thus, the value $c(C_m)$ is known for every possible coalition which can be formed from the set of actors $A = \{A_1,\ldots,A_5\}$. The values $c(C_m)$ are given in Table 1 for all $2^5-1=31$ coalitions $C_m$ with $m=0,1,2,\ldots,30$.

## Calculation of the τ-value for the Supply Network Distribution Game

A prerequisite for the calculation of the τ-value as a solution $\underline{v}_\tau$ with $\underline{v}_\tau = (v_{1,\tau},\ldots, v_{N,\tau})$ for the supply network distribution game is that the values of the characteristic function c for all three types of coalitions are available. That is, $c(C_0)$ has to be available for the grand coalition $C_0 = \{A_1,\ldots,A_5\}$, while $c(MC_n)$ is required for each marginal coalition $MC_n$ with $n = 1,\ldots,5$ and $c(AC_{n.q})$ has to be known for each outsider coalition $AC_{n.q}$.

The value $c(C_0) = 100{,}000$ for the grand coalition $C_0$ is immediately available from table 1, since, according to the efficiency condition, the whole efficiency gain $G = 100{,}000$ has to be distributed exactly among all 5 actors $A_1,\ldots,A_5$ in the supply network.

The values $c(MC_n)$ for the marginal coalitions $MC_n$ with $n = 1,\ldots,5$ can be determined with the aid of the definition $MC_n = C_0 \setminus \{A_n\}$. The results are given in Table 2.

The values $c(AC_{n.q})$ for the outsider coalitions $AC_{n.q}$ with $n = 1,\ldots,5$ can be obtained immediately from Table 1. However, the calculation of these values $c(AC_{n.q})$ requires a tremendous amount of work, since 75 $(= 5 \bullet 15 = n \bullet q)$ feasible outsider coalitions have to be considered. Therefore, this calculation is omitted owing to space restrictions.

The components $v_{n.max}$ of the upper bound UB (ideal point) are calculated with equation (3) on the basis of the values $c(C_0)$ and $c(MC_n)$. The value $c(C_0)$ is immediately given by the efficiency gain

*Table 2.*

| Values of the characteristic function c for all marginal coalitions $MC_n$ | |
|---|---|
| $MC_n$ | $c(MC_n)$ |
| $MC_1$ | $c(\{A_1,...,A_5\} \setminus \{A_1\}) = c(\{A_2,A_3,A_4,A_5\}) = 90{,}000$ |
| $MC_2$ | $c(\{A_1,...,A_5\} \setminus \{A_2\}) = c(\{A_1,A_3,A_4,A_5\}) = 80{,}000$ |
| $MC_3$ | $c(\{A_1,...,A_5\} \setminus \{A_3\}) = c(\{A_1,A_2,A_4,A_5\}) = 75{,}000$ |
| $MC_4$ | $c(\{A_1,...,A_5\} \setminus \{A_4\}) = c(\{A_1,A_2,A_3,A_5\}) = 65{,}000$ |
| $MC_5$ | $c(\{A_1,...,A_5\} \setminus \{A_5\}) = c(\{A_1,A_2,A_3,A_4\}) = 60{,}000$ |

*Table 3.*

| Components $v_{n.max}$ of the upper bound UB of the τ-value | |
|---|---|
| $A_n$ | $v_{n.max}$ |
| $A_1$ | $c(C_0)$-$c(MC_1)$ = 100,000 - 90,000 = 10,000 |
| $A_2$ | $c(C_0)$-$c(MC_2)$ = 100,000 - 80,000 = 20,000 |
| $A_3$ | $c(C_0)$-$c(MC_3)$ = 100,000 - 75,000 = 25,000 |
| $A_4$ | $c(C_0)$-$c(MC_4)$ = 100,000 - 65,000 = 35,000 |
| $A_5$ | $c(C_0)$-$c(MC_5)$ = 100,000 - 60,000 = 40,000 |

G to be distributed: $c(C_0) = G = 100{,}000$. Thus, the components $v_{n.max}$ of the upper bound UB of the τ-value are those in Table 3.

The components $v_{n.min}$ of the lower bound LB (threat point) of the τ-value are calculated with equation (5) for each of the 5 actors $A_1$ to $A_5$ in the supply network. This calculation is shown exemplarily for $A_4$:

$$v_{4.\min} = \max\left\{c_{4.1}; c_{4.2}; 0\right\} = \max\left\{5{,}000; 5{,}000; 0\right\} = 5{,}000$$

because of (see Exhibit 4)

It follows from the components $v_{n.max}$ of the upper bound UB and from the components $v_{n.min}$ of the lower bound LB calculated above, that the standard case for the calculation of the τ-valuc with $\sum_{n=1}^{N} v_{n.\max} \neq \sum_{n=1}^{N} v_{n.\min}$ applies.

According to equation (8b) the linear factor $\gamma$ is as follows: (see Exhibit 5)

Then, components $v_{n.\tau}$ of the τ-value (Table 4) are calculated with equation (8a) as the linear combination of the upper bound (ideal point) and the lower bound (threat point).

In the end, there is exactly one τ-Value $\underline{v}_\tau$ = $^1/_{23}$ • (170,000; 340,000; 425,000; 625,000; 740,000) as a unique solution for the distribution problem and thus for the supply network distribution game.

*Table 4.*

| Components $v_{n.\tau}$ of the τ-value for the supply network distribution game | |
|---|---|
| $A_n$ | $v_{n.\tau}$ |
| $A_1$ | $^{17}/_{23}$•10,000 + $^{6}/_{23}$•0 = $^{1}/_{23}$•170,000 |
| $A_2$ | $^{17}/_{23}$•20,000 + $^{6}/_{23}$•0 = $^{1}/_{23}$•340,000 |
| $A_3$ | $^{17}/_{23}$•25,000 + $^{6}/_{23}$•0 = $^{1}/_{23}$•425,000 |
| $A_4$ | $^{17}/_{23}$•35,000 + $^{6}/_{23}$•5,000 = $^{1}/_{23}$•625,000 |
| $A_5$ | $^{17}/_{23}$•40,000 + $^{6}/_{23}$•10,000 = $^{1}/_{23}$•740,000 |

## CONCLUSION, LIMITATIONS AND NEEDS FOR FURTHER RESEARCH

For managerial purposes, implementing the τ-value based solution concept presented in this paper in Supply Chain Management Software Systems or Enterprise Resource Planning Systems with a Supply Chain Management module could be considered. Software systems could then calculate a recommendation for allocating the efficiency gain among the actors of a supply network. Yet such a calculation of the shares of the efficiency gains for the actors can be taken as the basis for discussion only. This calculation at best conceals aspects of rational discourse about a fair distribution of efficiency gains. But this rational discourse, in turn, depends on the central prerequisite that the actors in a supply network accept the τ-value based solution concept for the above mentioned generic distribution problem as fair. This paper has presented good reasons for the fairness of the τ-value based distribution of the efficiency gains in supply networks. Moreover,

*Exhibit 4.*

$$c_{4.1} = c\left(\left\{A_4\right\} \mid AC_{4.1}\right) = c\left(\left\{A_4\right\}\right) \quad \textit{for} \quad AC_{4.1} = \left\{A_4\right\} = 5,000$$

$$c_{4.2} = \max\left\{c\left(\left\{A_4\right\} \mid AC_{4.q}\right) = c\left(AC_{4.q}\right) - \sum_{m \in \left(IN_{4.q} \setminus \{4\}\right)} v_{m.\max} \;\middle|\; q = 2,\dots,15\right\} = 5,000$$

*Exhibit 5.*

$$\gamma = \frac{G - \sum_{n=1}^{N} v_{n.\min}}{\sum_{n=1}^{N} v_{n.\max} - \sum_{n=1}^{N} v_{n.\min}}$$

$$= \frac{100,000 - \left(0+0+0+5,000+10,000\right)}{\left(10,000+20,000+25,000+35,000+40,000\right) - \left(0+0+0+5,000+10,000\right)} = \frac{17}{23} \approx 0,74$$

five requirements were introduced as evaluation criteria to assess the suitability of the τ-value solution concept for the distribution of efficiency gains in supply networks.

It would be far removed from reality, however, to assume that disputes about the fairness of proposals for the distribution of efficiency gains are conducted in rational discourse in management practice. Instead, other factors such as „political" strategies when staging a distribution conflict, the power of the involved actors to enforce their interests as well as personal motives such as vanity or plans for revenge can play major roles in management practice. These additional factors are outside rationality and, for the most part, cannot be covered by the game theoretic concepts discussed in this paper. Only aspects of power are partially integrated in the τ-value solution concept by means of the threat point. Since most of the additional factors mentioned previously are not captured, it would be naive to assume that definitive solutions for an efficiency gain distribution problem could be determined with the aid of a game theoretic solution concept, or a software based on such concepts, in actual management practice. Instead, solutions obtained by the application of game theoretic solution concepts can serve only as recommendations for problem solutions which can be derived with good reasons under the assumption of rational actors. If, and to what extent, the recommendations are followed (provided the assumption of rationality of action is shared at all) can neither be answered in this paper nor by game theory.

Moreover, some basic limitations for the application of the τ-value in practice must be considered. Firstly, it is assumed that managers of supply networks have information about the efficiency gain of the whole supply network realized by the grand coalition as well as the values of the characteristic function for all marginal and outsider coalitions. In management practice, it is often difficult to gather this information, or it is at least difficult to procure it in the desired accuracy. The τ-value solution concept can only be applied if the required information is actualy available. This is a significant restriction for the applicability of the τ-value solution concept. Nevertheless, this restriction must be regarded with some reservations.

Firstly, the restriction applies to all solution concepts of cooperative game theory. It is no unique deficiency of the τ-value. The critique that applicability can always be questioned based on doubts pertinent to the procurability of information is something applicable to other approaches to the operationalization of fairness. If such information is missing, it is in all cases impossible to operationalize fairness. Secondly, two ways were shown by which the efficiency gain of a supply network could be determined in management practice. If the effort of comparative modeling is eschewed, rough estimations based on expert opinions can be considered. There are good reasons for both ways of determining an efficiency gain. But the trade off between high accuracy at high effort and low accuracy at low effort for the determination of information cannot be eliminated. Thirdly, it has been shown that the τ-value is in principle characterized by lower informational prerequisites than other game theoretic solution concepts such as the Shapley value and the Nucleolus. This owes to the fact that the calculation of the τ-value only requires the values of the characteristic functions for all marginal and outsider coalitions, while the alternative solution concepts presuppose that the values of the characteristic function are known for all combinatorial possible coalitions. Although the τ-value is superior owing to its relatively low informational prerequisites, it can entail tremendous effort to obtain the values of the characteristic function for all outsider coalitions. Future research should therefore analyze strategies to reduce the great number of outsider coalitions in order to reduce the information premises as well as the calculation effort of the τ-value. For this purpose, plausibility considerations of how to remove the outsider coalitions that are not economically plausible from the set of all theoretically possible outsider coalitions are promising approaches. Advancements of the τ-value solution concept will be considered in a future paper.

A further problem in the application of the τ-value in management practice is the question of who determines the values $v_{n.max}$ and $v_{n.min}$ for the ideal point and the threat point respectively – both of which play very important roles in the calculation of the τ-value according to the equivalent equations (7a,b) and (8a,b). The τ-value solution concept does not make a clear statement concerning this problem. This also holds true for all other solution concepts of cooperative game theory. It is nevertheless desirable to answer this question for the practical acceptance of each game theoretic solution concept for efficiency gain distribution problems. There are two alternative answers to this question.

Firstly, one may have a supply network coordinated by a focal company. Such focal networks are widespread in practice, for example in the automobile industry and in the computer industry. In this case, it is obvious that the values $v_{n.max}$ and $v_{n.min}$ for the ideal point and the threat point respectively are estimated by the focal company. If the companies do not accept these estimations, their only way out is to leave the focal network due to their low network power. Secondly, an impartial third party trusted by all actors of the supply network could be contracted as a consultant to estimate the values $v_{n.max}$ and $v_{n.min}$ for the ideal point and the threat point respectively. This approach might rarely be realizable in practice, since recourse to an impartial third party is usually only practiced in exceptional conflict situations such as an unsuccessful collective bargaining. Moreover, recourse to an impartial third party involves the presupposition that the actors in the supply network are in conflict about the values $v_{n.max}$ and $v_{n.min}$ and thus it stands in contrast to the widely accepted notion that supply networks could only survive successfully in the long-run if they are based on mutual trust. Thirdly, it is conceivable to formulate the exchange of opinions between the actors about the values $v_{n.max}$ and $v_{n.min}$ which they regard as "realistic" as a non cooperative coalition formation game. Such approaches do exist in the game-theory literature. But in comparison to the τ-value and other solution concepts from cooperative game theory, such games require still more information about the opinions of the actors with respect to the values assumed to be known. Furthermore, it cannot be ensured that such non cooperative coalition formation games lead to a balance of opinions about the relevant values $v_{n.max}$ and $v_{n.min}$ of all actors in the supply network. This is why this approach from the non cooperative game theory is not viewed as promising for management practice and was thus not discussed in detail in this paper.

All in all, at this point of time the question of who estimates the values $v_{n.max}$ and $v_{n.min}$ for the ideal point and the threat point respectively cannot be answered in a satisfying or at least universally valid way. Except for the case of focal networks no convincing concept to answer this question is known which would be broadly accepted in management practice. Thus, there is considerable demand for further research to develop game theoretic solution concepts for management practice.

Furthermore, a basic limitation of the game theoretic solution concept on basis of the τ-value proposed here is that it can only serve to distribute already realized efficiency gains in an already existing supply network. It is, in other words, a static solution concept. A dynamization, however, is conceivable in two directions. Firstly, models from principal agent theory or findings from applied psychology could be utilized to analyze how incentives can be used in an existing supply network to increase the realizable efficiency gain. Secondly, one may consider analyzing under which conditions supply networks can be composed of actors with at least partially conflicting interests and how these supply networks fall apart. For this purpose, recourse to the type of coalition formation games mentioned above is recommended.

Such dynamizations would require integration of the τ-value solution concept in more comprehensive models from principal agent theory and/or from non cooperative game theory. Such complex modelings are not known up to now and require considerable additional research effort.

In addition, one fundamental critique about such integrative dynamical modelings can be voiced: while the scope of the addressed real problems is extended, the theoretical assumptions and the information requirements are considerably compounded at the same time. In keeping with the questions posed above regarding the applicability of the τ-value solution concept in management practice it is exceptionally difficult to substantiate the practicability of such integrative dynamical modelings under realistic operating conditions.

## REFERENCES


Bilbao, J. M., Jiménez-Losada, A., Lebrón, E., & Tijs, S. H. (2002). The τ-value for games on Matroids. *Top (Madrid)*, *10*(1), 67–81. doi:10.1007/BF02578941

Branzei, R., Dimitrov, D., & Tijs, S. (2005). *Models in Cooperative Game Theory – Crisp, Fuzzy, and Multi-Choice Games*. Berlin: Springer.

Croson, R., & Donohue, K. (2006). Behavioral causes of the bullwhip effect and the observed value of inventory information. *Management Science*, *52*(3), 323–336. doi:10.1287/mnsc.1050.0436

Curiel, I. (1997). *Cooperative Game Theory and Applications – Cooperative Games Arising from Combinatorial Optimization Problems*. Boston: Kluwer Academic Publishers.

Derks, J., & Tijs, S. (2000). On merge properties of the Shapley value. *International Game Theory Review*, *2*(4), 249–257. doi:10.1142/S0219198900000214

Driessen, T. (1987). The τ-value: a survey . In Peters, H. J. M., & Vrieze, O. J. (Eds.), *Surveys in Game Theory and Related Topics* (pp. 209–213). Amsterdam, The Netherlands: Stichting Mathematisch Centrum.

Driessen, T., & Tijs, S. (1982). The τ-value, the nucleolus and the core for a subclass of games . In Loeffel, H., & Stähly, P. (Eds.), *Methods of Operations Research 46* (pp. 395–406). Königstein, Germany: Verlagsgruppe Athenäum Hain Hanstein.

Driessen, T. S. H. (1985). *Contributions to the Theory of Cooperative Games: The τ-Value and k-Convex Games.* Unpublished doctoral dissertation, University of Nijmegen, The Netherlands.

Driessen, T. S. H., & Tijs, S. H. (1983). *Extensions and Modifications of the τ-Value for Cooperative Games* (Report No. 8325). Nijmegen, The Netherlands: University of Nijmegen, Department of Mathematics.

Driessen, T. S. H., & Tijs, S. H. (1985). The τ-value, the core and semiconvex games. *International Journal of Game Theory*, *14*(4), 229–247. doi:10.1007/BF01769310

Fehr, E., & Schmidt, K. M. (1999). A theory of fairness, competition, and cooperation. *The Quarterly Journal of Economics*, *114*(3), 817–868. doi:10.1162/003355399556151

Fromen, B. (2004). *Fair Distribution in Enterprise Networks – Solution Concepts Stemming from Cooperative Game Theory*. Doctoral dissertation, University of Duisburg-Essen, Germany. Wiesbaden, Germany: Gabler.

Krol, B., Keller, S., & Zelewski, S. (2005). E-logistics ocercome the bullwhip effect. *International Journal of Operations and Quantitative Management*, *11*(4), 281–289.

Kuhn, H. W., Harsanyi, J. C., Selten, R., Weibull, J. W., Damme, E., van Nash, J. F. Jr, & Hammerstein, P. (1996). The work of John Nash in game theory – Nobel seminar, december 8, 1994. *Journal of Economic Theory*, *69*(1), 153–185. doi:10.1006/jeth.1996.0042

Lee, H. L., Padmanabhan, V., & Whang, S. (1997). Information distortion in a supply chain: The bullwhip effect. *Management Science*, *43*(4), 546–558. doi:10.1287/mnsc.43.4.546

Mahdavi, I., Mohebbi, S., Cho, N., Paydar, M. M., & Mahdavi-Amiri, N. (2008). Designing a dynamic buyer-supplier coordination model in electronic markets using stochastic Petri nets. *International Journal of Information Systems and Supply Chain Management*, *1*(3), 1–20.

McCullen, P., & Towill, D. (2002). Diagnostics and reduction of bullwhip in supply chains. *Supply Chain Management*, *7*(3), 164–179. doi:10.1108/13598540210436612

Meertens, M. A., & Potters, J. A. M. (2006). The nucleolus of trees with revenues. *Mathematical Methods of Operations Research*, *64*(2), 363–382. doi:10.1007/s00186-006-0088-y

Metters, R. (1997). Quantifying the bullwhip effect in supply chains. *Journal of Operations Management*, *15*(2), 89–100. doi:10.1016/S0272-6963(96)00098-8

Pazner, E. A. (1977). Pitfalls in the theory of fairness. *Journal of Economic Theory*, *14*(2), 458–466. doi:10.1016/0022-0531(77)90146-6

Schmeidler, D. (1969). The nucleolus of a characteristic function game. *SIAM Journal on Applied Mathematics*, *17*(6), 1163–1170. doi:10.1137/0117107

Shapley, L. S. (1953). A value for n-person games. In H. W. Kuhn & A. W. Tucker (Eds.), *Contributions to the Theory of Games – Volume II, Annals of Mathematics Studies 28* (pp. 307-317). Princeton, NJ: Princeton University Press.

Tijs, S. H. (1981). Bounds for the core and the τ-value. In Moeschlin, O., & Pallaschke, D. (Eds.), *Game Theory and Mathematical Economics* (pp. 123–132). Amsterdam, The Netherlands: North Holland Publishing Company.

Tijs, S. H. (1987). An axiomatization of the tau-value. *Mathematical Social Sciences*, *13*(2), 177–181. doi:10.1016/0165-4896(87)90054-0

Tijs, S. H., & Driessen, T. S. H. (1983). *The τ-Value as a Feasible Compromise Between Utopia and Disagreement (Report 8312)*. Nijmegen, The Netherlands: University of Nijmegen, Department of Mathematics.

Tijs, S. H., & Driessen, T. S. H. (1986). Game theory and cost allocation problems. *Management Science*, *32*(8), 1015–1028. doi:10.1287/mnsc.32.8.1015

Varian, H. L. (1976). Two problems in the theory of fairness. *Journal of Public Economics*, *5*(3/4), 249–260. doi:10.1016/0047-2727(76)90018-9

Wu, F., Zsidisin, G. A., & Ross, A. D. (2007). Antecedents and outcomes of e-procurement adoption: An integrative model. *IEEE Transactions on Engineering Management*, *54*(3), 576–587. doi:10.1109/TEM.2007.900786

Xiao, T., Luo, J., & Jin, J. (2009). Coordination of a supply chain with demand stimulation and random demand disruption. *International Journal of Information Systems and Supply Chain Management*, *2*(1), 1–15.

Zhang, D. (2006). A network economic model for supply chain versus supply chain competition. *Omega – The International Journal of Management Science, 34*(3), 283-295.

Zokaei, A. K., & Simons, D. W. (2006). Value chain analysis in consumer focus improvement – A case study of the UK red meat industry. *International Journal of Logistics Management*, *17*(2), 141–162. doi:10.1108/09574090610689934

## APPENDIX: PROOF OF FORMULAS (7A) AND (7B)

In the pertinent literature, the τ-value according to equation (8a) is usually regarded as linear combination of the ideal point ($v_{n.max}$) and the threat point ($v_{n.min}$) for each actor $A_n$ with the linear factor $\gamma$ and $0 \leq \gamma \leq 1$:

$$\forall n = 1,...,N : v_{n.\tau} = \gamma \bullet v_{n.\max} + (1-\gamma) \bullet v_{n.\min} \tag{9}$$

Under consideration of the efficiency condition according to formula (2) it follows for the τ-value:

$$\begin{aligned} & \left(\forall n = 1,...,N : v_{n.\tau} = \gamma \bullet v_{n.\max} + (1-\gamma) \bullet v_{n.\min}\right) \wedge \sum_{n=1}^{N} v_{n.\tau} = G \\ \Leftrightarrow\ & \sum_{n=1}^{N} \left(\gamma \bullet v_{n.\max} + (1-\gamma) \bullet v_{n.\min}\right) = G \\ \Leftrightarrow\ & \gamma \bullet \left(\sum_{n=1}^{N} v_{n.\max} - \sum_{n=1}^{N} v_{n.\min}\right) + \sum_{n=1}^{N} v_{n.\min} = G \quad // \quad \sum_{n=1}^{N} v_{n.\max} \neq \sum_{n=1}^{N} v_{n.\min} \\ \Leftrightarrow\ & \gamma = \frac{G - \sum_{n=1}^{N} v_{n.\min}}{\sum_{n=1}^{N} v_{n.\max} - \sum_{n=1}^{N} v_{n.\min}} \end{aligned} \tag{10}$$

It must be noted that the abovementioned equivalence transformations only hold under the condition $\sum_{n=1}^{N} v_{n.\max} \neq \sum_{n=1}^{N} v_{n.\min}$. Thus, for the proof of the equations (a,b) a case differentiation is required. Initially, only the standard case $\sum_{n=1}^{N} v_{n.\max} \neq \sum_{n=1}^{N} v_{n.\min}$ is considered. Usually this is fulfilled, since, in the solution space $R_{\geq 0}^{N}$, the ideal point lies above the threat point: $\sum_{n=1}^{N} v_{n.\max} \geq \sum_{n=1}^{N} v_{n.\min}$ with $v_{n.max} > v_{n.min}$ for at least one actor $A_n$. The special case $\sum_{n=1}^{N} v_{n.\max} = \sum_{n=1}^{N} v_{n.\min}$ is regarded later. The third possible case $\sum_{n=1}^{N} v_{n.\max} \leq \sum_{n=1}^{N} v_{n.\min}$ with $v_{n.max} < v_{n.min}$ for at least one actor $A_n$ is ruled out by the integrity condition according to equation (6).

**Case A):** $\sum_{n=1}^{N} v_{n.\max} \neq \sum_{n=1}^{N} v_{n.\min}$

By the representation of the τ-value for all involved actors $A_n$ with n=1,…,N the τ-value is a solution point $L_{v.\tau}$ in the N-dimensional non-negative real solution space $R_{\geq 0}^{N}$:

$$L_{v.T} = \gamma \bullet \begin{pmatrix} v_{1.\max} \\ \dots \\ v_{N.\max} \end{pmatrix} + (1-\gamma) \bullet \begin{pmatrix} v_{1.\min} \\ \dots \\ v_{N.\min} \end{pmatrix} \quad \wedge \quad \gamma = \frac{G - \sum_{n=1}^{N} v_{n.\min}}{\sum_{n=1}^{N} v_{n.\max} - \sum_{n=1}^{N} v_{n.\min}}$$

$$\Leftrightarrow \quad L_{v.T} = \frac{G - \sum_{n=1}^{N} v_{n.\min}}{\sum_{n=1}^{N} v_{n.\max} - \sum_{n=1}^{N} v_{n.\min}} \bullet \begin{pmatrix} v_{1.\max} \\ \dots \\ v_{N.\max} \end{pmatrix} + \left(1 - \frac{G - \sum_{n=1}^{N} v_{n.\min}}{\sum_{n=1}^{N} v_{n.\max} - \sum_{n=1}^{N} v_{n.\min}}\right) \bullet \begin{pmatrix} v_{1.\min} \\ \dots \\ v_{N.\min} \end{pmatrix}$$

$$\Leftrightarrow \quad L_{v.T} = \frac{G - \sum_{n=1}^{N} v_{n.\min}}{\sum_{n=1}^{N} v_{n.\max} - \sum_{n=1}^{N} v_{n.\min}} \bullet \frac{\sum_{n=1}^{N} v_{n.\max}}{G} \bullet \begin{pmatrix} \frac{v_{1.\max}}{\sum_{n=1}^{N} v_{n.\max}} \bullet G \\ \dots \\ \frac{v_{N.\max}}{\sum_{n=1}^{N} v_{n.\max}} \bullet G \end{pmatrix} + \dots \tag{11}$$

$$\frac{\sum_{n=1}^{N} v_{n.\max} - G}{\sum_{n=1}^{N} v_{n.\max} - \sum_{n=1}^{N} v_{n.\min}} \bullet \frac{\sum_{n=1}^{N} v_{n.\min}}{G} \bullet \begin{pmatrix} \frac{v_{1.\min}}{\sum_{n=1}^{N} v_{n.\min}} \bullet G \\ \dots \\ \frac{v_{N.\min}}{\sum_{n=1}^{N} v_{n.\min}} \bullet G \end{pmatrix}$$

By the definitional introduction of the factors α and β with:

$$\alpha = \frac{G - \sum_{n=1}^{N} v_{n.\min}}{\sum_{n=1}^{N} v_{n.\max} - \sum_{n=1}^{N} v_{n.\min}} \bullet \frac{\sum_{n=1}^{N} v_{n.\max}}{G} \qquad \beta = \frac{\sum_{n=1}^{N} v_{n.\max} - G}{\sum_{n=1}^{N} v_{n.\max} - \sum_{n=1}^{N} v_{n.\min}} \bullet \frac{\sum_{n=1}^{N} v_{n.\min}}{G} \tag{12}$$

equation (11) can be represented in a simplified but equivalent form:

$$L_{v.\tau} = \begin{pmatrix} v_{1.\tau} \\ \ldots \\ v_{N.\tau} \end{pmatrix} = \alpha \bullet \begin{pmatrix} \frac{v_{1.\max}}{\sum_{n=1}^{N} v_{n.\max}} \bullet G \\ \ldots \\ \frac{v_{N.\max}}{\sum_{n=1}^{N} v_{n.\max}} \bullet G \end{pmatrix} + \beta \bullet \begin{pmatrix} \frac{v_{1.\min}}{\sum_{n=1}^{N} v_{n.\min}} \bullet G \\ \ldots \\ \frac{v_{N.\min}}{\sum_{n=1}^{N} v_{n.\min}} \bullet G \end{pmatrix} \quad (13)$$

$$= \alpha \bullet \begin{pmatrix} v_{1.\max} \\ \ldots \\ v_{N.\max} \end{pmatrix} \bullet \frac{G}{\sum_{n=1}^{N} v_{n.\max}} + \beta \bullet \begin{pmatrix} v_{1.\min} \\ \ldots \\ v_{N.\min} \end{pmatrix} \bullet \frac{G}{\sum_{n=1}^{N} v_{n.\min}}$$

For the components $v_{n.\tau}$ of the τ-value $\underline{v}_{\tau}$ and the corresponding solution point $L_{v.\tau}$ in the solution space $R^{N}_{\geq 0}$ follows immediately:

$$\forall n = 1,\ldots,N: \; v_{n.\tau} = \alpha \bullet \frac{v_{n.\max}}{\sum_{m=1}^{N} v_{m.\max}} \bullet G + \beta \bullet \frac{v_{n.\min}}{\sum_{m=1}^{N} v_{m.\min}} \bullet G \quad (14)$$

Proposition (14) conforms to equations (7a) and (7b) for the standard case $\sum_{n=1}^{N} v_{n.\max} \neq \sum_{n=1}^{N} v_{n.\min}$. Thus, it has been proven that the *new type* of τ-value formula according to equations (7a) and (7b) conforms to the usual representation of the τ-value as a linear combination of the ideal point and the threat point to equation (8a). Strictly speaking, this applies only for the standard case assumed here $\sum_{n=1}^{N} v_{n.\max} \neq \sum_{n=1}^{N} v_{n.\min}$.

**Case B):** $\sum_{n=1}^{N} v_{n.\max} = \sum_{n=1}^{N} v_{n.\min}$

In this special case the ideal point and the threat point have to be the same by virtue of the fulfillment of equation (6) and $v : A \rightarrow R_{\geq 0}$. It has to hold $v_{n.\min} = v_{n.\max}$ for all n=1,…,N. Thus, the ideal point and the threat point lie on the same point on the hyperplane H in the solution space $R^{N}_{\geq 0}$. This means for the representation of the τ-value as a linear combination of the ideal point and the threat point according to equation (9) that for the linear factor γ with $0 \leq \gamma \leq 1$ each of the both values $\gamma = 0$ und $\gamma = 1$ can be chosen arbitrarily based on the coincidence of the ideal point and the threat point. From this it follows for the τ-value in the special case considered here:

$$\begin{aligned} &\forall n = 1,\ldots,N: \; v_{n.\tau} = \gamma \bullet v_{n.\max} + (1-\gamma) \bullet v_{n.\min} \quad \wedge \quad v_{n.\min} = v_{n.\max} \\ &a)\; \gamma = 0: \quad v_{n.\tau} = 0 \bullet v_{n.\max} + (1-0) \bullet v_{n.\min} = v_{n.\min} = v_{n.\max} \\ &b)\; \gamma = 1: \quad v_{n.\tau} = 1 \bullet v_{n.\max} + (1-1) \bullet v_{n.\min} = v_{n.\max} = v_{n.\min} \end{aligned} \quad (15)$$

The same result $v_{n.\tau} = v_{n.min} = v_{n.max}$ is assumed in the representation of τ-value according to formula (7a) and (7b) for the special case $\sum_{n=1}^{N} v_{n.\max} = \sum_{n=1}^{N} v_{n.\min}$. This is why the *new type* of $\tau$-value formula according to equation (7a) and (7b) corresponds exactly to the usual representation of the $\tau$-value as a linear combination of the ideal point and the threat point according to equation (8a) once more.

*This work was previously published in International Journal of Information Systems and Supply Chain Management, Volume 3, Issue 2, edited by John Wang, pp. 1-24, copyright 2010 by IGI Publishing (an imprint of IGI Global)*

# Section 3
# Production Planning and Inventory Management

# Chapter 10
# Dynamic Price and Quantity Postponement Strategies

**Yohanes Kristianto**
*University of Vaasa, Finland*

## ABSTRACT

*This paper studies duopolistic competition under dynamic price and production quantity postponement for two differentiable products, which share common components from one supplier at a certain degree of substitution. Both price and quantity postponement is benchmarked according to the Bertrand and Cournot Stackelberg game. In addition, system dynamic is applied to show the long term effect of both strategic decisions (price and production quantity) on profit and against demand uncertainty. The results show that price postponement is appropriate for high modular products (make-to-stock) and production quantity postponement for special orders (make-to-order). The final part of the paper concludes with results and outlines future research directions.*

## INTRODUCTION

Price and production capacity are two strategic decisions which product managers face over time. Kreps and Scheinkman (1983) showed that if firms choose capacities before engaging in Bertrand-like price competition, then the Cournot outcome is the result if the given capacities are at Cournot levels, or they should be rationed when the capacity cannot meet market demand. Davidson and Doneckere (1986), however, argued against this investigation and showed that the alternative rationing rule can eliminate idle capacity because the players agree to compete at higher equilibrium capacity.

Because the products undertaken by order-based firms are characterized by uniqueness, uncertainty and complexity, however, the Kreps and Scheinkman or Davidson and Doneckere rationing rules are difficult to apply to this type of firm. One reason is that order-based firms are different from

DOI: 10.4018/978-1-4666-0918-1.ch010

mass production-based firms in many respects. These differences extend to their requirements with respect to product substitutability because consumer preferences are diversified among the available brands (Perloff & Salop, 1985). Since a homogenous product gives no options to consumers, most discussions of price or production postponement focus on their appropriateness, depending on the single firm demand uncertainty (Fine & Freund, 1986; Miegham, 1998; Miegham & Dada, 1999), while the product substitutability effect is often considered exogenous, so that such models may underestimate the benefit of production and price postponement to mass customized products (Kristianto & Helo, 2010).

The effectiveness of product substitutability degrees has been extensively studied in a large number of contributions (Spence, 1976; Singh & Vives, 1984; Katz & Shapiro, 1985; Perloff & Salop, 1985; Martin, 1995; Colombo, 2002; Lambertini et al., 2004; Panchal et al., 2007). Some of them, for example Singh and Vives (1984) analyse the dominant strategy between price and quantity predetermined contract in a differentiated duopoly. On the other hand, Cellini and Lambertini (2002) and Lambertini and Mantovani (2004) investigated a long term join venture in Research and Development (R&D) to optimize the product differentiation degree (hereafter called product substitutability degree) of cooperating and non-cooperating firms according to the competition, according to Cournot (Kristianto & Helo, 2009).

With respect to previous efforts in product substitutability degree investigation, so far, few serious attempts have been made to investigate the effect of product substitability degree on price and production quantity dynamics instead of their values at any given of time. However, the dynamic property is important with regard to the optimum price and quantity postponement decision, at which every player has no reason to change his price or production quantity decision. Our effort in this paper broadly follows Singh and Vives (1984), except that we take into account the possible effects of long term price and production quantity postponement strategic decision, resulting from the presence of product substitutability, as a result of common product platform application, and Dr. C.F Ross (1925) in terms of the possible effect of change in the rate of price and production quantity resulting from demand variety. In particular, unlike most of the existing literature on repeated games under product differentiation, we explicitly model those demand uncertainty effects which affect firms' production quantities as well as prices.

In addition to recent literature, the open loop water tank analogy is a special case of price and production postponement for a continuous product substitutability distribution. This new approach is quite different to previous methods in the differentiated duopoly game (Singh and Vives, 1984) or price and production postponement (Fine & Freund, 1986; Miegham, 1998; Miegham & Dada, 1999), where the decision is assumed to depend merely on the price or production quantity at any given time, without considering whether the price or production quantity is increasing or decreasing at this time. Even this new approach also quite different to Lambertini et al. (2002), where competition is assumed under Cournot solely, without considering production quantity competition under the Bertrand game or impact of price postponement to capacity and flexible investment (Biller et al., 2005), where price postponement is used to balance between available supply and demand, without considering time the demand is increasing or decreasing at this time or Birge et al. (1988) where price and capacity postponement is used to substitutable product, without considering the demand variety. Indeed, in order to comprehend price and production quantity postponement application appropriately, we compare price and production quantity postponement in terms of their profitability at several product substitutability degrees and under varied demand.

The following sections first introduce related literature on dynamic analysis in competition, product substitutability in duopoly competition and the research area of this paper (Related Literature Section). Analytical model section is started by price postponement analysis using the Cournot game model, which continues with production quantity postponement (hereafter called production postponement) by applying the Bertrand game model. Discussion section presents and discusses the simulation results, which are concluded in Conclusion and further research, which explores the information behind the simulation results in the previous section and discusses some future research opportunities.

## RELATED LITERATURE

Dynamic analysis in competition was firstly presented by Ross (1925) and it was rediscussed further by Smithies and Savage (1940). Dynamic analysis was used to represent a decision maker who intends to plan his capacity in advance according to the present situation. It is clear that production capacity needs long term planning and the paper addresses a problem of two competitors adapting to a new demand function with the goal of profit stability in the future. In contrast, Dudey (1992) argues against both papers by introducing dynamic edgeworth-bertrand competition in order to solve dynamic competition under capacity constraint, which causes Nash equilibrium inexistance. The Dudey model assumes that customers come to the market at different times and the firm's price can be reset at any time with an opportunity that at least one of the duopolists can sell all the units it is able to produce. When this game is a duopoly, the payoff function of each firm maps the duopolists' strategy choices into the firm's total expected profit. Even though the Dudey model used dynamic pricing, this model presents price as a short term decision, which can be changed at any time. However, our model posits price and production quantity as two strategic decisions, which are fixed at a certain finite time in order to handle demand change.

Similar to the approaches of Smithies and Savage (1940) or Dudey (1992) on price and quantity, Singh and Vives (1984) focus their analysis on flexible capacity / price appropriateness to hedge against predetermined price/quantity contracts. Their analysis adapts to a new demand or price after making a price or delivery quantity contract formerly under the absence or presence of product substitutability degree. In contrast to that paper, we contribute to this literature by adding dynamic behavior onto the Singh and Vives (1984) duopoly model by analyzing the impact of demand uncertainty on the firm´s profitability by applying dynamic price or production quantity postponement strategy. In particular, the Cournot duopoly model (Singh & Vives, 1984) is a special case of price postponement and the Bertrand duopoly model a special case of production postponement. In conclusion, our contribution is focused on the dynamic analysis of the Singh and Vives (1984) model. With a different objective to the Smithies and Savage (1940) or Dudey (1940) models, this paper uses dynamic analysis to investigate the price and production quantity postponement effect on supply chain profit.

## ANALYTICAL MODEL

Suppose now that two firms can agree on only two types of contracts: the price contract and quantity contract. Singh and Vives (1984) use predetermined price or quantity to supply customer demand at any levels. From this point on, we refer our discussion on price postponement as make-to-stock and production postponement as make-to-order. The reasons are that the price postponement manufacturer holds inventory at his final manufacturing stage and the price is determined after customer demand (hereafter called production quantity) is known, while the production quantity postponement manufacturer never produces any products before an exact order specification based price is known. To focus the discussion, this paper uses Singh and Vives demands and their reverse function by assuming that both products are perfect substitutes. The effects of this assumption are products having a sticky price and quantity, which enforces equal cost application. Sticky price and quantity in this case are a situation where both duopolists have no reason to change their decisions on price or production quantity. To gather general understanding for this concept, both postponement concepts will be discussed separately and then the general concept will be developed. Beforehand, some notations are introduced to guide the following discussion.

## Notations

| | |
|---|---|
| $Q$ | Equilibrium quantity |
| $p_1$ | Retailer 1 price |
| $p_2$ | Retailer 2 price |
| $q_1$ | Retailer 1 quantity |
| $q_2$ | Retailer 1 quantity |
| $c$ | Supplier price |
| $\gamma$ | Product substitutability degree |
| $a$ | Market price |
| $b$ | Market size |
| $\pi_p$ | Supply chain profit according to the Cournot game |
| $\pi_q$ | Supply chain profit according to the Bertrand game |

## Model Description for Price Postponement

In this model we consider a Cournot duopoly model (see Cournot, 1960; Gibbons, 1992) with price function for retailers given by

$$P(Q) = a - Q \qquad (1)$$

Where $Q$ is total production quantity from both retailers (retailer 1 and 2). In the Cournot game firms choose their own quantity to maximize their profit by taking their opponent's quantity as a given, and in the Bertrand game they choose their price to maximize their profit by taking their opponent's price as a given. This means that they are going to sacrifice price in the Cournot game and quantity in the Bertrand game. We thus propose a methodology to avoid price sacrifice by applying the Dynamic Stackelberg game just after a predetermined quantity or price game.

To illustrate, we suppose two firms can make two types of contracts, namely price and quantity contracts. If the firms choose price postponement, then they must hedge against production fluctuation as a result of demand uncertainty. If the firms choose production quantity postponement, then they must hedge against price fluctuation. Firms first choose what type of contract and afterwards they compete on the chosen type of contracts by considering selling and material prices. Restricting attention to the subgame perfect of this two stage game, we shall see that if firms choose price postponement, then predetermined quantity is used to optimize the selling price, where it is finally used by the supplier to optimize his material price. Both retailers do not have any benefits by shifting from their optimum point, while the supplier also does not have any reasons to threaten retailers. From this point on, the game is started from stage 2, where both retailers decide their capacity

## Stage 2 Retailers Decide their Capacity According to the Cournot Game

In this stage retailers 1 and 2 simultaneously choose production quantity to maximize their profit, taking supplier price (c) as a given. That is, maximize $q = q_1 = q_2$, where $q$ is a function of c, derived from profit equation below

$$\underset{q2}{Max\pi} = (a - q_1 - q_2 - c)q_2 \tag{2}$$

By assuming equal costs function (c) and incorporating a degree of product substitability (γ), so (2) can be modified according to Cournot duopoly inversion (see Singh & Vives, 1984) as follows

$$\underset{q2}{Max\pi} = (\frac{a}{1+\gamma} - \frac{1}{1-\gamma^2}q_1 - \frac{\gamma}{1-\gamma^2}q_2 - c)(q_2) \tag{3}$$

The first order condition (FOC) for supplier-1 is then

$$\left(\frac{a}{1+\gamma} - \frac{q_1}{1-\gamma^2} - 2\frac{\gamma.q_2}{1-\gamma^2} - c\right) = 0 \tag{4}$$

Similarly, the FOC for supplier-2 is

$$\left(\frac{a}{1+\gamma} - \frac{q_2}{1-\gamma^2} - 2\frac{\gamma.q_1}{1-\gamma^2} - c\right) = 0 \tag{5}$$

We can solve (4) and (5) simultaneously to be

$$q_2 = q_1 = \frac{\left(1-\gamma^2\right)\left(\frac{a.\left(1-2.\gamma\right)}{\left(1+\gamma\right)} - c.\left(1-2.\gamma\right)\right)}{1-4.\gamma^2} \tag{6}$$

We see it slope down for the increasing of material price (c). This relationship also supports the subgame perfect principle with both retailers taking the material price as given and at the same time the supplier maximizes his profit by taking the retailer´s order quantity as given.

Since two-stage games can solve only individual contracts,, this game uses production quantity and price contracts consecutively. The reason is we assume that the products are perfect substitutes. We shall use this upcoming section to make price contracts for both retailers.

## Stage 1 Price Decision

Singh and Vives (1984) use a two-stage game, in which the firm will have to supply the amount the consumers demand at a predetermined price or quantity. This paper applies a similar principle to Singh and Vives (1984), except that we take into account the price or quantity at infinite time in order to optimize the postponed decision resulting from the presence of long term price or production quantity contract. Different to the Stackelberg game of Ferstman and Kamien (1987), Fujiwara (2006) or Clemhout et al. (1971), this stage is developed by using time and Laplace domain dynamics, which describes price or quantity response against production quantity or price contract decision. These approaches describe a natural response instead of optimal setting, even though we can guarantee their optimality by assigning optimal production or price quantity. This approach is described as follows

$$\dot{p}\left(t\right) = \left(\frac{a}{1+\gamma} - \frac{1}{1-\gamma^2}q_1 - \frac{\gamma}{1-\gamma^2}q_2 - p\left(t\right)\right);$$
$$p(0) = p_0 \tag{7}$$

By assuming sticky quantity and prices, then (7) can be reformulated as

$$\dot{p}\left(t\right) = \left(\frac{a}{1+\gamma} - \frac{1+\gamma}{1-\gamma^2}q - p\left(t\right)\right) \tag{8}$$

Equation (8) describes price dynamics, which is caused by quantity decision. This paper uses the analogy of water level in a tank: for instance, if production quantity is increased then price is automatically reduced. The same case applies in a water tank: if the water outflow is increased then the tank level is automatically reduced.

*Figure 1. Dynamic price postponement*

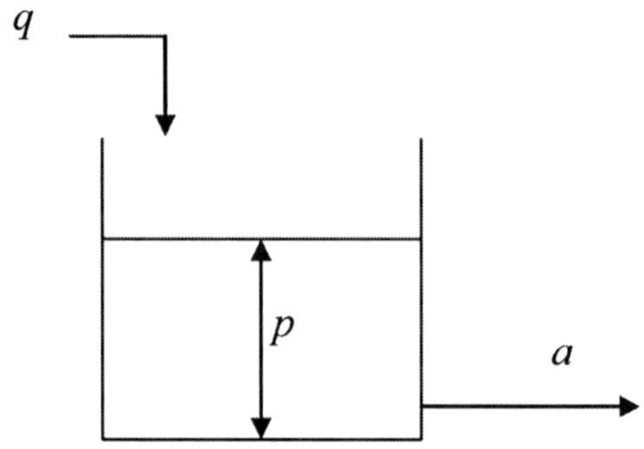


Figure 1 is taken from waterflow analogy. Inflow describes a steady state quantity and it is assumed to be constant. Outflow describes the actual demand and it is assumed to be dynamic. The price decision naturally follows according to demand. We can rearrange (8) according to

$$\dot{p}(t) + p(t) = \frac{a}{1+\gamma} - \frac{1+\gamma}{1-\gamma^2} q \tag{9}$$

Equation (9) is a first order price dynamics. By manipulating it into time domain price dynamics, Laplace domain price dynamics is obtained as follows

$$\frac{p(s)}{q(s)} = \frac{-(1+\gamma)}{(s+1)(1-\gamma^2)} \tag{10}$$

Equation (10) is formed by excluding $\left(\frac{a}{1+\gamma}\right)$ in (9) and a step response (1/s), which represents demand variety, can be attached to (10) so that we have $\frac{p(s)}{q(s)} = \frac{-(1+\gamma)}{s.(s+1)(1-\gamma^2)}$. This Laplace domain is finally converted to time domain price dynamics as follows

$$p(t) = \frac{a}{1+\gamma} - \left(1 + \frac{(1+\gamma)}{1-\gamma^2}.e^{-.t}\right).q(t) \tag{11}$$

At a steady state, the price function can be simplified to be $p = \frac{a}{1+\gamma} - .q$

The result of equation (10) is used by the supplier to set his selling price to retailers as the following relationship

$$\max_c \left(b - p_1 + \gamma p_2\right).c \tag{12}$$

By combining (6) and (12) then we have

$$\max_c \left( b + (\gamma - 1)\frac{a}{1+\gamma} - . \frac{(1-\gamma^2)\left(\frac{a.(1-2.\gamma)}{(1+\gamma)} - c.(1-2.\gamma)\right)}{1-4.\gamma^2} \right).c \tag{13}$$

$$c = \frac{1-4.\gamma^2}{2(1-2.\gamma)(1-\gamma^2)} \left( \frac{(1-\gamma^2)\frac{a.(1-2.\gamma)}{(1+\gamma)}}{(1-4.\gamma^2)} + (1-\gamma)\frac{a}{1+\gamma} - b \right). \tag{14}$$

Equation (14) describes a strong relationship among product substitutability (γ), retailer´s quantity and supplier price setting. We can see that supplier price is a concave function of product substitutability (γ).

## Model Description for Production Postponement

Consider a Bertrand duopoly model with price function (see Gibbon, 2002; Kristianto & Helo, 2009) for retailers given by

$$Q = b - p_i + \gamma.p_j \tag{15}$$

*Figure 2. Dynamic quantity postponement*

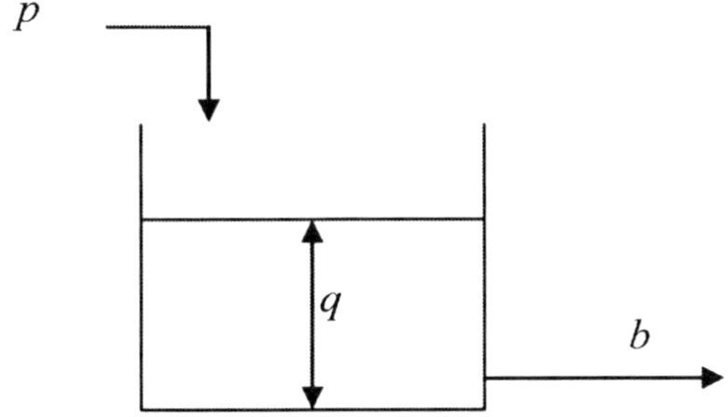


Where $p_i$ and $p_j$ is price of product 1 and 2.

In the Bertrand game firms choose their own price to maximize their profit by taking their opponent's price as a given. We thus propose a methodology which is similar to the previous Cournot game, except that we take into account the quantity at infinite time in order to optimize the postponed decision resulting from the presence of long term price contract

This game decides the equilibrium price first before capacity and it can be described as follows:

Figure 2 is taken from waterflow analogy. Inflow describes a steady price state and it is assumed to be constant. The managerial price related decision is located at the output and it is assumed to be dynamic. Quantity automatically follows, whatever the price pattern, as the water level is also controlled by its flow in a storage tank. These two situations are identical to one another.

## Stage 2 Retailers Decide Their Price According to the Bertrand Game

In this stage, retailers 1 and 2 simultaneously choose the product price to maximize their profit, taking the supplier price (c) as given. That is, maximize $p = p_1 = p_2$, where $q$ is a function of c, derived from the profit equation below

$$\max_{p1}\left(b - p_1 + \gamma.p_2\right)\left(p_1 - c\right) \tag{16}$$

The first order condition is

$$b - 2p_1 + \gamma.p_2 + c = 0 \tag{17}$$

Similarly, the FOC from the second product variant is

$$b - 2p_2 + \gamma.p_1 + c = 0 \tag{18}$$

Solving these two equations simultaneously, one obtains

$$P_2 = P_1 = \frac{\left(\gamma + 2\right)\left(c + b\right)}{\left(4 - \gamma^2\right)} \tag{19}$$

We see it slope up for the increasing of material price (c). This relationship supports the Bertrand principle, where both retailers and suppliers maximize the price by sacrificing production quantity.

Since two-stage games can solve only an individual contract, this game uses product price and quantity contracts consecutively. The reason is we assume that the products are perfect substitutes.

## Stage 1 Leader Decides His Own Profit Function

At the first stage we can find the material price by optimizing supplier profit function as

$$\max_{c}\left(b - p_1 + \gamma p_2\right).c \tag{20}$$

Find c by insert (19) into (20), so we get

$$\max_{c}\left(b + \left(\gamma - 1\right)\frac{\left(\gamma.c + \gamma.a + 2.c + 2.a\right)}{\left(4 - \gamma^2\right)}\right).c \tag{21}$$

$$c = \frac{\left(4 + \gamma - 2\right).b}{\left(1 - \gamma\right)\left(2.\gamma + 4\right)} \tag{22}$$

Equation (22) describes a strong relationship among product substitutability (γ), retailer quantity and supplier price setting. We can see that supplier price is a concave function of product substitutability (γ). We shall use this upcoming section to make a quantity contract for both retailers.

## Capacity Decision

In the same manner as price postponement, quantity postponement uses time and Laplace domain dynamics as follows

$$\dot{q}\left(t\right) = \left(b - p_1 + \gamma.p_2 - q\left(t\right)\right); p\left(0\right) = p_0 \tag{23}$$

Equation (23) describes quantity dynamics, which is caused by pricing decision. This paper uses the analogy of water level in a tank as well as the price postponement model (see Figure 2)

By assuming sticky price and quantity, then (23) can be rewritten as

$$\dot{q}\left(t\right) - q\left(t\right) = \left(b - \left(1 - \gamma\right)p\right) \tag{24}$$

Equation (24) is a first order price dynamics, which describes firm effort to achieve optimum production quantity level against pricing strategy. By manipulating it into time domain price dynamics, Laplace domain price dynamics is obtained as follows

$$\frac{q\left(s\right)}{p\left(s\right)} = \frac{-\left(1 - \gamma\right)}{\left(s + 1\right)} \tag{25}$$

Equation (25) is formed by excluding $b$ in (24) and a step response (1/s), which represents demand variety, can be attached to (25) so that we have $\frac{q\left(s\right)}{p\left(s\right)} = \frac{-\left(1 - \gamma\right)}{s.\left(s + 1\right)}$. This Laplace domain is finally converted to time domain quantity dynamics by incorporating $b$ as follows

$$q\left(t\right) = b - \left[1 + \frac{1}{1 - \gamma}.e^{-k.t}\right].p\left(t\right) \tag{26}$$

Equation (26) is a quantity postponement as a result of price predetermined contract.

## DISCUSSION

Studies on price and production postponement have been able to shed light on the supply chain as a dynamic system. In addition, they have underscored the importance of such long term stability as values, meanings and commitments and paved the way for more elaborate research on interface between supply chain and revenue management.

What happens if both firms choose the price contract? In that case firms choose production quantity to maximize their profit, taking $p$ as given. This yields the profit reaction which corresponds to the Bertrand reaction. Notice that it is downward sloping for the increasing of product substitutability degree. The reverse action is shown for production quantity contract, where firms choose price to maximize profit. These discrepancies inform us that price postponement is an appropriate choice for higher substitutability degree (see Figure 3).

To examine the case with erratic demands, suppose firms strictly predetermine their price or quantity settings, which is strictly revised in their profitability. Now if we randomize $a$ or $b$ value to represent demand variety, then we get $\pi_{(p)}$ or $\pi_{(q)}$ as a function of $q$, $p$ and $\gamma$; hence, the results can be drawn as follows (Figure 4 and Figure 5).

*Figure 3. Product substituability*

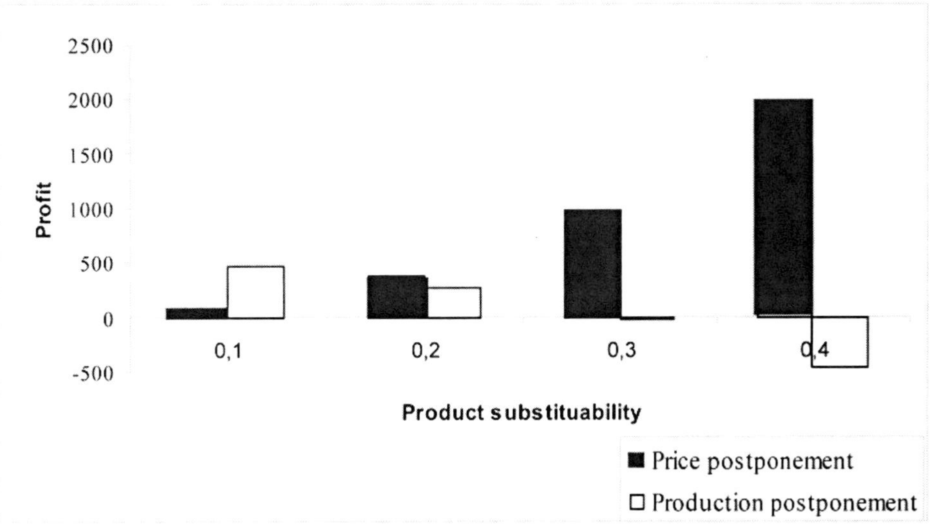


*Figure 4. Profit dynamics*

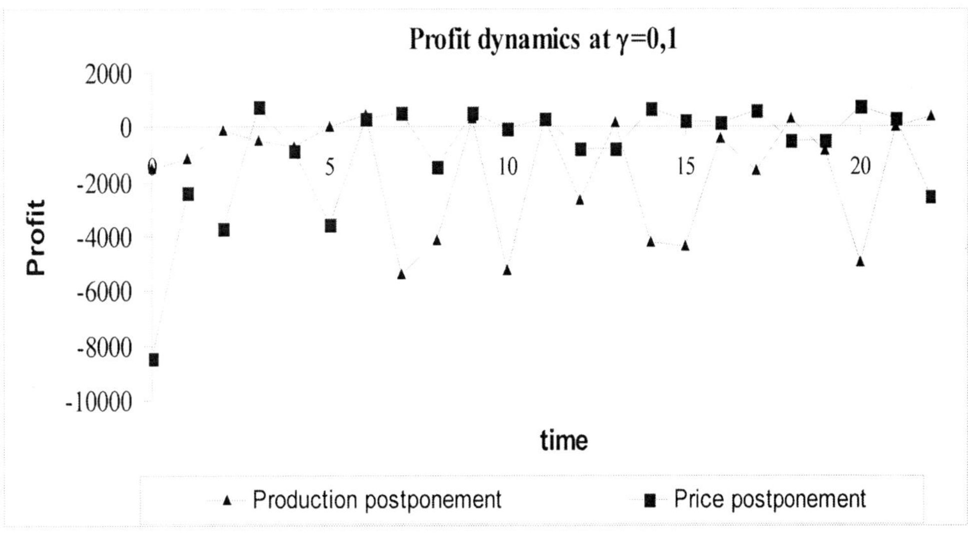


Figure 4 and 5 clearly show that price postponement is more stable at varied demands, which also shows further that price postponement gives higher stability at higher product substitutability degrees. These results again ensure that price postponement reaction is appropriate at higher compatible products.

We know that profit stability is a common goal of price and production quantity contracts. High profit stability ensures two firms can cooperate without worrying about any losses. Below is a comparison between two postponement types, which is gathered by finding profit at different price or quantity settings. Now if we measure

*Figure 5. Profit dynamics at γ=0,9 for price and production postponement*

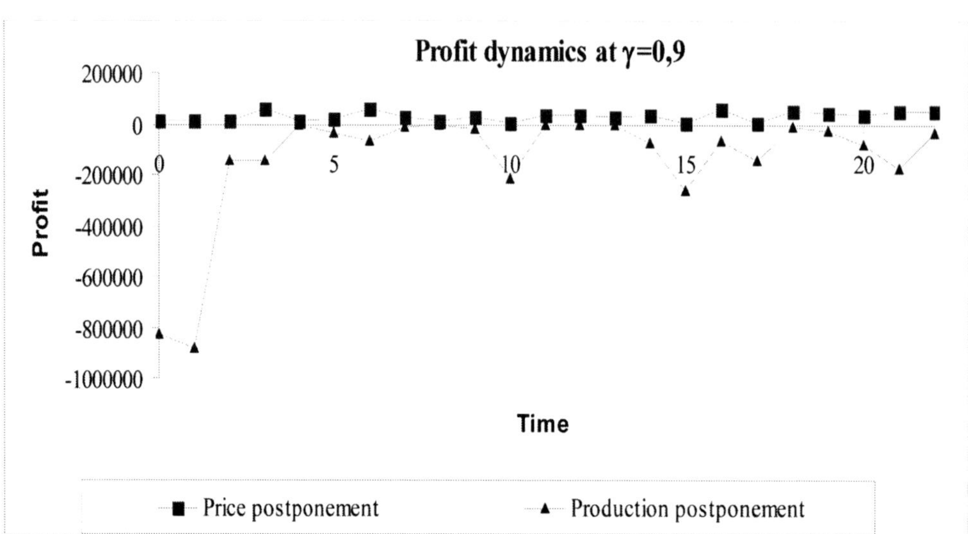


*Figure 6. Profit deviation at γ=0,1 and 0,9 for price and production postponement*

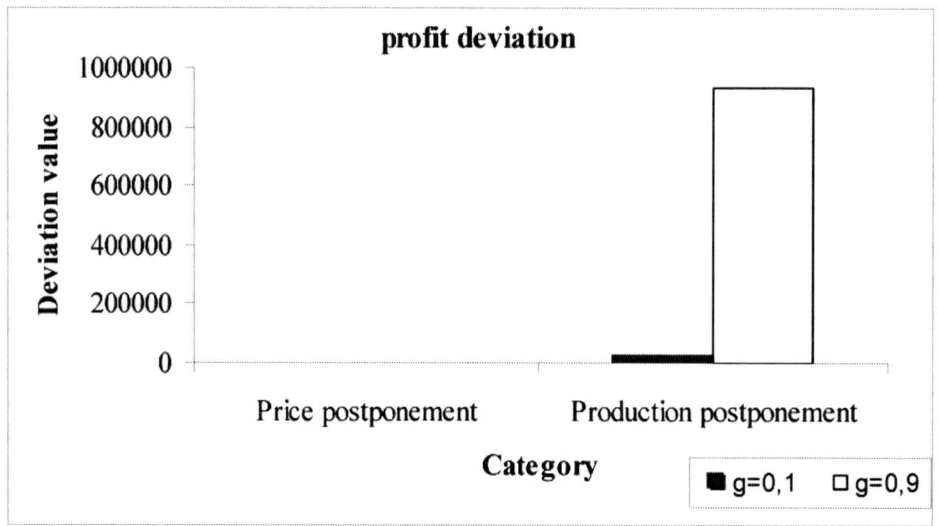


their deviation from optimum profit value (replace $q$ or $p$ by our own assumption), then we get $\pi_{(p)}$ or $\pi_{(q)}$ as a function of $q$, $p$ and $\gamma$. Benchmarking them against optimal value, the results can be exhibited in Figure 6.

In addition to the managerial implication, price postponement is also significantly superior to production postponement from the customer service level point of view (see Figure 7). That figure shows that price postponement can cover market demand at a higher level than production postponement. This result supports the managerial policy of product commonality, where price can be determined later after final customization is completed.

*Figure 7. Service level at different postponement types*

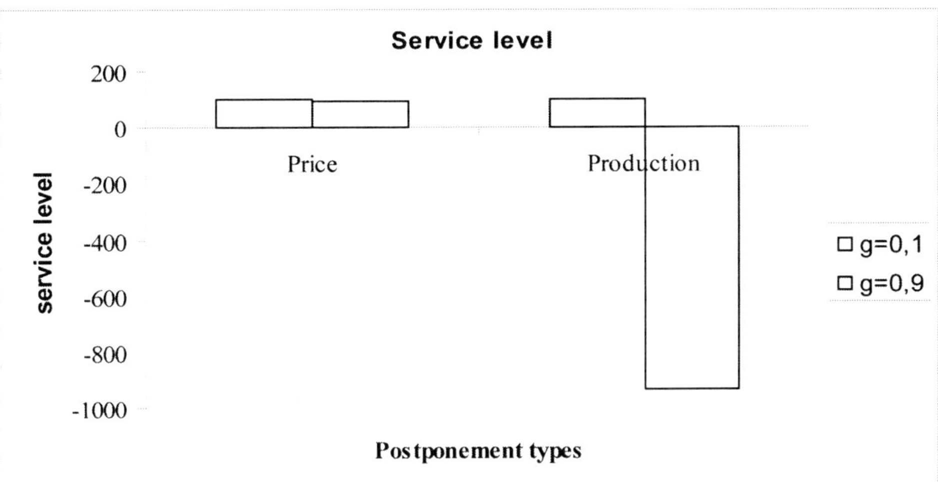


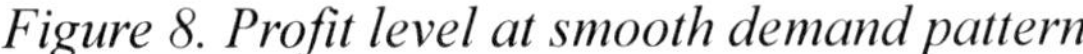

*Figure 8. Profit level at smooth demand pattern*

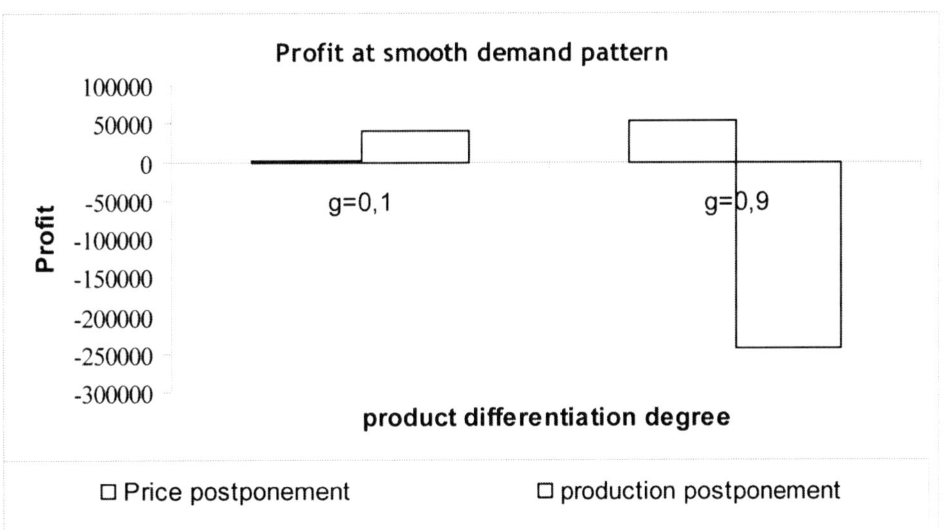


What happens if demand pattern is smoothed? In that case, firms choose production postponement to maximize their profit, taking $p$ as given (see Figure 8).

Furthermore, price postponement is still superior over production postponement at a higher product substitutability degree. This result once more supports the view that make-to-order (production postponement) is only appropriate to highly unique products. The reverse result is shown for price postponement, where firms develop a common platform to maximize profit. Those discrepancies also inform us that price postponement is an appropriate choice for higher substitutability degree.

## CONCLUSION AND FURTHER RESEARCH

This paper revisited the Singh and Vives model on price and production postponement by considering the dynamic behavior of demands. We may summarize the results derived from the model, as follows.

1. Price postponement is superior to production postponement at many respects. This type of contract guarantees profit stability and at the same time supports the product standardization effort
2. Price postponement is also a dominant strategy for substitutable products. This conclusion is at odds with the previous Singh and Vives conclusion (Singh & Vives, 1984). This discrepancy is caused by the Singh and Vives model perhaps assuming that in Bertrand price-like competition, the quantity setting will avoid both firms having to reduce their production quantity further. On the contrary, this paper assumes sticky prices and quantities, where it pushes both firms to cooperate at higher levels. By sticky prices and quantities, this paper is more appropriate for common platform based products instead of two widely differentiated products.
3. Production quantity postponement (make-to-order) is a dominant strategy for highly differentiable products. This conclusion supports the article of Miegham and Dada (1999), who discusses postponement strategies differentiation according to their applicability.

The dynamic behavior analysis in this paper helps decision makers to decide their long term postponement policy with regard to their manufacturing types, namely make-to-stock or make-to-order. The analysis results also support both modularity and customization principles in mass customized products, where decision uncertainty can be reduced by making closer customer order decoupling, point to sales point. This paper suggests that product developers design common platform products and decide the price according to customer specific requirements.

In terms of future research direction, the oligopoly model should be considered for development according to future market demand, which is determined by how close customer requirements are met, so in future the oligopoly model quantity and price can be replaced with some parameters such as inventory and lead times. From this result, a sequence between lead times and inventory can be determined and the outcome will be a decision about whether agility or efficiency is more important for a company, so that the outcome can be used by top management to compose their business strategy. Finally, future research should accommodate strategic and tactical level alignment in order to develop comprehensive decision analysis.

## REFERENCES

Biller, S., Muriel, A., & Zhang, Y. (2006). Impact of price postponement on capacity and flexible investment. *Production and Operations Management*, *15*(2), 198–214. doi:10.1111/j.1937-5956.2006.tb00240.x

Birge, J. J., Drogosz, I., & Duenyas, I. (1998). Setting single-period optimal capacity levels and prices for substitutable products. *International Journal of Flexible Manufacturing Systems*, *10*, 407–430. doi:10.1023/A:1008061605260

Cellini, R., & Lambertini, L. (2002). A differential game approach to investment in product differentiation. *Journal of Economic Dynamics & Control*, *27*, 51–62. doi:10.1016/S0165-1889(01)00026-4

Clemhout, S., Leitmann, G., & Wan, H. Y. Jr. (1971). A differential game model of duopoly. *Econometrica*, *39*(6), 911–938. doi:10.2307/1909667

Colombo, L. (2002). *Product differentiation and cartel stability with stochastic demand market*. Societa Italiana di Economia Publica.

Cournot, A. (1960). *Researches into the mathematical principtes of the theory of weatth* (N. T. Bacon trans., English ed. of Coumot, 1838). New York: Kelley.

Davidson, C., & Deneckere, R. (1986). Long-run competition in capacity, short-run competition in price, and the cournot model. *The Rand Journal of Economics*, *17*, 404–415. doi:10.2307/2555720

Dudey, M. (1992). Dynamic Edgeworth-Bertrand competition. *The Quarterly Journal of Economics*, *107*(4), 1461–1477. doi:10.2307/2118397

Fershtman, C., & Kamien, M. I. (1987). Dynamic duopolistic competition with sticky prices. *Econometrica*, *55*(2), 1151–1164. doi:10.2307/1911265

Fine, C., & Freund, R. (1986). Optimal investment in product-flexible manufacturing capacity. *Management Science*, *36*(4), 449–466. doi:10.1287/mnsc.36.4.449

Fujiwara, K. (2006). A Stackelberg game model of dynamic duopolistic competition with sticky prices. *Economic Bulletin*, *12*(12), 1–9.

Gibbons, R. (1992). *A Primer in Game Theory*. Upper Saddle River, NJ: Pearson Education Limited.

Katz, M. L., & Shapiro, C. (1985). Network externalities, competition, and compatibility. *The American Economic Review*, *75*(3), 424–440.

Kreps, D., & Scheinkman, J. (1983). Quantity precommitment and Bertrand competition yield Cournot outcomes. *The Bell Journal of Economics*, *14*, 326–337. doi:10.2307/3003636

Kristianto, Y., & Helo, P. (2009). Strategic thinking in supply and innovation in dual sourcing Procurement. *Int. J. Applied Management Science*, *1*(4), 401–419.

Kristianto, Y., & Helo, P. (2010). Built-to-order supply chain: response analysis with control model. *Int. J. Procurement Management*, *3*(2), 181–198. doi:10.1504/IJPM.2010.030734

Lambertini, L., & Mantovani, A. (2004). *Process and product innovation. A differential game approach to product life cycle*. Retrieved January 5, 2008, from http://www2.dse.unibo.it/seminari/Mantovani

Martin, S. (1995). R & D joint ventures and tacit product market collusion. *European Journal of Political Economy*, *11*, 733–741. doi:10.1016/0176-2680(95)00026-7

Mieghem, J. A. V. (1998). Investment strategies for flexible resources. *Management Science*, *44*(8), 1071–1078. doi:10.1287/mnsc.44.8.1071

Mieghem, J. A. V., & Dada, M. (1999). Price versus production postponement: Capacity and competition. *Management Science*, *45*(12), 1631–1649.

Panchal, J. H., Fernandez, M. G., Paredis, C. J. J., Allen, J. K., & Mistree, F. (2007). An interval-based constraint satisfaction (IBCS) method for decentralized, collaborative multifunctional design. *Concurrent Engineering, Research and Application*, *15*(3), 309–323.

Perloff, J. M., & Salop, S. C. (1985). Equilibrium with product differentiation. *The Review of Economic Studies*, *52*(1), 107–120. doi:10.2307/2297473

Ross, C. F. (1925). A mathematical theory of competition. *American Journal of Mathematics*, *47*(3), 163–175. doi:10.2307/2370550

Singh, N., & Vives, X. (1984). Price and quantity competition in a differentiated duopoly. *The Rand Journal of Economics*, *15*, 546–554. doi:10.2307/2555525

Smithies, A., & Savage, L. J. (1940). A dynamic problem in duopoly. *Econometrica*, *8*(2), 130–143. doi:10.2307/1907032

Spence, M. (1976). Product differentiation and welfare. *The American Economic Review*, *66*(2).

*This work was previously published in International Journal of Information Systems and Supply Chain Management, Volume 3, Issue 4, edited by John Wang, pp. 70-82, copyright 2010 by IGI Publishing (an imprint of IGI Global)*

# Chapter 11
# Determining Optimal Price and Order Quantity Under the Uncertainty in Demand and Supplier's Wholesale Price

**Seong-Hyun Nam**
*University of North Dakota, USA*

**Hisashi Kurata**
*International University of Japan, Japan*

**John Vitton**
*University of North Dakota, USA*

**Jaesun Park**
*University of North Dakota, USA*

## ABSTRACT

*Manufacturers in a high-tech durable product industry may have to make operational decisions in the presence of uncertainties associated with product demand and supplier's wholesale price. In this paper, the authors investigate the impact of such uncertainties on the activities of a manufacturer and its supplier and develop an optimization model that describes how the manufacturers should reflect the uncertainty issues in their pricing and order quantity policy to achieve a desirable profit. In the modeling process, three important managerial problems are discussed, i.e., the effect of coordination between a manufacturer and its supplier in dealing with uncertainties on product demand and supplier's wholesale price, strategies for mitigating both errors in demand forecasting and supply risk, and modeling frameworks to determine the optimal solution for price and order quantity based on the varying levels of coordination. To identify best operational decisions under market uncertainty, the authors use the stochastic optimal control theory.*

DOI: 10.4018/978-1-4666-0918-1.ch011

## INTRODUCTION

In recent years, manufacturers in the high-tech durable product industry (e.g., personal computers, digital music devices, plasma and LCD TVs) have experienced some emerging changes in the environment. Due to the rapid advancements in technology, the life cycle of high-tech durable products tends to be shorter and the obsolescence rate of these products gets faster, similar to fashion items. In order to respond properly to the shifting markets, manufacturers have to continuously bring innovative product ideas into the markets and make an effort to quickly employ the newly available technology for reducing operational costs. The recent advent of Internet technology, for example, helps find cost-effective suppliers on a global basis, which enables manufacturers to offer better prices to their customers. This means that the customers in this particular market segment become increasingly sensitive to price and the competition of high-tech durable products at the retail level has become more complex and tougher than ever before.

One of the important factors that determine a manufacturer's competitiveness in the high-tech durable product industry is product availability at a reasonable price. However, it is difficult for manufacturers to decide the most profitable levels of production and pricing because of the uncertainty associated with the supplier's wholesale price as well as product demand. Suppliers may have to change their unit wholesale price to be profitable as the factors in the external environment change. A supplier's unit wholesale price $S(t)$ at time $t$, for instance, can be affected by uncertain factors, such as the exchange rate at the time when the manufacturer buys the materials, the volatility of sourcing prices from foreign suppliers, the price changes of raw commodities, and fluctuation in manufacturer's demand. This also leads to manufacturers' retail price adjustments because the changes in the supplier's wholesale price $S(t)$ affect the production costs. Therefore, without building a close long-term strategic partnership with key suppliers, manufacturers can hardly satisfy an ever-changing marketplace and create values for customers.

In this paper, we are concerned with the two major types of uncertainty that are associated with product demand and in supplier's wholesale price. After we investigate how such uncertainties can disrupt the activities of a manufacturer and a supplier, we attempt to develop an optimization model that describes how the manufacturer should reflect the uncertainty issues in their pricing and order quantity policy to achieve a desirable profit. In the modeling process, three important managerial problems are discussed: 1) the effect of coordination between a manufacturer and its supplier in dealing with uncertainties on product demand and supplier's wholesale price; 2) strategies for mitigating both errors in demand forecasting and supply risk; 3) modeling frameworks to determine the optimal solution for price and order quantity based on the varying levels of coordination in the given time. To identify best operational decisions under market uncertainty, we use the stochastic optimal control theory.

The remainder of this paper is organized as follows: Section 2 contains a brief literature review. Section 3 studies decision-making problems and demonstrates the models for optimal stocking and pricing policies. The numerical examples that support the model and the analytical discussions are presented in Section 4. Finally, Section 5 summarizes conclusions reached.

## LITERATURE REVIEW

Most of the supply chain management literature (Hwang et al., 1967; Bensoussan et al., 1974; Pekelman, 1974; Hartl & Sethi, 1984; Feichtinger & Hartl, 1985; Gaimon, 1988) has focused on understanding how to balance production, inventory, and price under demand uncertainty. Levy (1995) examined the demand uncertainty and supplier reliability, Lee and Tang (1998 a, b) examined the trade-off between the consignment and turnkey arrangement under demand uncertainty, and Iyer and Berger (1997) were interested in the effect of demand uncertainty in the apparel industry and studied how Quick Response in the manufacturer–retailer channel could solve decision problems. In Iyer and Berger's model, two dimensions of demand uncertainty, initial demand and post-initial time demand were used. Iyer and Berger assumed normal distribution of the two types of demand and measured the uncertainties of demand by the variances. Emmons and Gilbert (1998) examined a multiplicative model of demand uncertainty and demonstrated how uncertainty could lead to an increase in retail prices. By using the scenario approach, Tsiakis et al. (2001) explained the impact of variances in demand uncertainty on operational decisions. Aviv (2001) considered two different demand forecasting models, which focused on manufacturer–supplier collaboration and noncollaboration in a two-stage supply chain consisting of two members. Their study is relevant to products with long life cycles and a relatively low level of demand uncertainty.

Chen and Lee (2004), on the other hand, used discrete scenarios with given probabilities and uncertain product prices based on fuzzy set theory to examine the simultaneous optimization of multiple conflict objectives under demand uncertainty. For multiproduct manufacturing plants, Gupta and Maranas (2003) suggested a multiperiod planning and scheduling approach to reduce the uncertainty resulting from changing demand. To hedge against demand uncertainty, Gupta and Maranas (2000a, 2003) and Gupta et al. (2000b) assumed a normal probability distribution, while Jung et al. (2004) used safety stock.

Bitran and Caldentey (2003) were concerned with the development of dynamic pricing models from the perspective of revenue management. Fan et al. (2005) developed a numerical method for the dynamic pricing model by using the optimal control approach. In theory, the majority of existing pricing models assume that cost and market parameters are known equally and perfectly to both a manufacturer and a retailer, which is known as information symmetry. In practice, however, the retailer's knowledge of manufacturing costs is often the province of the manufacturer. Lau et al. (2006) highlighted in their study that manufacturing cost is the most likely to be asymmetrically known and showed that the explicit consideration of knowledge asymmetry in manufacturing costs leads to interesting and surprising deviations from earlier symmetrical manufacturing knowledge results. Their findings indicated that under a linear demand curve, a manufacturer should overstate manufacturing costs, whereas under an iso-elastic demand curve, the manufacturer benefits both itself and the entire system by understating manufacturing costs.

In addition to demand uncertainty, any uncertain change or force emerging in any stage of the supplying process is also an important dimension that a manufacturer must consider when making operational decisions. However, not many studies have investigated how supply-side uncertainty could influence the manufacturer's production decisions. Gupta and Maranas (2003) and Liu and Sahinidis (1998) recognized the effect of sudden or unexpected changes in commodity price and supplier-side costs. They argued that such uncertain variables should be incorporated into the production planning process. Gurnani and Tang (1999) considered both demand and supplier cost uncertainty and developed an optimal ordering policy using a two-period dynamic programming tool. Kogut (1985), Kogut and Kulatilaka (1994), and Huchzermeier and Cohen (1996) developed a global supply chain network model considering currency exchange rate uncertainty. Parlar and Perry (1996), Ciarallo et al. (1994), and Wang and Gerchak (1996) developed a supply chain network model considering the uncertainty of supply capacity. Li and Kouvelis (1999) developed a supply chain network model considering the uncertainty of wholesale price using the geometric Brownian motion with drift term. The reader is referred to Tang (2006) for an extensive review of supply chain risk management.

Recognizing that research with a focus of supply-side uncertainty seems to be limited in the supply chain management field, we attempt to examine the impact of supplier's wholesale price uncertainty on a manufacturer's operational decisions and propose a coordination-based solution to handle the uncertainty related problems.

## MODEL

Manufacturers make products based on demand forecast. However, it is difficult to determine the accurate future demand, especially for manufacturers who operate in a highly volatile market. The rapid changes occurring in the marketplace can impede the forecasting capability of a manufacturer, leading to meaningful errors in managing both production and the supply chain. Moreover, manufacturers using an extensive set of foreign suppliers may have an increased risk in their operational management. The exchange rate, for example, constantly fluctuates in an unknown direction and thus causes the changes in foreign suppliers' wholesale prices. This event can create a serious problem for manufacturers if their pricing is heavily dependent on a supplier's wholesale price. Therefore, manufacturers must consider how the uncertainty could intervene in their operational processes and find ways to mitigate the risks associated with it. In this section, we discuss the effects of the uncertainty related to demand forecasting and supplier's wholesale price and develop a decision-making model for price and optimal order quantity toward the manufacturer's profit maximization. Proofs and a list of some important notation and symbols used in this paper are in Appendix.

## Consumer Demand with Respect to Price

For the manufacturer, matching actual demand with forecasted demand is a particularly challenging task in volatile markets. To capture the effect of price on consumers' actual demand, let $P_U$ be the highest price a customer can pay and $M(P) = M\{P(t)\}$ denote the actual market demand based on price $P(t)$ at a time *t*, which consists of a deterministic part $\{D(P(t)), t \geq 0\}$ and a random part (White noise; $\varepsilon_W(t), t \geq 0$) (Basu, 2003). In a real scenario, when the manufacturer needs to decide product price and order quantity before time *t*, it is realistic to use forecasted demand. To study the optimal pricing strategy based on the accuracy of the demand forecast, the deterministic part of market demand $D\{P(t)\}$ is assumed to be a deterministic-bounded, continuous, real-value function with respect to time *t*, which is unknown to the manufacturer and the supplier before time *t*. Therefore,

$$M(P) = D(P) + \varepsilon_W(t) > 0 \text{ for } P(t) \leq P_U \text{ and}$$
$$M(P) = D(P) + \varepsilon_W(t) = 0 \text{ for } P(t) > P_U$$

Let $\{Q(t), t \geq 0\}$ be the forecasted demand based on $M(P)$. If a deviation of $M(P) - Q(t)$ occurs at time *t*, an adjustment must be made in order to meet the actual demand. The manufacturer continuously monitors and dynamically adjusts for the deviation. It is assumed that the deviation can be corrected (a *measurable* control variable) through regulatory action, which involves coordination between a manufacturer and a supplier.

In an infinitesimal time interval after the adjustment is applied, the resulting improvement is expected to be $n(t)\{M(P) - Q(t)\}dt$, where $n(t) \in [0, \infty)$ represents the speed of revision that is controlled by the level of coordination between a manufacturer and a supplier at time *t* (Nam et al., 2009). The proposed model indicates that high coordination could lead to a more accurate demand forecast. By maintaining close collaborative relations with the supplier, the manufacturer can take corrective action as needed. Note that the forecast is assumed imprecise and that the longer the forecast horizon, the more imprecise the forecast. In order to measure the degree of the uncertainty, a disturbance term is incorporated into the model. The disturbance is assumed to be a Wiener type in the form of $\sigma_Q dW(t)$. The disturbance is assumed to be mitigated through coordination between the manufacturer and the supplier. The resulting improvement by taking corrective action can be expressed as $n(t)\cdot\{M(P)-Q(t)\}dt$. Therefore, the underlying stochastic process associated with the corrected level of coordination is obtained as follows:

$$dQ(t) = n(t)\{M(P) - Q(t)\}dt\ ,$$

and

$$dQ(t) = n(t)\{D(P) + \varepsilon_W(t) - Q(t)\}dt$$

Hence,

$$dQ(t) = n(t)\{D(P) - Q(t)\}dt + \varepsilon_W(t)dt\ .$$

Let $W(t)$ be a Weiner process. Then, $\varepsilon_W(t) = (\sigma_Q)dW / dt$ (Arnord, 1974; Dixit & Pindyck, 1994), and

$$dQ(t) = n(t)\{D(P) - Q(t)\}dt + \sigma_Q \cdot dW\ ,$$

where $dW(t)$ is a Weiner process (Brownian motion).

The simplest generalization of the underlying stochastic process $Q(t)$ associated with the speed of correction can be defined as follows (Nam et al., 2009):

$$dQ(t) = n(t)[D(P) - Q(t)]dt + \sigma_Q dW(t) \tag{1}$$

Since (1) is the continuous-time version of the first-order autoregressive process in discrete time $\sigma_Q$ that is the coefficient of variance rate change of $Q(t)$, then the buyer can estimate $\sigma_Q$ of (1) using discrete-time data by running the regression $Q(t) - Q(t-1) = \alpha + bQ(t-1) + \varepsilon(t)$ (for more details on the this process, see Arnold, 1974; Dixit & Pindyck, 1993). Assume that $Q_0$ is either normally distributed or constant. If the value of $Q(t)$ is currently $Q_0 = [Q(t = t_0)]$ for $t_0 \le x \le t \le T$ and $Q(t)$ follows (1), then the integral (in the Ito sense) of the infinitesimal diffusion as defined in (1) can be obtained as follows (using the method given by Theorem 8.4.2 in Arnold, 1974):

$$Q(t) = e^{-N(t)}\{Q_0 + \int_{t_0}^{t} n(x)D(P(x))e^{N(x)}dx + \sigma_Q \int_{t_0}^{t} e^{N(x)}dW(x)\} \tag{2}$$

where $N(x) = \int_{t_0}^{t} n(x)dx$ (i.e., cumulative adjustment through coordination).

By letting $K(t) = E[Q(t)]$ and $G(t) = E[Q^2(t)]$, *K(t)* and *G(t)* can be obtained from (2) as follows(Nam et al., 2009):

$$K(t) = E[Q(t)] = e^{-N(t)}\{K_0 + \int_{t_0}^{t} n(x)D\{P(x)\}e^{N(x)}dx\}$$ with $K_0 = E[Q_0]$ and

$$G(t) = e^{-2N(t)}[G_0 + \int_{t_0}^{t}\{2n(x)K(x)D(P(x)) + n^2(x)\}e^{2N(x)}dx]$$

with $G_0 = E[Q_0^2]$

Then, variance of future order quantity at time $t$ can be derived as follows:

$$Var[Q(t)] = V^Q(t) = E[Q(t) - K(t)]^2 = G(t) - K^2(t) \tag{3}$$

Using (3), the manufacturer measures the future risk of its order quantity and estimates inventory cost. For the purpose of evaluating the inventory cost that is affected by demand uncertainty, we use both holding and shortage costs. Set $C^H$ and $C^S$ as unit holding and shortage cost, respectively. The salvage value at the end of each period is assumed to be equal to zero. Then, the terms $C^S\{D(P) - Q(t)\}\ \ I_{\{D(P)>Q(t)\}}$ refer to penalties for stock outs (a loss of good will) and the terms $C^H\{Q(t) - D(P)\}\ \ I_{\{D(P)<Q(t)\}}$ imply overstock (inventory holding cost). The asymmetric loss function is appropriate when the loss differs for the value of $Q(t)$, which is the equidistance from the target $D(P)$. Therefore, the total penalty cost $l(t)$ using asymmetric loss function, is expressed as follows (Nam et al., 2009):

$$l(t) = \begin{cases} C^S\{D(P) - Q\} & \text{if } D(P) > Q \\ C^H\{Q - D(P)\} & \text{if } D(P) < Q \end{cases}$$

The expected loss (penalty cost) under the underline forecasting process is denoted as $L(t) = E[l(t)]$. In order to derive $L(t)$, the probability of underestimated forecasting must be known. Let $R(t) = \Pr[Q(t) < M(P)]$, then the process of the equation is explained as follows (Nam *et al.*, 2009):

Lemma 1.

$$R(t) = \Pr[Q(t) < M(P)] = \left\{1 / \sqrt{2\pi\ \sigma_R^2(t)}\right\}\left[\int_{-\infty}^{Y(t)} \exp\{-x^2 / 2\sigma_R^2(t)\}dx\right] \tag{4}$$

*where* $\sigma_R^2(t) = \sigma_Q^2(t - t_0) + \exp\{-2N(t)\}\sigma_Q^2\int_{t_0}^{t}\exp\{2N(x)\}dx$ and

$$Y(t) = [D(P) - \exp\{-N(t)\}Q_0 - \exp\{-N(t)\}\int_{t_0}^{t} n(x)D(P(x))\exp\{N(x)\}dx.$$

Lemma 1 explains how the varied levels of coordination can influence the forecasting accuracy. By using (4) in Lemma 1, the manufacture can evaluate the future errors on forecasting -the probability of whether the future forecasting at time $t$ might be overestimated or underestimated at the present time. Therefore, a manufacturer can manage the desirable future levels of coordination within its organizational capability. Since $R(t)$ is known, the *expected loss function* $L(t)$ can be derived as follows (Nam et al., 2009):

Lemma 2.

$$L(t) = E[l(t)] = C^h \cdot \{A(t) - D(P)\} \cdot \{1 - R(t)\} + C^S \cdot R(t) \cdot \{D(P) - B(t)\} \tag{5}$$

*where*

$$V^R(t) = \sqrt{G(t) - K^2(t) + \sigma_Q^2(t)},$$
$$u(t) = \{D(P) - K(t)\} / V^R(t),$$

$$\varphi\{u(t)\} = (1 / \sqrt{2\pi})\exp\{-u^2(t) / 2\},$$
$$\Phi(u) = \Phi\{u(t)\} = \int_{-\infty}^{u(t)} \varphi(x)dx,$$
$$A(t) = K(t) + V^R(t)\left[\varphi(u(t)) / \{1 - \Phi(u(t))\}\right],$$ and
$$B(t) = K(t) - V^R(t)[\varphi(u(t)) / \Phi(u(t))].$$

Based on the output of (5), the manufacturer selects the optimized coordination level that generates the operational flexibility toward desirable profit attainment.

## Supplier's Wholesale Price Uncertainty

We assume that all supplier candidates are able to meet the manufacturer's needs and the manufacturer is able to choose a supplier who offers the best wholesale price. The geometric Brownian motion is one of the most frequently used models to deal with uncertainty in supplier wholesale price (Dixt & Pindyck, 1993; Li & Kouvels, 1998). By using the geometric Brownian motion, the supplier wholesale unit price *S(t)* at time *t* is defined as follows:

$$dS(t) = \alpha(t)S(t)dt + \sigma_S(t)S(t)dW(t) \tag{6}$$

where $dW$ is the increment of a Weiner process.

In (6) the current price is known, but future prices are lognormally distributed with a variance that grows linearly with the time horizon. Assuming that $S_0$ is either normally distributed or constant, if $S(t)$ is defined as the process of (6), then the integral (in the Ito sense) of the infinitesimal diffusion in (6) can be obtained as follows (using the method given by Corollary 8.4.3 in Arnold, 1974):

$$S(t) = S_0\left[\exp\left\{\int_{t_0}^{t}\alpha(x)dx - (\sigma_S^2/2)(t-t_0) + \sigma_S^2\{W(t)-W(t_0)\}\right\}\right]$$

The expected value and variance of $S(t)$ at time *t* can be derived as follows:

Lemma 3.

$$J(t) = E\left[S(t)\right] = \left[J_0 \cdot \exp\left\{\int_{t_0}^{t}\alpha(x)dx\right\}\right], \; with\, J_0 = E\left[S_0\right]$$

$$V^S(t) = E\{S(t) - J(t)\}^2 = \left[\exp\left\{2\int_{t_0}^{t}\alpha(x)dx\right\}\right]\left[M_0 \cdot \exp\{\sigma_S^2(t-t_0)\} - (J_0)^2\right]$$

$$with\, M_0 = E[S_0^2]$$

## Production Cost under Supplier Wholesale Price Uncertainty

When the supplier's frequent wholesale price changes affect the manufacturer's unit production cost, consumer markets may react negatively to price increases. Years ago, Rubbermaid (Supplier of plastic goods, e.g., laundry baskets) tried to pass along price increases in plastics due to oil price increase. Walmart refused to accept higher prices for the plastic product, and Rubbermaid had to withdraw the price increase. As the supplier's price normally increases, retail price must be adjusted in order to make a profit when demand decreases. When demand decreases, a manufacturer needs to change product quantities according to a new demand scenario. As a result, the manufacturer's need to change production capacity results from additional production costs. Hence, the additional cost (AC) per unit including $S(t)$ per unit can be defined as follows:

$$AC(t) = \left[\delta[S(t) - J(t)]^2 + S(t)\right]$$

Then, the expected cost per unit can be expressed as follows:

$$C(t) = E[AC(t)] = \left[\delta \cdot \exp\{2\int_{t_0}^{t}\alpha(x)dx\}\left\{M_0 \cdot \exp\{\sigma_S^2(t-t_0)\} - (J_0)^2\right\} + J(t)\right] = \delta V^S(t) + J(t) \tag{7}$$

The ratio of variance of $S(t)$ reflects the additional cost per unit based on the supplier's wholesale price change, which is determined as a $\delta \in (0, \infty)$ If the supplier's frequent wholesale price change has a significant impact on production cost, then the value of $\delta$ is considered as high. If the supplier's wholesale price change does not affect any production cost, then $\delta = 0$. *Expected Price Change under Supplier's Wholesale Price Change*

The rate of price change at time $t + dt$ is determined by the current price at time $t$ and supplier's changed wholesale price. The manufacturer may need to adjust the price based on the determined fractional rate of the supplier's wholesale price change. Let $\beta(t) \in (-\infty, \infty)$ be the fractional rate of the supplier's unit wholesale price change that is reflected in the new price. The price rate at time $t + dt$ is expressed as follows:

$$P(t + dt) = P(t) + \beta(t)[S(t + dt) - S(t)]$$

Then

$$\lim_{dt \to 0}[\{P(t + dt) - P(t)\} / dt] = \lim_{dt \to 0}[\beta(t)\{[S(t + dt) - S(t)]\} / dt]$$

Therefore, we obtain the equation as follows:

$$\dot{P}(t) = \beta(t) \cdot \dot{S}(t)$$

Let $\dot{P}^E(t) = E[\dot{P}(t)]$, then the expected change rate of the price at time $t$ is expressed as $\dot{P}^E(t) = \beta(t)\dot{J}(t)$, since

$$\dot{P}^E(t) = E[\dot{P}(t)] = \beta(t)E[[\dot{S}(t)] = \beta(t)\dot{J}(t).$$

## Formula

The profit of the manufacturer at time $t$ is expressed as $P(t)D(P) - AC(t)Q(t) - l(t)$

Thus, the expected profit can be formulated as follows:

$$E[P(t)]D[E\{P(t)\}] - E[AC(t)]E[Q(t)] - E[l(t)] = P^E(t)D\{P^E\} - C(t)K(t) - L(t) \quad (8)$$

Henceforth, we use $P$ instead of $P^E$ for the sake of simple exposition unless the need exists to distinguish between $P$ and $P^E$ Equation (8) can be rewritten as follows:

$$P^E(t)D\{P^E\} - C(t)K(t) - L = P(t)D\{P(t)\} - C(t)K(t) - L(t)$$

Let $P_L$ and $P_U$ be the lower and upper bound of price. The initial price and initial demand are given as $P_0 = P(t_0)$ and $D_0 = D[P_0]$. Based on the optimal control theory, the price regulation (PR) decision is obtained as follows:

$$\underset{P}{Max}\int_0^T \begin{bmatrix} P(t)D[P(t)] - C(t)K(t) - \\ L\{G(t), K(t), P(t)\} \end{bmatrix} dt$$

s.t $\dot{G}(t) = -2n(t)G(t) + 2n(t)K(t)D\{p(t)\} + \sigma_Q^2$

$$\dot{K}(t) = n[D\{p(t)\} - K(t)]$$

$$\dot{P}(t) = \beta(t)\dot{J}(t)$$

$$\dot{J}(t) = \alpha(t)J(t)$$

$$P_L \le P(t) \le P_U$$

Then, the time index $t$ can be deleted for the sake of simple exposition. Here, we discuss the theoretical properties of the optimal control policies of a PR problem. With co-states $\lambda$, the corresponding Hamiltonian can be derived as follows:

$$H = PD(P) - C(J)K - L(G,K,P) + \lambda_1\{-2nG + 2nKD(P) + \sigma_Q^2\} + \lambda_2 n\{D(P) - K\} + \alpha J(\lambda_3\beta + \lambda_1)$$

The Lagrangian with multipliers $w_1$ and $w_2$ is expressed as follows:

$$\Gamma = H + w_1(P_U - P) + w_2(P - P_L)$$

Then, $\lambda_i(t)$ $(i = 1,2,3, and 4)$ satisfies the following equations:

$\dot{\lambda}_1(t) = -(\partial H / \partial G) = L + 2n\lambda_1$ , with $\lambda_1(T) = 0$

$\dot{\lambda}_2(t) = -(\partial H / \partial G) = C + L_K - 2nD\lambda_1 + n\lambda_2$, with $\lambda_2(T) = 0$

$\dot{\lambda}_3(t) = -(\partial H / \partial P) = -\{D + PD_P - L_P + nD_P(2\lambda_1 K + \lambda_2)\}$, with $\lambda_3(T) = 0$ and

$\dot{\lambda}_4(t) = -(\partial H / \partial J) = -\{C_J K + \alpha(\lambda_3\beta + \lambda_4)\}$, with $\lambda_4(T) = 0$

where

$L_G = \partial L / \partial G = C^H\{1 - R(t)\}A_G(t) - C^S \cdot, R(t) \cdot B_G(t)$

$L_K = \partial L / \partial K = C^H\{1 - R(t)\}A_K(t) -, C^S R(t) B_K(t)$

$L_P = \partial L / \partial P = C^S R_P(D - B) - C^H R_P(A - D) + C^S R(D_P - B_P) + C^H(1 - R)(A_P - D_P)$

Then,

$$\lambda_1(t) = -\exp(2nt)\int_t^T L_G(x)e^{-2nx}dx ,$$

$$\lambda_2(t) = -\exp(nt)\int_t^T \left\{ \begin{matrix} C(x) + L_K(x) - \\ 2nD\lambda_1(x) \end{matrix} \right\} \exp(-nx)dx$$

$\lambda_3(t) =$ and $\int_t^T \left\{ \begin{matrix} D(P) + P(x)D_P(P) - L_P(x) + \\ n(x)D_P(P)\{2\lambda_1(x)K(x) + \lambda_2(x)\} \end{matrix} \right\} dx$

$$\lambda_4(t) = \exp(-\int_{t_0}^t \alpha(x)dx)\int_t^T \left\{ \begin{matrix} \alpha(x)\beta(x)\lambda_3(x) + \\ C_J(x)K(x) \end{matrix} \right\} \cdot \exp(\int_{t_0}^X \alpha(y)dy)dx$$

Therefore, the necessary conditions for optimal solution are characterized as follows:

$$\partial\Gamma / \partial P = D + PD_P - L_P + \lambda_1(2nKD_P) + \lambda_2 nD_P + w_2 - =0 \tag{9}$$

$$\dot{w}_1 = \lambda_3(t) + w_2 - w_1$$

$$w_1 \geq 0,\ w_1(P_U - P^*) = 0, w_2 \geq 0,\ \text{and}\ w_2(P^* - P_L) = 0 \tag{10}$$

From (9) and (10), the optimal solution for price can be obtained as follows:

Lemma 4.

*For a given* $J(t)$ *the optimal price* $P^*(t)$ is as follows:

$$P^*(t) = \begin{cases} P_L & \text{if } \dot{\lambda}_3(t) > 0 \\ P^0 & \text{if } \dot{\lambda}_3(t) = 0 \\ P_U & \text{if } \dot{\lambda}_3(t) < 0 \end{cases},$$

Where

$$P^0 = \left[ P \in (P_L, P_U) \middle| \begin{matrix} D(P) + PD_p(P) - \\ L_p(P) + nD_p(P)\{2\lambda_1(P)K(P) + \lambda_2(P)\} = 0 \end{matrix} \right].$$

For a given expected wholesale price *J(t)*, the manufacturer formulates the pricing strategy based on the derivative of the marginal profit with respect to price. If the derivative of the marginal profit with respect to price is positive, the manufacturer should choose the minimum price policy; if it is negative, the maximum price policy must be taken as an optimal strategy. However, if it is equal to zero, the manufacturer should select the price ($P^0$) that exists between the low limit ($P_L$) and upper limit ($P_U$).

For notational purposes, the realized demand under retail price $P(t)$ at time $t$ is expressed as $D(t|P)$, and the realized values under price $P(t)$ at time $t$ are expressed as $I_1(t|P)$, $I_2(t|P)$, $\lambda_1(t|P)$, $\lambda_2(t|P)$, $\lambda_3(t|P)$, $L_P(t|P)$ and $D_P(t|P)$

Let $I_1(t|P) = D(t|P) + D_P(t|P)P(t) - L_P(t|P)$ and then

$$\dot{\lambda}_3(t|P) = -\begin{bmatrix} I_1(t|P) + \\ nD_P(t|P)I_2(t|P) \end{bmatrix}$$

The manufacturer must decide the introductory price when launching its product in the marketplace. Theorem 1 demonstrates how the manufacturer can determine the optimal introductory price with the given expected supplier's initial wholesale price.

**Theorem 1.** *For a given expected initial supplier's wholesale price* $J(t_0)$, *the optimal* introductory price $P_0^* = P^*(t = t_0)$ can be determined as follows:

(a) *If* $I_1(t_0|P_U) > 0$ and $I_2(t_0|P_U) \le 0$ then $P_0^* = P_U$

(b) *If* $I_1(t_0|P_L) < 0$ and $I_2(t_0|P_L) \ge 0$, then $P_0^* = P_L$

(c) *If there exists* $P_0 \in (P_L, P_U)$ such that

   (i) $I_1(t_0|P_0) < 0$, $I_2(t_0|P_0) < 0$ and $I_1(t_0|P_0) + nD_P(t_0|P_0)I_2(t_0|P_0) = 0$

   or

   (ii) $I_1(t_0|P_0) > 0$ $I_2(t_0|P_0) > 0$ and $I_1(t_0|P_0) + nD_P(t_0|P_0)I_2(t_0|P_0) = 0$

   or

   (iii) $I_1(t_0|P_0) = 0$ and $I_2(t_0|P_0) = 0$,

   *then* $P_0^* = P_0$.

(d) Otherwise

   (i) *if* $\dot{\lambda}_3(t_0|P_L) > 0$ then $P_0^* = P_L$

   (ii) *if* $\dot{\lambda}_3(t_0|P_U) < 0$ then $P_0^* = P_U$

Note that $I_1(t_0|P_U)$ refers to the outcome of three functions-demand, marginal demand with respect to price, and marginal expected loss with respect to price. $I_2(t_0|P_U)$ implies the proportion of marginal profit with respect to $K$ and $G$ If the profit margin is high and the forecasting uncertainty is low (i.e., the condition of [a]), the manufacturer should select an introductory price as high as possible. However, if the profit margin is low and forecasting uncertainty is high (i.e., the condition of [b]), the manufacturer should set the lowest introductory price. Otherwise (i.e., the condition of [c]), the optimal introductory price should be obtained between the low limit ($P_L$) and upper limit ($P_U$).

In Theorem 2 and Theorem 3, we derive the conditions for establishing the optimal pricing strategy considering future forecasting errors, such as overestimation and underestimation. For notational simplicity, let

$$Ws(t) = n\exp(nt)\int_t^T C(x)\exp(-nx)dx + C^S$$

and

$$Wh(t) = n\exp(nt)\int_t^T C(x)\exp(-nx)dx - C^H$$

**Theorem 2.** *For a given coordination level* $n(t) = n$ with an absolute error term of $|D(t) - K(t)| > 9\sqrt{V^R(t)}$ for $t \in |t_1, t_2|$

(a) *If the demand forecast is underestimated* $\{i.e., R(t) = 1\}$, then $P_L > Ws - \{D(P_L) / D_P(P_L)\}$ and $P^*(t) = P_L$ for $t \in |t_1, t_2|$.

(b) *If the demand forecast is overestimated* $\{i.e., R(t) = 0\}$ then $P_L > Wh - \{D(P_L) / D_P(P_L)\}$ and $P^*(t) = P_L$ for $t \in |t_1, t_2|$.

Based on the assumption that the absolute error is large enough, if either underestimated or overestimated forecast with $n$ level of coordination during $t_1$ and $t_2$ is derived from $Q(t)$, the manufacturer should obtain the lowest pricing strategy based on the value of $Ws$ and $Wh$. If the probability that demand forecast is always underestimated is expected to be high, the manufacturer should obtain the lowest price during $t_1$ and $t_2$. However, the determined lowest price must be greater than the value of $Ws - \{D(P_L) / D_P(P_L)\}$. If the probability that the demand forecast is always overestimated is expected to be high, the manufacturer should also obtain the lowest price during $t_1$ and $t_2$. The determined lowest price must be greater than the value of $Wh - \{D(P_L) / D_P(P_L)\}$.

Theorem 3 and Theorem 4 present optimally determined decisions on price and order quantity based on the given supplier's expected wholesale price and the terminal condition.

**Theorem 3.** *For a given expected wholesale price* $J(t)$ we assume that the terminal condition is defined as $I_1(T|P_L) < 0$ and one of the following conditions holds for $t \in (t_0, T)$:

(a) $I_1(t|P_L) + nD_P(t|P_L)I_2(t|P_L) < 0$

(b) $I_1(t|P_L) < 0$, $L_G(t|P_L) < 0$ and $\{C(J) + L_K(t|P_L)\} < 2nD(t|P_L)\lambda_1(t|P_L) < 0$

*Then,* $P^*(t) = P_L$ and

$$K^*(t) = e^{-nt}\left[K_0 + n\int_{t_0}^t D(P_L)e^{nx}dx\right]$$

for all $t \in [t_0, T]$.

The condition of $I_1(T|P_L)$ indicates that the outcome of the three functions-demand, marginal demand with respect to price, and marginal expected loss with respect to price-is expected to be negative with the lowest price at terminal time $T$. This means that the product is losing its popularity near the end of its product life cycle so that any retail price setting at $T$ leads to the firm's profit loss. If either condition (a) or (b) is held, the manufacturer must charge the lowest price over the entire planned time horizon. Based on this information, the expected optimal order quantity $K^*(t)$ is determined.

**Theorem 4.** *For a given expected wholesale price* $J(t)$ we assume that the terminal condition is defined as $I_1(T|P_U) > 0$ and one of the following conditions holds for $t \in [t_0, T)$:

(a) $I_1(T|P_U) + nD_P(t|P_U)I_2(t|P_U) > 0$

(b) $I_1(t|P_U) > 0, L_G(t|P_U) < 0$ and $0 < 2nD(t|P_U)\lambda_1(t|P_U) < \{C(J) + L_K(t|P_U)\}$.

*Then,* $P^*(t) = P_U$ and

$$K^*(t) = e^{-nt}\left[K_0 + n\int_{t_0}^{t} D(P_U)e^{nx}dx\right]$$

for all $t \in [t_0, T]$.

Theorem 4 is developed under the condition of

$$P^*(t) = \begin{cases} P_U = \$990 & \text{for } 0 \le t < 0.77 \\ P^0 \in (\$823.22, \$990) & \text{for } 0.77 \le t \le 5.99 \\ P_L = \$823.22 & \text{for } 5.99 < t \le 6 \end{cases}$$

.

The condition of $I_1(T|P_U) > 0$ means that the outcome of three functions of demand, marginal demand with respect to price, and marginal expected loss with respect to price is expected to be positive with the highest price at terminal time *T.* This implies that the product will still possess competitiveness until the end of its product life cycle. If either condition (a) or (b) is held, the manufacturer must charge the highest price over the entire planned time horizon. Based on this information, the expected optimal order quantity $K^*(t)$ is determined.

**Theorem 5.** *Suppose* $I_1(T|P_L) < 0$ If $\ddot{\lambda}_3(t) \ge 0$ or if $\dot{\lambda}_3(t)$ is a convex or concave function for all t, and if one of the following conditions is satisfied,

(a) $I_1(t_0|P_U) > 0$ and $I_2(t_0|P_U) < 0$,

(b) $I_1(t_0|P_U) + nD_P(t_0|P_U)I_2(t_0|P_U) > 0$

*then Opportunity Pricing Policy is optimal for a given* $J(t)$ and the optimal price and order quantity are formulated as follows:

$$P^*(t) = \begin{cases} P_U & \text{for } t_0 < t < \tau_1 \\ P^0 & \text{for } \tau_1 \le t \le \tau_2 \text{ and} \\ P_L & \text{for } \tau_2 < t < T \end{cases}$$

$$K^*(t) = \begin{cases} K(t|P = P_U) & \text{for } t_0 < t < \tau_1 \\ K(t|P = P^0) & \text{for } \tau_1 \le t \le \tau_2 \\ K(t|P = P_L) & \text{for } \tau_2 < t \le T \end{cases}$$

*where*

$$\tau_1 = Min\left[t_0 < t \le T \middle| \dot{\lambda}_3(t|P = P^0) = 0\right],$$

$$\tau_2 = Max\left[\tau_1 \le t \le T \middle| \dot{\lambda}_3(t|P = P^0) = 0\right] \text{ and}$$

$P^0 \in (P_L, P_U)$.

Theorem 5 takes into account the initial and terminal time of the product life. In this theorem, we are concerned with optimally combined price-order quantity decisions especially for the situation where the product starts with high popularity but demand declines over time. Theorem 5 suggests that the manufacturer should start with a high price until time $\tau_1$ but gradually decrease the price during $\tau_1$ and $\tau_2$ and after that, keep a low price until the ending time. This is an opportunity pricing strategy that can promote profit maximization of the manufacturer. The optimal order quantity is determined with the optimal price under the same time interval.

**Theorem 6.** *Suppose that* $I_1(T|P_L) < 0$, $\dot{\lambda}_3(t)$ is a convex function for all t, and there exists $t_1 \in (t_0, T)$ such that $\dot{\lambda}_3(t_1|P_U) < 0$

And also assume that one of the following conditions is held:

(a) $I_1(t_0|P_U) < 0$ and $I_2(t_0|P_U) > 0$

(b) $I_1(t_0|P_U) + nD_P(t_0|P_U)I_2(t_0|P_U) < 0$

*Then, Umbrella Pricing Policy is optimal for a given expected wholesale price* $J(t)$ and the optimal price and order quantity are formulated as follows:

$$P^*(t) = \begin{cases} P_L & \text{for } 0<t<\tau_1 \\ P^0 & \text{for } \tau_1 \le t \le \tau_2 \\ P_U & \text{for } \tau_2 < t < \tau_3 \text{ and} \\ P^0 & \text{for } \tau_3 \le t \le \tau_4 \\ P_L & \text{for } \tau_4 < t \le T \end{cases}$$

$$K^*(t) = \begin{cases} K(t|P = P_L) & \text{for } 0 \le t<\tau_1 \\ K(t|P = P^0) & \text{for } \tau_1 \le t \le \tau_2 \\ K(t|P = P_U) & \text{for } \tau_2 < t < \tau_3 \\ K(t|P = P^0 & \text{for } \tau_3 \le t \le \tau_4 \\ K(t|P = P_L & \text{for } \tau_4 < t \le T \end{cases}$$

where

$\tau_1 =$

$$Min[0 < t \le t_1 | \dot{\lambda}_3(t|P = P^0) = 0]$$

$\tau_2 =$

$$Max[\tau_1 \le t < t_1 | \dot{\lambda}_3(t|P = P^0) = 0]$$

$\tau_3 =$

$$Min[t_1 < t \le T | \dot{\lambda}_3(t|P = P^0) = 0]$$

$\tau_4 =$

$$Max[\tau_3 \le t \le T | \dot{\lambda}_3(t|P = P^0) = 0]$$

Theorem 6, similar to Theorem 5, considers both the initial and terminal sales of the product. When the product receives little attention during its introduction phase, gradually gains popularity, but the product demand declines again near the terminal stage, the manufacturer should use a low initial price and then gradually increase the price as time progresses, until the product reaches the maximum price. After the maximum price is reached, the manufacturer gradually decreases the price for the reminder of the planned time horizon.

## NUMERICAL EXAMPLES

In this section, numerical examples are presented based on the theoretical results of the previous section. The optimal policy models developed in the previous section can be applied under a sequence of events as follows. $D(P)$ in this paper is the linear demand function as follows:

$$D(P) = a - bP(t)\ (a > 0, b > 0)$$

The current time horizon is six months: $t_0 = 0 \le t \le T = 6$

First, a numerical example is shown for initial prices based on the actual, forecasted demand, and the variance of the forecast at the initial time with the following assumptions and parameters. The optimal price policy is investigated at the initial time based on an effect of $a$ on the initial demand without changing parameter $b$. Assumptions were made that $Q(t = 0) = Q_0$ is normally distributed with mean $K_0 = D(P_0) + 1000$ and standard deviation of 60, and $S(t = 0) = S_0$ is also normally distributed with mean $J_0 = 60$ and standard deviation of 2.

Let the wholesale price range be:

$$\$60 \le J(t) \le \$80.99,$$

$$P_L = 1.5(\$60) = \$90$$

*Table 1. Optimal introductory price (P*)*

| $a$ | 60000 | 70000 | 75000 | 80000 | 90000 | 100000 | 110000 | 200000 |
|---|---|---|---|---|---|---|---|---|
| $D(P_d)$ | 6000 | 16000 | 19548 | 22046 | 27048 | 32044 | 37106 | 127106 |
| $P^*$ | \$90.00 | \$90.00 | \$92.42 | \$96.59 | \$104.92 | \$113.26 | \$121.49 | \$121.49 |

$$P_U = 1.5(\$80.99) = \$121.49$$

and other parameters are as follows:

$$K_0 = Q_0$$
$$V^{Q_0} = (60)^2$$
$$G_0 = V^{Q_0} + K_0^2 = (60)^2 + K_0^2$$
$$V^{S_0} = 2$$
$$C^S = \$30$$
$$C^H = \$5$$
$n = 1$ and
$$M_0 = V^{S_0} + (J_0)^2 = 3602$$

The intercept a of linear demand function varies eight different values (e.g., 60,000, 70,000, 75,000, 80,000, 90,000, 100,000, 110,000, 200,000}, and slope b of linear demand function is 200, and apply Theorem 1. Then,

$$D(P_0) = \alpha - 600P_0$$

and results are summarized in Table 1.

Table 1 indicates that the higher the initial demand, the higher the initial price, and that if the initial demand is lower than 19548, then the minimum price, $P_L = 90$ is the optimal price. However, if the initial demand is higher than 32044, then the maximum price, $P_U = 121.49$ is the optimal price at the beginning. Otherwise, the optimal initial prices increase as initial demand increases.

Secondly, variances, $V^{Q_0}$ of the initial forecast are studied to determine their affect on the optimal price at the initial period. Assume:

*Table 2. Optimal introductory price based on standard deviation*

| $\sqrt{V^{Q_0}}$ | 10 | 60 | 100 | 150 | 200 |
|---|---|---|---|---|---|
| $P^*$ | \$ 96.59 | \$ 96.59 | \$ 96.59 | \$ 90 | \$ 90 |

$$D(P) = 80000 - 600P_0$$

and every parameter is the same as the previous example, except the parameters of $\sqrt{V^{Q_0}}$ that varies $\{10, 60, 100, 150, 200\}$. Using Theorem 1 (b) and (d) results in the following optimal price, summarized in Table 2.

Table 2 implies that an optimal initial price should be minimum, $P_L$ , when the standard deviation of the initial forecast is relatively high. This indicates that if the variance of the initial forecast is high, then it is better to charge the minimum optimal price.

Thirdly, the accuracy of the forecast is determined through supply chain coordination and the variance of forecast influence $\tau_1$ and $\tau_2$. We assume that $Q(t = 0) = Q_0$ is normally distributed with mean $K_0$ standard deviation $W_0 = 100$ $K_0 = D(P_0) - 10W_0$ and $G_0 = (K_0)^2 + 10$ $S(t = 0) = S_0$ is also distributed normally with mean $J_0$ and standard deviation of 2. Suppose that the supplier with the initial charge of \$600 ( $J_0 = \$600$ ) gradually decreases its price as time progresses. Hence, let $\alpha(t) = -0.05$ then $J_0 = \$600$ and $J_6 = \$444.49$ The manufacturer sets:

*Table 3. and $\tau_2$ change based on $n$ = 1, 2, or 3*

| $n$ | 1 | 2 | 3 |
|---|---|---|---|
| $\tau_1$ | 0.77 | 0.261 | 0.112 |
| $\tau_2$ | 5.99 | 6 | 6 |

$P_U = \$990$
$P_L = \$823.22$ and
$\$823.22 \le P(t) \le \$990$

The other parameters are as follows:

$D(P) = 300,000 - 300P(t)$
$\sigma_S = 0.02$
$C^S = \$400$
$C^H = \$100$ and
$n(t) = 1$

Applying Theorem 5 and 6, the optimal price can be determined as follows:

$$P^*(t) = \begin{cases} P_U = \$990 & \text{for } 0 \le t < 0.77 \\ P^0 \in (\$823.22, \$990) & \text{for } 0.77 \le t \le 5.99 \\ P_L = \$823.22 & \text{for } 5.99 < t \le 6 \end{cases}$$

Because as the manufacturer's coordination with the supplier increases, the accuracy of the forecast improves, we investigated how these changes affect the optimal price policy and order quantities in two different ways. First, all the parameters are considered to be the same except for the parameter $n(t) = n$ that takes three different values: $n$ = 1, 2, or 3. The results are summarized in Table 3.

From Table 3, as the coordination level and accuracy of forecast improve, $\tau_1$ decrease and $\tau_2$ increase. The manufacturer needs to charge the maximum price for an extended time as the level of $n$ increases. Next, we set all the parameters the same except for the parameter $\sigma_Q$ which takes four different values; $\sigma_Q$=100, 250, 500, or 1000. The results are summarized in Table 4.

*Table 4. and $\tau_2$ change based on $\sigma_Q$ with $n = 1$*

| $\sigma_Q$ | 100 | 250 | 500 | 1000 |
|---|---|---|---|---|
| $\tau_1$ | 0.77 | 0.2621 | 0 | 0 |
| $\tau_2$ | 5.99 | 5.98 | 5.96 | 5.91 |

Table 4 indicates that the manufacturer with $\sigma_Q = 100$ should charge the maximum price, \$990, during the first 0.77 months, but the manufacturer with $\sigma_Q = 250$ will charge the maximum price during the first 0.2621 months. A manufacturer with a lower variance can charge longer period than a manufacturer with a larger variance. However, the manufacturer with a lower variance can charge a minimum price for a shorter period than the manufacturer with a larger variance. Table 4 implies that both $\tau_1$ and $\tau_2$ should be decreased as forecast variance increases; in other words, the manufacturer with a smaller forecast variance can charge the maximum price longer than a manufacturer with a larger forecast variance. Also, the duration of the minimum price charge for the manufacturer with a lower forecast variance is shorter than one with a larger forecast variance. Overall, the numerical examples shown support the theories developed in the previous section as being reliable.

## CONCLUSION

A manufacturer may have to make decisions amid the uncertainties associated with product demand and supplier's wholesale price. One way for a manufacturer to deal with these uncertainties is to improve its decision-making ability by establishing a well-defined coordination mechanism that maintains close collaborative interactions with suppliers and to obtain optimized operational decisions. In this study, we attempted to integrate into the model the two major variables of forecasting error associated with demand uncertainty and risk incurred from unpredictable changes in supplier's wholesale price. In the modeling process, we showed the stochastic steps to reach optimal decisions on price and order quantity based on the different levels of coordination between the manufacturer and its supplier. Theorem 1of the model proposes the optimal introductory price under the given supplier's initial wholesale price. Our analysis showed that if the profit margin is expected to be high and forecasting uncertainty is assumed to be low (i.e., high level of coordination), then selecting the high introductory price is optimal, whereas if the profit margin is expected to be low and forecasting uncertainty high, choosing the lowest introductory price is optimal. Otherwise, the optimal introductory price should be obtained between the low limit ( $P_L$ ) and upper limit ( $P_U$ ). Theorem 2 demonstrates the useful guidelines for how to determine the lowest optimal price in the case of under stocking or overstocking of supply driven by unexpected product demand. Interestingly, even though it is anticipated that demand exceeds supply, our optimization model suggested the low-retail-pricing strategy as an optimal solution. The optimization models presented in Theorem 3, 4, 5, and 6 characterize the three functional conditions of demand, marginal demand with respect to price, and marginal expected loss with respect to price that the manufacturer must evaluate over the product life cycle and explain the optimally combined solutions for price and order quantity in the given time by balancing marginal revenue with marginal cost. The findings of those Theorems suggested that, in order to promote profit maximization, the manufacturer should define a policy that specifies when and how to strategically change the retail price as the popularity and sales potential of the product changes over its product lifetime and that the expected optimal order quantity must be obtained based on the determined optimal retail price.

Our study has two important managerial implications. First, we find that there may be significant performance improvements to manufacturers from creating a collaborative work relationship with their suppliers. If the use of a coordination mechanism mitigates the uncertainties associated with product demand and supplier's wholesale price, then manufacturers may be able to have increased power to make the right operational decisions. Although high levels of coordination require an investment cost, an important advantage of the close interaction with suppliers may be the manufacturer's superior decision-making capability. Second, our findings provide a modeling framework for decision making that guides the manufacturer in evaluating the coordination levels needed to deal with demand and supply uncertainty and in formulating optimal pricing and ordering strategy toward profit maximization. If the optimization policy with a quality coordination mechanism can lead to improved operational performance associated with product availability, then manufacturers may be able to offer better customer service, thus improving sales.

This paper contains several research limitations. First, we considered a single supplier in the modeling process. The inclusion of multiple suppliers in the model is needed for practical usage. Second, it is considered that manufacturers may have more than one type of supply uncertainty, but we mainly focused on only the uncertainty associated with the supplier's wholesale price. Finally, the theoretical findings of the model should be tested with real-world data.

## REFERENCES


Arnold, L. (1974). *Stochastic differential equations: Theory and applications*. New York: John Willey & Sons.

Ash, R. B. (1972). *Real analysis and probability*. New York: Academic Press.

Aviv, Y. (2001). The effect of collaborative forecasting on supply chain performance. *Management Science, 47*(10), 1326–1343. doi:10.1287/mnsc.47.10.1326.10260

Basu, A. K. (1993). *Introduction to stochastic process*. Oxford, UK: Alpha Science International Ltd.

Bensoussan, A., Hurst, E. G. Jr, & Naslund, B. (1974). *Management Application of Modern Control Theory*. New York: Elsevier.

Bitran, G., & Rene, C. (2003). An overview of pricing models for revenue management. *Manufacturing & Service Operations Management, 5*(3), 203–229. doi:10.1287/msom.5.3.203.16031

Chen, C., & Lee, W. (2004). Multi-objective optimization of multi-echelon supply chain networks with uncertain product demands and prices. *Computers & Chemical Engineering, 28*, 1131–1144. doi:10.1016/j.compchemeng.2003.09.014

Dixit, A., & Pindyck, R. (1993). *Investment under uncertainty*. Princeton, NJ: Princeton University Press.

Emmons, H., & Gilbert, S. M. (1998). Note: The role of return policies in pricing and inventory decisions for catalogue goods. *Management Science, 44*(2), 276–283. doi:10.1287/mnsc.44.2.276

Fan, Y. Y., Bhargava, H. K., & Natsuyama, H. (2005). Dynamic pricing via dynamic programming. *Journal of Optimization Theory and Applications, 127*(3), 565–577. doi:10.1007/s10957-005-7503-z

Feichtinger, G., & Hartl, R. F. (1985). Optimal pricing and production in an inventory model. *European Journal of Operational Research, 19*, 45–56. doi:10.1016/0377-2217(85)90307-8

Gaimon, C. (1988). Simultaneous and dynamic price, production, inventory, and capacity decisions. *European Journal of Operational Research, 35*, 426–441. doi:10.1016/0377-2217(88)90232-9

Gupta, A., & Maranas, C. D. (2000a). A two-stage modeling and solution framework for multisite midterm planning under demand uncertainty. *Industrial & Engineering Chemistry Research, 39*(10), 3799–3813. doi:10.1021/ie9909284

Gupta, A., & Maranas, C. D. (2003). Managing demand uncertainty in supply chain planning. *Computers & Chemical Engineering, 27*, 1219–1227. doi:10.1016/S0098-1354(03)00048-6

Gupta, A., Maranas, C. D., & McDonald, C. M. (2000b). Midterm supply chain planning under demand: Customer demand satisfaction and inventory management. *Computers & Chemical Engineering, 24*, 2613–2621. doi:10.1016/S0098-1354(00)00617-7

Hartl, R. F., & Sethi, S. P. (1984). Optimal control problem with differential inclusions: Sufficiency conditions and an application to a production-inventory model. *Optimal Control Applications & Methods, 5*(4), 289–307. doi:10.1002/oca.4660050403

Huchzermeier, A., & Cohen, M. (1996). Valuing operational flexibility under exchange rate risk. *Operations Research, 44*, 100–113. doi:10.1287/opre.44.1.100

Hwang, C. L., Fan, L. T., & Erickson, L. E. (1967). Optimal production planning by the maximum Principle. *Management Science, 13*, 750–755. doi:10.1287/mnsc.13.9.751

Iyer, A. V., & Bergen, M. E. (1997). Quick response in manufacturer-retailer channels. *Management Science, 43*, 559–570. doi:10.1287/mnsc.43.4.559

Jung, J. Y., Blau, G., & Pekny, J. F. (2004). A simulation based optimization approach to supply chain management under demand uncertainty. *Computers & Chemical Engineering, 28*, 2087–2106. doi:10.1016/j.compchemeng.2004.06.006

Kogut, B. (1985). Designing global strategies: profiting from operational flexibility. *Sloan Management Review*, 27–38.

Lau, A., Lau, H., & Zhou, Y. (2006). Considering asymmetrical manufacturing cost information in a two-echelon system that uses price-only contracts. *IIE Transactions, 38*, 253–271. doi:10.1080/07408170590961148

Lee, H., & Tang, C. S. (1998a). Variability reduction through operations reversal. *Management Science, 44*, 162–173. doi:10.1287/mnsc.44.2.162

Lee, H. L., & Tang, C. S. (1998b). Managing supply chains with contract manufacturing . In Lee, H. L., & Ng, S. M. (Eds.), *Global supply chain and technology management*. Orlando, FL: Production and Operations Management Society Publishers.

Levy, D. (1995). International sourcing and supply chain stability. *Journal of International Business Studies, 26*, 343–360. doi:10.1057/palgrave.jibs.8490177

Li, C. L., & Kouvelis, P. (1999). Flexible and risk-sharing supply contracts under price uncertainty. *Management Science, 45*(10), 1378–1398. doi:10.1287/mnsc.45.10.1378

Liu, M. L., & Sahinidis, M. V. (1998). Robust process planning under uncertainty. *Industrial & Engineering Chemistry Research, 37*, 1883–1899. doi:10.1021/ie970694t

Nam, S. H., Vitton, J., & Kurata, H. (2009). Determining the optimal number of suppliers utilized by contractors. *International Journal of Production Economics*.

Pekelman, D. (1974). Simultaneous price-production decision. *Operations Research, 22*, 788–794. doi:10.1287/opre.22.4.788

Ryan, T. P. (2000). *Statistical methods for quality improvement*. New York: John Wiley & Sons.

Tang, C. H. (2006). Perspectives in supply chain risk management. *International Journal of Production Economics, 103*, 451–488. doi:10.1016/j.ijpe.2005.12.006

Tsiakis, P., Shah, N., & Pantelides, C. (2001). Design of multi-echelon supply chain networks under demand uncertainty. *Industrial & Engineering Chemistry Research, 40*, 35–85. doi:10.1021/ie0100030

## APPENDIX

*List of notation and symbols used in the paper*

| | |
|---|---|
| *P(t)* | Price at time *t* |
| $P_L, P_U$ | Lower and upper bound of price, respectively |
| *M(P)* | $:= M(P) = D(P) + \varepsilon_W(t)$, actual market demand based on price *P(t)* at time $t$ |
| *D(P)* | Deterministic part of market demand |
| $\varepsilon_W(t)$ | White noise: $\varepsilon_W(t) = (\sigma_Q)dW / dt$ |
| *dW(t)* | Standard Brownian motion (Weiner process) |
| *S(t)* | Supplier's unit wholesale price at time *t* and follow geometric Brownian motion |
| *dS(t)* | $= \alpha(t)S(t)dt + \sigma_S(t)S(t)dW(t)$, geometric Brownian motion |
| *Q(t)* | Forecasted demand based on demand *M(P)* at time *t* with $Q_0 = Q(t = t_0)$ |
| $\sigma_Q$ | Coefficient of variance rate change of *Q(t)* |
| *dQ(t)* | $n(t)\left[D(P) - Q(t)\right]dt + \sigma_Q dW(t)$ |
| *n(t)* | The speed of revision that is controlled by the level of coordination between manufacturer and supplier at time *t* |
| $\varepsilon(t)$ | Error term in the regression $Q(t) - Q(t-1) = \alpha + bQ(t-1) + \varepsilon(t)$ |
| $C^S, C^H$ | Unit shortage and holding cost, respectively |
| *R(t)* | $= \Pr\left[Q(t) < M(P)\right]$, the probability of forecasted demand, which is less than the actual demand at time *t* |
| *L(t)* | $= E[l(t)]$, expected loss at time *t*, including inventory holding and stock-out cost, where *l(t)* is the inventory penalty at time *t* |
| $V^R(t)$ | $= \sqrt{G(t) - K^2(t) + \sigma_Q^2(t)}$, standard deviation of *R(t)* |
| $\delta$ | Supplier's price change rate and $\delta \in [0, \infty)$ |
| $\beta(t)$ | The fractional rate of the supplier's unit price change that actually reflects the new price and $\beta(t) \in (-\infty, \infty)$ |

**Proof of Lemma 1.**

Since $M(P) = D(P) + \varepsilon_W(t)$ and $dM(P) = \dot{D}(P)dt + \sigma_Q dW(t)$, the continuous version of $M(P)$ can be formulated as $M(P) = D(P) + \int_{t_0}^{t} \sigma_Q dW(s)$, $E[M(P)] = D(P)$ and $Var[M(P)] = \sigma_Q^2(t - t_0)$. Let

$$Y(t) = \begin{bmatrix} D(P) - \exp\{-N(t)Q_0 - \exp\{-N(t)\} \\ \int_{t_0}^{t} n(x)D(P)\exp\{N(x)\}dx \end{bmatrix}.$$

Accordingly,

$$R(t) = \Pr[Q(t) < M(P)] =$$
$$\Pr\left[\int_{t_0}^{t} \{\sigma_Q \exp(N(x) - N(t)) - \sigma_Q\}dW(x) < Y(t)\right]$$
$$= \left\{1 / \sqrt{2\pi\sigma_R^2(t)}\right\}\left[\int_{-\infty}^{Y(t)} \exp\left\{-x^2 / 2\sigma_R^2(x)\right\}dx\right] \text{ and}$$

$$\sigma_R^2(t) = \sigma_R^2(t - t_0) +$$
$$\exp\{-2N(t)\}\sigma_R^2 \int_0^t \exp\{2N(x)\}dx$$

by theorem 9.2.3 in Arnold (1974).

**Proof of Lemma 2.**

$$L(t) = E[l(t)] = C^S \bullet E[(M - Q)|M > Q] \bullet P[M > Q] + C^h \bullet E[(Q - M)]|M < Q] \bullet P[M < Q] \quad \text{(Chung, 1974).}$$

Since $E[M(P)] = D(P)$ and $Var[M(P)] = \sigma_M^2 t$, then

$E[M(P) - Q(t)] = D(P) - K(t)$ and

$Var[M(P) - Q(t)] = G(t) - K^2(t) + \sigma_Q^2 t$.

Let

$$V^R(t) = \sqrt{G(t) - K^2(t) + \sigma_Q^2 t}$$

$$u(t) = [D(P) - K(t)] / V^R(t)$$

$A(t) = K(t) + V^R(t)[\varphi(u(t)) / \{1 - \Phi(u(t))\}]$, and

$B(t) = K(t) - V^R(t)[\varphi(u(t)) / \Phi(u(t)]$.

Using the truncated normal distribution (Ryan, 2000), the *expected value* is expressed as

$$E[M(t) - Q(t) | M(t) - Q(t) > 0] = D(t) - K(t) + V^R(t)[\varphi(u(t)) / \{1 - \Phi(u(t))\}],$$ and

$$E[Q(t) - M(t) | Q(t) - M(t) > 0] = K(t) - D(t) + V^R(t)[\varphi(-u(t)) / \{1 - \Phi(u(t))\}].$$

Since $\varphi(-u) = \varphi(u)$ and

$\{1 - \Phi(-u)\} = \Phi(u)$

$$L(t) = E[l(t)] = C^h \bullet \{A(t) - D(P)\} \bullet \{1 - R(t)\} +, C^S \bullet R(t) \bullet \{D(P) - B(t)\}$$

**Proof of Lemma 3.**

By a similar calculation as in Arnold (1974, 1992), the function $J(t)$ must satisfy the following:

$\dot{J}(t) = \alpha(t)J(t)$, with initial condition of $J_0 = E[S_0]$.

Hence, $J(t) = E[S(t)] = [J_0 \cdot \exp\{\int_{t_0}^{t} \alpha(x)dx\}]$,

with $J_0 = E[S_0]$ and the function $M(t) = E[S^2(t)]$ must satisfy the following:

$\dot{M}(t) = [2\alpha(t) + \sigma_S^2]M(t)$,

with initial condition of $M_0 = E[S_0^2]$ Hence,

$M(t) =$
$\left[M_0 \cdot \exp\left\{2\int_{t_0}^{t} \alpha(x)dx + \sigma_S^2(t-t_0)\right\}\right]$ and
$V^S(t) = E\{S(t) - J(t)\}^2$
$\left[\exp\left\{2\int_{t_0}^{t} \alpha(x)dx\right\}\right]\left[M_0 \cdot \exp\{\sigma_S^2(t-t_0)\} - (J_0)^2\right]$, with $M_0 = E[S_0^2]$.

**Proof of Lemma 4.**

If $\dot{\lambda}_3(t) > 0$, then (5) requires $w_1 = 0$; hence, from (6), $P^* = P_L$.

If $\lambda_3(t) < 0$ then (5) requires $w_2 = 0$; hence, from (6), $P^* = P_U$.

If $\lambda_3(t) = 0$ cannot be sustained over an interval of time, then the control is "bang-bang"; it is at its maximum level, $P^* = P_U$,

if $\dot{\lambda}_3(t) < 0$, while it is at its minimum level, $P^* = P_L$,

if $\lambda_3(t) > 0$. Hence, $\lambda_3(t) < 0$ implies $P^* = P_U$,

$\dot{\lambda}_3(t) = 0$ implies $P_L < P(t) < P_U$, and $\dot{\lambda}_3(t) > 0$ implies $P^* = P_L$.

If $\dot{\lambda}_3(t) = 0$ then $P^* = P^0$ and

$$P^0 = \left[P \in (P_L, P_U) \middle| \begin{array}{l} D(P) + PD_P(P) - L_P(P) + \\ nD_P(P)\{2\lambda_1(P)K(P) + \lambda_2(P)\} = 0 \end{array}\right].$$

For notational simplicity, let $D(t|P) = D\{P(t), t\}$ be the realized demand under retail price $P(t)$ at time $t$. Similarly, $I_1(t|P)$, $I_2(t|P)$, $\lambda_1(t|P)$, $\lambda_2(t|P)$, $\lambda_3(t|P)$, $L_P(t|P)$, and $D_P(t|P)$ are the realized values under retail price $P(t)$ at time $t$.

Let $\begin{array}{l} I_1(t|P) = D(t|P) + \\ D_P(t|P)\,P(t) - L_P(t|P) \end{array}$ and

$\begin{array}{l} I_2(t|P) = \\ 2K(t|P)\,\lambda_1(t|P) + \lambda_2(t|P) \end{array}$, then

$\dot{\lambda}_3(t|P) = -\{I_1(t|P) + nD_P(t|P)\, I_2(t|P)\}$.

**Proof of Theorem 1.**

Because $nD_P(t|P) \le 0$ by hypothesis:

(a) if $I_1(t_0|P_U) > 0$ and $I_2(t_0|P_U) < 0$, then $\dot{\lambda}_3(t_0|P_U) < 0$ and $P_0^* = P_U$ by Lemma 1;

(b) suppose $I_1(t_0|P_L) < 0$ and $I_2(t_0|P_L) > 0$, then $\dot{\lambda}_3(t_0|P_U) > 0$ and $P_0^* = P_L$ by Lemma1.

Suppose either $I_1(t_0|P_0) > 0$ , $I_2(t_0|P_0) < 0$ or $I_1(t_0|P_0) < 0$, $I_2(t_0|P_0) > 0$ , then $\dot{\lambda}_3(t|P) \ne 0$; hence c) holds; otherwise d) holds .

**Proof of Theorem 2.**

(a) If $R(t) = 1$, then $Q(t) < M(P)$ almost everywhere and $K(t) < D(P)$ (Ash,1972) and implies that the buyer's forecast of demand is always underestimated. The assumption $D(t) - K(t) > 9\sqrt{V^R(t)}$ indicates that $\varphi(u) = 0$ because $\varphi(u) = 0$, where $u = (D - K) / \sqrt{V^R} > 9$.
If $|D(t) - K(t)| > 9\sqrt{V^R(t)}$, then $\varphi(u) = 0$ and $A_K = B_K = 1$ and $A_G = B_G = 0$. Then $L(t) = C^S[D(P) - K]$ and hence $L_K = -C^S$, $L_G = 0$, and $L_P = C^S D_P$. Therefore $\lambda_1(t) = 0$.
For (a), $\lambda_2(t) = -c^{nt} \cdot \int_t^T C(x)c^{-nx}dx \le 0$ for all *t*, and
$\lambda_3(t) =$
$\int_t^T \left[D(P) + D_P\{P(x) - Ws(x)\}\right]dx$.
If $P_L > Ws - (D / D_P)$, then $P > Ws - (D / D_P)$ and $D + D_P(P - Ws) < 0$. Hence $\lambda_3(t) < 0$ and $\dot{\lambda}_3(t) > 0$ and $P^*(t) = P_L$ for $t \in [t_1, t_2]$ by Lemma 4.

(b) If $R(t) = 0$, then $Q(t) > M(P)$ almost everywhere and $K(t) > D(P)$ (Ash, 1972) and implies that the buyer's forecast of demand is always overestimated. The assumption $D(t) - K(t) > 9\sqrt{V^R(t)}$ indicates that $\varphi(u) = 0$ because $\varphi(u) = 0$ where $u = (D - K) / \sqrt{V^R} > 9$. If $|D(t) - K(t)| > 9\sqrt{V^R(t)}$, then $\varphi(u) = 0$ and $A_K = B_K = 1$ and $A_G = B_G = 0$. Then $L(t) = C^H[K - D(P)]$ and hence $L_K = C^H$, $L_G = 0$, and $L_P = -C^H D_P$. Therefore $\lambda_1(t) = 0$, $\lambda_2(t) = -e^{nt} \cdot \int_t^T C(x)e^{-nx}dx \le 0$ for $t \in [t_1, t_2]$, and

$\lambda_3(t) =$
$\int_t^T \left[D(P) + D_P\{P(x) - Wh(x))\}\right]dx$.

If $P_L > Wh - (D / D_P)$, then $P > Wh - (D / D_P)$ and $D + D_P(P - Wh) < 0$. Hence $\lambda_3(t) < 0$ and $\dot{\lambda}_3(t) > 0$ and $P^*(t) = P_L$ for $t \in [t_1, t_2]$ by Lemma 4.

**Proof of Theorem 3.**

Since $\lambda_1(T) = \lambda_2(T) = 0$ by assumption, $\dot{\lambda}_3(T|P_L) = -I_1(T|P_L)$.

If $I_1(T|P_L) < 0$, then $\dot{\lambda}_3(T|P_L) > 0$.

Suppose

$$I_1(t|P_L) + nD_P(t|P_L)I_2(t|P_L) < 0$$

for all $t \in [t_0, T)$ and $I_1(T|P_L) < 0$, then $\dot{\lambda}_3(t|P_L) > 0$ for all $t \in [t_0, T]$,

and the optimal price is $P^*(t) = P_L$ for all $t \in [t_0, T]$ by Lemma 1.

Suppose $L_G(t|P_L) > 0$ for $t \in [t_0, T)$, then $\lambda_1(t|P_L) < 0$ for $t \in [t_0, T)$.

If

$$C(J) + L_K(t|P_L) - 2nD(t|P_L)\lambda_1(t|P_L) < 0$$

for $t \in [t_0, T)$, then $\lambda_2(t|P_L) > 0$ for all $t \in [t_0, T)$.

Hence if

$$0 > 2nD(t|P_L)\lambda_1(t|P_L) > [C(J) + L_K(t|P_L)]$$

holds for all $t \in [t_0, T)$, then both $\lambda_1(t|P_L) > 0$ and $\lambda_2(t|P_L) > 0$.

Therefore

$$I_2(t|P) = 2K(t|P)\lambda_1(t|P) + \lambda_2(t|P) > 0$$

and $2nD_P(t|P_L)I_2(t|P_L) \leq 0$ because $D_P(t) \leq 0$.

Hence

$\dot{\lambda}_3(t|P) =$
$-\{I_1(t|P) + nD_P(t|P)I_2(t|P)\} > 0$
and $P^* = P_L$ for all $t \in [t_0, T]$ by Lemma 4.

And the optimal order quantity is $K^*(t) = e^{-nt}\left[K_0 + n\int_{t_0}^{t} D(P_L)e^{nx}dx\right]$.

**Proof of Theorem 4.**

Theorem 4 can be proved by the same logic as the proof of Theorem 3.

**Proof of Theorem 5.**

To solve this theorem, a conjecture about the structure of the solution was made in order to seek a path to satisfy the conditions. First of all, it is assumed that there is an initial period, say $t_0 \le t \le \tau_1$, with $P^*(t) = P_L$. Price changes after at $t = \tau_1$. Thus the hypothesis is:
$P^*(t) = P_U$ for $t_0 < t < \tau_1$, $P^*(t) = P^0$ for $\tau_1 < t < \tau_2$, and $P^*(t) = P_U$ for $\tau_2 < t < T$.

Since $\lambda_1(T) = \lambda_2(T) = 0$ by assumption and $\dot{\lambda}_3(T|P_U) = -I_1(T|P_U)$.

If $I_1(T|P_U) > 0$, then $\dot{\lambda}_3(T|P_U) < 0$. Since $D_P(t) \le 0$, if $I_1(t_0|P_U) > 0$ and $I_2(t_0|P_U) < 0$, then
$I_1(t_0|P_U) +$
$nD_P(t_0|P_U)I_2(t_0|P_U) > 0$

and $\dot{\lambda}_3(t_0) < 0$.

Suppose $I_1(t_0|P_U) + nD_P(t_0|P_U)I_2(t_0|P_U) > 0$ at $t = t_0$, then $\dot{\lambda}_3(t_0) < 0$ and $P^*(t) = P_U$ for $t = t_0$ by Lemma 1, and if $I_1(T|P_L) < 0$, then $\dot{\lambda}_3(T) > 0$ at $t = T$, and the optimal price is $P^*(t) = P_L$ at $t = T$ by Lemma 1.

If $\ddot{\lambda}_3(t) \ge 0$, then $\dot{\lambda}_3(t)$ is a nondecreasing function of time *t* with $\dot{\lambda}_3(t_0) < 0$ and $\dot{\lambda}_3(T) > 0$. At least one transition point between 0 and $T$ should exist where $\dot{\lambda}_3(\tau_1|P^0) = 0$ and $\dot{\lambda}_3(\tau_2|P^0) = 0$, such as

$\tau_1 = \min\{0 < t \le T | \dot{\lambda}_3(t|P^0) = 0\}$

with $P^0 \in (P_L, P_U)$ and $\tau_2 = \max\{\tau_1 \le t \le T | \dot{\lambda}_2(t|P^0) = 0\}$ with $P^0 \in (P_L, P_U)$ by Lemma 1.

Suppose $\dot{\lambda}_3(t)$ that is a convex or concave function for all $t$ with $\dot{\lambda}_3(t_0) < 0$ and $\dot{\lambda}_3(T) > 0$, then at least one transition point should exist between 0 and $T$, where $\dot{\lambda}_3(\tau_1)$ will be zero such as $\tau_1 = \min\{0 < t \leq T \mid \dot{\lambda}_3(t \mid P^0) = 0\}$ with $P^0 \in (P_L, P_U)$ by Lemma 1.

**Proof of Theorem 6.**

Theorem 6 can be proved by applying the same logic of the proof of Theorem 5 to the conditions of Theorem 6.

*This work was previously published in International Journal of Information Systems and Supply Chain Management, Volume 3, Issue 4, edited by John Wang, pp. 1-24, copyright 2010 by IGI Publishing (an imprint of IGI Global)*

# Chapter 12
# Simulation of Inventory Control System in a Supply Chain Using RFID

**Ibrahim Al Kattan**
*American University of Sharjah, UAE*

**Taha Al Khudairi**
*American University of Sharjah, UAE*

## ABSTRACT

*This paper employs a simulation model in a Supply Chain Management (SCM) system. This study is one of the first to present simulation model of inventory control system in supply chain management using barcode and Radio Frequency Identification (RFID). The main objective of this model is to compare two inventory systems in a supply chain, one using RFID, versus the barcode. The model will help company to consider moving from a barcode system to the RFID application. A quantitative analysis based on a simulation model is developed. The model runs for both systems using ARENA simulation software with a comparison between the two systems. Furthermore, the simulation model is tested by applying three different types of demand for both scenarios. The results have shown that regardless of demand distribution pattern and customer order rate, the outcomes of the model are consistent and provide promising RFID technology adoption to improve inventory control of the entire supply chain system. The installation and unit cost of RFID implementation were estimated and considered to be the main barrier. Such model can offer the policymakers insight into how RFID might improve SCM system performance. Additional test has been conducted for demand with normal and triangular distributions using real data provided by ABC-Dubai Company. The results obtained from running the two models for these distributions are consistent with the original results.*

DOI: 10.4018/978-1-4666-0918-1.ch012

## INTRODUCTION

Globalization encourages businesses to make large investments in Supply Chain Management (SCM) applications. The advent of telecommunications and transportation technologies has motivated continuous development of supply chain applications. However, automated data capture and information tracking in real-time have created a major bottleneck, affecting the ability of organizations to optimize their investments in supply chain systems. RFID technology has received considerable attention for its ability to help in the tracking of items through the whole supply chain system. This technology is different from barcode technology in two ways: firstly, it does not require a line-of-sight, and secondly, RFID tags have unique codes. The RFID system consists of tags, readers and radio waves to communicate with all chains in the organization system. RFID is able to identify and deliver a whole range of benefits across a variety of supply chains, the main ones being the following: supplier; manufacturer; distributor; retailer and end user (customer), Al Khudairi (2007).

Many pioneer organizations and companies, such as Wal-Mart, Tesco and the United States Department of Defense, have invested in RFID technology. The potential benefits arise from an increase in supply chain visibility, an increase in efficiency and a decrease in total costs. Thus, RFID promises to have a major impact on supply chains, allowing trading partners to collaborate more effectively and achieve new levels of efficiency and responsiveness. Among many proposed applications of RFID, inventory management in manufacturer-retail stores has been commonly identified as an imperative application, Min et al. (2002). Although business consultants and academic researchers are highly interested in the estimation and assessment of the benefits and values of RFID as deployed in retail stores, most of their figures and claims are based on certain simple assumptions rather than on numerical simulation models, Al Kattan and Al Khudairi (2007).

RFID as an emerging technology has generated an enormous amount of interest in the supply chain as stated by Lee et al. (2005). Inventory accuracy is significantly affected when RFID technology is not employed. Without this accuracy, the supply chain has incorrect information which in turn affects the whole network. Inventory cost also has a great impact on the supply chain inventory. The sharing of inventory information between suppliers and retailers not only improves the supply chain fill rate but also reduces inventory levels. The RFID technology enhanced the information system of the inventory to be tracked more accurately in real-time. More considerably, the complete integration of inventory data throughout the whole supply chain drivers, from the manufacturer's shop-floor to warehouses to retail stores, brings prospects for improvement in reducing processing time and labor cost.

The main focus of this research is to compare the benefits of using RFID on total inventory cost throughout the entire supply chain for two systems - one with Barcode as most company currently practicing, and the other with RFID technology. A quantitative analysis based on a simulation model will be presented, and a comparison between the two systems based on simulation results will be discussed. To achieve a good system comparison, two scenarios are modeled and simulated in such a way as to give output that offers a significant change in total inventory cost throughout the entire supply chain. The objective of this research is to focus on a simulation model that will be analyzed for both systems, and to find the effect of total inventory cost by using RFID technology. The data for this quantitative model is obtained by having a company called "ABC-Dubai Company," and simulation results will be compared between two systems using the same data. The difference between them is that one uses RFID technology and the other use barcode. Simulation block diagrams for all Supply Chain (SC) locations will

*Figure 1. RFID systems (Maloni 2006)*

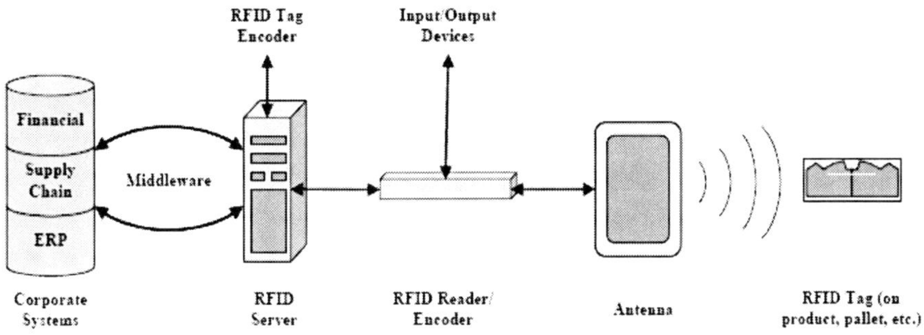


also be implemented, Al Khudairi (2007). The main components of RFID are: a tag, a reader, and data processing equipment, as illustrated in Figure 1, Maloni (2006).

An introduction to the research is presented in this section. A supportive literature review is presented in section II. Section III discusses design of supply chain simulation models embedded with and without RFID. A proposed simulation model using ARENA with two the scenarios is explored in section IV. Analysis of the results with the application of the ABC-Dubai Company using numerical values is presented in section V. Finally the conclusion and future work are summarized in section VI.

## LITERATURE REVIEW

Over the years, many firms have focused their attention on the effectiveness and efficiency of separate business functions. A growing number of firms have begun to realize the strategic importance of planning, controlling and designing the supply chain. Min et al. (2002), synthesized past supply chain modeling efforts and identified key challenges and opportunities associated with supply chain modeling. RFID is an emerging technology, the growth of which has expanded exponentially over the last decade, as stated by Wyld (2006). RFID, which uses radio waves to identify objects, is projected to rapidly replace barcode technology as the principal means of identifying items in the supply chain, and in a wide variety of other applications. Sakamura, (2001) had examine the use of RFID technology in commercial applications and identified its application in various industries. The author concluded that successful implementation of RFID systems requires a long-term strategic plan, careful planning at the tactical level to roll out deployment throughout the organization, and a change in operational business processes.

RFID is considered to be a major breakthrough in supply chain management, Pradhan (2005). The authors described Hewlett-Packard (HP) initiative to enhance the framework to provide safe, secure and adaptive supply chain solutions. They concluded that by using HP's adaptive approach, coupled with understanding of physically-based infrastructure, needs to become a key differentiator as they challenge and transform existing business practices and operations within supply chain industries. Some work has been done to demonstrate the impact of RFID on cost efficiency and agility, Heikkila et al. (2006). The authors used three key concepts to illustrate agility as the goal for supply chain management. They showed that by applying advanced manufacturing technologies, management practices, and adopting the possibilities already provided by the modern information technology, it is possible to simultaneously improve both agility and cost efficiency. In addition, there has also been some investigative work into the pros and cons of RFID technology in supply chain management, Michael (2005).

RFID is utilized as a tool for more effective Supply Chain Management in the retail industry, especially in food and grocery retailing, Prater et al.. (2005). The authors indicated that the adoption of RFID may further increase structural concentration within the retail sector of the economy and have a major impact on retail operations at shop-floor level and on the customers' shopping experience. They concluded that companies should move forward and study the possibility of improving grocery supply chain effectiveness with the utilization of RFID. Some work has been done on the adoption of RFID technology in the warehousing industry, Vijayaraman (2006). This study has been conducted in order to examine whether empirical data supports the role of RFID. RFID technology also has an important role in supply chains in reducing total inventory cost Khan (2006). Khan has summarized the business drivers which are triggering the adoption of RFID in addition to many benefits such as reducing labour costs and improving inventory control.

Most recent publications are qualitative studies that provide business cases for RFID deployments. Research on the impact of RFID on supply chains using analytical approaches is still at an early stage. Some researchers have provided a quantitative analysis to demonstrate the potential benefits of RFID technology in inventory reduction and service level improvement, Lee et al. (2005). The authors have developed a simulation model to study how RFID can improve supply chain performance by modeling the impact of RFID technology in a manufacturer-retailer supply chain environment. In their work, they showed a supply chain model that consists of a supplier and a retailer, focusing on analyzing the effects of three factors that are influenced by RFID implementation. These factors are: inventory accuracy, shelf replenishment policy, and inventory visibility.

Another study has dealt with how RFID technology can improve supply chain performance in fast moving consumer goods industry, Telkamp (2006). The author developed two analytical models that illustrate the potential impact of RFID on product availability through higher inventory accuracy by redesigning of the replenishment from the backroom store. Modeling and simulation of average inventory level and service level in supply chain systems have been carried out in literature by Vieira (2004). The author has proposed the use of computer simulation in modeling and evaluation of supply chain performance using ARENA simulation software. The author has mentioned that few works have been actually tried to simulate a whole supply chain and the interaction between its stages such as overall inventory management, associated costs and service level. Altiok et al. (2001) tried to build a simple inventory model that has a single production machine and a single warehouse using ARENA simulation software. In addition, in order to create more efficient simulation models, Seppanen (2000) has explained the process by which Visual Basic for Application (VBA) can be utilized in the development of industrial strength of ARENA simulation models, Kelton (2007). The author has shown how to transfer the data between ARENA and Microsoft EXCEL using VBA.

## CHALLENGES OF RFID ADOPTION

The lack of clarity of costs and benefits represents just one of a considerable number of RFID risks and challenges. Cost and the payback of new technology implementation is one of the most important factors facing a company in the competitive market. The following are selected RFID risks and challenges organized by areas such as: technology, Return on Investment (ROI), privacy/security, supply chain, and implementation as summarized from Maloni, (2006).

### Technology

The technological risks and challenges involved with RFID stem from many different sources; a strong dependency on technology is one of the most prominent issues. The readability for tag users has been significantly lower than 100%, due to deviations in various reader brands, the defect rate of chips/antennas, signal distortion/ absorption/ reflection as well as issues related to signal collision from numerous tags and readers.

### Return on Investment

As the true cost-efficiency of RFID hardware and services is still undetermined, users who utilize RFID solutions may discover it to be a more costly resolution, especially when compared to tag costs. As a result, it may be uncertain as to the direct return on investment for RFID solutions. Furthermore, the costs and benefits of are not evenly distributed among the supply chain and that outcomes may differ according to industry, supporting products with greater significance.

### Privacy and Security

A main concern for consumers and companies that implement RFID has been related to the privacy and security of RFID data. Aside from principal protection methods of consumer and company information, companies utilizing the system must also protect themselves from illegitimate readings of tags from criminals or even competitors. Because tags still operate throughout and after the process, public attention has arisen over the possible ways to record customer geographic location, product utilization, spending habits, and any other types of confidential information associated with the tags and each customer. This can particularly become an issue when customers are unaware of the way in which RFID and tags use and transmit data.

### Supply Chain

One of the most obvious obstacles for RFID solutions implementation is a need for management and recognition from members of the supply chain. Another difficulty is that members of bigger supply chains hold a benefit of more resources to implement and investigate RFID, and may improve their strategy and control inside the chain itself. Furthermore, these members may be anxious to find that RFID increases control given that vendors are able to continue to access detailed and advantageous information.

### Implementation

Many issues surrounding the implementation of RFID solutions including resistance from company users as well as costs associated with implementation. It is also crucial to sustain dual processes and systems to avoid any complications that may arise while using RFID solutions and have high costs associated with errors, etc.

## SIMULATION OF SUPPLY CHAIN MODEL

This section is considering supply chain system design with five main chains of organizations or drivers: Customer, Retailer, Distribution Center, Manufacturer, and Supplier, as shown in Figure 2, (Kilton 2003 and Vieira 2004). The simulation model in this research starts from customer demand, to retailer, passing through retailer-shelf, backroom, warehouse to distribution center-warehouse, and finally to the manufacturer's warehouse and plant, Lee et al. (2005) and Al Khudairi (2007). The Supply Chain (SC) cycle starts when the customer sets an order for product item(s) to the retailer. All orders of a certain item could be considered as an annual demand. Product items could be packaged in boxes, pallets and containers. The other end of the SC cycle is

*Figure 2. Proposed supply chain systems (Kilton 2003 and Vieira 2004)*

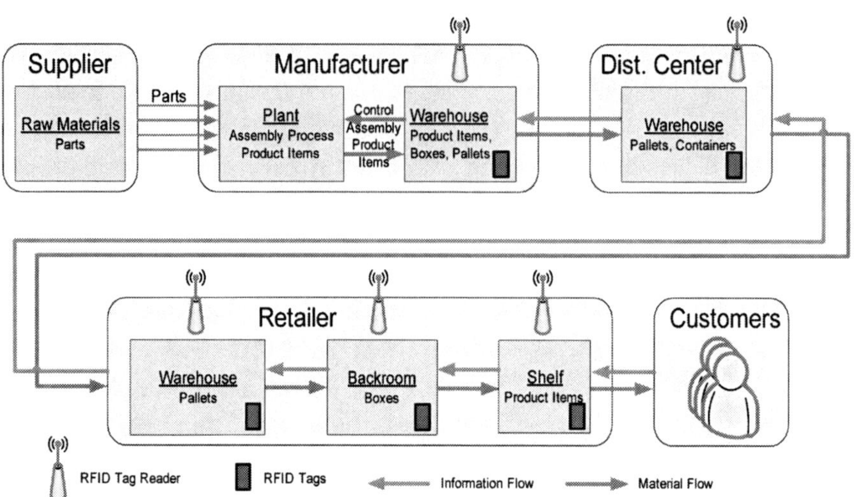


the manufacturing plant, where the final product is made using raw materials and/or assembling semi-finished items from different suppliers. The cycle is long and distanced from each chain of organization. Using a barcode is cheap but not necessarily efficient. The process could be accelerated by implementing RFID. A schematic figure of the proposed simulation model is shown in Figure 2, which includes the locations of RFID readers and tags in red color. RFID tags are used on all packages and items so that it could be scanned easily by tag readers through the supply chain system. There are two directions of flow within the supply chain. First there is the information flow, the pull process which is initiated by the customer(s) demand, transmitted through the supply chain and ends as an input to the manufacturing production plant. This direction is from right to left as shown in Figure 2. The second flow is the material flow, the push process (production process) which travels in the opposite direction to the first one. This flow represents supply chain reaction to customer demand by ordering raw and semi-finished material, pushing the final product items forward to satisfy customer demand. Push and pull processes should be controlled to have product availability with minimum total inventory cost in all chains. This is the main objective of the SCM system that looks at all chains rather than having low inventory in one chain over the other. The simulation model used an (R, Q) inventory policy in all chains of SCM storage. The main entity in the simulation model is called *product item* (PI). This entity is created at the manufacturer's block (raw materials and parts assembly) and transferred through the system stages until the end user or customer's block. The following sections show the flow of the information from customers to the manufacturer.

## Retailer Storage

The retailer block consists of three storage locations: shelves, a backroom store and a warehouse as indicated in Figure 2, (Kilton 2003 and Vieira 2004). Each block has its own inventory and RFID tag readers. The identification code of a product item is used for the information flow and to interact with all other blocks between demand and supply. A common type of packaging is when individual product items are packed into cardboard boxes, which are then gathered on a pallet. Depending on the storage size of each block the packaging volume is assigned accordingly. For example, product items are used in retailer's shelf, boxes are used in retailer's backroom and pallets are used in retailer's warehouse. The number of product items per box or number of boxes per pallet can be set according to the packing system used. The simulation model considers all storages using product items only. However, the loading and unloading time with and without RFID scanning is taken into consideration when items are handled between retailer locations. To simulate the model of retailer storage, the following assumptions are used:

Shelf location; the assumed variables consist of: customer arrival and demand distribution; initial inventory level; stock level; barcode scanning time; and finally lead-time from the retailer's backroom.

1. Backroom store location; the assumed variables consist of: shelf annual demand; initial inventory level; stock level; reorder level; lead-time from retailer's warehouse and barcode scanning time.
2. Warehouse location; the assumed variables consist of: backroom store annual demand; initial inventory level; reorder level; stock level; lead-time from DC's warehouse and barcode scanning time.

## Distribution Center (DC)

DC warehouse supports multiple retailers, where the assumption used is for one retailer. The lead-time between DC and manufacturer is longer than that from DC to retailer. The assumed variables consist of: the retailer's warehouse annual demand; initial inventory level; reorder level; stock level; lead-time from manufacturer to DC warehouse and barcode scanning time.

Demand (D): the amount required by $j^{th}+1$ storage location and it represents $j^{th}+1$ order quantity.

Stock Level (SL): the maximum capacity of the storage space.

Order Quantity (Q): denotes to the amount of items that can be calculated by using the minimum total inventory cost equation of the $j^{th}$ storage location to set an order for the $j^{th}-1$ storage location. The relation among these variables is illustrated by the following general equations:

Inventory Level = Inventory Level − Demand
$$IL_{jth} = IL_{jth} - D_{jth+1} \quad (1)$$

Order Quantity = Stock Level $Q_{jth} = SL_{jth}$ (2)

Inventory Level = Inventory Level + Order Quantity $IL_{jth} = IL_{jth} + D_{jth}$ (3)

## Manufacturing Plant

The simulation model includes two main stores product items at each stage with its inventory policy. For each storage location an inventory mechanism is needed to control the flow of products items from one location to the other. An (R, Q) inventory policy is used for controlling the inventory of all locations in the supply chain system. There are four variables in this inventory control system: inventory level, stock level, order quantity and annual demand. The following is the definition of those variables and how they are used in inventory control equations: as shown in Figure 2. The simulation model includes the following assumed variables for:

## Inventory Control Mechanism

The supply chain system contains a place to Inventory Level (IL): the number of items blocks in this location: Manufacturer production line (plant) and Manufacturer storage and shipping area (warehouse) to be stored in a certain location.

The above equations are used for every storage location in the SC system and denoted by the prefix of the location in the ARENA simulation model. For example, the above equations can be re-written for the retailer's shelf inventory as follows:

RET_shelf_IL=RET_shelf_IL−Customer_Demand_Size ... (4)

RET_shelf_Q = RET_shelf_SL .. (5)

RET_shelf_IL=RET_shelf_IL + RET_shelf_Q .. (6)

Note that in Equation (4), the customer is considered jth+1 block with respect to the retailer's shelf storage location. Also note that when dealing with different packaging volumes such as boxes, pallets and containers the general equation (1) can be re-written as follows:

$$IL_{jth} = IL_{jth} - \frac{D_{jth+1}}{CF} \quad \dots \qquad (7)$$

Where, (CF) is the Conversion Factor between two different packaging volumes. This should be applied to all storage locations which have unit packaging other than product items as shown in Figure 2. The simulation model assumes product items through the entire supply chain system.

## Cost Structure

There are four cost components associated with each storage location: purchasing cost, order cost, holding cost and shortage cost. There are other costs that are actually embedded in these costs but they are calculated at the point when there is a need to do so. The main cost equation for each storage location is the Total Cost. This equation consists of all costs that are mentioned above. This cost is calculated per unit time, for example days, weeks, months, years, etc. The unit time used in this work is 'days', so all costs are calculated per day unit time.

*Table 1. Normal and worst-case lead-times*

| Order Direction | Normal LT (days) | Worst-case LT (days) |
|---|---|---|
| RET_BR ► RET_shelf | 0.25 | 5.25 |
| RET_WH ► RET_BR | 0.5 | 5 |
| DC_WH ► RET_WH | 1 | 4.5 |
| MFG_WH ► DC_WH | 3 | 3.5 |
| MFG_PT ► MFG_WH | 0.5 | 0.5 |

Total Cost = Average Order Cost + Average Holding Cost + Average Shortage Cost (8)

Average Order Cost: Every time an order is placed, there is a cost incurred.

Average Holding Cost: Whenever there are items actually physically in inventory (IL>0), a holding cost is incurred. The average holding cost per day is the total cost divided by the number of working days per year.

Average Shortage Cost: Whenever there is a back-order (i.e. IL<0), a shortage cost is incurred,

The average shortage cost per day is this total divided by the number of working days per year.

## Lead-Time Effect

A constant lead-time is used and it is added between two storage locations. Table 1 shows the proposed lead-time values for the SC system. In Table 1, there are two types of lead-times, normal (NLT) and worst case (WCLT). NLTs are used in normal operation when there is enough stock at the time of setting the order between jth and jth-1 storage locations. WCLTs are used when there is a stock outage at the time of order between jth and jth-1 storage locations. This stock outage may be propagated through the supply chain, therefore Table 1 shows the maximum or worst case scenario when there is a stock outage through all supply chain locations. For example, consider that the retailer's warehouse sends an order to the DC's warehouse.

*Table 2. Input data from ABC-Dubai company*

| System Variable | Description and Value |
|---|---|
| Customer arrival distribution (case 1) | Exponential with 20 customers/day |
| Customer demand distribution (case 1) | Uniform with min. of 5 and max. of 10 items |
| Customer arrival distribution (case 2) | Exponential with 20 customers/day |
| Customer demand distribution (case 2) | Normal with mean of 8 and standard deviation of 2. |
| Customer arrival distribution (case 3) | Exponential with 20 customers/day |
| Customer demand distribution (case 3) | Triangular with min.=6,mode=8 and max.=10 items |
| Demand of other stages | 20% of retailer's shelf demand |
| Inventory shortages | Allow shortages |
| Customers willing to wait | 50% satisfied, 33% partially satisfied, 17% lost sales |
| Lead-time distribution | Constant, using historical data |
| Backroom to shelf lead-time | 0.25 day |
| Warehouse to backroom lead-time | 0.5 day |
| DC to warehouse lead-time | 1 day |
| Manufacturer (MFG) to DC lead-time | 3 days |
| MFG plant to warehouse lead-time | 0.5 day |
| Setup cost | $100 |
| Unit holding cost | $5 /year |
| Unit shortage cost | $10 /year |
| Barcode scanning labor cost | $0.01 /second or $36 /hour |
| Barcode scanning time distribution | Uniform with min. of 10 and max. of 20 seconds |
| Lost sales penalty cost | $50 /customer |
| Manufacturer batch size | 500 items |

The DC's warehouse checks for inventory availability and finds that there is no available stock, so it orders from the Manufacturer's warehouse which in turn discovers that its inventory is out of stock. Moreover, it also sends an order to the manufacturer's plant. Now the lead-time for the manufacturer' plant to finish and submit the order to its warehouse is equal to half day. Then once the warehouse receives the order, it sends the item to the DC's warehouse which takes three days' lead-time. When the DC's warehouse receives the order, it immediately sends it to the retailer's warehouse with a lead-time of one day, so the total lead-time becomes 0.5 + 3 + 1 = 4.5 days. When measuring the annual demand for each storage location, WCLT for each storage location should be considered to evaluate the reorder level of that location.

## REAL DATA FOR CASE STUDY

To test the proposed model, real data are collected from ABC-Dubai Company. An (R, Q) inventory policy is used for inventory control system. Barcode labels are used on items in all locations of the SC system. Table 2 shows the input data used in this model. In addition to this data two demand distributions are applied to test the validity of the model for different demand distributions. Those distributions are included in Table 2 and represented by case 2 and case 3.

*Figure 3. Hierarchy of ARENA model (Kilton 2003 and Vieira 2004)*

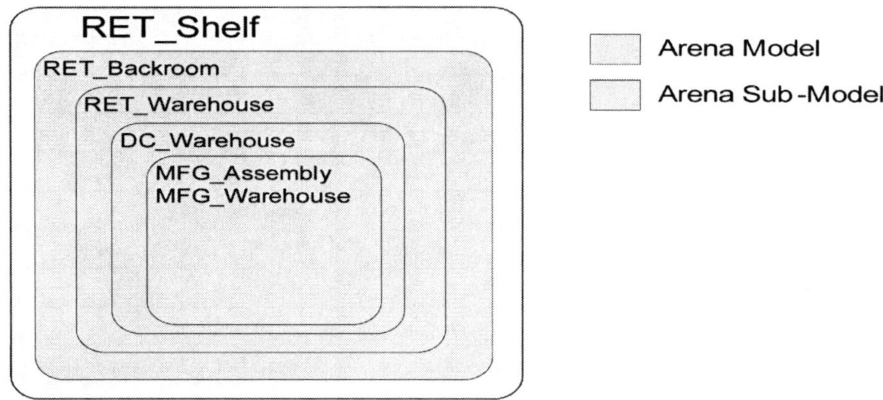


*Figure 4. Retailer-Shelf ARENA model for RFID (Kilton 2003 and Vieira 2004)*

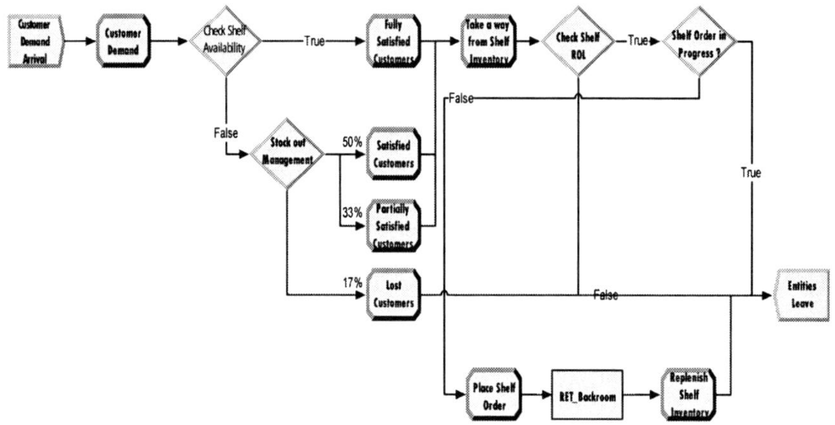


## PROPOSED ARENA MODEL

The ARENA simulation model hierarchy of the supply chain system is shown in Figure 3. The top level model is the retailer-shelf model which is initiated by customer demand. The concept of ARENA submodel is used here. The output of inventory system from the retailer-shelf model is used as an input to the next level which is the retailer-backroom and so on. Therefore the flow of inventory information is defined as in the following sequence: Customer order ⇒ Retailer-Shelf Inventory ⇒ Retailer-Backroom Inventory ⇒ Retailer-Warehouse Inventory ⇒ DC-Warehouse Inventory ⇒ Manufacturer-Warehouse Inventory, Manufacturer-Plant ⇒ Supplier.

### Retailer Shelf (RS) Model

Figure 4 illustrates the model for Retailer-Shelf (RS) location. This model and all other sub-models are built using Basic Process and Advanced Process panels inside ARENA software. This model is the top of the hierarchy in which calls to other submodels are executed when an order for a demand is placed.

## Retailer-Backroom (RB) Sub-Model

The ARENA sub-model of the retailer's backroom (RB), in this submodel the demand is basically the ordered quantity from the previous stage (in this case the retailer's shelf). The inventory control algorithm is the same for the retailer's warehouse and DC's warehouse submodels with some small modifications.

## Retailer-Warehouse (RW) Sub-Model

The ARENA sub-model of the retailer's warehouse (RW), in this submodel the demand is basically the ordered quantity from the retailer's backroom.

## Distribution Center (DC) Sub-Model

The ARENA sub-model of DC's warehouse, in this submodel the demand is basically the ordered quantity from the retailer's warehouse.

## Manufacturer (MFG) Sub-Model

The ARENA sub-model of the manufacturer's warehouse, in this submodel there are two main segments; warehouse segment and plant segment.

# ANALYSIS OF RESULTS

In this section, the ARENA output report of the supply chain simulation model is discussed. The model runs two scenarios. The first scenario uses ABC-Dubai Company data for SCM current system using barcodes (without RFID). The second scenario uses ABC-Dubai Company data for the SCM system using RFID. This section presents; the total annual inventory cost for all chain storages in SCM with and without RFID; the simulation model for SCM with RFID implementation eliminates the delay used by barcode; and the delay caused by barcode scanning adds an additional labor cost. In addition, a comparison of customer satisfaction for two different scenarios; with and without RFID implementation will be presented. Furthermore, more tests are carried out to see the effect of applying different demand distributions on both systems.

The ARENA model implemented using RFID is modified for the SCM system using barcode as shown in Figure 5. In order to simulate the system using barcodes, some modifications should be made on this model in order to work with barcodes. In addition, some new constants and variables are added to the model. Table 3 shows additional constants, expressions and variables used for the retailer's shelf.

The modifications done on the barcode system are:

1. With barcodes, there is a time delay when performing the scanning operation. This operation is done by the worker since he has to scan each item/box/pallet by applying the barcode scanner with a line-of-site on the barcode sticker, in order to get each products information. This delay time is set by ABC-Dubai Company to be a uniform distribution between 10 and 20 seconds per item as shown previously in Table 2. This delay value is stored in the expression RET_shelf_BST.
2. The above delay causes an additional labor cost since the worker takes some time to scan every item. This cost is set by ABC-Dubai Company to be $0.01/seconds or $36/hour as shown previously in Table 2. This cost value is stored in the constant BC_LCPS.

The above modifications are added to the simulation model by using the Delay module and assign module as highlighted in Figure 5 for the retailer's shelf model. These modifications are added in the position when the order quantity is shipped from backroom to shelf. This means that the use of barcode is performed when shelf inventory is replenished as shown in Figure 5. Barcode

*Figure 5. Retailer-shelf model with barcode modifications (Kilton 2003 and Vieira 2004)*

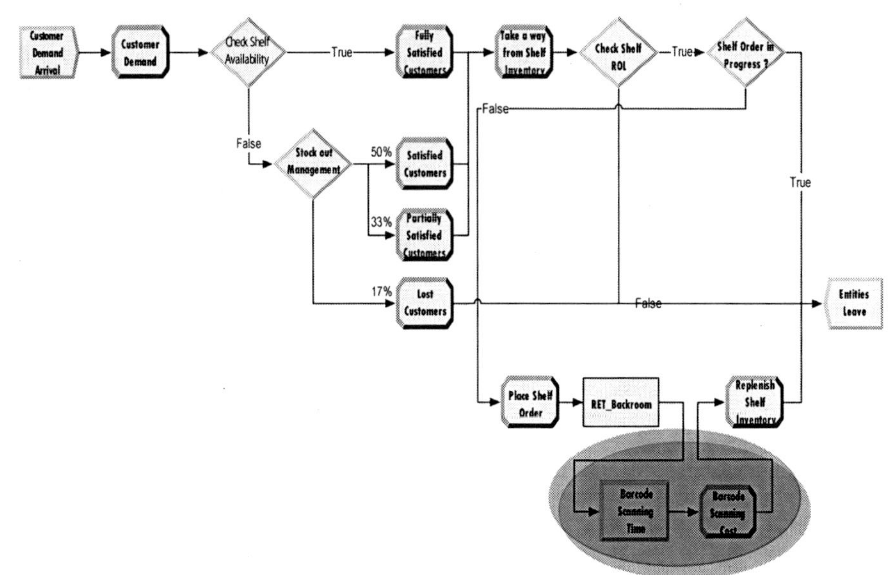

*Table 3. Additional data structure for ARENA model with barcodes*

| Mnemonic | Description |
|---|---|
| BC_LCPS | Barcode labor cost per second |
| BC_ST | Barcode scanning time |
| LSPC | Lost sales per customer |
| RET_shelf_BST | Retailer's shelf barcode scanning time |
| BC_TSC | Barcode total scanning cost |
| Shelf Lost Sales | Retailer's shelf total lost sales |
| Total System Cost | The total annual cost for all SCM locations |

modifications are inserted in all SCM locations (backroom, warehouse, DC, and MFG) so Figure 5 and Table 3 are provided as an example for the retailer's shelf location only. The other SCM locations are treated in the same way. In the retailer's shelf location, the system encountered lost customers. The penalty of losing one customer is set at \$50. This penalty is multiplied by the number of lost customers as part of lost sales opportunities. In addition to the cost results, it is interesting to observe customer satisfaction with each system. Customer satisfaction results are carried out at retailer's shelf model only.

When the system is implemented using RFID it has the same structure and implementation as the ARENA model presented in the previous section. By using RFID implementation, the effect of barcode scanning delay is eliminated since RFID readers are installed in each SCM location. They perform the scanning operation directly without human intervention and without direct line-of-sight, as it is the case with barcodes. Consequently in this scenario there is no scanning delay and hence no scanning cost.

## Total SCM Inventory Cost with and without RFID

The results from running the simulation model for all storages in the supply chain system are presented in Figure 6. The bar charts show the annual inventory cost for each storage location in the supply chain system for both scenarios. At each stage, the annual inventory cost is calculated by using standard inventory equations (economic order quantity for minimizing total inventory

*Figure 6. Comparison of total inventory cost for both scenarios*

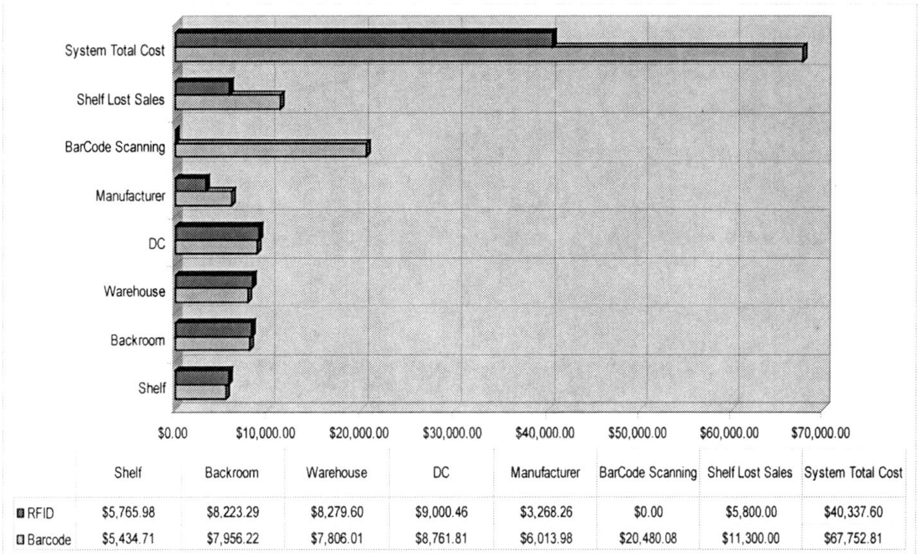


| | Shelf | Backroom | Warehouse | DC | Manufacturer | BarCode Scanning | Shelf Lost Sales | System Total Cost |
|---|---|---|---|---|---|---|---|---|
| RFID | $5,765.98 | $8,223.29 | $8,279.60 | $9,000.46 | $3,268.26 | $0.00 | $5,800.00 | $40,337.60 |
| Barcode | $5,434.71 | $7,956.22 | $7,806.01 | $8,761.81 | $6,013.98 | $20,480.08 | $11,300.00 | $67,752.81 |

cost). There is an additional cost component that comes up when RFID is not used, which is the barcode scanning cost. The system employing RFID implementation reduces this type of cost. Consequently, system encountered customers lost. The penalty of loosing one customer is set to be $50. This penalty is multiplied by the number of lost customers as part of lost sales opportunity.

By using barcode, there is an extra time for scanning each item so that this delay is multiplied by the order quantity and also by labor cost per unit time. This labor cost is accumulated through chain blocks so the total barcode scanning cost is the sum of scanning costs for all blocks. The total annual inventory cost is the sum of the total annual inventory cost for all storages plus the cost of the labor scanning cost. The total cost of the SCM system using barcodes is $67,752, as shown in Figure 6, while the total system cost using RFID is $40,337. The difference between the two is $27,415 and this cost is for single item with an average annual demand of 37,500 items. By dividing the total saving over the annual demand, the unit saving in the inventory cost is $0.73. With 50% of this saving a reasonable active RFID tag could be installed. The other 50% of saving could be used to cover the remaining RFID system component cost. For an application of FRID on 1000 different items in the supply chain system the payback could be in a short period.

## Estimate the Customer Satisfaction with and without RFID

The number of customers and their level of satisfaction are shown in Figure 7. Using RFID improves system responsiveness and the demand has been met with full customer satisfaction. It is clear that with RFID accuracy of 100% the system reaches its maximum performance. As indicated in Figure 9, the number of fully satisfied customers is 4335 using RFID, while this number is 3844 using barcodes. The number of lost customers in the case of RFID is 116, while 226 customers are lost using barcodes. According to ABC-Dubai policy, the penalty for losing one customer is estimated $50. This affects the total inventory cost in the system, as lost customers are considered lost sales. Although satisfied and partially satisfied customers are fewer when

*Figure 7. Comparison of customer satisfaction for both scenarios*

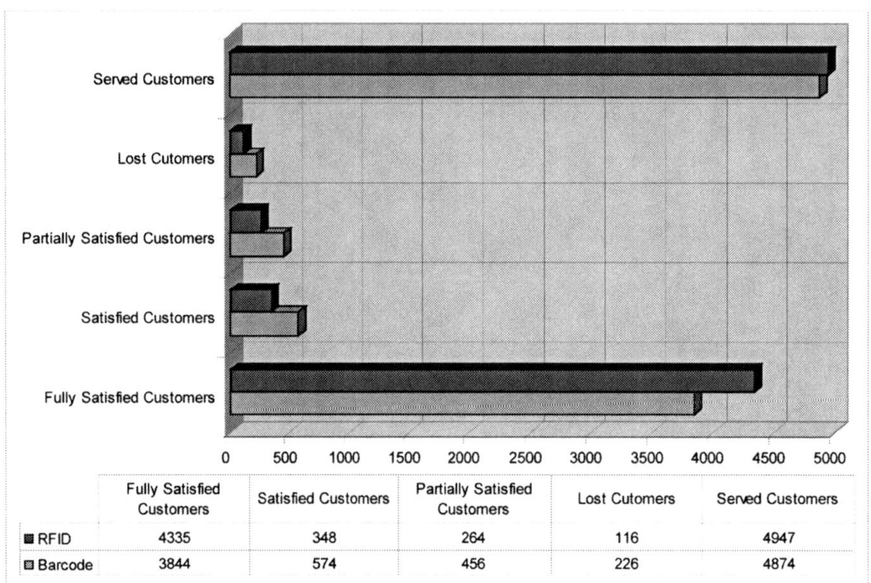


| | Fully Satisfied Customers | Satisfied Customers | Partially Satisfied Customers | Lost Cutomers | Served Customers |
|---|---|---|---|---|---|
| RFID | 4335 | 348 | 264 | 116 | 4947 |
| Barcode | 3844 | 574 | 456 | 226 | 4874 |

*Figure 8. Comparing demand distribution on systems total cost for both scenarios*

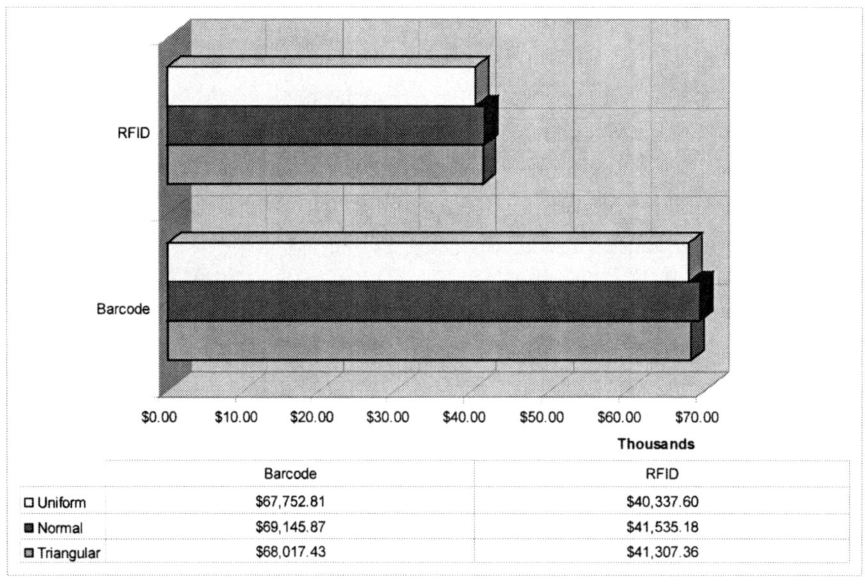


| | Barcode | RFID |
|---|---|---|
| Uniform | $67,752.81 | $40,337.60 |
| Normal | $69,145.87 | $41,535.18 |
| Triangular | $68,017.43 | $41,307.36 |

using RFID, the number of lost and fully satisfied customers, however, is the company's main concern. Table 4 shows a comparison of the four types of customer satisfaction between the RFID and Barcode. The improvement in number of customers and percentage for each customer is illustrated. Table 4 indicates that the total and the fully satisfied customers are improved using RFID.

*Figure 9. Comparison of customer lost for both scenarios*

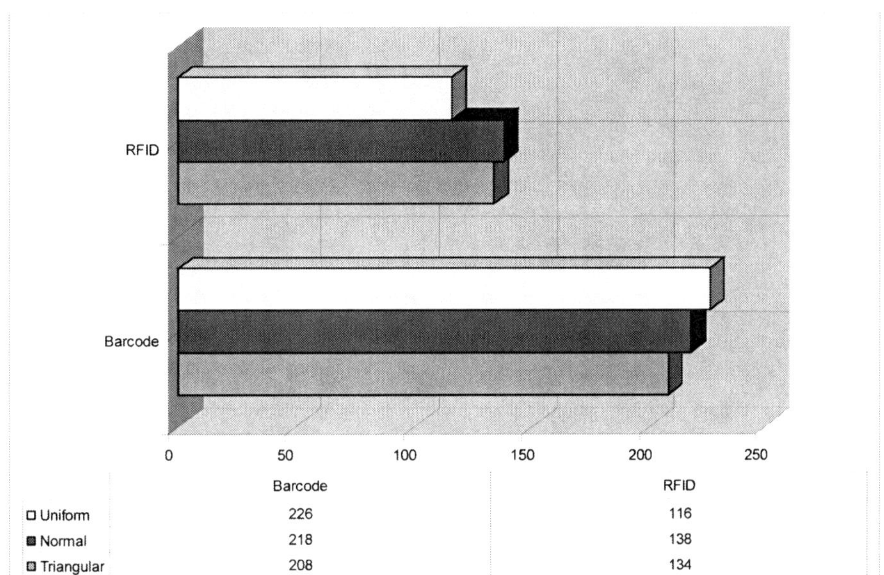


*Table 4. Comparison of customer satisfaction between barcode and RFID applications*

| Type | Barcode system | | RFID system | | Improvement No. of Customers | Improvement In percentage |
|---|---|---|---|---|---|---|
| | No. of Cust. | % | No of Cust. | % | | |
| Fully Satisfied Customers | 3844 | 0.7886746 | 4335 | 0.87628 | 491 | 0.08761406 |
| Satisfied Customers | 574 | 0.1177677 | 348 | 0.07034 | 226 | 0.047422083 |
| Partially Satisfied Customers | 456 | 0.09355765 | 264 | 0.05336 | 192 | 0.040191977 |
| Lost Customers | 226 | 0.04636 | 116 | 0.02344 | 110 | 0.022919931 |
| Total Served Customers | 4874 | | 4947 | | | |

## Testing Model Validity

In addition to the results obtained for both scenarios, other tests are carried out on both systems. For the purpose of this research the author has chosen two different demand distributions and customer order rate. The idea behind applying different demand distributions is to test the validity of the ARENA model for different kinds of data. Table 5 demonstrates the three different demand distributions applied to the system. The uniform distribution is the original distribution that has been discussed in previous sections. The other two demand distributions are normal and triangular. Figure 8 illustrates a comparison of the three types of distributions on total cost for both RFID and barcode scenarios. Furthermore, the number of lost customers is also carried out and the comparison for both scenarios is illustrated in Figure 8. It is clear that for all cases, the number of lost customers in the system using RFID is lower than in that using barcodes. Consequently,

*Table 5. Testing the model with three different demand distributions*

| Case no. | Demand Pattern | Parameters | Demand rate per day |
|---|---|---|---|
| 1 | Uniform | Min=5, Max=10 | 20 |
| 2 | Normal | Mean=8, SD=2 | 20 |
| 3 | Triangular | Min=6, Mode=8, Max=10 | 20 |

the results have shown that by applying different demand distributions, the output of the model remains consistent and provides lower total cost with less number of lost customers in the RFID implementation scenario.

## CONCLUSION AND FUTURE WORK

Throughout this research, opportunities for RFID technology to provide significant benefits in inventory cost in a supply chain management system are demonstrated. The author proposed using computer simulation modeling to evaluate supply chain system performance. Two models were simulated; one using RFID technology and the other using barcode. The result of the models using ARENA simulation software were analyzed and discussed in detail. Subsequently, several comparisons have been accomplished by measuring both the total inventory cost and customer satisfaction for the entire supply chain storages for both systems. Using the current system with barcode technology, the responsiveness of the system is much slower and more labor dependent. The barcode system encounters some inventory shortages and lost customers. These shortages occur because there are delays for barcode scanning and information transfer. This research considers tracking inventory at item-level. Therefore it is multiplied by the order quantity when replenishing the inventory at each storage location, causing the system to have back-orders.

The proposed system with RFID adoption has a vital role to play in various stages of the supply chain. This study has proven that RFID technology is more reliable identification tool for tracking than barcode technology. By using RFID, the inventory data becomes more accurate and much easier to share among all supply chain players, which helps to control the inventory level. As a result, decisions on replenishment as well as manufacturing plans may be improved substantially. Results have shown that the system with barcode technology has a total cost of $67,752 for the entire supply chain. This cost includes all supply chain inventory cost, lost sales due to lost customers and barcode scanning cost in all stages. In addition, the number of lost customers is 226 and the number of fully satisfied customers is 3,844. On the other hand, the proposed system with RFID technology adoption reduces system total cost to $40,337, the number of lost customers to 116, and increases the number of fully satisfied customers to 4,335. This improvement represents 40.46% saving from the current barcode model, with a net value of $27,415.

Further, the authors have investigated other opportunity to use different demand distribution behavior for other items using the data supplied by ABC-Dubai Company. Additional testing has been conducted for demand with normal and triangular distributions. The results obtained from running the two models for these distributions are consistent with the original results. In the case of barcodes, the total cost of $69,145 and $68,017 are for normal distribution demand, and triangular distribution demand, respectively. The number of lost customers is 218 and 208 for normal distribution and triangular distribution demand, respectively. On the other hand, the proposed system for RFID reduces the total cost to $41,535 in the case of normal distribution demand, and $41,307 in the case of triangular distribution demand. Moreover, the proposed system for RFID reduces the number of lost customers to 138 in the case of normal distribution demand, and 134 in the case of triangular distribution demand.

The above results are carried out for a single type of items. ABC-Dubai Company handles thousands of items. The company's plan requires the implementation of RFID for a wider range of items (say around 1000 different items). Hence, this leads to a reduction in total annual inventory cost of $27,415,000. Finally, the proposed system with RFID deployment, using the ARENA simulation model, has shown improvement in supply chain inventory control by reducing the total inventory cost, the number of lost customers and improving customer satisfaction.

Future work could be done to improve the findings of this research. There are some issues that could be taken into consideration for future work development. These issues include having RFID read accuracy of less than 100%, other supply chain configurations, and multi-product inventory model. The inventory model used in this research is a single-period, single-order, continuous review with lead-time less than cycle time. Other models, such as multi-period multi-order continuous review where lead-time is greater than cycle time, can also be carried out to observe the effect of RFID technology on total inventory cost and customer satisfaction.

## REFERENCES

Al Kattan, I., & Al Khudairi, T. (2007). Improving supply chain management effectiveness using RFID. *In IEEE International Engineering Management Conference,* Austin, Texas.

Al Khudairi, T. (2007). *Integrating RFID into supply chain management in process inventory control.* Unpublished Master thesis dissertation, American University of Sharjah, UAE.

Albright, B. (2006). *Ready or not RFID's coming, frontline solutions website*. Retrieved January 2007. from http://www.findarticles.com/p/articles/mi_m0dis/is_12_4/ai_112366601

Altiok, T., et al. (2001). *Simulation Modeling and Analysis with ARENA*. Cyber Research Inc.

Brain, M. (2006). How UPC bar codes work. *Howstuffworks website*. Retrieved May 2009 from http://electronics.howstuffwworks.comupc

Heikkila, J., & Holmstrom, J. (2006). The impact of RFID and agent technology innovation on cost efficiency and agility. *International Journal of Agile Manufacturing*, *9*(1), 1–6.

Jones, P., Clarke-Hill, C., Hiller, D., & Comfort, D. (2005). The benefits, challenges and impacts of radio frequency identification technology (RFID) for retailers in the UK. *Marketing Intelligence & Planning*, *23*(4), 395–402. doi:10.1108/02634500510603492

Journal, R. F. I. D. (2006). The cost of RFID equipment. *RFID Journal website*. Retrieved May 2009 from http://www.rfidjournal.com/faq/20: RFID Research Center. (2001). *The myths and realities of RFID.* University of Arkansas, USA: Information Technology Research Institute.

Khan, A., & Kurnia, S. (2005). *Explaining the Potential of RFID.* Department of Information Systems, University of Melbourne, Australia.

Khan, A., & Kurnia, S. (2006). *Exploring the potential benefits of RFID: a literature-based study.* Department of Information Systems, University of Melbourne, Australia (pp. 1-12).

Kilton, D., Sadowski, R., & Sturrock, D. (2003). *Simulation with Arena* (3rd ed.). McGraw-Hill, Series Industrial Engineering and Management Science.

Lee, Y., Cheng, F., & Leung, Y. (2005). A quantitative view on how RFID will improve a supply chain. *IBM Research Report* (pp. 1-45).

Maloni, M. (2006). *Understanding Radio Frequency Identification (RFID) and its impact on the supply chain.* Pen State Behrend: Black School of Business, PA, USA

Michael, K. (2005). The pros and Cons of RFID in supply chain management. In *Proceedings of the International Conference on Mobile Business, IEEE Computer Society* (pp. 1-7).

Min, H., & Zhou, G. (2002). Supply chain modeling: past, present, and future. *Computers & Industrial Engineering, 43*(2), 231–249. doi:10.1016/S0360-8352(02)00066-9

Pradhan, S. et al. (2005). RFID and sensing in the supply chain: challenges and opportunities. *HP Laboratories Palo Alto HPL* (pp. 1-11).

Prater, E. et al. (2005). Future impact of RFID on e-supply chains in grocery retailing. *Supply Chain Management.* ABI/INFORM Global (pp. 134-142).

Sakamura, K. (2001). Radio frequency identification and non-contact smart cards. *IEEE Micro,* 4–6. doi:10.1109/MM.2001.977752

Seppanen, M. (2000). Developing industrial strength simulation models using visual basic for applications (VBA). *Proceedings of the 2000 Winter Simulation Conference* (pp. 77-82).

Tellkamp, C. (2006). *The impact of Auto-ID technology on process performance RFID in the FMCG supply chain.* Unpublished doctoral dissertation, Graduate School of Business Administration, Economics, Law and Social Sciences (HSG) University of St. Gallen, Switzerland.

Vieira, G. (2004). Ideas for modeling and simulation of supply chains with arena. *Proceedings of the 2004 Winter Simulation Conference* (pp. 1418-1427).

Vijayaraman, B., & Osyk, B. (2006). An empirical study of RFID implementation in the warehousing industry. *The International Journal of Logistics Management, 17*(1), 6–20. doi:10.1108/09574090610663400

Wyld, D. (2006). RFID 101: the next big thing for management. *Management and Research News, 29*(4).

*This work was previously published in International Journal of Information Systems and Supply Chain Management, Volume 3, Issue 1, edited by John Wang, pp. 68-86, copyright 2010 by IGI Publishing (an imprint of IGI Global)*

# Chapter 13
# Two-Commodity Markovian Inventory System with Set of Reorders

**N. Anbazhagan**
*Algappa University, India*

**B. Vigneshwaran**
*Thiagarajar College of Engineering, India*

## ABSTRACT

*This article examines a two commodity substitutable inventory system—two different brands of super computers under continuous review. The demand points for each commodity are assumed to form independent Poisson processes. The reordering policy is to place orders for both the commodities when the total net inventory level drops to any one of the prefixed levels with prescribed probability distribution. Lost sales are assumed during the stock out period. The lead time for a reorder is exponentially distributed with parameter* $(\mu_k)$, *depending on the size of the ordering quantity. The limiting probability distribution for the joint inventory levels is also evaluated. Various operational characteristics and total expected cost rate are derived. Numerical examples are provided to find optimal reorder quantity and band width* $r$.

## INTRODUCTION

In dealing with multi-commodity inventory system in a single location, the joint and coordinated reordering policy have been given more attention than that of individual reorder for each commodity separately. There are many advantages for joint replenishment, as they share setup costs, quantity discounts, utilize the same transport facilities, etc.

In the real life situation, the global sales agencies deal with rare electronic products like super computers with almost same configuration and identical functioning. This motivates the

DOI: 10.4018/978-1-4666-0918-1.ch013

researcher to consider the substitutable commodities inventory control under joint reorder policy. In general these high tech machines are having substitutable nature because of its common number crunching behavior(General purpose computer). Keeping them in stock for sales purpose is high risk but yield high profit. Maintaining these "A" type items of ABC inventory classification, attracted many researchers in the past decade. We also assumed elongated production schedule and that the lead time distribution parameter depends on the number of items ordered.

In continuous review inventory systems, Ballintify (1964) and Silver (1974) have considered a can-order policy, which has a can order level and a reorder level. In this policy, an item must be ordered when its inventory level reaches the reorder level and when an item in the group is ordered, other items with inventory positions at or below their respective can-order levels are also ordered. Subsequently, many articles have appeared with models involving the above policy. Another article of interest is due to Federgruen et. al (1984), which deals with the general case of compound Poisson demands and non-zero lead time. A review of inventory models under joint replenishment is provided by Goyal and Satir (1989). Kalpakam and Arivarignan (1993) have introduced $(s, S)$ policy with a single reorder level $s$ defined in terms of the total number of items in the stock.

The work on methods to solve the joint replenishment problem throughout the years has been extensive. Readers are referred to the publications of Fung and Ma (2001), Goyal (1973, 1974, 1988), Goyal and Belton (1979), Kaspi and Rosenblatt (1991), Nilsson et al. (2007), Nilsson and Silver (2008), Olsen (2005), Silver (1976), Van Eijs (1993), Viswanathan (1996, 2002, 2007) and Wildeman et al. (1997) and references therein.

Xiao and Tiaojun et al. (2007) have developed a dynamic game model of a supply chain consisting of one manufacturer and one retailer to study the coordination mechanism and the effect of demand disruption on the coordination mechanism, where the market demand is sensitive to retail price and service. Shi and Kuiran and Xiao and Tiaojun (2008) analysed a supply chain consisting of a risk-neutral manufacturer selling a perishable product to a loss-averse retailer and presented the optimal ordering decision between the manufacturer and the retailer in a single period inventory with uncertain demand. Two types of contracts, buyback contract and markdown-price contract with the retailer's loss aversion consideration are investigated, respectively.

Anbazhagan and Arivarignan (2000) have considered a two commodity inventory system with coordinated reordering policy. Anbazhagan and Arivarignan (2001) have analysed a model with a joint ordering policy which places orders for both commodities whenever the total net inventory level drops to a prefixed level $s$. The demand points for each commodity form independent Poisson processes and the lead time is distributed as negative exponential. They have also assumed unit demands for both commodities.

Yadavalli et al. (2004, 2006) have analyzed two commodity inventory system under various ordering policies. Sivakumar et al. (2007) have considered a two commodity coordinated and individual ordering policies with renewal demands. In this policy, two reorder levels $r_i$ and $s_i (< r_i)$ are marked. An individual order at level $r_i$ for $i$-th commodity may stand cancelled whenever the inventory level is less than or equal to $s_i$ before replenishment and a new joint reorder for both commodities is placed.

The present work generalizes the work of Anbazhagan and Arivarignan (2001) by assuming the set of reorder levels with prescribed probability distribution for reordering. The rest of this paper is organized as follows: section 2 deals with problem formulation. Section 3 analyses part of the system and the system performance measures are computed in section 4. Section 5, total expectedcost rate is computed. Some numerical examples are considered to illustrate the model description in section 6. The last section is meant for conclusion.

## PROBLEM FORMULATION

Consider a two commodity inventory system with the maximum capacity $S_i$ units for $i$-th commodity $(i = 1,2)$. The demand points for each commodity are assumed to form independent Poisson processes with parameter $\lambda_i$, $i = 1,2$. The two commodities are assumed to be substitutable. That is, if the inventory level of one commodity reaches zero, any demand for this commodity will be satisfied by the item of the other commodity. If the total net inventory level drops to a prefixed level $s-k$, $k = 0,1,\cdots,r,$ where $r$ denote the reorder band width. An order will be placed for $Q^i_{s-k}(= S_i - s + k)$ units of i-th commodity with probability

$$p_k(\geq 0),\ k = 0,1,\ldots,\ r,\ \sum p_k = 1,$$

$$(0 \leq r \leq s-2 \text{ and } S_i - s + k > s+1), i = 1,2.$$

The lead time initiated at level $s-k$ is assumed to be distributed as exponential with parameter $\mu_k(>0)$ depends on the reorder quantity $Q^i_{s-k}$. The demands that occur during stock out periods are assumed to be lost sales.

Let $L_i(t)$ denotes the net inventory level of $i$-th commodity at time $t$. Then

$L = \{(L_1(t), L_2(t)), t \geq 0\}$ is a stochastic process with state space $E = \{0,1,\cdots,S_1\} \times \{0,1,\cdots,S_2\}$. The state space of inventory level process $L$ is shown in the Figure 1.

From the assumptions made on demand and replenishment processes, it follows that $L$ is a vector valued Markov process. To determine the infinitesimal generator

$$\tilde{A} = ((\ a((i,j);(k,l))\ )), \quad (i,j),(k,l) \in E,$$

we use the following arguments:

The demand for the first commodity takes the state of $L$ from $(i,j)$ to $(i-1,j)$ and the intensity of transition $a(i,j,i-1,j)$ is given by $\lambda_1$. The demand for second commodity takes the state from $(i,j)$ to $(i,j-1)$ with the intensity of transition given by $\lambda_2$. From the state $(i,j)$, when $i+j = s-k$ a replenishment takes the joint

*Figure 1. Space of inventory levels*

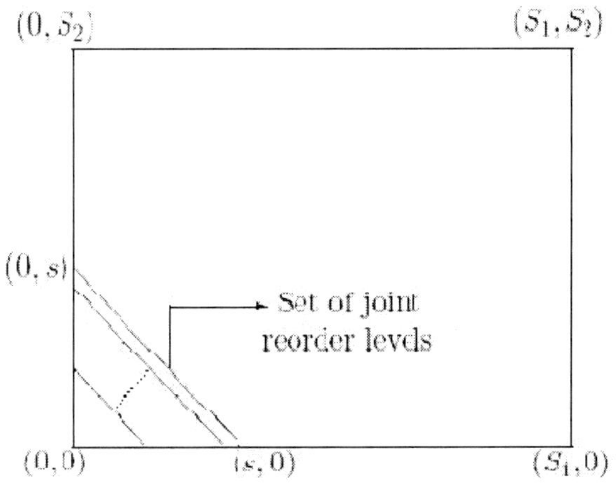

inventory level to $(i + Q^1_{s-k}, j + Q^2_{s-k})$ with probability $p_k$ and the intensity of transition is given by $\mu_k$, $k = 0,1,\cdots,r$. To obtain the intensity of passage $-a((i,j);(i,j))$ of state $(i,j)$, we make use of the identity

$$a((i,j);(i,j)) = -\sum_{k}\sum_{\substack{l \\ (i,j)\neq(k,l)}} a((i,j);(k,l)).$$

Hence, $a((i,j);(k,l))$ can be writtten as

$$\begin{cases}
\lambda_1 + \delta_{0j}\lambda_2, & k = i-1, \quad i = 1,2,\cdots,S_1, & l = j, \quad j = 0,1,\cdots,S_2 \\
\delta_{io}\lambda_1 + \lambda_2, & k = i, \quad i = 0,1,\cdots,S_1, & l = j-1, \quad j = 1,2,\cdots,S_2 \\
-(\lambda_1+\lambda_2), & k = i, \quad i = s+1,\cdots,S_1, & l = j, \quad j = 0,1,\cdots,S_2 \\
-(\lambda_1+\lambda_2), & k = i, \quad i = 0,1,\cdots,s, & l = j, \quad j = s-i+1,\cdots,S_2 \\
-(\lambda_1+\lambda_2+\sum_{t=0}^{s-i-j}\mu_t p_t), & k = i, \quad i = s-r,\cdots,s & l = j \quad j = 0,1\cdots,s-i \\
-((1-\delta_{0i}\delta_{0j})(\lambda_1+\lambda_2) + H[(s-r-1)-(i+j)]\sum_{t=0}^{r}\mu_t p_t + H[(i+j)-(s-r)]\sum_{t=0}^{s-i-j}\mu_t p_t), & k = i, \quad i = 0,1,\cdots,s-r-1 & l = j, \quad j = 0,1,\cdots,s-i \\
\mu_{s-m}p_{s-m} & k = i+Q^1_m, \quad i = m,m-1,\cdots,0 \quad \text{with } m = s,s-1,\cdots,s-r & l = j+Q^2_m \quad j = m-i,,\cdots,0 \\
0, & \text{otherwise.}
\end{cases}$$

In order to write down the infinitesimal generator $\tilde{A}$ in a matrix form, we arrange the states in lexicographic order and group $S_2+1$ states as

$$i = (((i,0),(i,1),\cdots,(i,S_2)), i = 0,1,\cdots,S_1.$$

Then the rate matrix $\tilde{A}$ has the block partitioned form with the following sub matrix $[\tilde{A}]_{ij}$ at the $i$- the row and $j$-th column position.

$$[\tilde{A}]_{ij} = \begin{cases}
B & \text{if} \quad j = i-1, & i = S_1,S_1-1,\cdots,1 \\
A & \text{if} \quad j = i, & i = S_1,S_1-1,\cdots,s+1 \\
A_{s+1-i} & \text{if} \quad j = i, & i = s,s-1,\cdots,1,0 \\
M_{[j-i-Q^1_s][S_1-j+1]} & \text{if} \quad j = S_1,S_1-1,\cdots,S_1-(s-i) & i = s,s-1,\cdots,s-r \\
M_{[j-i-Q^1_s][S_1-j+1]} & \text{if} \quad j = S_1-(s-i)+r,\cdots,S_1-(s-i) & i = s-r-1,\cdots,1,0 \\
0 & \text{otherwise.}
\end{cases}$$

Where,

$$[B]_{mn} = \begin{cases}
\lambda_1 & \text{if} \quad n = m, \quad m = S_2,S_2-1,\cdots,1 \\
\lambda_1+\lambda_2 & \text{if} \quad n = m, \quad m = 0 \\
0 & \text{otherwise.}
\end{cases}$$

$$[A]_{mn} = \begin{cases}
\lambda_2 & \text{if} \quad n = m-1, \quad m = S_2,S_2-1,\cdots,1 \\
-(\lambda_1+\lambda_2) & \text{if} \quad n = m, \quad m = S_2,S_2-1,\cdots,0 \\
0 & \text{otherwise.}
\end{cases}$$

$$[M_{ij}]_{mn} = \begin{cases}
p_i\mu_i & \text{if} \quad n = Q^2_s + i + m, \quad m = j-1,\cdots,1,0 \\
0 & \text{otherwise.}
\end{cases}$$

$$\textit{with } i = 0,1,\cdots,r \textit{ and } j = 1,2,\cdots,s+1-i$$

$$[A_i]_{mn} = \begin{cases}
\lambda_2 & \text{if} \quad n = m-1, \quad m = S_2,S_2-1,\cdots,1 \\
-(\lambda_1+\lambda_2) & \text{if} \quad n = m, \quad m = S_2,S_2-1,\cdots,i \\
-(\lambda_1+\lambda_2+\sum_{k=0}^{i-1-m}p_k\mu_k) & \text{if} \quad n = m, \quad m = i-1,i-2,\cdots,0 \\
0 & \text{otherwise.}
\end{cases}$$

$$with\ i = 1,2,\cdots,r+1$$

$$[A_i]_{mn} = \begin{cases} \lambda_2 & if \quad n = m-1, \quad m = S_2, S_2-1,\cdots,1 \\ -(\lambda_1+\lambda_2) & if \quad n = m, \quad m = S_2, S_2-1,\cdots,i \\ -(\lambda_1+\lambda_2+\sum_{k=0}^{i-1-m} p_k\mu_k) & if \quad n = m, \quad m = i-1, i-2,\cdots,i-r-1 \\ -(\lambda_1+\lambda_2+\sum_{k=0}^{r} p_k\mu_k) & if \quad n = m, \quad m = i-r-2,\cdots,1,0 \\ 0 & otherwise. \end{cases}$$

$$with\ i = r+2,\cdots,s$$

$$[A_{s+1}]_{mn} = \begin{cases} \lambda_1+\lambda_2 & if \quad n = m-1, \quad m = S_2, S_2-1,\cdots,1 \\ -(\lambda_1+\lambda_2) & if \quad n = m, \quad m = S_2, S_2-1,\cdots,s+1 \\ -(\lambda_1+\lambda_2+\sum_{k=0}^{s-m} p_k\mu_k) & if \quad n = m, \quad m = s, s-1,\cdots,s-r \\ -([1-\delta_{0m}](\lambda_1+\lambda_2)+\sum_{k=0}^{r} p_k\mu_k) & if \quad n = m, \quad m = s-r-1,\cdots,1,0 \\ 0 & otherwise. \end{cases}$$

Using the above rate matrix $\tilde{A}$, the steady state probabilities are evaluated which inturn give the system performance measures and total expected cost rate as described in the following sections.

## STEADY STATE RESULTS

It can be seen from the structure of $\tilde{A}$ that the homogeneous Markov process $\{(L_1(t), L_2(t)), t \geq 0\}$ on the finite state space $E$ is irreducible. Hence, the limiting distribution

$$\Phi = (\varphi^{(0)}, \varphi^{(1)}, \varphi^{(2)}, \ldots, \varphi^{(S_1-1)}, \varphi^{(S_1)})$$

with $\varphi^{(q)} = (\varphi^{(q,S_2)}, \varphi^{(q,S_2-1)}, \cdots, \varphi^{(q,0)})$, where $\varphi^{(i,j)}$ denotes the steady state probability for the state $(i,j)$ of the inventory level process exists and is given by

$$\Phi\tilde{A} = 0 \ and \sum\sum_{(i,j)\in E} \varphi^{(i,j)} = 1$$

The first equation of the above yields the following set of equations:

$$\varphi^{(i)}B + \varphi^{(i-1)}A_{s-i+2} = 0, \qquad i = 1,2,\cdots,s+1$$

$$\varphi^{(i)}B + \varphi^{(i-1)}A = 0, \qquad i = s+2, s+3,\cdots,Q_s^1$$

$$\varphi^{(i)}B + \varphi^{(i-1)}A + \sum_{k=0}^{i-Q_s^1-1} \varphi^{(k)} M_{[i-Q_s^1-1-k][Q_s^1+s+2-i]} = 0,$$
$$i = Q_s^1+1,\cdots,Q_s^1+r+1$$

$$\varphi^{(i)}B + \varphi^{(i-1)}A + \sum_{k=i-Q_s^1-r-1}^{i-Q_s^1-1} \varphi^{(k)} M_{[i-Q_s^1-1-k][Q_s^1+s+2-i]} = 0,$$
$$i = Q_s^1+r+2,\cdots,S_1$$

$$\varphi^{(S_1)}A + \sum^{s} \varphi^{(k)} M_{[s-k][1]} = 0,$$

After a great deal of simplifications, the first four of above equations become

$$\varphi^{(i)} = (-1)^i \varphi^{(0)} \mathop{\Omega}_{m=s+1}^{s+2-i} A_m B^{-1}, \quad i = 1,2,\cdots,s+1$$

$$= (-1)^i \varphi^{(0)} \left( \mathop{\Omega}_{m=s+1}^{1} A_m B^{-1} \right) (AB^{-1})^{i-s-1}, \quad i = s+2,\cdots,Q_s^1$$

$$= \varphi^{(0)} \Bigg\{ (-1)^i \left( \mathop{\Omega}_{m=s+1}^{1} A_m B^{-1} \right) (AB^{-1})^{i-s-1} +$$

$$\sum_{l=Q_s^1+1}^{i} \sum_{k=0}^{l-Q_s^1-1} (-1)^{i-l+1} [\delta_{k0} + (1-\delta_{k0})(-1)^k \mathop{\Omega}_{m=s+1}^{s+2-k} A_m B^{-1}]$$

$$M_{[l-Q_s^1-1-k][Q_s^1+s+2-l]} B^{-1}(AB^{-1})^{i-l} \Bigg\}, \quad i = Q_s^1+1,\cdots,Q_s^1+r+1$$

$$=\varphi^{(0)}\Bigg\{(-1)^{i}\left(\underset{m=s+1}{\overset{1}{\Omega}} A_{m}B^{-1}\right)(AB^{-1})^{i-s-1} +$$

$$\sum_{l=Q_s^1+1}^{Q_s^1+r+1}\sum_{k=0}^{l-Q_s^1-1}(-1)^{i-l+1}[\delta_{k0}+(1-\delta_{k0})(-1)^{k}\underset{m=s+1}{\overset{s+2-k}{\Omega}} A_{m}B^{-1}]$$

$$M_{[l-Q_s^1-1-k][Q_s^1+s+2-l]}B^{-1}(AB^{-1})^{i-l} +$$

$$\sum_{l=Q_s^1+r+2}^{i}\sum_{k=l-Q_s^1-r-1}^{l-Q_s^1-1}(-1)^{i-l+k+1}[\underset{m=s+1}{\overset{s+2-k}{\Omega}} A_{m}B^{-1}]$$

$$M_{[l-Q_s^1-1-k][Q_s^1+s+2-l]}B^{-1}(AB^{-1})^{i-l}\Bigg\} \quad i=Q_s^1+r+2,\cdots,S_1$$

where $\varphi^{(0)}$ can be obtained by solving,

$$\varphi^{(S_1)}A+\sum_{k=s-r}^{s}\varphi^{(k)}M_{[s-k][1]}=0 \text{ and}$$

$$\sum_{i=0}^{S_1}\varphi^{(i)}\,e=1,$$

that is

$$\varphi^{(0)}\Bigg\{\Bigg[(-1)^{S_1}\left(\underset{m=s+1}{\overset{1}{\Omega}} A_{m}B^{-1}\right)(AB^{-1})^{S_1-s-1} +$$

$$\sum_{l=Q_s^1+1}^{Q_s^1+r+1}\sum_{k=0}^{l-Q_s^1-1}(-1)^{S_1-l+1}[\delta_{k0}+(1-\delta_{k0})(-1)^{k}\underset{m=s+1}{\overset{s+2-k}{\Omega}} A_{m}B^{-1}]\times$$

$$M_{[l-Q_s^1-1-k][Q_s^1+s+2-l]}B^{-1}(AB^{-1})^{S_1-l} +$$

$$\sum_{l=Q_s^1+r+2}^{S_1}\sum_{k=l-Q_s^1-r-1}^{l-Q_s^1-1}(-1)^{S_1-l+k+1}[\underset{m=s+1}{\overset{s+2-k}{\Omega}} A_{m}B^{-1}]M_{[l-Q_s^1-1-k][Q_s^1+s+2-l]}B^{-1}(AB^{-1})^{S_1-l}\Bigg]A+$$

$$\sum_{k=s-r}^{s}[(-1)^{k}\underset{m=s+1}{\overset{s+2-k}{\Omega}} A_{m}B^{-1}]M_{[s-k][1]}\Bigg\}=0,$$

and

$$\varphi^{(0)}\Bigg\{I+\sum_{i=1}^{s+1}(-1)^{i}\left(\underset{m=s+1}{\overset{s+2-i}{\Omega}} A_{m}B^{-1}\right)+$$

$$\sum_{i=s+2}^{Q_s^1}(-1)^{i}\left(\left(\underset{m=s+1}{\overset{1}{\Omega}} A_{m}B^{-1}\right)(AB^{-1})^{i-s-1}\right)+$$

$$\sum_{i=Q_s^1+1}^{Q_s^1+r+1}\Bigg((-1)^{i}\left(\underset{m=s+1}{\overset{1}{\Omega}} A_{m}B^{-1}\right)(AB^{-1})^{i-s-1} +$$

$$\sum_{l=Q_s^1+1}^{i}\sum_{k=0}^{l-Q_s^1-1}(-1)^{i-l+1}[\delta_{k0}+(1-\delta_{k0})(-1)^{k}\underset{m=s+1}{\overset{s+2-k}{\Omega}} A_{m}B^{-1}]\times$$

$$M_{[l-Q_s^1-1-k][Q_s^1+s+2-l]}B^{-1}(AB^{-1})^{i-l}\Bigg)+$$

$$\sum_{i=Q_s^1+r+2}^{S_1}\Bigg((-1)^{i}\left(\underset{m=s+1}{\overset{1}{\Omega}} A_{m}B^{-1}\right)(AB^{-1})^{i-s-1} +$$

$$\sum_{l=Q_s^1+1}^{Q_s^1+r+1}\sum_{k=0}^{l-Q_s^1-1}(-1)^{i-l+1}[\delta_{k0}+(1-\delta_{k0})(-1)^{k}\underset{m=s+1}{\overset{s+2-k}{\Omega}} A_{m}B^{-1}]\times$$

$$M_{[l-Q_s^1-1-k][Q_s^1+s+2-l]}B^{-1}(AB^{-1})^{i-l} +$$

$$\sum_{l=Q_s^1+r+2}^{i}\sum_{k=l-Q_s^1-r-1}^{l-Q_s^1-1}(-1)^{i-l+k+1}[\underset{m=s+1}{\overset{s+2-k}{\Omega}} A_{m}B^{-1}]\times$$

$$M_{[l-Q_s^1-1-k][Q_s^1+s+2-l]}B^{-1}(AB^{-1})^{i-l}\Bigg)\Bigg\}e=1.$$

## SYSTEM PERFORMANCE MEASURES

In this section, some performance measures of the system are derived as follows:

### Mean Inventory Level

Let $\beta_1$ denote the average inventory level of the first commodity in the steady state. Then

$$\beta_1 = \sum_{i=1}^{S_1} i \left( \sum_{j=0}^{S_2} \varphi^{(i,j)} \right). \tag{1}$$

Let $\beta_2$ denote the average inventory level of the second commodity in the steady state. Then the following is obtained

$$\beta_2 = \sum_{j=1}^{S_2} j \left( \sum_{i=0}^{S_1} \varphi^{(i,j)} \right). \tag{2}$$

### Mean Reorder Rate

Let $\beta_3$ denote the mean reorder rate then

$$\beta_3 = \sum_{i=s-r}^{s} \sum_{k=0}^{i} [(\lambda_2 + \delta_{ko}\lambda_1) p_{s-i} \varphi^{(k,i+1-k)} + (\lambda_1 + \delta_{ko}\lambda_2) p_{s-i} \varphi^{(i+1-k,k)}]. \tag{3}$$

### Mean Shortage Rate

When $\beta_4$ denote the mean shortage rate,

$$\beta_4 = (\lambda_1 + \lambda_2)\varphi^{(0,0)}. \tag{4}$$

## COST ANALYSIS

We assume a specified cost structure for the proposed inventory system as follows:

Under the above cost structure, the total expected cost per unit time(total expected cost rate) in the steady state for this model is defined to be

$$TC(S_1, S_2, s, r) = h_1\beta_1 + h_2\beta_2 + k\beta_3 + c\beta_4.$$

By Substituting the values for $\beta_i$'s we can compute the value of $TC(S_1, S_2, s, r)$.

## NUMERICAL ILLUSTRATION

As the total expected cost rate is obtained in a complex form, the convexity of the total expected cost rate cannot be studied by analytical methods. Hence, numerical search procedures are used to find the local optimal values for $(S_1, S_2)$ with fixed $(s, r)$, $s$ with fixed $(S_1, S_2, r)$ and $r$ with fixed $(S_1, S_2, s)$. With a large number of numerical examples it is found that the total expected cost rate in the long run is either convex function of both variables $S_1$ and $S_2$ or an any one of the variables $r$ and $s$.

In Table 1 gives the total expected cost rate as a function of $S_1$ and $S_2$ by fixing constant values for the other variables and costs. After obtaining the local optima, $S_1^*$ and $S_2^*$, of $S_1$ and $S_2$ respectively, the sensitivity analysis are carried out to see how the changes in $S_1$ and $S_2$ affect the total expected cost rate(refer Figure 2). For this the values of $\dfrac{TC(S_1, S_2, 7, 5)}{TC(S_1^*, S_2^*, 7, 5)}$ by fixing the parameters and costs as

$$\lambda_1 = 1.97; \lambda_2 = 2.3; p_i = (0.7)(0.3)^i, i = 0, 1, \cdots, r-1;$$
$$p_r = 1 - \sum_{i=0}^{r-1} p_i; \mu_i = 2.1, i = 0, 1, \cdots, r;\ h_1 = 3.85; h_2 = 3.85;$$

$k = 1900; c = 13.2$ are computed.

*Table 1. Sensitivity of $S_1$ and $S_2$ on total expected cost rate*

| $S_2$ $S_1$ | 26 | 27 | 28 | 29 | 30 | 31 | 32 |
|---|---|---|---|---|---|---|---|
| 37 | 1.003936 | 1.002484 | 1.002456 | 1.003937 | 1.006957 | 1.011483 | 1.017387 |
| 38 | 1.004172 | 1.001817 | 1.000664 | 1.000818 | 1.002349 | 1.005286 | 1.009593 |
| 39 | 1.005912 | 1.002910 | 1.000902 | **1.000000** | 1.000297 | 1.001856 | 1.004700 |
| 40 | 1.008842 | 1.005405 | 1.002776 | 1.001064 | 1.000371 | 1.000780 | 1.002349 |
| 41 | 1.012711 | 1.009014 | 1.005959 | 1.003652 | 1.002193 | 1.001674 | 1.002170 |
| 42 | 1.017322 | 1.013502 | 1.010185 | 1.007464 | 1.005435 | 1.004192 | 1.003818 |
| 43 | 1.022522 | 1.018688 | 1.015237 | 1.012253 | 1.009824 | 1.008037 | 1.006978 |

*Table 2. Effect of s values on total expected cost rate*

| $(S_1,S_2)$ $s$ | (30,30) | (30,39) | (39, 39) | (39, 30) |
|---|---|---|---|---|
| 2 | 228.716014 | 213.661212 | 218.354931 | 229.809047 |
| 3 | 212.059065 | 200.519941 | 204.358882 | 214.430411 |
| 4 | 200.202119 | 192.132618 | 194.187768 | 203.214017 |
| 5 | 192.269202 | 187.682272 | 187.028237 | 195.282891 |
| 6 | 187.599127 | **186.527344** | 182.247822 | 189.947379 |
| 7 | **185.708059** | 188.157373 | 179.360957 | 186.671357 |
| 8 | 186.251015 | 192.167226 | 177.999702 | 185.043106 |
| 9 | 188.981934 | 198.252234 | **177.888945** | **184.751244** |
| 10 | 193.714656 | 206.220952 | 178.825403 | 185.565518 |
| 11 | 200.294259 | 216.014495 | 180.659775 | 187.321893 |

*Figure 2. Effect of $S_1$ and $S_2$ on total expected cost rate*

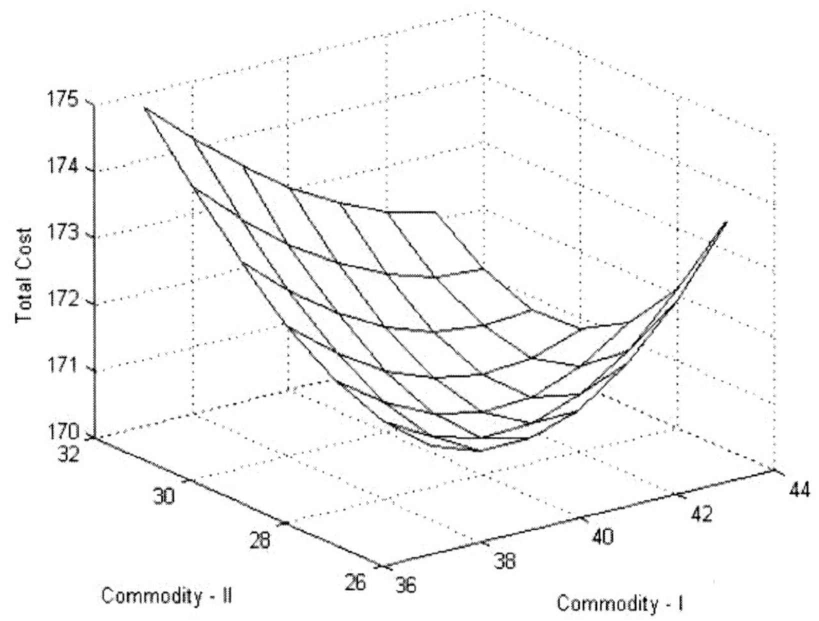

*Figure 3. Effect of s values on total expected cost rate*

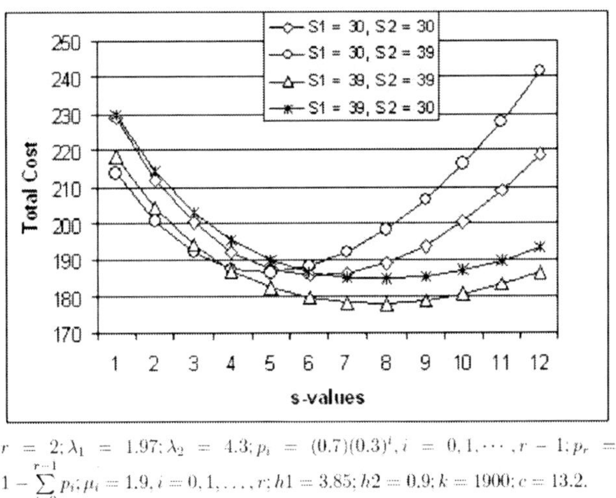


$r = 2; \lambda_1 = 1.97; \lambda_2 = 4.3; p_i = (0.7)(0.3)^i, i = 0, 1, \cdots, r-1; p_r = 1 - \sum_{i=0}^{r-1} p_i; \mu_i = 1.9, i = 0, 1, \ldots, r; h1 = 3.85; h2 = 0.9; k = 1900; c = 13.2.$

*Figure 4. Effect of r values on total expected cost rate*

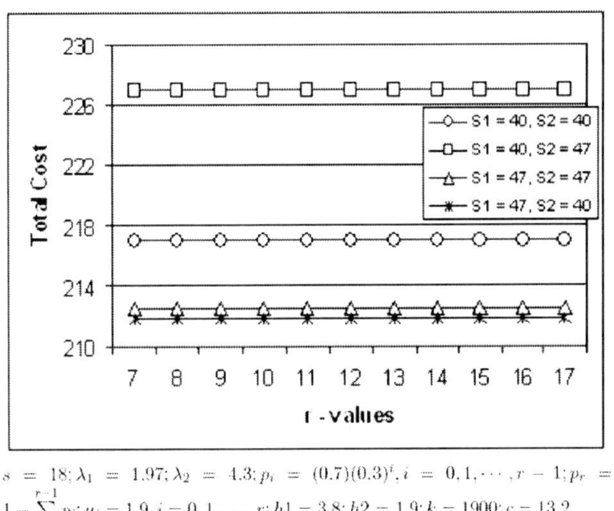


$s = 18; \lambda_1 = 1.97; \lambda_2 = 4.3; p_i = (0.7)(0.3)^i, i = 0, 1, \cdots, r-1; p_r = 1 - \sum_{i=0}^{r-1} p_i; \mu_i = 1.9, i = 0, 1, \ldots, r; h1 = 3.8; h2 = 1.9; k = 1900; c = 13.2.$

Here $S_1^* = 39$ and $S_2^* = 29$ and $TC(39, 29, 7, 5) = 170.9521$. It appears that the total expected cost rate is more sensitive to the changes in $S_2$ than that of in $S_1$.

Fixing all parameters and other cost values except $s$, the total expected cost rates are computed as shown in Table 2. The four curves in Figure 3 correspond to $(S_1, S_2) = (39, 30)$, $(S_1, S_2) = (39, 39)$, $(S_1, S_2) = (30, 39)$ and $(S_1, S_2) = (30, 30)$ represent different convex functions of $s$.

In Table 3 the total expectedcost rates by fixing constant values for all variables and costs except $r$ are presented. The four curves in Figure 4 correspond to $(S_1, S_2) = (40, 40)$, $(S_1, S_2) = (40, 47)$, $(S_1, S_2) = (47, 47)$ and

*Table 3. Effect of r values on total expected cost rate*

| $(S_1 S_2)$ r | (40,40) | (40,47) | (47, 47) | (47, 40) |
|---|---|---|---|---|
| 7 | 16.9522690994 | 226.9452680535 | 212.5174925549 | 211.8771321337 |
| 8 | 216.9521472578 | 226.9451933866 | 212.5173454789 | 211.8769527539 |
| 9 | 216.9521274595 | 226.9451845726 | 212.5173152901 | 211.8769145636 |
| 10 | 216.9521249370 | **226.9451844201** | 212.5173093197 | 211.8769066175 |
| 11 | **216.9521248537** | 226.9451848044 | 212.5173082060 | 211.8769050203 |
| 12 | 216.9521249548 | 226.9451849869 | 212.5173080188 | 211.8769047166 |
| 13 | 216.9521250069 | 226.9451850503 | 212.5173079940 | 211.8769046643 |
| 14 | 216.9521250258 | 226.9451850699 | **212.5173079930** | 211.8769046571 |
| 15 | 216.9521250318 | 226.9451850756 | 212.5173079941 | **211.8769046568** |
| 16 | 216.9521250336 | 226.9451850772 | 212.5173079946 | 211.8769046571 |
| 17 | 216.9521250341 | 226.9451850777 | 212.5173079949 | 211.8769046573 |

*Table 4. Effect of holding costs $h_1$ and $h_2$ on optimal values*

| $h_2$ $h_1$ | 0.7 | | 0.8 | | 0.9 | | 1.0 | | 1.1 | |
|---|---|---|---|---|---|---|---|---|---|---|
| 3.75 | 169.6795 | | 170.7347 | | 171.7899 | | 172.7817 | | 173.7267 | |
| | 41 | 31 | 41 | 31 | 41 | 31 | 40 | 29 | 40 | 29 |
| 3.8 | 170.8946 | | 171.9520 | | 172.9612 | | 173.9704 | | 174.9206 | |
| | 41 | 31 | 40 | 30 | 40 | 30 | 40 | 30 | 40 | 29 |
| 3.85 | 172.1097 | | 173.1649 | | 174.2201 | | 175.1582 | | 176.1146 | |
| | 41 | 31 | 41 | 31 | 41 | 31 | 40 | 30 | 40 | 29 |
| 3.90 | 173.3015 | | 174.3275 | | 175.3368 | | 176.3460 | | 177.3085 | |
| | 40 | 31 | 40 | 30 | 40 | 30 | 40 | 30 | 40 | 29 |
| 3.95 | 174.4823 | | 175.5153 | | 176.5245 | | 177.5307 | | 177.3085 | |
| | 40 | 31 | 40 | 30 | 40 | 30 | 39 | 29 | 40 | 29 |

$(S_1, S_2) = (47, 40)$ represent different convex functions of r(see Table 3). It is noted that when $S_1 = 40$, for two different values of $S_2$, say $S_2^{(i)}$, $i = 1,2$ the associated total cost $TC^{(i)}$ satisfies the relation: $S_2^{(1)} < S_2^{(2)} \Rightarrow TC^{(1)} < TC^{(2)}$ and the same relationship holds for $S_1 = 47$ also.

Next the impact of holding costs $h_1$ and $h_2$ on the optimal values $(S_1^*, S_2^*)$ and the corresponding total expected cost rate are studied. For the fixed parameters and the probability distribution $s = 7; r = 5; \lambda_1 = 1.97; \lambda_2 = 2.3; p_i = (0.7)(0.3)^i$, $i = 0,1,\cdots,r-1; p_r = 1 - \sum_{i=0}^{r-1} p_i$; $\mu_i = 1 + i/2, i = 0,1,\cdots,r; k = 1900; c = 13.2$, we observed that the total cost rate increases when $h_1$ and $h_2$ increases (see Table 4).

The impact of ordering cost per order and the shortage cost per unit item on the optimal values $(S_1^*, S_2^*)$ and the corresponding total expected cost rate are studied by fixing the parameters and the probability distribution:

*Table 5. Effect of ordering cost (k) and shortage cost (c) on optimal values*

| c k | 13.0 | | 13.1 | | 13.2 | | 13.3 | | 13.4 | |
|---|---|---|---|---|---|---|---|---|---|---|
| 1800 | 170.2455 | | 170.2477 | | 170.2499 | | 170.2521 | | 170.2543 | |
| | 39 | 29 | 39 | 29 | 39 | 29 | 39 | 29 | 39 | 29 |
| 1850 | 172.2150 | | 172.2172 | | 172.2193 | | 172.2214 | | 172.2235 | |
| | 40 | 30 | 40 | 30 | 40 | 30 | 40 | 30 | 40 | 30 |
| 1900 | 174.1448 | | 174.1469 | | 174.1490 | | 174.1511 | | 174.1532 | |
| | 40 | 30 | 40 | 30 | 40 | 30 | 40 | 30 | 40 | 30 |
| 1950 | 176.0745 | | 176.0766 | | 176.0787 | | 176.0808 | | 176.0829 | |
| | 40 | 30 | 40 | 30 | 40 | 30 | 40 | 30 | 40 | 30 |
| 2000 | 177.9472 | | 177.9492 | | 177.9513 | | 177.9533 | | 177.9553 | |
| | 41 | 31 | 41 | 31 | 41 | 31 | 41 | 31 | 41 | 31 |

$$s = 7; r = 5; \lambda_1 = 1.97; \lambda_2 = 2.3; p_i = (0.7)(0.3)^i,$$
$$i = 0,1,\cdots,r-1; p_r = 1 - \sum_{i=0}^{r-1} p_i;$$
$$\mu_i = 1 + i/2, i = 0,1,\cdots,r; h_1 = 3.85; h_2 = 0.9.$$

From the Table 5 it is noted that the total cost rate increases when $k$ and $c$ increases.

## CONCLUSION

In this paper, a two commodity substitutable inventory system with set of reorder levels with band width $r$ have been dealt. This model is most suitable for two different rare items which are substitutable(for "A" type items in ABC inventory classifications). The joint probability distribution of the inventory levels in the steady state and the stationary measures of system performances have been derived. Illustration has also been provided to prove the existence of local optima when the total cost function is treated as a function of only two variables $S_1$ and $S_2$ or a single variable $s$ or $r$. The authors are working in the direction of generalizing the demand process as renewal type.

## ACKNOWLEDGMENT

N. Anbazhagan's research is supported by the DST Fast Track Scheme grant for 'Young Scientists' through research project SR/FTP/MS-04/2004. The authors would like to thank the anonymous referees for their perceptive comments and valuable suggestions on a previous draft of this paper to improve its quality.

## REFERENCES

Anbazhagan, N., & Arivarignan, G. (2000). Two-commodity continuous review inventory system with coordinated reorder policy. *International Journal of Information and Management Sciences*, *11*(3), 19–30.

Anbazhagan, N., & Arivarignan, G. (2001). Analysis of two-commodity markovian inventory system with lead time. *Journal of Applied Mathematics & Computing*, *8*(2), 519–530.

Ballintify, J. L. (1964). On a basic class of inventory problems. *Management Science*, *10*, 287–297. doi:10.1287/mnsc.10.2.287

Federgruen, A., Groenvelt, H., & Tijms, H. C. (1984). Coordinated replenishment in a multi-item inventory system with compound Poisson demands. *Management Science*, *30*, 344–357. doi:10.1287/mnsc.30.3.344

Fung, R. Y. K., & Ma, X. (2001). A new method for joint replenishment problems. *The Journal of the Operational Research Society*, *52*, 358–362. doi:10.1057/palgrave.jors.2601091

Goyal, S. K. (1973). Determination of economic packaging frequency of items jointly replenished. *Management Science*, *20*, 232–235. doi:10.1287/mnsc.20.2.232

Goyal, S. K. (1974). Determination of optimal packaging frequency of jointly replenished items. *Management Science*, *21*, 436–443. doi:10.1287/mnsc.21.4.436

Goyal, S. K. (1988). Economic ordering policy for jointly replenished items. *International Journal of Production Research*, *26*, 1237–1240. doi:10.1080/00207548808947937

Goyal, S. K., & Belton, A. S. (1979). On a simple method of determining order quantities in joint replenishments for deterministic demand. *Management Science*, *25*, 604.

Goyal, S. K., & Satir, T. (1989). Joint replenishment inventory control: Deterministic and stochastic models. *European Journal of Operational Research*, *38*, 2–13. doi:10.1016/0377-2217(89)90463-3

Kalpakam, S., & Arivarignan, G. (1993). A coordinated multicommodity (s,S) inventory system. *Mathematical and Computer Modelling*, *18*, 69–73. doi:10.1016/0895-7177(93)90206-E

Kaspi, M., & Rosenblatt, M. J. (1991). On the economic ordering quantity for jointly replenished items. *International Journal of Production Research*, *29*(1), 107–114. doi:10.1080/00207549108930051

Nilsson, A., Segerstedt, A., & van der Sluis, E. (2007). A new iterative heuristic to solve the joint replenishment problem using a spreadsheet technique. *International Journal of Production Economics*, *108*(1-2), 399–405. doi:10.1016/j.ijpe.2006.12.022

Nilsson, A., & Silver, E. A. (2008). A simple improvement on Silver's heuristic for the joint replenishment problem. *The Journal of the Operational Research Society*, *59*(10), 1415–1421. doi:10.1057/palgrave.jors.2602446

Olsen, A. L. (2005). An evolutionary algorithm to solve the joint replenishment problem using direct grouping. *Computers & Industrial Engineering*, *48*, 223–235. doi:10.1016/j.cie.2005.01.010

Shi, K., & Xiao, T. (2008). Coordination of a supply chain with a loss-averse retailer under two types of contracts. *Int. J. Information and Decision Sciences*, *1*(1), 5–25. doi:10.1504/IJIDS.2008.020033

Silver, E. A. (1974). A control system of coordinated inventory replenishment. *International Journal of Production Research*, *12*, 647–671. doi:10.1080/00207547408919583

Silver, E. A. (1976). A simple method of determining order quantities in joint replenishments under deterministic demand. *Management Science*, *22*, 1351–1361. doi:10.1287/mnsc.22.12.1351

Sivakumar, B., Anbazhagan, N., & Arivarignan, G. (2007). Two commodity inventory system with individual and joint ordering polices and Renewal demands. *Stochastic Analysis and Applications*, *25*(6), 1217–1241. doi:10.1080/07362990701567314

Van Eijs, M. J. G. (1993). A note of the joint replenishment problem under constant demand. *The Journal of the Operational Research Society*, *44*(2), 185–191.

Viswanathan, S. (1996). A new optimal algorithm for the joint replenishment problem. *The Journal of the Operational Research Society*, *47*, 936–944.

Viswanathan, S. (2002). On optimal algorithms for the joint replenishment problem. *The Journal of the Operational Research Society*, *53*, 1286–1290. doi:10.1057/palgrave.jors.2601445

Viswanathan, S. (2007). An algorithm for determining the best lower bound for the stochastic joint replenishment problem. *Operations Research*, *55*(5), 992–996. doi:10.1287/opre.1070.0401

Wildeman, R. E., Frenk, J. B. G., & Dekker, R. (1997). An efficient optimal solution method for the joint replenished problem. *European Journal of Operational Research*, *99*, 433–444. doi:10.1016/S0377-2217(96)00072-0

Xiao, T., Luo, J., & Jin, J. (2007). Coordination of a supply chain with demand stimulation and random demand disruption. *International Journal of Information Systems and Supply Chain Management*, *2*(1), 1–15.

Yadavalli, V. S. S., Anbazhagan, N., & Arivarignan, G. (2004). A two-commodity stochastic inventory system with lost sales. *Stochastic Analysis and Applications*, *22*(2), 479–497. doi:10.1081/SAP-120028606

Yadavalli, V. S. S., Arivarignan, G., & Anbazhagan, N. (2006). Two commodity coordinated inventory system with Markovian demand. *Asia - Pacific Journal of Operational Research, 23*(4), 497-508.

## APPENDIX

Notations used in this paper

$0$ : zero matrix.

$e^T$ : $(1,1,\ldots,1)$.

$I_N$ : identity matrix of order $N$.

$\delta_{ij}$ : Kronecker delta

$$\sum_{j=0}^{i} a^j = \begin{cases} a^0 + a^1 + \cdots + a^i, & \textit{if } i \textit{ is nonnegative integer} \\ 0, & \textit{otherwise} \end{cases}$$

$[A]_{ij}$ : $(i,j)$ – th element of the matrix $A$.

$$\underset{i=j}{\overset{k}{\Omega}} c_i = \begin{cases} c_j c_{j-1} \cdots c_k & \textit{if } j \geq k \\ 1 & \textit{if } j < k \end{cases}.$$

k: ordering cost per order.
$h_i$ : holding cost for the i-th commodity per unit item per unit time (i=1,2).
c: shortage cost per unit item.

*This work was previously published in International Journal of Information Systems and Supply ChainManagement, Volume 3, Issue 2, edited by John Wang, pp. 52-67, copyright 2010 by IGI Publishing (an imprint of IGI Global)*

# Section 4
# Logistics

# Chapter 14
# A New Look at Selecting Third-Party Reverse Logistics Providers

**Reza Farzipoor Saen**
*Islamic Azad University-Karaj Branch, Iran*

## ABSTRACT

*The use of Data Envelopment Analysis (DEA) in many fields is based on total flexibility of the weights. However, the problem of allowing total flexibility of the weights is that the values of the weights obtained by solving the unrestricted DEA program are often in contradiction to prior views or additional available information. Also, many applications of DEA assume complete discretionary of decision making criteria. However, they do not assume the conditions that some factors are nondiscretionary. To select the most efficient third-party reverse logistics (3PL) provider in the conditions that both weight restrictions and nondiscretionary factors are present, a methodology is introduced. A numerical example demonstrates the application of the proposed method.*

## INTRODUCTION

Many manufacturers have understood that their core competences are not in the logistics-field, and have therefore progressively sought to buy logistics services and functions from third-party reverse logistics (3PL) provider (Bottani & Rizzi, 2006). The outsourcing of non core processes and activities makes it possible to focus on core manufacturing activities, while, at the same time, 3PL providers have specific logistics core competences, and they can manage logistics processes more efficiently than their customers.

3PL providers play a role in helping organizations in closing the loop for products offered by those organizations. Traditionally, reverse logistics

DOI: 10.4018/978-1-4666-0918-1.ch014

is an activity within organizations delegated to the customer service function, where customers with warranted or defective products would return them to their supplier.

One of the uses of data envelopment analysis (DEA) can be 3PL provider selection. In original DEA formulations the assessed decision making units (DMUs) can freely choose the weights or values to be assigned to each input and output in a way that maximizes its efficiency, subject to this system of weights being feasible for all other DMUs. This freedom of choice shows the DMU in the best possible light, and is equivalent to assuming that no input or output is more important than any other.

The free imputation of input-output values can be seen as an advantage, especially as far as the identification if inefficiency is concerned. If a DMU (3PL provider) is free to choose its own value system and some other 3PL provider uses this same value system to show that the first 3PL provider is not efficient, then a stronger statement is being made. The advantages of full flexibility in identifying inefficiency can be seen as disadvantages in the identification of efficiency. An efficient 3PL provider may become so by assigning a zero weight to the inputs and/or outputs on which its performance is worst. This might not be acceptable by decision makers (DMs) as well as by the analyst, who after spending time in a careful selection of inputs and outputs sees some of them being completely neglected by 3PL providers.

DMs may have in 3PL provider selection problems value judgments that can be formalized *a priori*, and therefore should be taken into account in 3PL provider selection. These value judgments can reflect known information about how the factors used by the 3PL providers behave, and/or "accepted" beliefs or preferences on the relative worth of inputs, outputs or even 3PL providers. For example, in 3PL provider selection problem in general, one input (price) usually overwhelms all other inputs, and ignoring this aspect may lead to biased efficiency results. 3PL providers might also supply some outputs that require considerably more resources than others and this marginal rate of substitution between outputs should somehow be taken into account when selecting a 3PL provider. To avoid the problem of free (and often undesirable) specialization, input and output weights should be constrained in DEA.

On the other hand, discretionary models for evaluating the efficiency of DMUs assume that all criteria are discretionary, i.e., controlled by the management of each DMU and varied at its discretion. Thus, failure of a DMU to produce maximal output levels with minimal input consumption results in a decreased efficiency score. In any realistic situation, however, there may exist exogenously fixed or nondiscretionary criteria that are beyond the control of a management. Banker & Morey (1986) illustrate the impact of exogenously determined inputs that are not controllable in an analysis of a network of fast food restaurants. In their study, each of the 60 restaurants in the fast food chain consumes six inputs to produce three outputs. The three outputs (all controllable) correspond to breakfast, lunch, and dinner sales. Only two of the six inputs, expenditures for supplies and expenditures for labor, are discretionary. The other four inputs (age of store, advertising level, urban/rural location, and presence/absence of drive-in capability) are beyond the control of the individual restaurant manager. Their analysis clearly demonstrates the value of accounting for the nondiscretionary character of these inputs explicitly in the DEA models they employ; the result is identification of a considerably enhanced opportunity for targeted savings in the controllable inputs and targeted increases in the outputs. In the case of 3PL provider selection, location and industry experience are generally considered nondiscretionary criterion.

The objective of this article is to propose a model for selecting 3PL providers in the presence of both weight restrictions and nondiscretionary factors. This article depicts the 3PL provider selection process through a DEA model, while allowing for the incorporation of DM's preferences and nondiscretionary factors simultaneously. The chief advantage of the proposed model is that it does not demand exact weights from the decision maker and considers nondiscretionary factors.

This article proceeds as follows. In the next Section, literature review is presented. Then the proposed method for 3PL provider selection is discussed. Next, numerical example is illustrated. Finally, concluding remarks are discussed.

## LITERATURE REVIEW

Some mathematical programming approaches have been used for 3PL provider selection in the past. Meade & Sarkis (2002) addressed the need for a strategic decision-making model to assist management in determining with which 3PL to partner in the reverse logistics process. They determined the following decision criteria to select the best 3PL:

- Location of product in its lifecycle
- The organization's strategic performance criteria
- The reverse logistics process functions required by the organization
- Organization (environmental/customer service) role of reverse logistics

They applied analytic network process (ANP) for 3PL provider selection. Göl & Çatay (2007) presented hand-on experiences of a pilot project conducted at a leading Turkish automotive company to redesign its logistics operations and to select a global logistics service provider, using analytic hierarchy process (AHP) for multi-criteria decision making. Efendigil et al. (2008) developed a conceptual framework integrating fuzzy logic and artificial neural networks as a tool while including the environmental factors in the design of a reverse logistics provider selection process. The framework presented in their study was developed through an analysis of environmental management practices and performance measurement metrics. Fuzzy AHP and ANN are employed to incorporate both quantitative and qualitative attributes of suppliers leading to a more effective and realistic 3PL selection process.

However, AHP and ANP have two main weaknesses. First subjectivity of AHP and ANP is a weakness. The decision maker provides the values for the pairwise comparisons and, therefore, the model is very dependent on the weightings provided by the decision maker[1]. Second the time necessary for completion of such a model is a weakness. The number of pairwise comparisons required could become cumbersome. Meanwhile, when the number of alternatives and criteria grows, the pairwise comparison process becomes difficult, and the risk of generating inconsistencies grows, hence jeopardizing the practical applicability of AHP and ANP.

Bottani & Rizzi (2006) provided a comprehensive and organic framework, encompassing the different attributes and criteria to be traded-off in the 3PL selection process. The framework is based on the analysis of the relevant literature related to logistics outsourcing. Then, they derived a quantitative methodology from the selection taxonomy which can be used effectively in the evaluation stage to quantitatively rank different alternatives. Their methodology is based on a fuzzy TOPSIS (Technique for Order Preference by Similarity to Ideal Solution) approach, and is therefore effective for dealing with both qualitative and quantitative selection criteria.

However, all of the abovementioned references do not consider weight restrictions and nondiscretionary factors. A technique that can deal with both weight restrictions and nondiscretionary factors is needed to better model such situation.

Recently, Farzipoor Saen (2009) proposed an imprecise DEA method for selecting 3PL providers in the conditions that both ordinal and cardinal data are present. However, his proposed approach, not only does not consider weight restrictions, but also does not consider nondiscretionary factors. As well, Haas et al. (2003) applied an analytical method to help logistics managers manage reverse logistics systems. They used DEA for selecting reverse logistics channels and utilized DEA to the reverse flows involved in municipal solid waste management systems to determine if the method could be used to aid managers of reverse logistics channels in running their distribution systems.

To the best of author's knowledge, there is not any reference that discusses 3PL provider selection in the presence of both weight restrictions and nondiscretionary factors. The approach presented in this article has some distinctive features.

- The proposed model does not demand exact weights from the decision maker[2].
- The proposed model considers nondiscretionary factors for 3PL provider selection
- The proposed model considers weight restrictions for 3PL provider selection
- Weights restrictions and nondiscretionary factors are considered simultaneously

## PROPOSED METHOD FOR 3PL PROVIDER SELECTION

One serious drawback of DEA applications in many real world problems have been the absence of DM judgment, allowing total freedom when allocating weights to input and output data. This allows DMUs to achieve artificially high efficiency scores by indulging in inappropriate input and output weights.

The most widespread method for considering judgments in DEA models is, perhaps, the weight restrictions inclusion. Weight restrictions allow for the integration of managerial preferences in terms of relative importance levels of various inputs and outputs. The idea of conditioning the DEA calculations to allow for the presence of additional information arose first in the context of bounds on factor weights in DEA's multiplier side problem. This led to the development of the cone-ratio (Charnes et al., 1989) and assurance region models (Thompson et al., 1990). Both methods constrain the domain of feasible solutions in the space of the virtual multipliers.

Weights restrictions may be applied directly to the DEA weights or to the product of these weights with the respective input or output level, referred to as *virtual input* or *virtual output*. Restrictions on virtual weights were proposed first by Wong & Beasley (1990). Sarrico & Dyson (2004) suggest, in line with Thompson et al. (1990), the use of virtual assurance regions, concluding that they can overcome problems of infeasibility as well as interpretation of target and efficiency scores, whilst retaining the benefit of the natural representation of preference structures.

The discussions in this article are provided with reference to the original DEA formulation by Charnes et al. (1978) below, which assumes constant returns to scale and that all input and output levels for all DMUs are strictly positive. The CCR model measures the efficiency of DMU$_o$ relative to a set of peer DMUs:

$$\begin{aligned}
&\max \quad \frac{\sum_{r=1}^{s} u_r y_{ro}}{\sum_{i=1}^{m} v_i x_{io}}, \\
&s.t. \\
&\quad \frac{\sum_{r=1}^{s} u_r y_{rj}}{\sum_{i=1}^{m} v_i x_{ij}} \le 1, \qquad j = 1,\dots,n, \\
&\quad u_r, v_i \ge \varepsilon \qquad \forall r \quad and \quad i
\end{aligned} \tag{1}$$

where there is a set of $n$ peer DMUs, {DMU$_j$: $j$ = 1, 2,…, $n$}, which produce multiple outputs $y_{rj}$ ($r$ = 1, 2, …, $s$), by utilizing multiple inputs $x_{ij}$ ($i$ = 1, 2, …, $m$). DMU$_o$ is the DMU under consideration. DMU$_o$ consumes $x_{io}$ ($i$=1, …, $m$), the amount of input $i$, to produce $y_{ro}$ ($r$=1, …, $s$), the amount of output $r$. $u_r$ is the weight given to output $r$ and $v_i$ is the weight given to input $i$. $\varepsilon$ is a positive non-Archimedean infinitesimal. DMU$_o$ is said to be efficient, if no other DMU or combination of

DMUs can produce more than $DMU_o$ on at least one output without producing less in some other output or requiring more of at least one input.

In (2) the various types of weight restriction that can be applied to multiplier models are shown.

$$\left.\begin{array}{l}\text{Absolute weight restrictions}\\ \qquad \delta_i \le v_i \le \tau_i \quad (g_i) \qquad \rho_r \le u_r \le \eta_r \quad (g_o)\\ \text{Assurance region of type I (relative weight restrictions)}\\ \qquad \alpha_i \le \dfrac{v_i}{v_{i+1}} \le \beta_i \quad (h_i) \qquad \theta_r \le \dfrac{u_r}{u_{r+1}} \le \zeta_r \quad (h_o)\\ \text{Assurance regions of type II (input-output weight restrictions)}\\ \qquad \gamma_i v_i \ge u_r \quad (l)\end{array}\right\} \tag{2}$$

The Greek letters ($\delta_i, \tau_i, \rho_r, \eta_r, \alpha_i, \beta_i, \theta_r, \zeta_r, \gamma_i$) are user-specified constants to reflect value judgments the DM wishes to incorporate in the assessment. They may relate to the perceived importance or worth of input and output factors. The restrictions *(g)* and *(h)* in (2) relate on the left hand side to input weights and on the right hand side to output weights. Constraint *(l)* links directly input and output weights. Absolute weight restrictions are the most immediate form of placing restrictions on the weights as they simply restrict them to vary within a specific range. Assurance region of type I, link either only input weights ($h_i$) or only output weights ($h_o$). The relationship between input and output weights are termed assurance region of type II.

Restrictions on virtual inputs/virtual outputs assume the form in (3), where the proportion of the total virtual output of $DMU_j$ accounted for by output *r* is restricted to lie in the range [$a_r, b_r$] and the proportion of the total virtual input of $DMU_j$ accounted for by input *i* is restricted to lie in the range [$c_i, d_i$].

$$\begin{array}{ll} a_r \le \dfrac{u_r y_{rj}}{\sum_{r=1}^{s} u_r y_{rj}} \le b_r, & r = 1,\dots,s \\ c_i \le \dfrac{v_i x_{ij}}{\sum_{i=1}^{m} v_i x_{ij}} \le d_i, & i = 1,\dots,m \end{array} \tag{3}$$

Assuming weights restrictions on inputs, the multipliers formulation, with the assurance region of type I, is as below:

$$\begin{array}{lll} \max & \sum_{r=1}^{s} y_{ro} u_r, & \\ s.t. & & \\ & \sum_{i=1}^{m} x_{io} v_i = 1, & \\ & \sum_{r=1}^{s} u_r y_{rj} - \sum_{i=1}^{m} v_i x_{ij} \le 0 & \forall j, \\ & \alpha_i v_{i+1} \le v_i & \\ & v_i \le \beta_i v_{i+1} & \\ & v_i \ge \varepsilon & \forall i, \\ & v_{i+1} \ge \varepsilon & \forall i, \\ & u_r \ge \varepsilon & \forall r. \end{array} \tag{4}$$

where $\varepsilon$ is the non-Archimedean infinitesimal. At this juncture, suppose that the input variables may be partitioned into subsets of discretionary (D) and nondiscretionary (N) variables. Thus,

$$I = \{1,2,\dots,m\} = I_D \cup I_N, \quad I_D \cap I_N = \Phi$$

Finally, the proposed model is given as below

$$\begin{array}{lll} \max & \sum_{r=1}^{s} u_r y_{ro} - \sum_{i \in N} v_i x_{io}, & \\ s.t. & & \\ & \sum_{i \in D} v_i x_{io} = 1, & \\ & \sum_{r=1}^{s} u_r y_{rj} - \sum_{i \in N} v_i x_{ij} - \sum_{i \in D} v_i x_{ij} \le 0 & \forall j, \\ & \alpha_i v_{i+1} \le v_i & \\ & v_i \le \beta_i v_{i+1} & \\ & v_i \ge \varepsilon & i \in D, \\ & v_{i+1} \ge \varepsilon & i \in D, \\ & v_i \ge 0 & i \in N, \\ & v_{i+1} \ge 0 & i \in N, \\ & u_r \ge \varepsilon & \forall r. \end{array} \tag{5}$$

As can be seen, the nondiscretionary but not the discretionary inputs, enter into the objective

*Table 1. Related attributes for eighteen 3PL providers and efficiency scores*

| 3PL provider o. (DMU) | Inputs | | | Outputs | | | Efficiency scores |
|---|---|---|---|---|---|---|---|
| | UOP ($) $x_{1j}$ | EE (1000 $) $x_{2j}$ | D (km) $x_{3j}$ | R (1000 $) $y_{1j}$ | VE (100000 $) $y_{2j}$ | RC (10000 kg) $y_{3j}$ | |
| 1 | 253 | 197 | 249 | 187 | 90 | 2 | .934 |
| 2 | 268 | 198 | 643 | 194 | 130 | 13 | .913 |
| 3 | 259 | 229 | 714 | 220 | 200 | 3 | 1 |
| 4 | 180 | 169 | 1809 | 160 | 100 | 3 | 1 |
| 5 | 257 | 212 | 238 | 204 | 173 | 24 | 1 |
| 6 | 248 | 197 | 241 | 192 | 170 | 28 | 1 |
| 7 | 272 | 209 | 1404 | 194 | 60 | 1 | .854 |
| 8 | 330 | 203 | 984 | 195 | 145 | 24 | .801 |
| 9 | 327 | 208 | 641 | 200 | 150 | 11 | .799 |
| 10 | 330 | 203 | 588 | 171 | 90 | 53 | 1 |
| 11 | 321 | 207 | 241 | 174 | 100 | 10 | .72 |
| 12 | 329 | 234 | 567 | 209 | 200 | 7 | .862 |
| 13 | 281 | 173 | 567 | 165 | 163 | 19 | .876 |
| 14 | 309 | 203 | 967 | 199 | 170 | 12 | .837 |
| 15 | 291 | 193 | 635 | 188 | 185 | 33 | 1 |
| 16 | 334 | 177 | 795 | 168 | 85 | 2 | .666 |
| 17 | 249 | 185 | 689 | 177 | 130 | 34 | 1 |
| 18 | 216 | 176 | 913 | 167 | 160 | 9 | 1 |

(5). The multiplier values associated with these nondiscretionary inputs may be zero, but the other variables must always be positive.

In the next section, a numerical example is presented.

## NUMERICAL EXAMPLE

The data set for this example is partially taken from Talluri & Baker (2002) and contains specifications on eighteen 3PL providers. The 3PL data set is selected from a case study that utilized DEA for monitoring customer–supplier relationship. For more details, please see Talluri & Baker (2002).

The 3PL provider inputs considered are unit operation cost (UOP)[3], environmental expenditures (EE), and distance (D). UOP is the cost spent for one unit operation of transportation. EE is the cost of environmental activities. The outputs utilized in the study are recycling capacity (RC), value of equipments (VE), and revenue from the sale of recyclables (R). D is generally considered as a nondiscretionary input variable. Also, RC is generally considered as a nondiscretionary output variable. Table 1 depicts the 3PL provider's attributes. According to the decision of DM, the importance of UOP must be greater than EE. Assume that UOP is, at least, twice as important as EE. The positive non-Archimedean infinitesimal, $\varepsilon$, has been set to 0.0001.

Applying model (5), the efficiency scores of 3PL providers (DMUs) have been presented in Table 1.

Model (5) identified 3PL providers 3, 4, 5, 6, 10, 15, 17, and 18 to be efficient with a relative efficiency score of 1. The remaining ten 3PL providers with relative efficiency scores of less

*Table 2. Sensitivity of efficiency scores to the $\alpha$*

| 3PL provider No. (DMU) | $\alpha = 2$ | $\alpha = 4$ | $\alpha = 8$ | $\alpha = 16$ |
|---|---|---|---|---|
| 1 | .934 | .927 | .923 | .92 |
| 2 | .913 | .896 | .886 | .88 |
| 3 | 1 | 1 | 1 | 1 |
| 4 | 1 | 1 | 1 | 1 |
| 5 | 1 | 1 | 1 | 1 |
| 6 | 1 | 1 | 1 | 1 |
| 7 | .854 | .837 | .825 | .819 |
| 8 | .801 | .776 | .761 | .753 |
| 9 | .799 | .771 | .754 | .745 |
| 10 | 1 | 1 | 1 | 1 |
| 11 | .72 | .7 | .687 | .68 |
| 12 | .862 | .844 | .832 | .826 |
| 13 | .876 | .85 | .833 | .824 |
| 14 | .837 | .812 | .796 | .788 |
| 15 | 1 | .991 | .979 | .973 |
| 16 | .666 | .629 | .606 | .594 |
| 17 | 1 | 1 | 1 | 1 |
| 18 | 1 | 1 | .996 | .994 |

than 1 are considered to be inefficient. Therefore, DM can choose one or more of these efficient 3PL providers.

Sensitivity analysis involves investigating the effects on the solutions of making possible changes in the values of the model parameters. To this end, the sensitivity of the results to the weights ( $\alpha$ ) are discussed. In the context of DEA, sensitivity analysis has been one of the widely studied topics. Since DEA is data based, it is very useful to assess changes of user-specified constants to reflect value judgments.

Table 2, describes the sensitivity of the relative efficiencies by changing the $\alpha$ for the model (5). It can be noted from the Table 2 that when $\alpha$ is increased, the efficiency score of most of DMUs and number of efficient DMUs are reduced. For example, when $\alpha$ is set to be 2, $DMU_{15}$ has an efficiency score of 1. When $\alpha$ is set to be 4, the efficiency score decreases to 0.991. Similarly, increasing $\alpha$, decreases the efficiency score of $DMU_{15}$.

Therefore, it can be concluded that, increasing the lower bounds in weights restrictions, will increase discrimination among efficiency scores of DMUs.

## CONCLUDING REMARKS

This article has provided a model for selecting 3PL providers in the presence of both weight restrictions and nondiscretionary factors.

The problem considered in this study is at initial stage of investigation and further researches can be done based on the results of this article. Some of them are as follows:

Similar research can be repeated for dealing with ordinal data and bounded data in the conditions that both weight restrictions and nondiscretionary factors exist. Other potential extension to the methodology includes the case that some of the 3PL providers are slightly non-homogeneous. One of the assumptions of all the classical models of DEA is based on complete homogeneity of DMUs (3PL providers), whereas this assumption in many real applications cannot be generalized. In other words, some inputs and/or outputs are not common for all the DMUs occasionally. Therefore, there is a need to a model that deals with these conditions.

## ACKNOWLEDGMENT

The author wishes to thank the anonymous reviewers for their valuable suggestions and comments.

## REFERENCES

Banker, R. D., & Morey, R. C. (1986). Efficiency analysis for exogenously fixed inputs and outputs . *Operations Research, 34*(4), 513–521. doi:10.1287/opre.34.4.513

Bottani, E., & Rizzi, A. (2006). A fuzzy TOPSIS methodology to support outsourcing of logistics services. *Supply Chain Management: An International Journal, 11*(4), 294–308. doi:10.1108/13598540610671743

Charnes, A., Cooper, W. W., & Rhodes, E. (1978). Measuring the efficiency of decision making units. *European Journal of Operational Research, 2*(6), 429–444. doi:10.1016/0377-2217(78)90138-8

Charnes, A., Cooper, W. W., Wei, Q. L., & Huang, Z. M. (1989). Cone-ratio data envelopment analysis and multi-objective programming. *International Journal of Systems Science, 20*(7), 1099–1118. doi:10.1080/00207728908910197

Farzipoor Saen, R. (2009). A mathematical model for selecting third-party reverse logistics providers. *International Journal of Procurement Management, 2*(2), 180–190. doi:10.1504/IJPM.2009.023406

Sarrico, C. S., & Dyson, R. G. (2004). Restricting virtual weights in data envelopment analysis. *European Journal of Operational Research, 159*(1), 17–34. doi:10.1016/S0377-2217(03)00402-8

Talluri, S., & Baker, R. C. (2002). A multi-phase mathematical programming approach for effective supply chain design . *European Journal of Operational Research, 141*(3), 544–558. doi:10.1016/S0377-2217(01)00277-6

Thompson, R. G., Langemeier, L. N., Lee, C. T., Lee, E., & Thrall, R. M. (1990). The role of multiplier bounds in efficiency analysis with application to Kansas farming . *Journal of Econometrics, 46*(1/2), 93–108. doi:10.1016/0304-4076(90)90049-Y

Wong, Y. H. B., & Beasley, J. E. (1990). Restricting weight flexibility in data envelopment analysis . *The Journal of the Operational Research Society, 41*(9), 829–835.

## ENDNOTES

1 A critical issue of traditional approaches is the correct choice of the weights. These must be assigned by the decision maker or a decision committee and are often very subjective measures. The basic idea of DEA is that the weights are chosen by an optimization procedure and not by the decision maker. Weights are assigned optimally for every input and output attribute. This makes the approach more robust against human inference.

2 Note that in traditional models of 3PL provider selection, the weights are allocated in

a crisp value, while in the proposed model; weights are defined in an interval. It is clear that interval definition of the weights for the decision maker is easier than the crisp weight assignment.

3 The inputs and outputs selected in this article are not exhaustive by any means, but are some general measures that can be utilized to evaluate 3PL providers. In an actual application of this methodology, decision makers must carefully identify appropriate inputs and outputs measures to be used in the decision making process.

*This work was previously published in International Journal of Information Systems and Supply Chain Management, Volume 3, Issue 1, edited by John Wang, pp. 58-67, copyright 2010 by IGI Publishing (an imprint of IGI Global)*

# Chapter 15
# Management of Logistics Planning

**Bjørnar Aas**
*Molde University College, The Norwegian School of Logistics, Norway*

**Stein W. Wallace**
*Lancaster University, England*

## ABSTRACT

*Logistics problems are gradually becoming more complex and a better understanding of logistics management as a subject is a key to deal with the new challenges. A core element of logistics management is logistics planning, which substitutes for low customer service levels, high waste, and the use of buffers and slacks in the execution of logistic activities. Furthermore, the availability of information and problem-solving capabilities are established as the core parts of logistics planning. Based on this, in this paper, a conceptual model for the management of logistics planning is proposed and discussed. In this regard, the model is built on ideas from microeconomics.*

## INTRODUCTION

Over the last few decades, developments in business life and business management in general, have led to changed and more complex logistics problems. This has made it harder for managers in organizations where logistics is important to be consistent in relating to and dealing with the many logistics management activities that arise.

DOI: 10.4018/978-1-4666-0918-1.ch015

Our practical experience from working with many different companies for a number of years is that many managers have rather low formal logistics competence. Further, that mastering quantitative logistics management activities is challenging since they focus on methods and techniques which are difficult to learn through experience. Nonetheless, such methods have often yielded large savings and are frequently promoted by commercial logistics companies. Hence, sound knowledge about such logistics management activities is necessary for logistics managers.

Quantitative logistics management covers activities such as routing, inventory management, and production planning necessary to create plans for executing logistics activities mainly at an operational level. Such plans will represent value since they improve the execution of logistics activities which again usually will create cost savings or other positive effects. In the literature, such planning, which is a core activity in most quantitative logistics management activities, is often called "logistics planning".

In order to conduct logistic planning it is necessary to have access to relevant logistics information and a capability to transform this information into an intelligent plan. However, these issues could be challenging and demanding in terms of resources. A logistics manager will have to trade off costs and benefits (more details later) and make sure that the necessary premises for efficient execution at the desired level are fulfilled. We regard these as important strategic or tactical (situation dependent) logistics management activities which concern the efficient allocation of scarce resources. Further, we believe that understanding these aspects of logistics management is of great importance for logistics managers as well as for academics.

Remark that management of logistics planning differs from logistics planning as such, since the focus now is on how to organize the ability to perform logistics planning.

In our opinion, management of logistics planning is presently given too little attention in research and education and we agree with Ghiani et al. (2004, p. xiv) who claim that "logistics planning tends to be treated either integrated and qualitative, or mathematical and very specific".

When taking the latter approach, all necessary logistics information is usually assumed to be available and the focus is on problem solving to obtain solutions. Managerial aspects beyond the obligatory chapters of why the logistics management activity in question is important are not given much attention, see e.g., Silver et al. (1998) and Nahmias (2005).

Similarly, these issues are not given much attention in logistics research and textbooks that take a more integrated and qualitative approach, whether they declare themselves as supply chain management-founded or not. Information availability and problem-solving capability are widely regarded as being important since they are assumptions for being able to exercise logistics planning. However, trade-off issues are rarely mentioned and in particular, not much attention is paid to the related costs of information collection and problem solving, and how they should be balanced relative to each other, e.g., Christopher (1998), Chopra and Meindel (2001), and Ballou (2004).

There are also several contributions that focus on information technology and information systems and their importance for enhancing the competitiveness of logistics. However, the focus is generally on the information aspect, and the problem-solving capability is not emphasized, e.g., Hammant (1995), Gustin et al. (1995), and Closs and Goldsby (1997).

A consequence of this divided and detached research and presentation of quantitative logistics management activities is that it becomes difficult for logistics mangers to get a full picture of what logistics planning includes on strategic or tactical levels. Furthermore, it becomes difficult to see the common features, and subsequently common management issues which exist across the different logistics management activities.

This article will focus on the area of logistics management that is related to the management of efficient logistics planning using quantitative logistics management tools. More specifically, this means that we will explore the general objective of logistics planning and the coherence between this and its two main components; problem solving and information collection. In doing this we will propose a conceptual model that covers these issues.

The conceptual model is used as a tool to reach the main purpose of this article, namely to contribute to a greater understanding and awareness of what management of efficient logistics planning entails with respect to quantitative logistics tools.

Our aim is to contribute to a further development of logistics management as a subject and subsequently, in the long run, contribute to increased quality of logistics planning in organizations.

The article is conceptual and based on a review of relevant literature and builds on our experience from participating in numerous industry projects.

The rest of this paper is organized as follows. In the following section we review the background of our research. Thereafter, in the two subsequent sections, we present and discuss our conceptual model. Conclusions and suggestions for further research are presented in the last section.

## BACKGROUND

To understand what management of efficient logistics planning entails, we first need to explore the concepts of logistics management and logistics planning.

A definition of logistics management commonly used is provided by the Council of Supply Chain Management Professionals, CSCMP (2007): "Logistics management is that part of supply chain management that plans, implements, and controls the efficient, effective forward and reverse flow and storage of goods, services and related information between the point of origin and the point of consumption in order to meet customers' requirements". Together with the supplementary information provided by CSCMP, see CSCMP (2007), we conclude that it is emphasized that logistics management is an integrating function which coordinates and optimizes all logistics activities, as well as integrates logistics activities with other business functions.

This elaboration clearly shows that logistics management comprises the management of various logistics activities and that logistics planning is a main responsibility for any logistics manager. Christopher (1998, p. 14) goes even further by describing logistics as "fundamentally a planning concept".

Although the two terms "logistics management" and "logistics planning" to a large extent are overlapping, it is important to notice that the first term comprises more than just the creation of logistics plans. By regarding the terms as more or less equivalent, an important perspective will disappear; the understanding of what management of efficient logistics planning is.

Even though we will focus on the management part, it is necessary to first examine the basic nature of logistics planning. Without sufficient knowledge about the planning activity itself, it is difficult to understand how to manage it efficiently.

The purpose of logistic planning is to enable logistics managers to make good logistics decisions. Where to route the vehicle? How much should we produce? How much should we stock? These are all traditional questions within logistics and the answers to those questions have usually been found by combining the available information with some sort of problem-solving capability, seeking to obtain rational solutions. We will now proceed by taking a look into history to find the origin of this approach.

The first scattered attempts to develop quantitative solution techniques started early, and a common approach was to explore how established scientific disciplines, like mathematics and statistics could be used to improve the execution of logistics activities. Descriptions of early contributions in the development of logistics methods are found in Woodbury (1960), Erlenkotter (1990), and Wagner (2002).

Today's awareness of the importance of logistics was triggered by amongst others the seminal contributions of Culliton et al. (1956) and Drucker (1962). According to Ballou (2007), it can be claimed that logistics management first becomes defined as a discipline in management science and practice in the 1950s or 1960s. Further he claims that until that time, the typical enterprise performed logistics purely on a functional basis and not much effort was put into analysis and research. "There was little attempt to integrate and balance the activities later to be known as logistics activities that were in cost/or service conflict", Ballou (2007 p. 334). During the following decades, the focus on developing logistics models to decrease logistics costs was high, leading to today's existence of numerous models and approaches (Ballou, 2007).

From the beginning, the main motivation for developing logistics planning methods was an incipient understanding that logistics planning could function as a substitute for high inventories and overcapacity in transportation and production. Or simply, it could be a substitute for the poor solutions that are often obtained when problems are solved without any planning or when insufficient efforts are put into planning.

As an offspring of disciplines like mathematics and statistics, where finding provably good or optimal solutions is essential, the early quantitative logistics models also had this focus. The early developments were mainly focused on how improvements could be realized by improving the calculations done ahead of the actual execution of a logistics activity. The outcome of the logistics planning was therefore an important input to the decision making process and sometimes the logistics planning process itself represented the entire decision-making process.

This link between decision making and logistics planning is important and in 1947, Simon published a seminal work concerning decision making and the concept of rationality in general which soon gained popularity (Simon, 1957).

In his work "Administrative Behavior", Simon explores the features of rational decisions and states that to claim that a decision is rational we must have satisfied the following criteria:

- We must have identified and listed all alternatives (collected all relevant information)
- We must have determined the consequences resulting from each alternative (processed all the collected information)
- We must have compared the accuracy and efficiency of each of these sets of consequences (made an objective comparison)

Simon asserts that if we manage to fulfill those criteria, we have acted like an economic man. Furthermore, he claims that to fulfill those criteria is difficult, in practice impossible, and contrasts the administrative man whose goal is to find satisfying solutions which are "good enough", rather than maximizing solutions. He claims this is more in line with real life decision making.

In many ways the development of logistics planning for the last sixty years has been in line with the ideas from Simon's work, resulting in a more conscious and diversified approach to logistics planning. The rationality criteria pinpointed by Simon have in different ways strongly influenced the development of most of the logistics planning methods used today. His idea has been a main justifier both for the further use and development of exact methods based on simplifications, and the development of logistics planning approaches with even less focus on proving optimality.

A continuation of Simon's work is the work of Kahneman and Tversky who clearly showed that to rely on the judgment of the individual when dealing with uncertainty is risky, see e.g. Kahneman et al. (1982). This finding, together with a gradually more complex business environment, which makes it more and more difficult to see whether a solution is rational or not, has caused more standardized methods, or even automated planning tools to avoid or reduce the problems of flawed intuitive decision making.

Actually, an important purpose of today's modern information systems is to contribute to rational decision making and therefore is designed to contribute in that respect. This is easy to discover by studying the structure of modern information systems which usually are divided into two main components: execution and planning systems. While the execution systems provide data, transaction processing, user access and infrastructure for running the company, the planning systems usually use data generated by the execution systems to support decision making on a strategic, tactical and operational level (Simchi-Levi et al., 2003).

A trend is that these information systems become more and more sophisticated and the newest Enterprise Resource Planning (ERP) systems are also capable of (at least to a certain extent) addressing SCM optimization problems as e.g., SAP's Advanced Planner and Optimizer solution (SAP, 2007). An important issue for these types of comprehensive software solutions is to unify information collection and problem solving in one application. This is looked upon as a cost-efficient way of managing the logistics information system structure and as a way to improve logistics decisions (Helo & Szekely, 2005).

The fact that logistics planning is gradually being carried out by the use of a company's information system has dramatically altered the logistics manager's tasks. Earlier, problem solving was given most of the attention, now the management of information (Hammant, 1995; Christopher, 1998), and the design of information systems / logistics systems must be regarded as equally important managerial activities (Closs & Savitskie, 2003; Ballou, 2004; Helo & Szekely, 2005; Feng & Yuan, 2006).

Summarizing, within a company, a core issue of logistics management has from the beginning been to perform logistics planning in a way that ensures an efficient execution of logistics activities. Planning methods developed through history to ensure this have been guided by a wish for rational solutions and today, a company's information collection and problem-solving capabilities constitute the basic elements of its logistics planning. Consequently a logistics manager must now be able to contribute to solutions where both the information availability and the problem-solving capability are treated in an efficient way. This view is the basis for our proposed model which will be presented in the next section. As stated earlier, this is not about logistics planning but about management of logistics planning, a subject so far scarcely addressed in the body of logistics literature.

## CONCEPTUAL MODEL

The conceptual model is based on ideas from microeconomics and more specifically on production theory. A basic assumption in all economic models, production models included, is that economic agents pursue their self-interest (Eaton & Eaton, 1991). In production theory different inputs are employed and transformed into outputs and in our case, every company aims, according to the assumption of pursuing self-interest, to find the best way of doing this.

In this article we will deal with the production of logistics plans and the economic and technical concerns related to this. Economic concerns include how much of a service to produce and what combination of input factors to use. Technical concerns relate to the process of transforming inputs to outputs, subject to technical rules specified by the production function. Both dimensions are important for our model and will be discussed with regard to logistics planning.

### Logistics Planning

Logistics planning is an activity where the goal is to find an intelligent way to coordinate and guide the use of logistics resources. Poor logistics plans usually lead to inefficient use of logistics resources and should subsequently be avoided.

Occasionally logistics inefficiencies are easy to discover, and reduce or eliminate, by improved logistics planning, but it is not always possible or desirable to do so. Firstly, at least in the short run, logistics planning has its limitations. It is not always possible to increase the planning effort (we will return to this soon). Secondly, due to cost-efficiency concerns, it is sometimes better to accept the presently best achievable solution than to spend extensive amounts of resources to obtain one that is objectively better.

The latter issue is an economic balancing / efficiency problem similar to what we find in production theory. We claim that a main objective of logistics planning is to substitute low customer service levels, high waste, and the use of buffers and slacks in the execution of logistics activities. This idea is usually referred to in the logistics literature, though often somewhat imprecisely. An often mentioned example of this is the statement "information can be used as a substitute for inventory", see for instance Mukhopadhyay and Cooper (1993), Dudley and Lasserre (1998), and Christopher (1998). As indicated by Lumsden et al. (1998), this is based on the assumption that it is actually possible to utilize information in such a way that inventory levels can be reduced. In other words, logistics planning capability is needed and it is therefore more correct to say that "logistics planning can be used as a substitute for inventory".

Often low customer service levels, high waste, buffers and slacks are caused by not putting sufficient effort into logistics planning. A good logistics plan can for instance be a substitute for not only too high inventories, but also for such as low utilization of vehicles or poor service if those are results of poor planning. However, since logistics planning has its technical limitations and is subject to economic trade-offs, the main issue is to find a level of logistics planning which is both possible and at the same time efficient.

Even though it is often possible to substitute logistics planning with for instance a high inventory level, those inputs cannot usually be regarded as perfect substitutes. It is reasonable to assume that in general there will be a decreasing marginal effect. This provides us with a convex production possibility curve where it becomes progressively more difficult to substitute one input for the other as we move away from the origin (see the upper part of Figure 2).

Sometimes the level of logistics planning is limited by a budget constraint. If so, finding the highest possible achievable level of logistics planning under the budget constraint is important. The optimal production level is found at the point where the isocost line corresponding to the budget is tangent to a production possibility curve. At this point the marginal returns per unit cost are the same for all inputs and the budget is binding (see Eaton and Eaton (1991) for more details about microeconomics issues).

## Logistics Planning Resources

If we assume that the level of logistics planning we want to produce is set, the level of inputs used must be determined. In Section 2 we have stated that the two main ingredients of logistics planning are logistics information availability and problem-solving capability and we regard these as the main inputs in the production of logistics planning.

Contrary to the situation in the previous section where the inputs are substitutes, the inputs are now complements. "Information is a complementary good with respect to productive processes" West and Courtney (1993, p. 230). Relevant logistics information cannot be substituted by problem-solving capacity and vice versa; the inputs must be consumed jointly in order to produce a logistics plan.

In microeconomics this phenomena is explained by what is called a negative cross price elasticity. This means that if the price of one input increases, the demand for the other input will also decrease, Eaton and Eaton (1991, p. 218). In practice this means that to collect more logistics information than can be used, or to have an overcapacity with regard to problem-solving capacity would represent a waste (unless it comes

*Figure 1. Relationships in logistics planning*

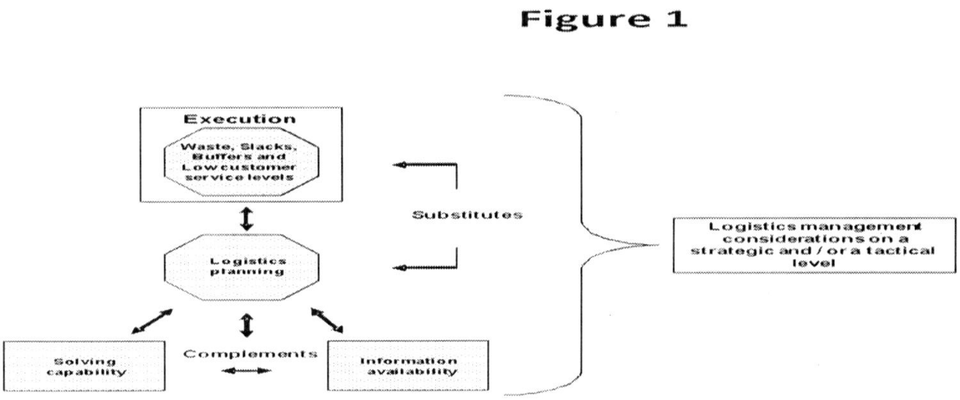

**Figure 1**

for free). The logistics planning cannot be better than what is imposed by the "weakest" link.

## The Proposed Model

So far, we have separately described two economic balance / efficiency issues which we regard as key managerial logistics planning activities. If we merge these, we get that efficient logistics planning is about efficient resource allocation across planning ability and information gathering, as well as waste, low customer service levels, buffers and slacks in the physical execution of the logistics services. Consequently; this is also the goal and field of responsibility of management of logistics planning.

The product of planning ability and relevant logistics information; the logistics plan could lead to less waste, improved customer service levels and less need for spare capacity and other forms of slacks and buffers in the execution of logistics, showing that logistics planning could function as a substitute. The elements and relationships described constitute the content of our conceptual model and are shown in Figure 1.

In the body of production literature, to determine the combinations of inputs that are technically possible is commonly described as a responsibility of entrepreneurs, e.g., Parkin (1997, p. 204). Similarly, issues of establishing the price / cost of the inputs and the benefits of the outputs are commonly described as activities outside the scope of microeconomics models. Logistics managers, however, must often deal with such issues. For instance, it is important to clarify how much relevant logistics information it is possible to collect, as well as the related costs. Likewise, it is part of the logistics manager's duties to determine alternative levels of logistics planning; the possible output-expansion path, and corresponding costs and benefits. Logistics planning therefore not only concerns adaptation issues, but also includes the design and dimensioning of planning resources. Occasionally, logistics planning also reveals that it is desirable to make changes with regard to the logistics resources used for executing the logistics.

All together this implies that logistics planning also concerns productivity issues. Productivity refers to the effective overall use of resources without making any assumptions about the production technology. This means that productivity is a function of being able to produce in accordance with the production possibility curve; static efficiency, as well as being at the right production possibility curve; dynamic efficiency (Shao & Lin, 2001).

*Figure 2. The case company's comprehension of "the way to go"*

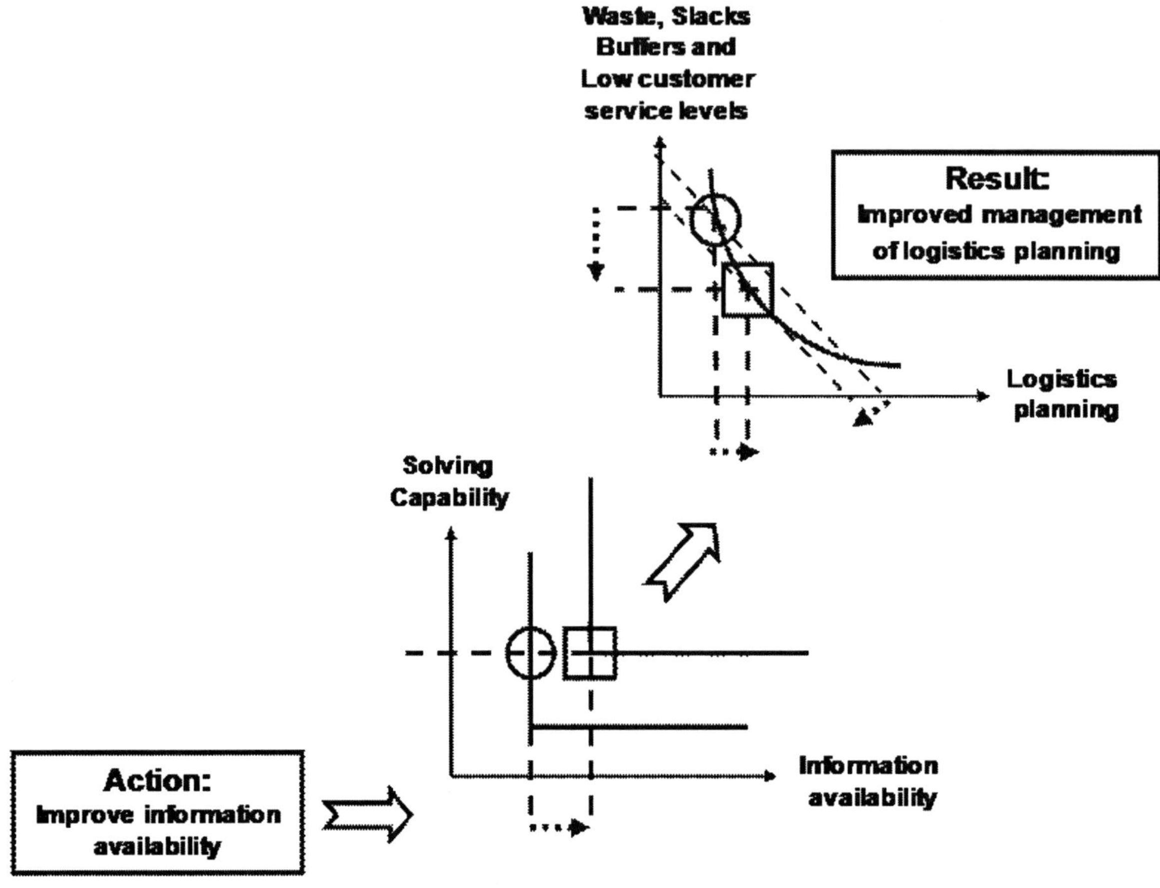


If an enterprise enters into a supply chain collaboration which drastically increases the information availability, or procures a new and sophisticated optimization program, the production possibility curve will shift. The underlying motivation for such actions should be to increase the overall productivity of the company. If this use of resources is suboptimal, or if the enterprise does not manage to exploit the new opportunities, this will result in a state of inefficiency.

Often improvements with regard to information availability, or problem-solving capability, come in large portions. For instance, a new information collection tool would rarely represent an increase in information availability that perfectly matches the (hopefully) existing lack of information. Often too much information will be collected leading to an overcapacity situation which could represent an even less efficient allocation between information availability and problem-solving capability. However, the overall productivity of the firm could have improved significantly due to the effects of moving to a higher level of logistic planning. Logistics managers need to be aware of this so that they do not forget that overall productivity measures are paramount local efficiency measures.

Another important issue is that it is seldom possible to obtain an ideal adoption, i.e. an ideal output level. The main reason for this is that the production cost for e.g., logistics planning and the cost of waste etc. will fluctuate over time. Often an immediate reaction to such changes is not possible or wanted if the changes are believed to be of temporary character. In the meantime the company would be in a state of inefficiency.

Due to the generic nature of our model, it will apply irrespective of which supply philosophy chosen (as for instance lean or agile). Modern supply philosophies require all a rather high level of logistics planning to realize the full benefits of the concepts. Still, due to the differences of the concepts, what is the preferable planning ability and relevant logistics information will vary across the concepts.

## MODEL DISCUSSION

We will now continue by elaborating on the practical application of the proposed model in the current business environment.

### Efficient Logistics Planning

Logistics is of utmost importance for most companies to stay competitive and considerable resources are used to procure, develop and maintain logistics systems (Closs & Savitskie, 2003; ELA, 2004; Ministry of Transport and Communication Finland, 2006). Consequently, to ensure a wise use of resources on logistics planning, the design of logistics systems is important for logistics managers. We believe that efficient logistics systems and subsequently efficient logistics planning can be obtained by following our proposed model.

#### The Different Levels of Information Collection and Solving Capability and Corresponding Costs

Given that an output-expansion path is determined, it is necessary to find the costs corresponding to the different levels of information collection and problem-solving capability. Unfortunately, not much research has been carried out to specifically reveal this. Some research contributions have reported logistics cost, but mainly on an aggregated level (ELA, 2004; Ministry of Transport and Communication Finland, 2006).

We claim that a more fine-meshed analysis of logistics costs is necessary to reveal the true cost of logistics planning and its components so that knowledge of cause-effect relationships can be obtained. This is consistent with research in informatics where finding the relationships between the costs and benefits /values of an information system is an important issue, see Mukhopadhyay and Cooper (1993), West and Courtney (1993), West (1994), Shao and Lin (2001), Closs and Savitskie (2003), and Lee (2008).

The resources used on information collection and problem solving have to be separated from other uses of the resources. For instance, resources used on collecting logistics information which is just passed on to customers as a service, should not be included. The human resources, IT resources etc. used for information collection, problem solving and management of the logistics planning should be included.

We believe that for a logistics manager it will not be difficult to at least approximately estimate the costs belonging to the different logistics planning levels.

#### The Different Levels of Information Collection and Problem-Solving Capability and Corresponding Value

The value of increased information collection and problem-solving capability is due to the fact that they are complements, linked to the output-expansion path. Each step on the output-expansion path has its corresponding value which reflects how well the level performs with regard to its ability to substitute low customer service levels, high waste and the use of buffers and slacks. Therefore, the value of the different substitution levels must be determined. This can be done by using optimization, simulation, or related tools. A technique aimed at revealing how much one should be willing to pay for perfect information --- "the expected value of perfect information" (EVPI) --- could also be applied. Alternatively, a pilot solution could be established for the purpose of comparisons against the existing solution.

After having determined the value of different levels of logistics planning, the best level can be found by cost-benefit comparisons across the levels.

It should also be noticed that the same information often can be used as input in more than one planning task. Or that the underlying process that must be established to create the needed information simultaneously can create other types of information which could be of value. Our conceptual model (Figure 1) will often be a simplification of the reality where in particular information gathering can be related to a multitude of planning, decision and service activities. This complicates the cost allocation and value assessment of this task.

## Example

To clarify and illustrate how the proposed model could function, we present a simple example in Figure 2. The example is based on well-known standard microeconomic ideas, and is in accordance with the earlier discussions. The starting point is that a company has discovered that it would be beneficial from a productivity viewpoint to improve the information availability. The figure assumes an established / defined marginal rate of substitution of "waste…etc" for given levels of logistics planning. It is also assumed that all points on the logistics production curve represent solving the same logistics problem.

The circles show today's situation, while the squares show the expected future position if the information availability gets improved.

The lower level figure shows the present unbalance caused by too little information available. By increasing the information availability to match the already existing level of solving capability, a more balanced (efficient) production of logistics plans will be the result. Simultaneously, a logistics plan of higher quality will be realized. However, this action will trigger a logistic plan that costs more to produce due to increased information collection costs. An example, partly illustrating this, can be found in Lee (2008).

If we turn to the upper figure, we will stay on the same logistics production curve (since we still are solving the same logistics planning problem). However, as we have lifted the logistic plan to a higher quality level, it is possible, in a profitable way, to increase e.g. the utilization of a warehouse.

Shifts in the logistics production curve will occur if the logistics problem is changed. By for instance including more vehicles, or by deciding to solve several logistics problems simultaneously, e.g., a combined routing and inventory problem.

Price changes affecting the cost of logistics planning, or "waste…etc", will influence the slope of the isocost line, and the tangent point at the logistics production line will change accordingly.

## LIMITATIONS

In this section we will return to the question of determining limitations for information collection and problem-solving capability, and by that clarify the joint product; the output-expansion path for logistics planning.

Limitations can have for example financial, technical, human, and inter- and intra-organizational causes. There will usually be some hard limitations and some which are possible to alter if enough resources are applied. We will pinpoint some sources that we consider relevant and significant. To do this, we have chosen to start by addressing the most important premise provider for logistics planning; the level of complexity. In an uncomplicated and predictable world, planning would be superfluous. In many ways, the aim of any planning is to cope with complexity.

According to Lumsden et al. (1998, p. 7), important aspects of complexity are; a large number of system states, heterogeneous systems, distributed decision making and uncertainty. It is those aspects that represent the main reasons for limitations with regard to information collection and problem solving and this will be a recurring theme in the two next sections.

## Limited Information Availability

Information is made up of meaningful data (Jessup & Valacich, 2008), and by summarizing previous research contributions, Mukhopadhyay and Cooper (1993) state that coverage and accuracy are the most important attributes of information. Coverage identifies how inclusively a description represents the relevant parts of the reality. The attribute covers an overview of the number of objects or events described, at what level they are described, the period that is covered, how often it is updated and the age of the description. Accuracy refers to the extent to which a description is in accordance with reality.

For companies, access to data has never been better (Wagner, 2002; Shapiro, 2007). Notwithstanding, it is also a fact that in general, the complexity of the business environment faced by most companies has increased significantly during the last few decades. For instance, globalization forces such as supply chaining and outsourcing have been, and still are, "flattening" the world (Friedman, 2004). Most of these globalization forces are related to new ways of dividing work between companies and consequently affect the companies' inter-organizational collaboration. For companies, this increased complexity of the business environment has generated a need for even more information to be able to handle more complex logistics planning problems.

Within SCM it is relatively well proven that sharing information can be beneficial for such as Vendor Managed Inventory and Efficient Consumer Response, Xu and Dong (2004) refer to Waller et al. (1999) and Xu et al. (2001). Even so, commentators have pointed out that many of the proposed benefits of the SCM concept are questioned, e.g., Disney et al. (2004). Research also shows that the theoretical scope of the SCM concept is scarcely practiced. A survey from 2002 shows that 47% of the companies in the survey only focus on improving their internal integration and not more than 8% try to integrate upstream and downstream in the supply chain, Ballou (2007) refers to Fawcett and Magnan (2002). This picture of the reality is supported by the findings of ELA (2004) and Ministry of Transport and Communication Finland (2006). Besides indicating that SCM integration is still in its infancy, this shows that internal integration is far from settled in most companies. It is not unusual that intra-organizational issues represent a significant hindrance for spreading information within a company, and hence for logistics planning. In particular, this is often a problem for logistics planning since logistics usually is not regarded as a core activity and consequently often becomes regarded as a less important planning task.

Compared to logistics planning which aims to optimize the benefits of one company, logistics planning with an SCM focus is more difficult. The reason is that collecting and sharing information throughout a supply chain brings up new types of problems which represent an increased complexity with regard to information coverage and accuracy. Hence, new types of limitations arise. We will briefly elaborate on some of the reasons for this.

One problem with the practice of SCM is that close relationships and extensive information sharing can lead to high transaction costs due to for instance the risk of opportunism (see Rindfleisch and Heide (1997) for more about the use of transaction costs theory). This is probably one of the main reasons why surveys show that many companies still do not share much information across the supply chain. Such practice in supply chains clearly affects the information coverage that is obtained by the members and functions as a limitation.

Information accuracy or undistorted transfer of information across the supply chain is also an important issue within logistics planning in supply chains. It can almost be regarded as a premise for being able to realize many of the proposed benefits of the SCM concept. However, this is not always easy to obtain. Wagner (2002) claims that with an individual company focus on inventory management, a general problem is that data often are dirty / have an insufficient accuracy. The problem with dirty data is also relevant in an SCM context where both more data and new types of dirty data appear. An example of the latter is the cases where actors in the supply chain try to manipulate demand data, see Xu and Dong (2004). In general, the appearance of dirty data and the need for "washing" require knowledge and resources which not all companies possess. It is clear that for a company, having to wash data or not being able to get accurate data would affect the coverage and accuracy of the information obtained.

Another, although related, challenge is that to collect and forward logistics information spread over a wide range of organizational units to the relevant decision makers in the chain, is still a major challenge. SCM architectures, which also often are trying to offer such services, easily become complex, unwieldy, difficult to alter, and costly (Jessup & Valacich, 2008). According to transaction cost theory, the procurement and maintenance of such architectures could lead to significant adaption, performance evaluation, and safeguarding costs. As a result, financial concerns can form limitations for information collection, see Rindfleisch and Heide (1997).

Extensive data exchange also often results in a problem of information overflow which already is experienced in many organizations. Shapiro (2007, p. 23) claims that a problem currently faced by managers is that "there is an overabundance of transactional data for the purposes of managerial decision making". For example, the recent popularity of RFID has brought with it significant challenges with regard to how to treat the huge amount of data collected (Bose & Lam, 2008).

It is clear that a company's ability to cope with the increased amounts of information could represent limitations with regard to what information coverage and accuracy are obtainable.

All in all, access to information has never been better. However, in line with the rapid development of the business environment, the criteria for what are satisfying information coverage and accuracy change accordingly. The result is that enterprises constantly meet hindrances which, at least in the short run, represent limitations.

## Limited Problem-Solving Capability

Many logistics planning problems are not solvable at all, or they are unsolvable within the given time frame. Not all companies have capability, or prioritize the use of resources to solve those that are solvable. Finally, not all problems are solved correctly. The first issue is related to technical limitation while the second issue usually is related to financial concerns. The last issue is often related to aspects of behavioural decision making.

How difficult a problem is to solve is closely related to its complexity. As for information availability, solving capability of logistics planning problems has been strongly influenced by the increased complexity of the business environment and the technological development within computer science. A result of this is that logistics problems have become gradually more complex, but at the same time it has become possible to solve more complex problems. See Shapiro (2007) for a broad and updated collection of the basic optimization methods which are commonly in use today.

Notwithstanding, limits still exist since many logistics problems remain unsolved. It is outside the scope of this article to describe all such problems. Instead, we focus on what we consider to be the main contributors to today's limitations in problem solving.

In OR/MS Today (2007), a series of essays on the topic "Great Unsolved Problems in O.R." has been published. These essays demonstrate some of the future challenges within this field and therefore also implicitly some of today's limitations in logistics planning.

One of the important challenges mentioned concerns how uncertainty is treated in decision support (Greenberg, 2007). Even though the risk of serious sub-optimality is usually emphasized when simplifying problems to make them solvable, deterministic assumptions are usually considered as necessary to reduce complexity. Very few real-life problems are deterministic and solving such problems with a deterministic approach often leads to solutions which are not optimal (Gal & Greenberg, 1997) or more typically, directly misleading (Wallace, 2000). Sensitivity analysis, which is recommended by most textbooks as a way to overcome this shortcoming, is seriously deficient, see Higle and Wallace (2003) for an outline and worked-out example. The deeper reasons for these difficulties are easier to understand in the light of options theory, see Wallace (2009) for details.

So, on one hand, deterministic logistics planning methods applied in real-life settings often have substantial limitations with regard to the output quality. On the other hand, the possibility to use stochastic methods is limited due to their lack of ability to cope with larger problems. It is clear that these limitations represent severe limitations with regard to solving capability. The severity of the limitations is situation dependent.

One result of this situation is that researchers focus gradually more on the uncertainty aspects, Bonfill et al. (2007) refer to Vin and Ierapetritou (2000), and that methods like Metaheuristics and Simulations are gaining popularity. In particular to combine those two methods is considered a promising approach (Fu, 2007). However, such methods have also shortcomings in the way that it is difficult to assess the exact quality of the output / how far the result diverges from optimality. In fact, defining optimality is far from straightforward. Another approach is to try to understand how optimal solutions to stochastic programs differ from their deterministic counterparts, and use that knowledge to create good solutions containing appropriate options see Lium et al. (2007) for such an example.

Another important challenge for problem solving, as for information availability, is related to the growth of the SCM concept. A core issue in SCM is that business functions should be managed from a supply chain perspective, and a major goal is to optimize on the supply chain level (Mentzer et al., 2001).

To optimize at SCM level is quite different from optimizing at company level since it usually means larger problems and increased uncertainty. Furthermore, to find goals which all the chain members can agree upon, whether decisions should be taken centralized or decentralized, and how to split the supply chain surplus, are challenging and usually quite difficult problems, see Ballou (2007) for more SCM challenges.

Centralized or decentralized decision making is particularly important to consider since the decision will constitute a premise for the choice and development of logistics planning models (Li & Wang, 2006). Li and Wang (2006) claim that much of the quantitative research concerning logistics planning showing that the SCM concept is beneficial has been based on centralized decision making and deterministic assumptions. Furthermore they state that this approach is insufficient since the idea of a centralized decision maker in a supply chain is often unrealistic and because uncertainty should definitively be taken into consideration.

In total, it is clear that the development of logistics planning tools capable of functioning properly in an SCM context is still in an initial phase, and that this represents an evident limitation with regard to solving capability.

As with information collection, problem-solving capability is also affected by human intervention. As mentioned in the Background section, the development of logistics planning tools to support logistics decision making has been guided by a wish to obtain gradually more rational decisions. However, even though there are many logistics planning methods today that in an excellent way support the decision making, most methods / tools still have different kinds of shortcomings and humans are still the designers, choosers and users of the tools. This implies that there is a risk that humans could construct and implement methods that do not solve the problem adequately, or even incorrectly. Shapiro (2007, p. 12): "Unfortunately, because most managers are not modelling experts, they can be deceived about the value of applying systems that translate input data into supply chain plans using mediocre models and methods". Furthermore, the answers from the tools could also be interpreted incorrectly.

A good example of the importance of choosing the right tool is the above discussion about using deterministic methods to solve stochastic problems. Often it is difficult for a logistics manger to recognize a setting / planning situation where deterministic planning tools are unsuitable. It is not an uncommon view that handling uncertainty can be postponed to the operational level, while decisions at the strategic and tactical levels are best handled using expected values and deterministic thinking with fairly simple models. As already discussed, this is completely erroneous, since the ability to handle uncertainty at operational level will depend critically on the strategic and tactical decisions: new machines can only be installed if there is enough room, supply chains often can only be changed at short notice if they are set up with change in mind.

Another important factor is that although the use of logistics planning methods and tools has increased, much problem solving is still done manually. This increases the risk of getting solutions influenced by judgmental biases; e.g., Kahneman et al. (1982). An important aspect is also that unexpected events will occasionally occur, which the tools or methods used are not capable of handling. Thus human interaction will be necessary.

All in all, this means that a company's resources regarding its employees' knowledge about logistics, planning, and decision making under uncertainty are of utmost importance for the level of problem-solving capability which can be obtained. Stated differently; a lack of human capital with regard to these issues could represent a major limitation for a company's solving capability.

## CONCLUSION

Logistics managers need to be able to assess the value of investments in logistics systems. To make investments without such assessments should be avoided. However, to assess the right value is not straightforward and the difficulties increase with the complexity of the logistics problem.

A relevant illustration is that one of the main arguments for the SCM concept is that information sharing is beneficial. However, so far, quantitative "proofs" which show the benefits of the SCM concept are relatively simple. Could it be that the "proofs" provided come from simplified situations with very few limitations, and that this not necessarily needs to be the situation in most supply chains? It might be true that improving information availability, generally speaking, is a good guideline. However, this assumes excess problem-solving capability. We believe that each logistics manager is in a unique situation and that it is important to realize that general statements do not necessarily apply.

Today there are many indications that companies struggle to obtain a sufficient level of solving capability to efficiently utilize the amount of information already available. There are reasons to believe that the inflation with regard to information availability, sooner or later will lead to a boost in the development of solving capability, as requested by Shapiro (2007), due to the complementary relationship of those inputs.

We have proposed a conceptual model for addressing the management of efficient logistics planning by using microeconomics thinking focusing on a rational use of resources. This way, we can balance information collection and problem-solving capability, and relate the level of logistics planning with overcapacity; all seen in the light of the overall productivity.

Our proposed model is consciously kept very general to be applicable to as many logistics planning situations as possible. It can be applied irrespective of whether the logistics planning is carried out by computers or humans. It is applicable for both small and large companies, simple and complex problems and is suitable for analysis of ex-post and ex-ante use of resources.

Further research should include empiric investigations of the practical applicability and usefulness of the proposed model. In particular, implementation issues should be addressed.

One interesting approach would be to reveal the relevant logistics cost parameters and compare across companies and businesses. Results could be used by individual companies and businesses to move towards a more efficient allocation of resources. A strongly related research opportunity is to develop a more detailed framework for managerial support assessing the costs and benefits of logistics planning in accordance with our proposed model. Additionally, our elaboration on limitations in this article is by no means complete, and research to reveal and map the limitation would be beneficial.

## ACKNOWLEDGMENT

The authors would like to acknowledge the contributions from the Norwegian oil company StatoilHydro ASA which have contributed significantly to this paper. We would also like to thank Stewart Clark at the Norwegian University of Science and Technology for reading and giving valuable comments to this article.

## REFERENCES

Ballou, R. H. (2004). *Business logistics management: Planning, organizing, and controlling the supply chain* (5th ed.). Upper Saddle River, NJ: Pearson.

Ballou, R. H. (2007). The evolution and future of logistics and supply chain management. *European Business Review, 19*(4), 332–348. doi:10.1108/09555340710760152

Bonfill, A., Espuña, A., & Puigjaner, L. (2007). Decision support framework for coordinated production and transport scheduling in SCM. *Computers & Chemical Engineering*. doi:. doi:10.1016/j.compchemeng.2007.04.020

Bose, I., & Lam, C. W. (2008). Facing the challenges of RFID data management. *International Journal of Information Systems and Supply Chain Management, 1*(4), 1–19.

Chopra, S., & Meindl, P. (2001). *Supply chain management – strategy, planning and operation*. Upper Saddle River, NJ: Prentice Hall.

Christopher, M. (1998). *Logistics and supply chain management* (2nd ed.). Essex, UK: Pearson.

Closs, D. J., & Goldsby, T. J. (1997). Information technology influences on world class logistics capability. *International Journal of Physical Distribution & Materials Management, 27*(2), 4–17.

Closs, D. J., & Savitskie, K. (2003). Internal and external logistics information technology integration. *The International Journal of Logistics Management, 14*(1), 63–76. doi:10.1108/09574090310806549

CSCMP. (2007). *Council of Supply Chain Management Professionals' home page*. Retrieved November 1, 2007, from http://cscmp.org/

Culliton, J. W., Lewis, H. T., & Steele, J. D. (1956). *The role of air freight in physical distribution*. Boston: Harvard University.

Disney, S. M., Naim, M. M., & Potter, A. (2004). Assessing the impact of e-business on supply chain dynamics. *International Journal of Production Economics*, *89*(2), 109–118. doi:10.1016/S0925-5273(02)00464-4

Drucker, P. (1962, April). The economy's dark continent. *Fortune Magazine*, 103.

Dudley, L., & Lasserre, P. (1998). Information as a substitute for inventories. *European Economic Review*, *33*(1), 67–88. doi:10.1016/0014-2921(89)90038-X

Eaton, B. C., & Eaton, D. F. (1991). *Microeconomics* (2nd ed.). New York: W. H. Freeman and Company.

Erlenkotter, D. (1990). Ford Whitman Harris and the Economic Order Quantity Model of 1913. *Management Science*, *38*(6), 937–846.

European Logistics Association (ELA) & A. T. Kearney Management Consultants. (2004). *Differentiation for Performance. Results of the Fifth Quinquennial European Logistics Study, Excellence in logistics 2003/2004, ELA*. GER.

Feng, C. M., & Yuan, C.-Y. (2006). The impact of information and communication technologies on logistics management. *International Journal of Management*, *23*(4), 909–944.

Friedman, T. L. (2004). *The world is flat*. New York: Farrar, Straus and Giroux.

Fu, M. C. (2007). Are we there yet? The marriage between simulation & optimization. *OR/MS Today, 34*(3), 16-17.

Gal, T., & Greenberg, H. J. (1997). *Advances in sensitivity analysis and parametric programming*. Boston: Kluwer Academic.

Ghiani, G., Laporte, G., & Musmanno, R. (2004). *Introduction to logistics systems planning and control*. New York: John Wiley & Sons.

Greenberg, H. (2007). Representing uncertainty in decision support. *OR/MS Today, 34*(3), 14-16.

Gustin, C. M., Daugherty, P. J., & Stank, T. P. (1995). The effect of information availability on logistics integration. *Journal of Business Logistics*, *16*(1), 1–21.

Hammant, J. (1995). Information technology trends in logistics. *Logistics Information Management*, *8*(1), 32–38. doi:10.1108/09576059510102235

Helo, P., & Szekely, B. (2005). Logistics information systems –An analysis of software solutions for supply chain co-ordination. *Industrial Management + Data Systems, 105*(1), 5-18.

Higle, J., & Wallace, S. W. (2003). Sensitivity analysis and uncertainty in linear programming. *Interfaces*, *33*(4), 53–60. doi:10.1287/inte.33.4.53.16370

Jessup, L., & Valacich, J. (2008). *Information systems today – Managing in the digital world* (3rd ed.). Upper Saddle River, NJ: Pearson/Prentice Hall.

Kahneman, D., & Tversky, A. (1982). *Judgment under uncertainty: Heuristics and biases* (Kahneman, D., Slovic, P., & Tversky, A., Eds.). Cambridge, UK: Cambridge University Press.

Lee, Y. M. (2008). Balancing accuracy of promised ship date and IT cost. *International Journal of Information Systems and Supply Chain Management*, *1*(1), 1–14.

Li, X., & Wang, Q. (2006). Coordination mechanisms of supply chain systems. *European Journal of Operational Research*, *179*(1), 1–16. doi:10.1016/j.ejor.2006.06.023

Lium, A.-G., Crainic, T. G., & Wallace, S. W. (2007). Correlations in stochastic programming: a case from stochastic service network design. *Asia-Pacific Journal of Operational Research*, *24*(2), 161–179. doi:10.1142/S0217595907001206

Lumsden, K. R., & Hulthén, L. A. R. (1998). Outline for a conceptual framework on complexity in logistics systems. In *Proceedings of NOFOMA 98,* Helsinki, Finland.

Mentzer, J. T., DeWitt, W., Keebler, J. S., Min, S., Nix, N. W., & Zacharia, Z. G. (2001). Defining supply chain management. *Journal of Business Logistics*, *22*(2), 1–25.

Ministry of Transportation and Communication. (2006). *Finland state of logistics 2006*. Helsinki, Finland: Edita Publishing Ltd.

Mukhopadhyay, T., & Cooper, R. B. (1993). A microeconomic production assessment of the business value of management information systems: The case of inventory control. *Journal of Management Information Systems*, *10*(1), 33–56.

Nahmias, S. (2005). *Production & operations analysis* (5th ed.). New York: McGraw-Hill/Irwin.

*OR/MS Today*. (2007). *34*(1-6).

Parkin, M. (1997). *Microeconomics* (3rd ed.). Reading, MA: Addison-Wesley.

Rindfleisch, A., & Heide, J. B. (1997). Transaction Cost Analysis: past, present and future applications. *Journal of Marketing*, *61*(4), 30–54. doi:10.2307/1252085

SAP. (2007). *SAP's homepage*. Retrieved November 5, 2007 from http://www.sap.com/solutions/business-suite/scm/index.epx

Shao, B. B. M., & Lin, W. T. (2001). Measuring the value of information technology in technical efficiency with stochastic production frontiers. *Information and Software Technology*, *43*(7), 447–456. doi:10.1016/S0950-5849(01)00150-1

Shapiro, J. F. (2007). *Modeling the supply chain* (2nd ed.). Belmont, CA: Thomson Higher Education.

Silver, E. A., Pyke, D. F., & Peterson, R. (1998). *Inventory management and production planning and scheduling* (3rd ed.). New York: Wiley & Sons.

Simchi-Levi, D., Kaminsky, P., & Simchi-Levi, E. (2003). *Managing the supply chain: The definitive guide for the business professional*. New York: McGraw-Hill.

Simon, H. A. (1957). *Administrative behavior* (4th ed.). New York: Free Press.

Wagner, H. (2002). And then there were none. *Operations Research, 50*(1), 217–226. doi:10.1287/opre.50.1.217.17777

Wallace, S. W. (2000). Decision making under uncertainty: Is sensitivity analysis of any use? *Operations Research*, *48*(1), 20–25. doi:10.1287/opre.48.1.20.12441

Wallace, S. W. (2009). *Stochastic programming and the option of doing it differently*. Operations Research.

West, L. A. Jr. (1994). Researching the costs of information systems. *Journal of Management Inquiry*, *11*(2), 75–108.

West, L. A. Jr, & Courtney, J. F. (1993). The information problems in organizations: a research model for the value of information and information systems. *Decision Sciences*, *24*(2), 229–252. doi:10.1111/j.1540-5915.1993.tb00473.x

Woodbury, R. S. (1960). The legend of Eli Whitney and interchangeable parts. *Technology and Culture*, *1*(3), 235–253. doi:10.2307/3101392

Xu, K., & Dong, Y. (2004). Information gaming in demand collaboration and supply chain performance. *Journal of Business Logistics*, *25*(1), 121–144.

*This work was previously published in International Journal of Information Systems and Supply Chain Management, Volume 3, Issue 3, edited by John Wang, pp. 1-17, copyright 2010 by IGI Publishing (an imprint of IGI Global)*

# Chapter 16
# An Empirical Investigation of Third Party Logistics Providers in Thailand:
## Barriers, Motivation and Usage of Information Technologies

**Duangpun Kritchanchai**
*Mahidol University, Thailand*

**Albert Wee Kwan Tan**
*National University of Singapore, Singapore*

**Peter Hosie**
*University of Wollongong, Dubai, United Arab Emirates*

## ABSTRACT

*Third Party Logistics (3PL) in Asia emerged as an important trend in logistical management and Thailand continues to develop in this service rapidly. While a great deal has been written about the dissemination of information technology (IT), few empirical investigations address the use of IT in relation to 3PLs in Thailand. In this article, the authors use an empirical study to investigate the profiles of 3PLs in Thailand and their company strategies for providing logistics service and use of IT. Survey results show that Thailand's 3PL companies must expend more effort to strengthen basic IT and infrastructure to enhance competitiveness. IT capabilities in Thailand are increasing rapidly and its effective adoption has the potential to significantly enhance the competitiveness of small 3PLs. Still many barriers exist to the successful adoption of IT by these providers. Given the importance of such companies in supply chain management, these issues must be fully understood.*

DOI: 10.4018/978-1-4666-0918-1.ch016

## INTRODUCTION

A Third Party Logistics (3PL) service represents more than a subcontracting or outsourcing service. Typically, subcontracting or outsourcing covers only one product (or a family of products) or one function that is produced by an outside vendor. In contrast, the functions performed by 3PL providers cut across multiple logistics functions. These functions can encompass the entire logistics process or, more commonly, selected activities within that process.

Undoubtedly, the emergence of 3PL has become an important trend in logistical management during the 1990s. While estimates vary concerning the size of 3PL, it is evident that opportunities for 3PL services will continue to grow. The Asian logistics market is poised for robust expansion with its annual market growth rates projected at 15% in Asia (TLI, 2003). Such high growth rates in Asia indicate the high potential of supply chain development in this region, particularly in Thailand.

Thailand's 3PL industry is developing rapidly and its service is gaining demand by shippers. Such development is strongly encouraged by the government in Thailand, which aims to develop the country into a premium integrated transport and logistics hub. Leading edge terminal facilities combined with logistics management competence are strengthening Thailand's position as an international trading, automotive and business hub.

There are a number of factors driving the growth of 3PL. Declining margins and a tougher competitive environment, together with the recent positive attitude towards outsourcing and focusing on core activities, are regarded as the strongest drivers for the emergence of the 3PL industry. Many companies are outsourcing their logistics operations to reduce costs, improve products, shorten lead time and enhance competitiveness for their supply chain (Mentzer et al., 2001). Thus, companies need to link electronically with suppliers or logistics providers to forge sophisticated inter functional connections with key customers (Narasimhan & Jayaram, 1998).

Information Technology (IT) has become an important tool for implementing supply chain management (SCM). It provides many essential applications in improving performance in supply chain. Advances in IT have made it possible for companies to develop and maintain the flexibility to respond quickly to changing demands and conditions. Besides improving supply chain efficiency, IT including the application of hardware, software and networks, can enhance information flow and facilitate decision making in supply chain and logistics operations. The most dramatic and potentially powerful uses of IT involve networks spanning any boundaries that are capable of significantly enhancing the productivity, flexibility, and competitiveness of many companies (Jun, Cai, & Kim, 2008).

A recent issue of the McKinsey Quarterly (Kanakamedala, Ramsdell, & Srivatsan, 2003) reported that some companies making heavy investments in SCM information systems actually performed worse than companies who did not invest in the technology. This finding conflicts with the received wisdom that technological investment in SCM will increase efficiency. Also, it contradicts what managers have been repeatedly advised about the likelihood that IT will improve logistics operations. In order to successfully adopt IT, companies must first streamline supply chain their processes and fully understand how to leverage technology to improve its performance.

Although IT in large 3PLs has been widely investigated (Larson & Gammelgaard, 2001; van Hoek, 2000) there is still a shortage of research in the field of small 3PLs. Minimal empirical investigation has been undertaken to analyze the adoption of IT by these companies. An empirical investigation was warranted given the limited quantitative evidence available on the usage of IT by small 3PL companies in Thailand.

The purpose of this study is to investigate the status of IT in the Thailand logistics industry and identify IT usage trends in the region. It is undertaken to achieve three specific objectives:

1. To identify the profile of 3PL companies, their markets and services offered in Thailand
2. To identify the 3PL company strategies and the performance measurements used by them
3. To understand the types of IT used by 3PL, their motivations and barriers for deployment

Following this introduction, a brief literature review of IT usage by 3PL provides is undertaken. The subsequent sections provide a profile of the respondent and their markets, followed by an analysis of IT usage by these companies. A mail survey was used to investigate IT usage from a sample size of 72 3PL companies throughout Thailand. The main research findings from the data collected are then presented. The concluding section discusses the managerial implication on the adoption of IT for small 3PLs.

## LITERATURE REVIEW

3PL companies provide outsourced or logistics services to companies for an aspect, or sometimes all of their SCM functions. 3PL logistics providers typically specialize in integrated warehousing and transportation. These services are capable of being systematically scaled and customized to clients needs in response to market conditions which determine the demands and delivery service requirements for products and materials. Four categories of 3PL providers are identified by Hertz and Alfredsson (2003):

1. Standard 3PL providers are the most basic form of service offered. Activities performed include the most basic functions of logistics 'picking and packing', warehousing, and distribution. 3PL functions are not the main activity of these companies.
2. Service developers 3PL providers offer customers advanced value added services in the form of: 'tracking and tracing', cross docking, specific packaging, or a unique security system. A strong IT platform, combined with a focus on economies of scale and scope, enable this type of 3PL provider to undertake these tasks.
3. Customer adapters are a form of 3PL provider which responds to requests from a customer to take complete control of the company's logistics activities. New logistics services are not provided or substantially improved and the customer base is typically quite small.
4. Customer developers are the highest level 3PL provision with respect to its processes and activities. A 3PL provider integrates its activities with the customer and controls the entire logistics function. Extensive and detailed tasks are performed for a few providers.

A variety of benefits and risks in relation to 3PL have been reported in the literature. These can be classified as strategy, finance and operations related. Outsourcing non-strategic activities enables companies to focus on their core competences and exploit external logistical expertise (Sink & Langley, 1997). 3PL providers can also contribute to improved customer satisfaction and provide access to international distribution networks (Bask, 2001). The most often cited risks are associated with loss of control over the logistics function and loss of in house capability and customer contact (Ellram & Cooper, 1990). However, usually shippers employ a mixed strategy regarding logistics and retain important logistics activities (e.g, order management) in house (Wilding & Juriado, 2004). Users of 3PL reported enhanced flexibility with regard to market (investments) and demand (volume flexibility) changes. Lack of responsiveness to customer needs is also cited as a problem of outsourcing (van Damme & Ploos van Amstel, 1996).

The literature focuses on the demand side of 3PL; a large number of studies have investigated the extent of 3PL usage across specific countries/regions and industries. A series of annual surveys conducted in the USA by Lieb and colleagues (Lieb, 1992; Lieb & Bentz, 2004, 2005; Lieb et al., 1993; Lieb & Miller, 2002; Lieb & Randall, 1996) is a well known example of such an approach. The main issues examined by such studies include services used, usage rate, contract renewal rates, outsourcing costs and geographical spread of services (Lau & Ma, 2008). Generally speaking, findings indicate the prominence of transport, warehouse and administration related (e.g,, freight payment) services and confirm the continuing growth of logistics outsourcing (Ashenbaum et al., 2005; Lieb & Bentz, 2005; Murphy & Poist, 1998).

Research regarding 3PL usage also includes experience from specific countries or industries. Country specific studies stress the prominence of transport and warehousing services and also identify other activities with growth potential. Examples include: Australia (Dapiran et al., 1996; Sohal et al., 2002); China (Hong et al., 2004); Malaysia (Sohail & Sohal, 2003); Mexico–US border (Maltz et al., 1993); New Zealand (Sankaran et al., 2002); and Singapore (Bhatnagar et al., 1999).

IT systems are increasingly being used to offer real time information to clients in logistics industry to enhance visibility for supply network members (Tan, 1999; Lewis & Talalayersky, 2000; Piplani et al., 2004; Sauvage, 2003). Subsequently, this has a massive impact on all business areas, especially transportation and distribution (Lewis & Talalayevsky 2000; Milligan 2000). The reasons for the exponential growth in the use of IT are summarized below:

- Significant reduction in assets, such as inventories and equipment
- More effective management of information, products, and cash flows among supply chain partners; and
- Dramatic reduction in the cost of IT throughout the 1900s and from 2000

The number and types of technology that exist to facilitate logistics operations continues to grow at a rapid pace. However, certain technologies have become critical components of the logistics and SCM functions. For example, in the area of traffic management, companies use the web or the Internet for communicating by email, tracking and tracing shipments, obtaining industry and carrier news and information, and conducting database searches. More than 50% of all transactions between carriers and their customers are estimated to be dealt with over the Internet (Stock and Lambert 2001). It is fast becoming the primary interface mechanism for Business-to-Business (B2B) and Business-to-Customer (B2C) transactions because transportation companies can quickly offer broad service and market reach capabilities.

The Internet and related web services have become strategic tools in determining success of SCMs. Companies use Internet sites to provide potential customers with product and price information. Many carriers have developed Internet sites to allow customers to track their shipments. In addition, companies use Internet based systems to share demand and production forecasts. Issues of capacity and security are challenges remaining with the use of the Internet for transportation and distribution transactions.

Another example would be the logistics information systems which can be categorized into four groups (Haverly & Whelan 1996):

1. **Transactional analysis:** Allows management to monitor costs and service by providing historical reporting of key performance indicators, such as carrier performance, shipping modes, traffic lane use, premium freight usage, and backhauls;
2. **Traffic routing and scheduling:** Provides features such as the sequence and timing of vehicle stops, route determinations, ship-

ping paperwork preparation, and vehicle availability;

3. **Freight rate maintenance and auditing:** Maintains a database of freight rates used to rate shipments or to perform freight bill auditing; and,
4. **Vehicle maintenance:** Features commonly provided by these packages include vehicle maintenance scheduling and reporting.

Depending on the size of the 3PL companies, the level of technology use will vary between and within companies (Neureuther & Kenyon, 2008). Despite such variation, it is clear that the use of technology is expanding at a rapid pace in the area of transportation and distribution, and it will continue to grow well into the future (Sahay & Mohan 2006; Cabdoi 2003; Frost & Sullivan 2003). Despite the fact that much has been written about the dissemination of IT, there is still a shortage of quality research in the field of small 3PLs with little empirical investigation into the usage of IT by them. This article is intended to this growing contribute knowledge base and suggests some possible research directions.

## RESEARCH METHODOLOGY

Five hundred (500) questionnaires were selected randomly from a logistics association database and were sent to their members by mail with a pre paid envelope attached. There were 72 valid responses, collected making a response rate of 14%. Some of the respondents indicated they are operating in more than one core logistics business (three key areas namely; Transportation, Warehousing or Freight Forwarding) and thus resulted in the total number of responses exceeding the number of respondents as shown in Table 1.

*Table 1. Percentage of core logistics businesses operated (n = 72)*

| Core logistics business | No. of responses | Percentage |
|---|---|---|
| Transportation | 32 | 37.7 |
| Warehousing/ Distribution | 17 | 20.0 |
| Freight Forwarding | 36 | 42.3 |
| **Total** | **85** | **100** |

## PROFILE OF 3PL COMPANIES

### Turnover and Staff Size of 3PL Companies

Seventy one percent (71%) of the respondents had an operating revenue of less than 25 million in Thai baht (US$794,000) in year 2005. Close to 28% of the respondents have operating revenue of between 26 to 500 million in Thai baht (US$825,000 to US$15.8 million) and only 2% has more than 500 million baht of operating revenue. Thus, the respondents are skewed towards the small and medium size enterprise (SME) with exception of a few large companies. The majority of respondents (90%) were local companies with the rest emanating from the USA, Europe and Asia.

As shown in Table 2, close to 70% of respondents have less than 25 employees while 25% had 25 to 100 employees. Only a small percentage of the respondents have more than a 100 employees. Interestingly, 37% do not have any employee for IT and close to 61% have only 1 to 10 IT employees. This could be due to the fact that most of the respondents are SMEs.

### 3PL Services

From Figure 1, it is observed that almost all of 3PL companies attempt to provide as wide a range of 3PL services to their clients as possible. This can be seen from the numerous types of 3PL services offered by the companies. The most read-

*Table 2. Staff size of 3PL companies*

| Employment | Percentage |
|---|---|
| **Total number of employees** | |
| 1 – 25 employees | 69.6 |
| 26 – 100 employees | 24.6 |
| 101 – 200 employees | 4.4 |
| More than 200 employees | 1.4 |
| **Number of employees in IT function** | |
| None | |
| 1 – 3 employees | 37.7 |
| 4 – 10 employees | 40.5 |
| 11 – 20 employees | 20.1 |
| More than 20 employees | 1.4 |

ily available 3PL services offered include basis services such as: customer service (81%), traffic and transportation (77%), packaging (40%), logistics communication (36%), and warehousing and storage (29%). In contrast, the least readily available 3PL services on offer are: parts and service support (4%), order processing (5%), demand forecasting (6%), plant and warehouse site selection (8%) and material handling (9%).

## 3PL Industries and Customers

The range of industries that the 3PL companies are currently serving is shown in Figure 2. As observed, mechanical, chemical, textiles, electronics, consumer goods and beverages are the major industrial sectors that most 3PL companies render their services to. Paper and Health logistics are the minority industrial sectors serviced by these 3PL companies.

Finally in terms of customer composition, Figure 3 indicates that majority of the companies (50%) served between 20 to 99 customers while only 8% of respondents are serving more than 100 customers. This confirms the trend for 3PL to dedicate their services to a few customers and with the ultimate goal of eventually transforming themselves into 4PLs.

## COMPANY STRATEGIES AND METRICS FOR LOGISTICS SERVICE

The top three strategies adopted by the companies are: time based logistics (79%), logistics performance (79%) and strategic alignment of IT (72%) respectively as shown in Figure 4. In contrast, 49% does not agree on establishing partnership with rewards and risks. The current trend clearly favors players with strong solutions background, backed by strong process management and IT support infrastructure.

Fifty two percent (52%) of respondents have indicated that they measure their logistics performance. The top three logistics metrics are: on time delivery (74%), accurate, complete and damage free delivery (59%) and data accuracy (39%) respectively as shown in Figure 5. In contrast, order cycle time is not commonly measured by these companies.

## IT MOTIVATIONS, BARRIERS AND USAGE

Reacting to the trend towards greater use of e-commerce, most companies see the need to leverage IT, e-commerce and infrastructures. The top three prime motivators for adopting IT by the respondents are: reducing data entry errors (critical: 28%), decreasing labour cost (critical: 28%) and reducing order cycle time (critical: 28%) respectively as shown Figure 6. This could be due to growing pressure from customers for 3PL to install information systems to support their logistics services in order to increase data accuracy, cut costs, increase visibility and shorten delivery time.

The top three barriers to usage of IT for 3PL companies are: lack of education (agree: 71%), integration with legacy system (agree: 65%) and unaware of new technology (agree: 58%) respectively as shown in Figure 7. In contrast, 68% of the companies do not agree that IT is not a neces-

*Figure 1. Types of logistics services provided*

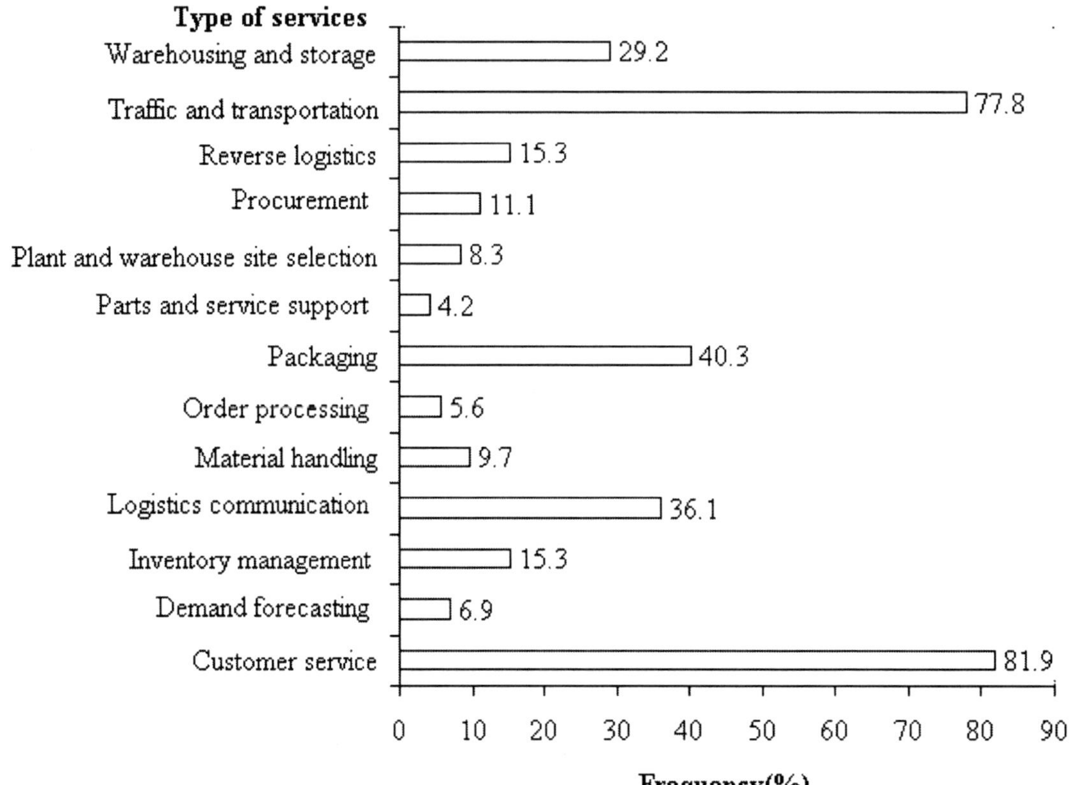


sity. Most organizations are treating IT as an essential tool for productivity.

In term of implementation of basic IT applications for transactional activities, 71% have implemented or are implementing financial management system, 44% have implemented or are implementing inventory management system and 40% have implemented or are implementing purchasing management systems as shown in Figure 8. The least implemented system is the production control system.

In the case of IT applications specializing on logistics operations, 72% have implemented or are implementing transportation management system, 43% have implemented or are implementing distribution resource planning system and 41% have implemented or are implementing warehousing management system as shown in Figure 9. It is not surprising that transportation management systems are the most commonly implemented system since 77% of the respondents are offering traffic and transportation services as shown in Figure 9.

In order to find out how ready the 3PL companies in Thailand are in embracing e-commerce, the respondents are asked to indicate their range of e-commerce services that they are currently offering or planning to offer to their clients. The majority of companies (80%) use e-commerce to monitor their delivery service, 40% uses e-commerce to track and trace service, and 30% use e-commerce to check stock status and to provide pre alert services as shown in Figure 10.

Respondents were asked to indicate their achievements from the use of their IT systems. The top three achievements indicated are: improve data quality (agree: 80%), cost reduced (agree: 75%) and more reliable delivery (agree: 75%) as

*Figure 2. Industries served by 3PL companies*

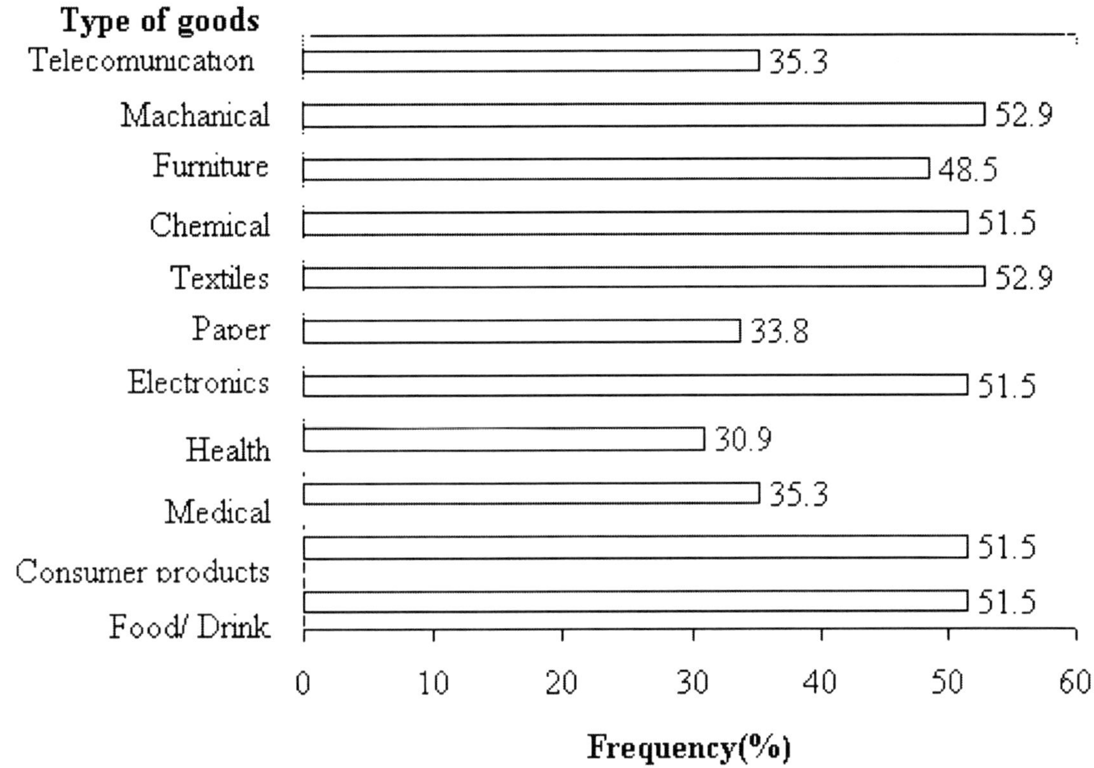


*Figure 3. Customer composition for each 3PL*

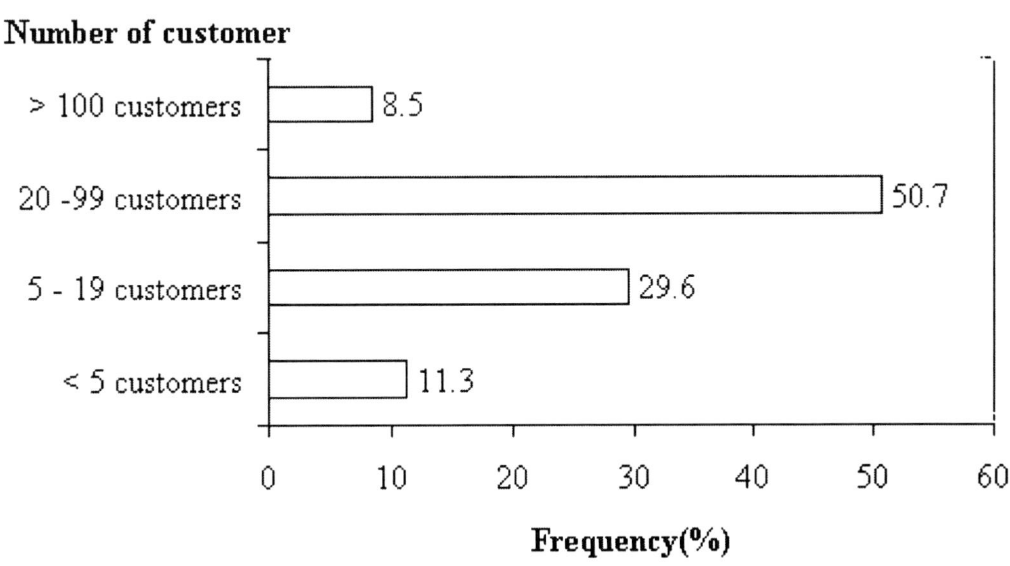

*Figure 4. Strategies adopted by companies*

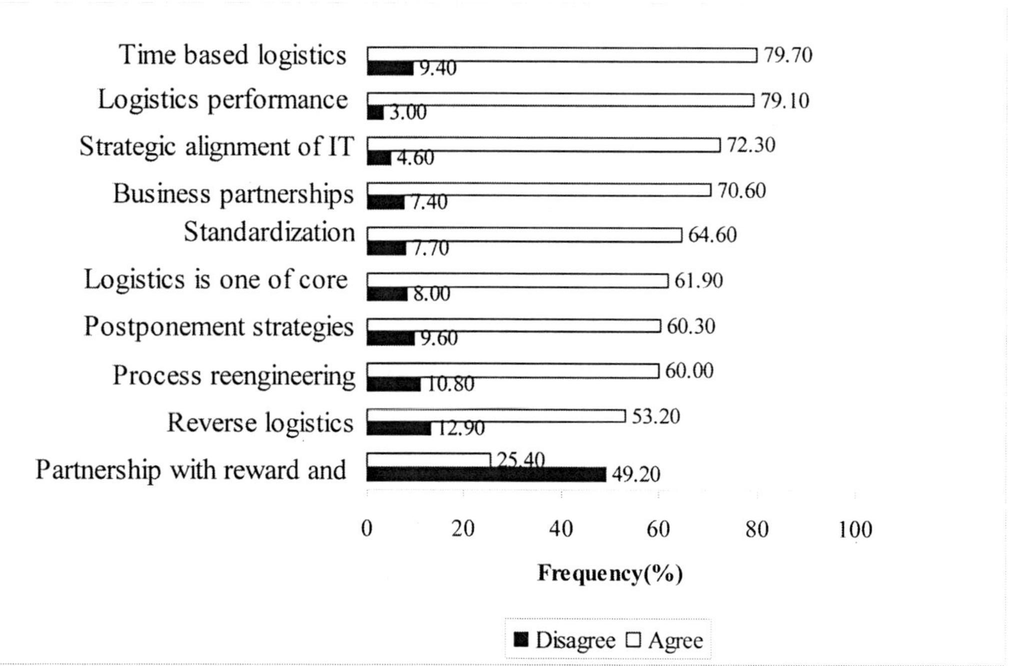


shown in Figure 11. With information sharing and exchange using the Internet, the data quality is expected to improve as compared to manual systems.

According to Lewis and Talalayersky (2000) most 3PL companies readily adopt IT but fail to realize that the right people and processes are ultimately the key critical success factors. The logistics industry is a service industry and therefore people are the crucial competitive factor since it is people who utilize IT tools. Figure 12 shows that top three important IT skills for personnel in 3PL companies to support their logistics operations. They are namely: word processing (agree: 78%), spreadsheet (agree: 78%) and operating system (OS) (agree: 78%). It is no surprise that word processing is listed as the most important IT skill since 3PL services will need to prepare a lot of documents for their shippers for export and import declaration. Spreadsheets are considered the next most important skill as the 3PL needs to consolidate information from many different IT systems in order to provide accurate information on the status of shipment details to shippers.

## SUMMARY OF FINDINGS

This research has provided an overview of IT utilization by the 3PL companies in Thailand. The readiness of industry in Thailand to adopt IT in SCM is moderate compared to other developed countries (Singapore, USA, etc). These findings show that IT is an essential tool in reducing operation cost, improving quality of data, and increasing the consistency of company's delivery. In fact, most of the respondents disagreed that IT is not a necessity in today's competitive environment where shippers expect real time information about their inventory and shipment status.

These findings also reveal that 3PL companies in Thailand have been investing in basic transactional and logistics information systems to enhance their business processes. Most of the respondents plan to improve their business processes and their logistics performances by using IT. However, most of the respondents either have no IT staff or only a few IT staffs to assist in the implementation of such systems. Unless they are expecting to outsource the IT implementation to

*Figure 5. Metrics for logistics performance*

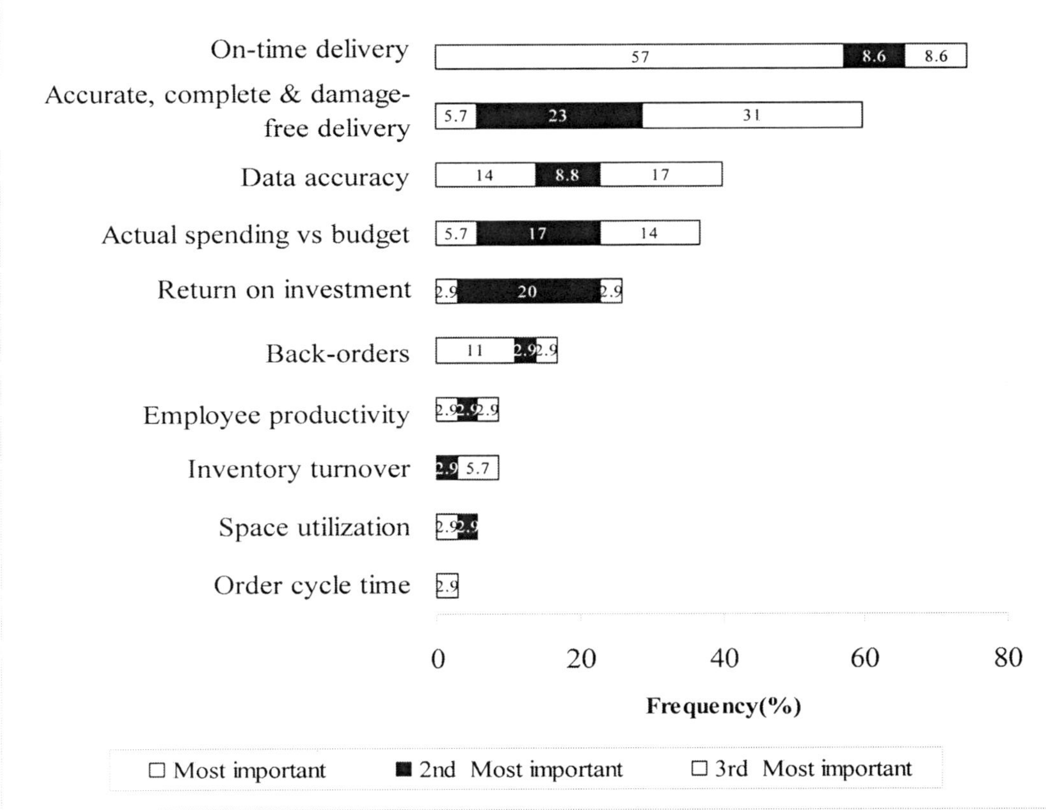


the IT vendor, it is hardly possible for these SME companies to be able to handle such large scale IT implementation projects. Furthermore, with a lack of IT knowledgeable and staffs, it is evident that most of the companies will not be capable of integrating their IT systems with other legacy systems as indicated by them as barriers for IT implementation.

Most logistics companies have several legacy computer systems to serve different needs, such as for order recording, order delivery, and so on. Currently, each client has a separate database that is operated independently of the information systems of other clients. The trend is to migrate to a more advanced and integrated IT system that will give 3PL companies the scope to expand their operations. The process of changing IT systems involves examining the business process and considering process reengineering before deciding on the architecture of the IT system for the companies.

The ultimate plan is to integrate the IT system with the web so that a web based logistics information system can be made available over time for their shippers. Most 3PL companies are in the processes of migrating to a web based information system with a common interface for all their clients. However, this standardization can be accomplished only with the cooperation of all clients.

## CONCLUSION

3PL is growing around the world as more and more corporations prefer to outsource their logistics operations to the 3PL or logistics service providers. The 3PL market in Thailand is highly

*Figure 6. Prime motivators for adopting IT*

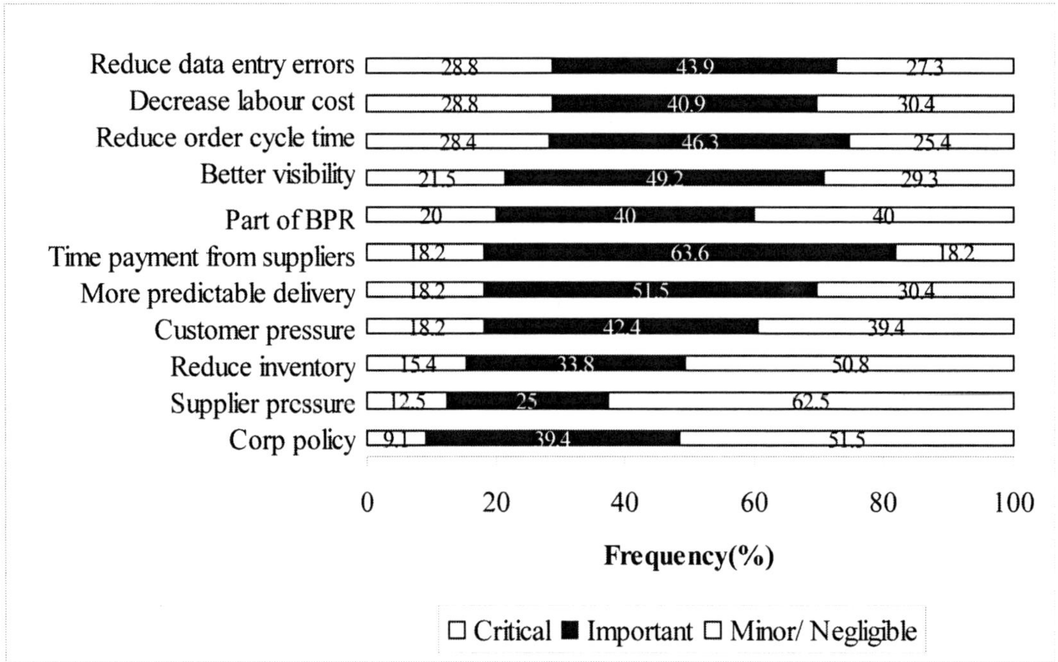


fragmented but there is also a high potential for growth of the market, which was evident from the research. More companies in Thailand are becoming aware of the benefits of 3PL and are outsourcing a part or whole of their logistics related activities to service providers.

This research has provided an overview of 3PL companies, mainly SME and their markets and IT implementation status. The majority of companies surveyed have implemented the basic IT systems but in order to compete with the larger companies, they will need to consider investing more in this area as well as in IT manpower. They should develop an IT system that is flexible and able to accommodate legacy data, Internet and related web based services logistics information systems. Web based information will help to reduce communications barriers such as complex logistics operational systems. Both financial and non financial factors, including tangible and intangible factors, should be considered when justifying investments in IT projects. Such investments should not simply be based on financial performance measures such as ROI.

Currently, 3PL companies in Thailand receive only some support for training and education. Government support is needed to train and educate employees in small logistics companies. In addition, educating the customers on how to use the logistics information system will certainly improve levels of customer satisfaction.

In conclusion, the competitive landscape for small 3PLs is continuously changing to reflect evolving customer requirements and other business pressures. The capability of emerging IT is increasing at a rapid rate and its effective adoption has the potential to significantly enhance the competitiveness of small 3PLs. However, it is clear that many barriers exist to the successful adoption of IT by these providers. Given the importance of such companies in contemporary supply chain configurations, it is important that these issues are fully understood and resolved where possible.

*Figure 7. Barriers to IT usage*

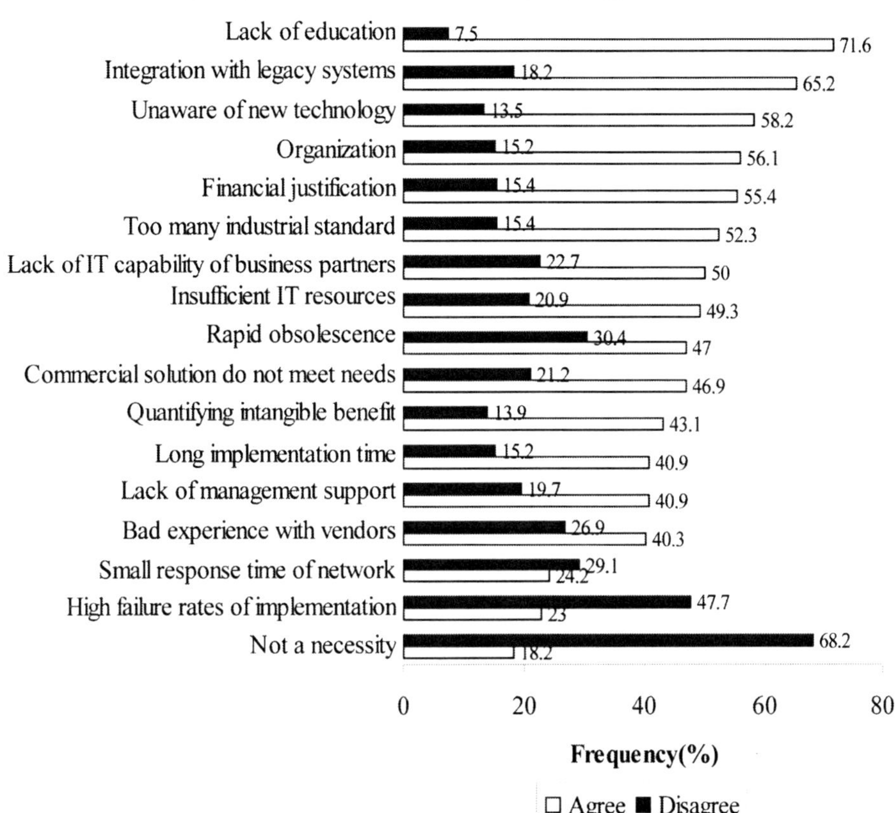


*Figure 8. Implementation of transaction systems in 3PL companies*

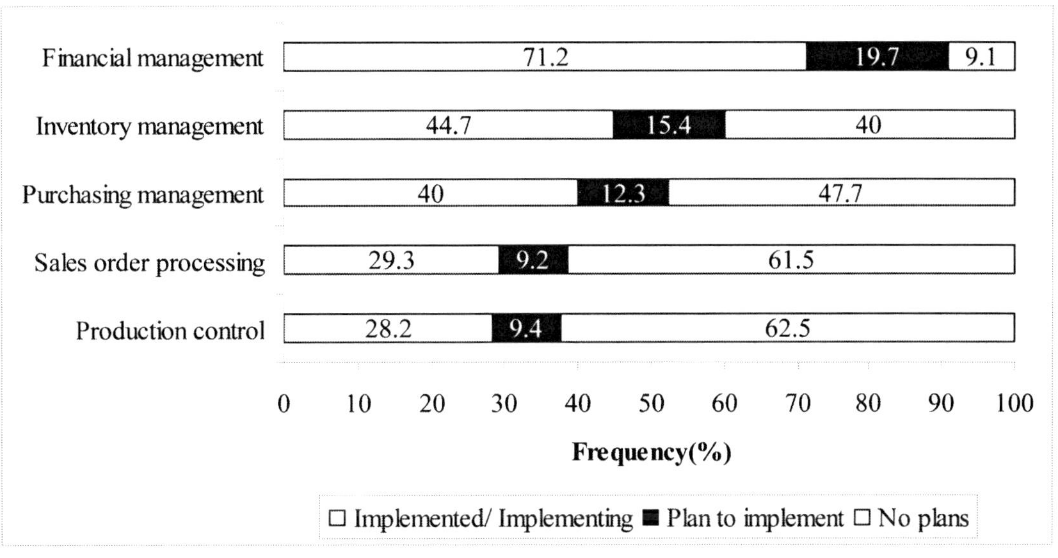

*Figure 9. Implementation of logistics systems in 3PL companies*

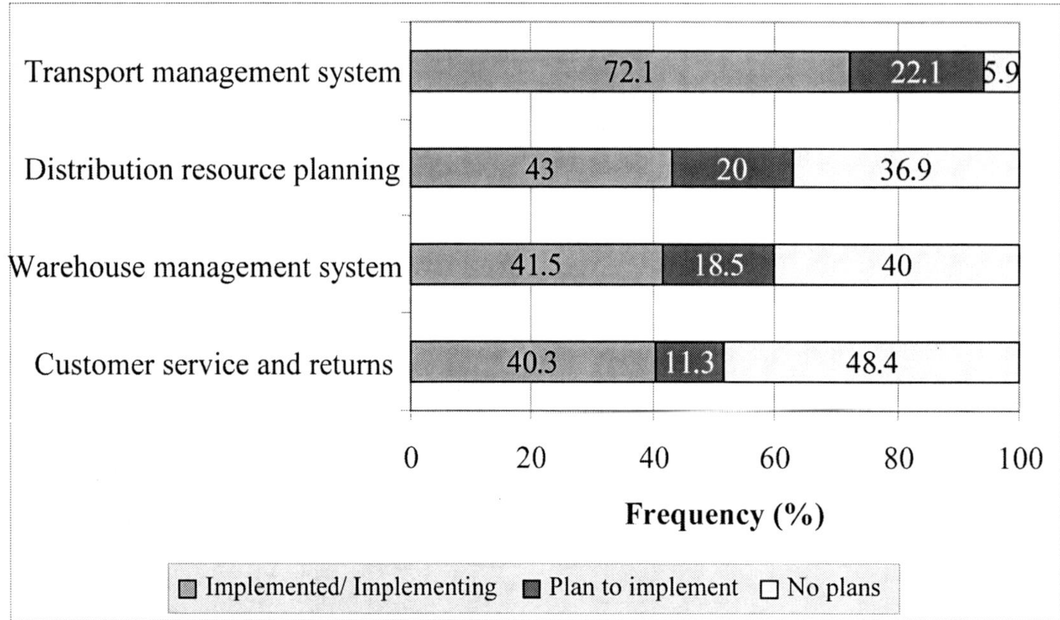


*Figure 10. Services using e-commerce*

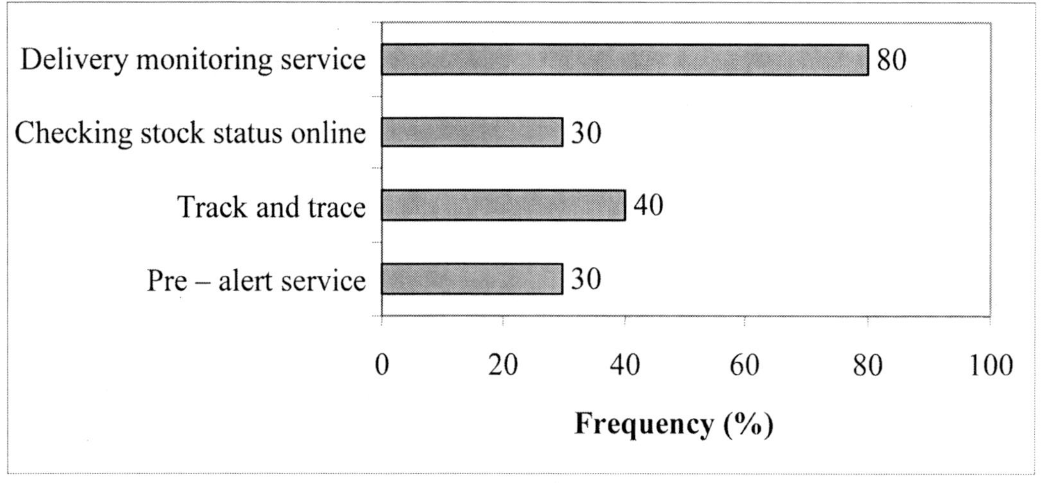

*Figure 11. Company achievements through implementing IT*

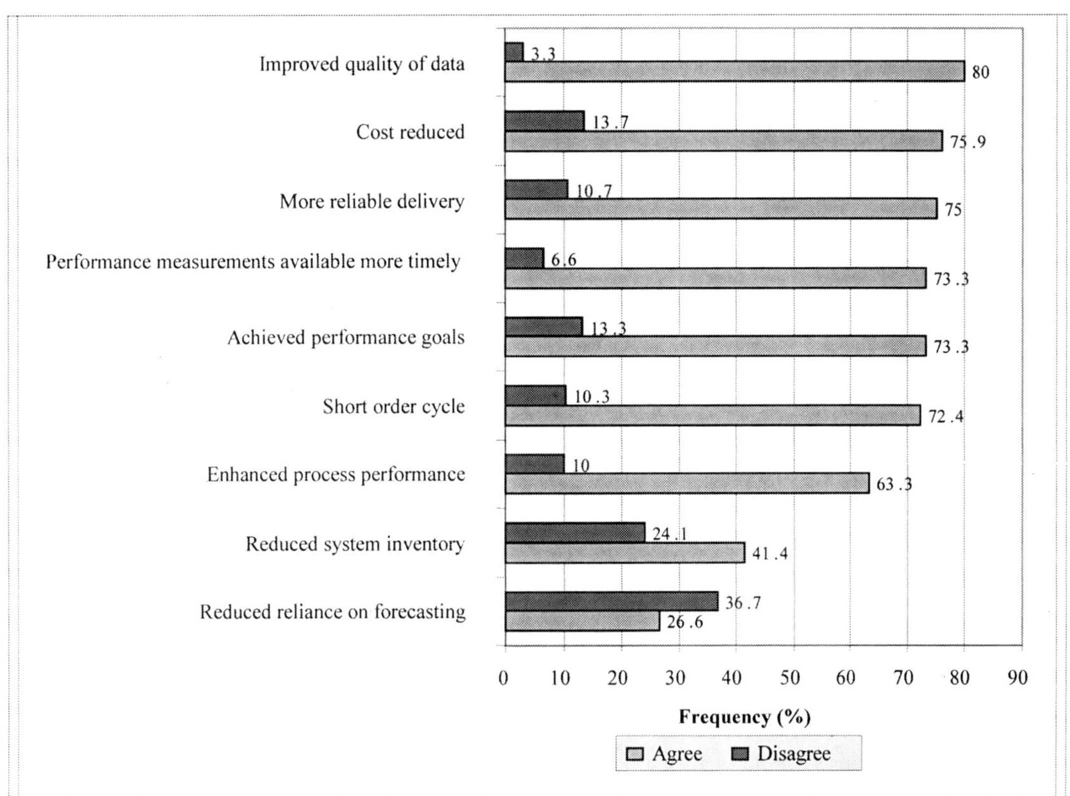


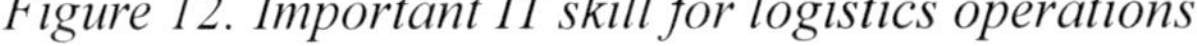

*Figure 12. Important IT skill for logistics operations*

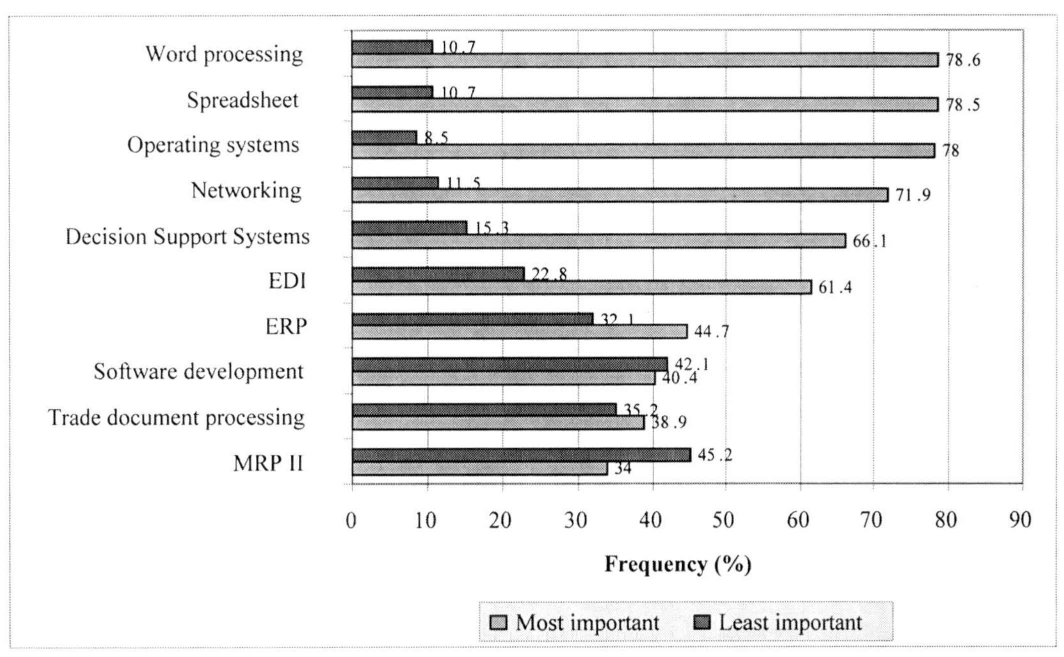

## REFERENCES

Ashenbaum, B., Maltz, A. B., & Rabinovich, E. (2005). Studies of trends in third party logistics usage: What can we conclude? *Transportation Journal*, *44*(3), 39–50.

Bask, A. H. (2001). Relationships between 3PL providers and members of supply chains – a strategic perspective. *Journal of Business and Industrial Marketing*, *16*(6), 470–486. doi:10.1108/EUM0000000006021

Bhatnagar, R., Sohal, A., & Millen, R. A. (1999). Third party logistics services: A Singapore perspective. *International Journal of Physical Distribution and Logistics Management*, *29*(9), 569–587. doi:10.1108/09600039910287529

Cabdoi, C. (2003). *Fourth party logistics market: A European perspective. 1 Dec 2003.* Retrieved March 3, 2009, from http://www.frost.com/prod/servlet/market-insight-top.pag?docid=8341069

Dapiran, P., Lieb, R. C., Millen, R. A., & Sohal, A. (1996). Third party logistics service usage by large Australian firms. *International Journal of Physical Distribution and Logistics Management*, *26*(10), 36–45. doi:10.1108/09600039610150442

Ellram, L. M., & Cooper, M. C. (1990). Supply chain management, partnerships and the shipper-third party relationship. *International Journal of Logistics Management*, *1*(2), 1–10.

Frost & Sullivan. (2004). F*ourth-party logistics: Turning a cost into a value proposition.* Retrieved March 3, 2009, from http//www.ebizq.net/topics/scm/features/3881.html

Hertz, S., & Alfredsson, M. (2003). Strategic development of third party logistics providers. *Industrial Marketing Management*, *32*, 139–149. doi:10.1016/S0019-8501(02)00228-6

Hong, J., Chin, A., & Lin, B. (2004). Logistics outsourcing by manufacturers in China: A survey of the industry. *Transportation Journal*, *43*(1), 17–25.

Jun, M., Cai, S., & Kim, D. (2008). The strategic implications of e-network integration and transformation paths for synchronizing supply chains. *International Journal of Information Systems and Supply Chain Management*, *1*(4), 39–59.

Kanakamedala, K., Ramsdell, G., & Srivatsan, V. (2003). Getting supply chain software right. *The McKinsey Quarterly*.

Larson, P., & Gammelgaard, B. (2001). Logistics in Denmark: A survey of the industry. *International Journal of Logistics: Research and Applications*, *4*(2), 191–205.

Lau, Kwok, H., & Ma, W. L. (2008). A supplementary framework for evaluation of integrated logistics service provider. *International Journal of Information Systems and Supply Chain Management*, *1*(3), 49–69.

Lewis, I., & Talalayersky, A. (2000). Third party logistics: Leveraging information technology. *Journal of Business Logistics*, *21*(2), 173–185.

Lieb, R. C. (1992). The use of third-party logistics services by large American manufacturers. *Journal of Business Logistics*, *13*(2), 29–42.

Lieb, R. C., & Bentz, B. A. (2004). The use of 3PL services by large American manufacturers: The 2003 survey. *Transportation Journal*, *43*(3), 24–33.

Lieb, R. C., & Bentz, B. A. (2005). The use of 3PL services by large American manufacturers: The 2004 survey. *Transportation Journal*, *44*(2), 5–15.

Lieb, R. C., Millen, R. A., & Van Wassenhove, L. N. (1993). Third party logistics: A comparison of experienced American and European manufacturers. *International Journal of Physical Distribution and Logistics Management*, *23*(6), 35–44. doi:10.1108/09600039310044894

Lieb, R. C., & Miller, J. (2002). The use of 3PL services by large American manufacturers: The 2000 survey. *International Journal of Logistics: Research and Applications*, *5*(1), 1–12.

Lieb, R. C., & Randall, H. L. (1996). A comparison of the use of third-party logistics services by large American manufacturers, 1991, 1994 and 1995. *Journal of Business Logistics*, *17*(1), 305–320.

Maltz, A. B., Riley, L., & Boberg, K. (1993). Purchasing logistics in the US-Mexico transborder situation: Logistics outsourcing in the US-Mexico co-production. *International Journal of Physical Distribution and Logistics Management*, *23*(8), 46–55. doi:10.1108/09600039310049835

Mentzer, J., Dewitt, W., & Keebler, J. (2001). Defining supply chain management. *Journal of Business Logistics*, *22*(2), 1–25.

Murphy, P. R., & Poist, R. F. (1998). Third-party logistics usage: An assessment of propositions based on previous research. *Transportation Journal*, *37*(4), 26–35.

Narasimhan, R., & Jayaram, J. (1998). Causal linkages in supply chain management: An exploratory study of North American manufacturing firms. *Decision Sciences*, *29*(3), 579–605. doi:10.1111/j.1540-5915.1998.tb01355.x

Neureuther, B. D., & Kenyon, G. N. (2008). The impact of information technologies on the US beef industry's supply chain. *International Journal of Information Systems and Supply Chain Management*, *1*(1), 48–65.

Sahay, B. S., & Mohan, R. (2006). 3PL practices: An Indian perspective. *International Journal of Physical Distribution and Logistics Management*, *36*(9), 666. doi:10.1108/09600030610710845

Sankaran, J., Mun, D., & Charman, Z. (2002). Effective logistics outsourcing in New Zealand: An inductive empirical investigation. *International Journal of Physical Distribution and Logistics Management*, *32*(8), 682–702. doi:10.1108/09600030210444926

Sauvage, T. (2003). The relationship between technology and logistics third-party providers. *International Journal of Physical Distribution and Logistics Management*, *33*(3), 236–253. doi:10.1108/09600030310471989

Sink, H. L., & Langley, C. J. (1997). A managerial framework for the acquisition of third-party logistics services. *Journal of Business Logistics*, *18*(2), 63–89.

Sohail, M. S., & Sohal, A. (2003). The use of 3PL services: A Malaysian perspective. *Technovation*, *23*, 401–408. doi:10.1016/S0166-4972(02)00003-2

Sohal, A., Millen, R. A., & Moss, S. (2002). A comparison of the use of 3PL services by Australian firms between 1995-1999. *International Journal of Physical Distribution and Logistics Management*, *32*(1), 59–68. doi:10.1108/09600030210415306

Stock, J. R., & Lambert, D. M. (2001). *Strategic Logistics Management* (4th ed.). New York: McGraw-Hill.

Tan (1999). The use of information technology to enhance supply chain management. *Production and Inventory Management Journal*, 40.

Transportation and Logistics Institute (TLI). (2003). *Report on the working group on logistics: Developing Singapore into a global integrated logistics hub.*

van Damme, D. A., & Ploos van Amstel, M. J. (1996). Outsourcing logistics management activities. *International Journal of Logistics Management*, 7(2), 85–95. doi:10.1108/09574099610805548

van Hoek, R. (2002). Using information technology to leverage transport and logistics service operations in the supply chain: An empirical assessment of the interrelation between technology and operation management. *International Journal of Information Technology and Management*, *1*(1), 115–130. doi:10.1504/IJITM.2002.001191

Wilding, R., & Juriado, R. (2004). Customer perceptions on logistics outsourcing in the European consumer goods industry. *International Journal of Physical Distribution and Logistics Management*, *34*(8), 628–644. doi:10.1108/09600030410557767

*This work was previously published in International Journal of Information Systems and Supply Chain Management, Volume 3, Issue 2, edited by John Wang, pp. 68-83, copyright 2010 by IGI Publishing (an imprint of IGI Global)*

# Section 5
# Supply Chain Monitoring and Performance Management

# Chapter 17
# Design and Development of an e-Platform for Supporting Liquid Food Supply Chain Monitoring and Traceability

**Dimitris Folinas**
*Alexander Technological Institute of Thessaloniki, Greece*

**Ioannis Manikas**
*Aristotle University of Thessaloniki, Greece*

## ABSTRACT

*In this paper, the deliverables of a research project are presented, which aims at the development of a web-based platform capable of supporting the traceability of liquid products like milk, wine and olive oil. First, it includes the design of a supply chain reference model and the identification of the data required for the efficient operation of the traceability system. The main elements of the proposed model defined in this paper are the entities, stages, events, and processes. The reference model consists of three distinct phases that represent stages of real-life supply chains. Each of these phases is defined by certain interactions between the above basic elements. Additionally, the proposed e-platform is based on the above reference model aiming to follow and register the production and distribution processes of the raw materials, semi-finals, and final products that are used in the examined industry.*

## INTRODUCTION

During the past decade, the credibility of Food Industry safety schemes was heavily challenged after a number of food crises, such as Bovine Spongiform Encephalopathy (BSE) and food-and-mouth disease. The necessity of sufficient traceability systems to tackle such crises brought into light the need for reassessing and updating traceability systems currently implemented in the

DOI: 10.4018/978-1-4666-0918-1.ch017

Figure 1. Generic food supply chain model

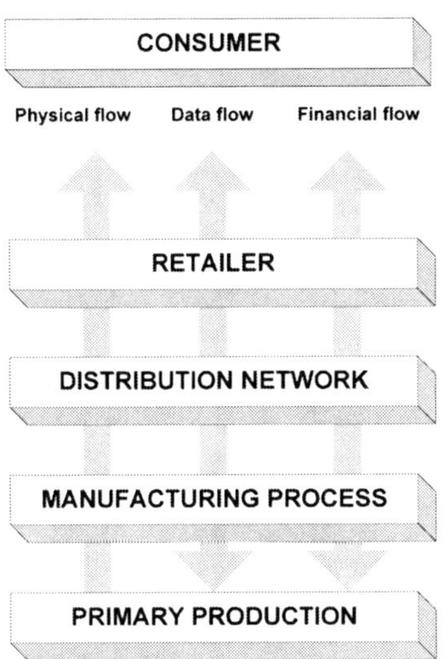


food sector. Thus, the successful control of the physical flow of the products along the supply chain and product safety assurance depends on the existence of an efficient traceability system (Giraud & Halawany, 2006; Morissey & Almonacid, 2005). This system must be able to identify each and every single unit produced and distributed from farm to fork.

The E.U. Regulation 178/2002 regarding the establishment of the European Food Safety Agency (EFSA) set the foundations towards more strict traceability requirements in the food sector and paved the way for further legal requirements at a national level assuring information flow transparency and efficient traceability in the Food Industry of each country-member of the E.U. According to the E.U. Regulation 178/2002, which took effect on January 1st 2005, "food and feed business operators shall be able to identify any person from whom they have been supplied with a food, a feed, a food-producing animal or any substance intended to be, or expected to be, incorporated into a food or a feed. Moreover, food business shall have in-place systems and procedures to identify the other businesses to which their products have been supplied. This information shall be made available to the competent authorities on demand".

This Regulation does not lay a specific methodology to be followed by all food business operators. Instead, food companies are free to choose those mechanisms that fit their needs and ensure efficient traceability for their products. According to ISO Quality Standards (1994), traceability is defined as: "the ability to trace the history, application, or location of an entity by means of recorded information" (ISO 8402:1994). Moreover, the same institution introduced at the beginning of 2006, two new standards that define the requirements for a traceability system within a food safety management system and the data that needs to be retained (ISO 22000:2005 - Food Safety Management Systems - Requirements, and ISO 22519 - Traceability System in the Agriculture Food Chain - General Principles for Design and Development). In the food chain, traceability means the ability to trace and follow a food, feed, food producing animal or substance through all stages of production and distribution (see Figure 1) (Beulens et al., 2005).

According to definitions at Article 3, (E.U. General Food Law Regulation): "stages of production and distribution refer to any stage including import, from and including the primary production of food, up to and including its sale or supply to the final consumer and, where relevant to food safety, the production, manufacture and distribution of feed".

The efficiency of a traceability system depends on the ability to collect safety and quality related information (Resende-Filho & Buhr, 2008; Sasazaki et al., 2004). According to Food Standards Agency (2002): "There is a wide range of traceability schemes currently used in food supply chains, from paper based to sophisticated IT-enabled. Moreover, as Wilson et al. (1998) notes IT-enabled systems have been developed and introduced over the recent years in the food sector, based on technologies implemented in more sophisticated industries, such as pharmaceuticals.

Salin (1998), Karkkainen (2003), and Roussos (2006), emphasize that Bar code and Radio Frequency Identification (RFID) technologies have been implemented in food chains, reducing errors associated with manual data handling, thus making tracking more feasible. While, at the same time, Regattieri et al., (2007) argued that some RFID properties limit traceability systems due of their costs, while (Feder, 2004; Fordice, 2004) pointed out the difficulties of management. Also, Wilson et al. (1998) and FSA (2002) indicates that the development of software systems and databases (data pools) increases the efficiency in collecting, transmitting, and analyzing larger volumes of safety and quality related data.

The main objective of the paper is to present the deliverables of a national project titled: "Design and development of an integrated traceability web-platform for the food industry". The project was funded by the National Program of the Development of Industrial Research and Technology (Call: 4.3, Action: 4.3.1), under the Third Community Support Program coordinated by the Greek Ministry of Development. The project involved two participants from Greece. The MIK3 Integrated Information Systems SA (a software house company) and the Department of Agricultural Economics (AE) of the School of Agriculture, Aristotle University of Thessaloniki. It lasted 18 months and the total budget was 210,000 €. The main objective of the project was to develop a web-platform that supports traceability in the food industry and especially in liquid products: milk, wine, and olive oil. This platform aims to follow and register the production and distribution processes of the raw materials, semi-finals, and final products that are used in the examined industry. The project consisted of four working phases:

1. Identification and categorization of the required data for the effective traceability.
2. Modeling of business processes of the Agribusiness Supply Chain.
3. Modeling traceability data in Physical Markup Language (PML) format.
4. Design and development of a web-platform for the monitoring of the required data in order to support the examined traceability process.

In this paper, all of the above working phases are presented. The first phase includes the design of a reference model for liquid food supply chain and the identification of data required for a traceability system to operate efficiently. The development of such a model requires the identification and analysis of each stage of the supply chain for each one of the product categories under study, from farm to fork, including all factors that affect quality (packaging materials, agrochemicals, antibiotics, fertilizers, climate, soil etc.). It also requires the modeling of the main entities of the proposed model in order to design and develop the web-platform in a more stable, reliable and effective manner.

## IDENTIFICATION AND CATEGORIZATION OF THE REQUIRED DATA FOR THE EFFECTIVE TRACEABILITY

Every traceability system is based initially on the identification and classification of the appropriate traceability data to become transparent in a supply chain.

A number of traceability data sources exist along a supply chain. Data are gathered from: 1) bar codes and RFID tags embedded or attached in raw materials, semi-final or final products those move across the food supply chain, 2) Sensors placed into machines, equipments, containers, etc. that are used in the various productions and logistics processes (warehousing, handling, distribution, packaging, etc.), and 3) Signals from other application systems that supply chain members maintain and use. Every item (physical object) in the supply chain is identified with the Electronic Product Code (EPC) unique number (e.g., an EPC is stored on a RFID tag). A number of readers are devices responsible for detecting when tags or other data sources enter their read range. They may also be capable of interrogating other sensors coupled to tags or embedded within tags.

Furthermore, data are generated from the various enterprise processes of supply chain members and maintain information to the smallest homogeneous product unit. It is also stored and presented as structured content (relational databases), unstructured content (plain text) or unstructured content with structured presentation (html files), in many and different forms (such as texts, charts, figures, or tables, etc.), as well as, in both typed and electronic format (plain or multimedia files). The above can be explained due to the following reasons: there is no standard way inside every enterprise to define how data should look or be represented, and the software applications vary from entity to entity in a food supply chain, and even if a standard software package is used, output formats commonly vary. Also, in the various logistics processes for example in the production process there are various information systems such as production management and enterprise resource planning systems, extension and food health service systems, third party control database management systems, public inspection database management systems, etc.

Traceability data can be distinguished to static and dynamic. Static data refer to product features that cannot change, such as retirement / catch date, country of origin, expiry date, size etc. Dynamic data refer to dynamic features that change over time while product is changing ownership while moving along the supply chain, such as lot / batch number, order ID, dispatch date, taste, content of chemical components etc. Traceability data can also be distinguished to mandatory and optional. Mandatory data shall be collected and archived from all members of the food supply chain, and most of them shall be communicated to the rest of the supply chain. An indicative categorization of traceability data to mandatory (Table 1) and optional ones (Table 2) for the examined sector are given (see Table 1 and Table 2).

## MODELING OF BUSINESS PROCESSES

The outcome of this work package is the development of a reference supply chain business model for supporting traceability in the examined sector. A number of references model have been proposed in order to support traceability. Bevilacqua et al. (2008) propose several models of traceability processes for a supply chain of fourth range vegetable products and to set up a computerized system for managing product traceability. Bertolini et al. (2006) design a model that aims to detect the possible critical points of a traceability system. Salomie et al. (2008) design a broker-based service oriented model that captures the main elements necessary to follow a product throughout its lifecycle, from manufacturing to the end consumer. Regattieri et al. (2006), provide a general framework for the identification of fundamental mainstays and functionalities in an effective traceability system.

This proposed model is explained in the next section in details. Specifically, in this package, the basic concepts and the main phases of the model are presented. Also, static modeling with Unified Modeling Language (UML) Class Diagrams and dynamic modeling with UML Activity Diagrams of the business processes that consists of the supply chain model is taken place. The next step is the development of eXtensible Markup Language (XML) Schemas from the UML diagrams.

### Basic Concepts and Main Phases of the Model

The phase is the main sub-system of the reference model. Each phase includes stages, which

*Table 1. An indicative list of mandatory data that constitute an efficient traceability system*

| Mandatory data | Data definition | Entry example |
|---|---|---|
| Lot number | A number or code assigned to uniquely identify a group or batch of products (inputs or outputs) | Lot number, Batch number<br>Production number, Pack date |
| Product ID | A number or code that uniquely identifies individual units of production (for fresh produce) | Bar code, RFID tag |
| Product description | A description of the product | 2 Kg oranges bag (net) |
| Supplier ID | A number or code that uniquely identifies the company that sells the product | EAN.UCC GLN |
| Quantity | Count, net weight, net volume of product. | 10, 500g |
| Unit of measure | Description of the units in which the quantity of product is being measured/ expressed | Kilo (Kg), Pound, bin, can etc |
| Buyer ID | A number or code that uniquely identifies the purchaser | EAN.UCC GLN<br>Internal Customer Number |

link together with actions, which are initiated by events. By breaking down the system to phases, we achieve a more realistic approach of the actual supply chain processes. The concept of stage refers to the process-taking place in limited space defined by a container of standard capacity in terms of product. According to the process type, the product properties can be altered in an extended or limited degree. The alterations can be described as extended when in this stage, the product is submitted to a certain transformation process, such as sterilization, pasteurization, fermentation, etc, and limited when in this stage, the product is simply stored into the container, such as storing, distribution, etc.

An event indicates the time when one or more actions should initiate. The event is realized instantly and triggers a respective action. An entity is defined as a distinguished real world object and is described in our model by a group of attributes. Each attribute refers to an elemental data that follows an entity. The values taken by each one of these attributes allow distinguishing between an entity and other similar ones. A group of entities includes a number of entities with similar attributes, such as all actors or all containers of the model. The categories of entities defined in our model are Container, Actor, Material, Sample, Document, Primary production unit, and Primary production support area.

In the proposed model there are three main phases as it is illustrated in Figure 2.

Phase 1: The natural environment is a system consisting of the primary production unit (vineyard, olive tree, milk-producing animal) and the primary production support area (farm, field, and stable). This phase

*Table 2. An indicative list of optional traceability data*

| Optional data | Data definition | Entry example |
|---|---|---|
| Byers' name | The name of the company that purchases the product | «XXXX» customer business name |
| Contact information | The company contact information | «XXXX»<br>phone number, fax number, email address |

*Figure 2. Reference supply chain business model for supporting traceability*

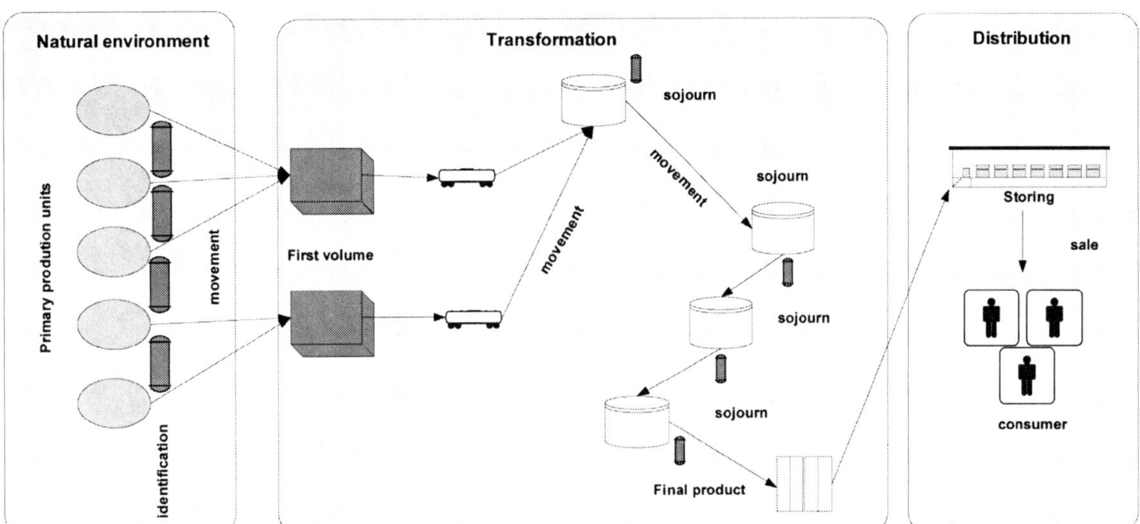


may represent natural systems such as tree and field, animal and farm, fish and the sea etc. The first phase is part of a wider horizontally integrated farming system that may include more than one of these systems, including support areas for several types of primary production units. The stage of creation refers to the production of the primary product (first volume) from the primary production unit. This is an intermediate stage of the natural life cycle of primary production unit, which is supported by all artificial and natural facilities comprising the primary unit support area and inputs, such as support materials for primary production unit. The efficient combination of the above will lead to the successful completion of the creation stage, with the production of the primary product, which in our model is raw milk, olive, or grape. With the event of sale, the ownership of the primary product is being passed to the next level of the supply chain, i.e., a distributor of the manufacturing company. Transition to the next stage is initiated by the event of identification that triggers the process of induction, while the primary product is being removed from the natural environment under proofed conditions and enters the phase of transformation.

Phase 2: The transformation phase includes all stages, events, and processes that lead in the production of the final product, as this is disposed in retail. This phase models the production process, having as inputs the primary product coming from the first phase and all materials for cleaning, disinfecting, and supporting the production process. This phase starts with the stage of storing/transformation of the primary product and ends with the packaging of the final product into retail units. The aim of this phase is to add value of shape. The stages of sojourn build the process line that lead in the production of the final product. While the processed material is passing through successive sojourn stages, this is being stored (into the storing containers) or transformed (into the transformation containers). The movement of the product from one stage to another is being initiated by the event of identification, which includes homogenization of the product volume, sampling, analysis, and decision for movement. The direction of movement depends on the results of the sample assessment and the production needs, business rules and other events related to the production process. The movement process refers to the disengagement of the product volume from one container/stage and consignment to another container, under controlled conditions. The initiation of the movement process and the direction of the product depend of the identification event that precedes this process. The transition to the next phase is realized after the final movement process where the final product disengages from the last stage, which is usually storing of the final retail packed product, or a packing line, when the final product is not stored in manufacturing facilities but is dispatched directly after it is packed. All the above are depicted in Figure 3, in the form of UML Activity Diagram.

Phase 3: The distribution phase includes mainly the storing of the final product to retailer warehousing facilities or to interceded stakeholders facilities such as third party logistics companies or distribution centers. The storing stages are linked with movement processes that are triggered by identification and sale events. The storing stages compose and model the distribution channel that links the manufacturing point with the point of sale. While the product is being moved along the stages of this phase, it is not transformed by any means. The main objective of this phase is to retain the quality of the product and to add value by disposing it to the correct point of retail sale at the correct time. According to the rules described above, the identification event, along with the sale event, trigger the movement process, from one storing stage to another, leading to the final event of retail sale. In this phase, the identification event does not include sampling and analysis but only the control of the documents that escort the product (shipping invoices). The final sale event disengages the product from the modeled supply chain, and passes the ownership of the product to the consumer who is charged with a price that quantifies the modeled value chain per product unit.

## Main Entities of the Model

The following entities of the proposed reference model are examined and modeled with UML Class Diagrams: 1) Container, 2) Actor, 3) Material, and 4) Sample.

At first, the Container (or Carrier) category refers to the entity that confines and carries the product, defining its volume, size and features that differentiate it from other volumes of the same product with similar features and adding value of shape, time, area and procurement. The added value is given to the product by the container through the main processes realized in it: transportation, storing, transformation, and distribution. According to these types of processes, the container entity category can be divided into the following sub-categories: 1) Transportation sub-category, 2) Storing sub-category, 3) Transformation sub-category, and 4) Distribution sub-category. The Class Diagram of the Container (Carrier) entity is presented in Figure 4.

The Actor entity category refers to all persons, natural or legal, that contribute by any means to any of the processes of the supply chain. This category can be distinguished to the following sub-categories: 1) Food handler, 2) Trader, 3) Auditor, and 4) Consumer. The Class Diagram of the entity Actor is presented in Figure 5.

The Material category refers to all materials introduced in any of the stages of the modeled supply chain. This category is distinguished in two sub-categories; products and supplementary materials (see Figure 6).

*Figure 3. Activity diagram for the movement*

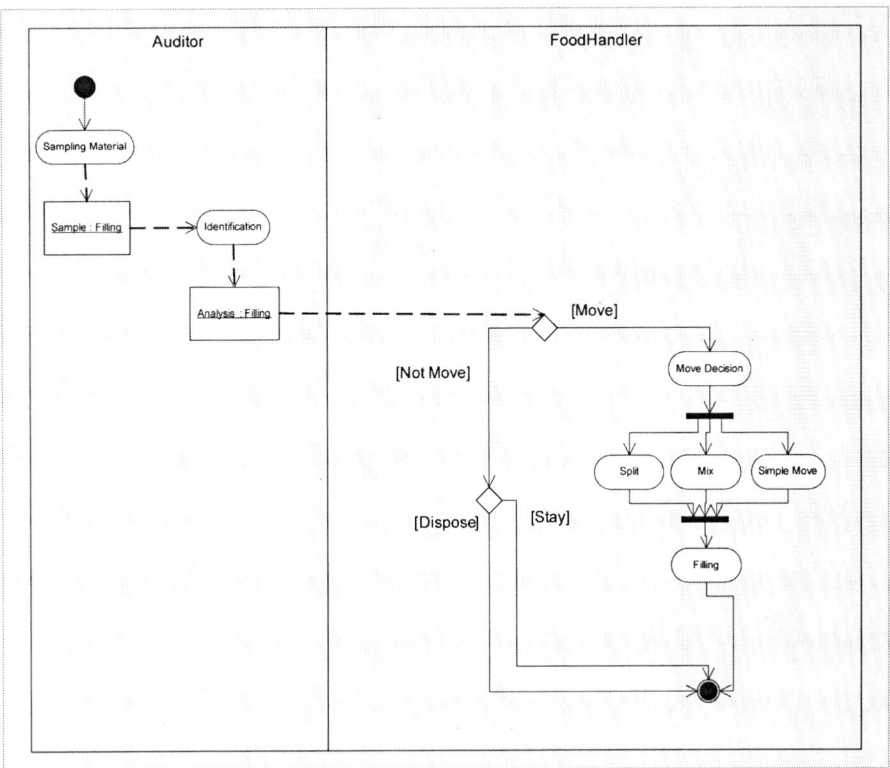


The Sample entity refers to a standard volume of product with such size that allows conduction of quality tests on it. Each sample represents a bigger volume of product with the same attributes as the sample. The documents entity refers to all certified or not certified documented information required for the realization of all the supply chain processes. Such documents are the invoices, the shipping notes, etc. The Class Diagram that models the Sample entity is presented in Figure 7.

All the entities described above exist in all the phases of the model, apart from two, that were developed to describe the first phase of the model. These are the Primary production unit and the Primary production support area. The Primary production unit referred to all natural entities, such as plants and animals, that produce the primary product (primary volume), as a result of processes realized during their natural life cycle. The Primary production support area refers to all natural and artificial installations that support the natural life cycle and satisfy all the needs of the primary production units. By establishing and modeling these basic concepts, we facilitate the development of systems and applications that support traceability in the food supply chain. The proposed reference model is the base for the development of a web application for traceability management for liquid food.

Finally, in this phase the extracted and transformed data are modeled, so as to allow their uniform management and utilization. Bevilacqua et al. (2008), points out that in a traceability system the need to share information requires the use of a standardized language. Several standard systems have been introduced in the market but the most promising is GS1 (GS1 Official Website, 2006), which it is considered as the most robust lot identification system in the world.

*Figure 4. Class Diagram for the entity carrier*

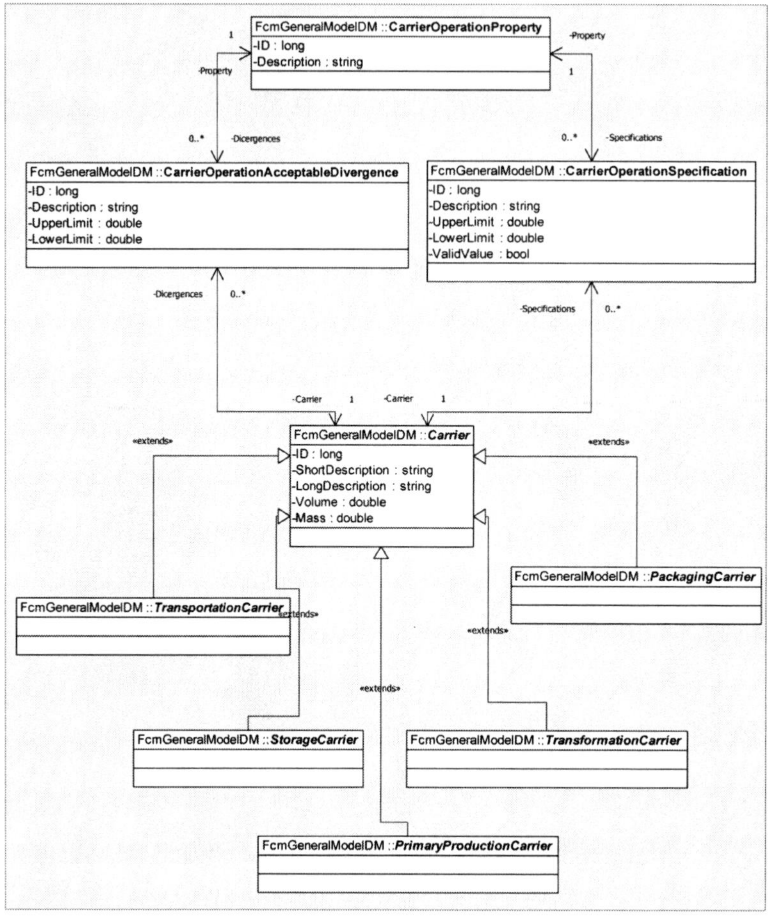


Moreover, Folinas et al. (2003) indicates that there is a need for common business vocabularies that describe the structure and the semantics of the traceability data. These vocabularies allow users to document these requirements (existing or proposed) in a neutral format that will act as a standard. In essence, a standard is just an agreed-upon set of elements, attributes, structure, semantics, and processes with which information can be used, exchanged or presented, in order to support the traceability process.

All the above can be easily represented in eXtensible Markup Language (XML) Schemas technology, which is the W3Consortium standard (W3C Specification). XML Schemas can be used to encode traceability information needed, to provide adequate technology for specifying standards and assuring that various documents prepared are valid. The last mentioned is very important due to the fact that in most other data formats, errors are not usually detected until something goes wrong leading to low information quality, which in turn leads to poor decision quality in food-crisis situations.

Due to its nature, XML is data-centric, unlike document-centric typed or electronic reports as pdfs or html pages, allowing information to be structured in a way that makes it readily accessible for the final users. As Laurent (1999) notes, it represents not only the information to be presented, but also the metadata encapsulating its meaning, and the structure of the information to

*Figure 5. Class Diagram for the entity actor*

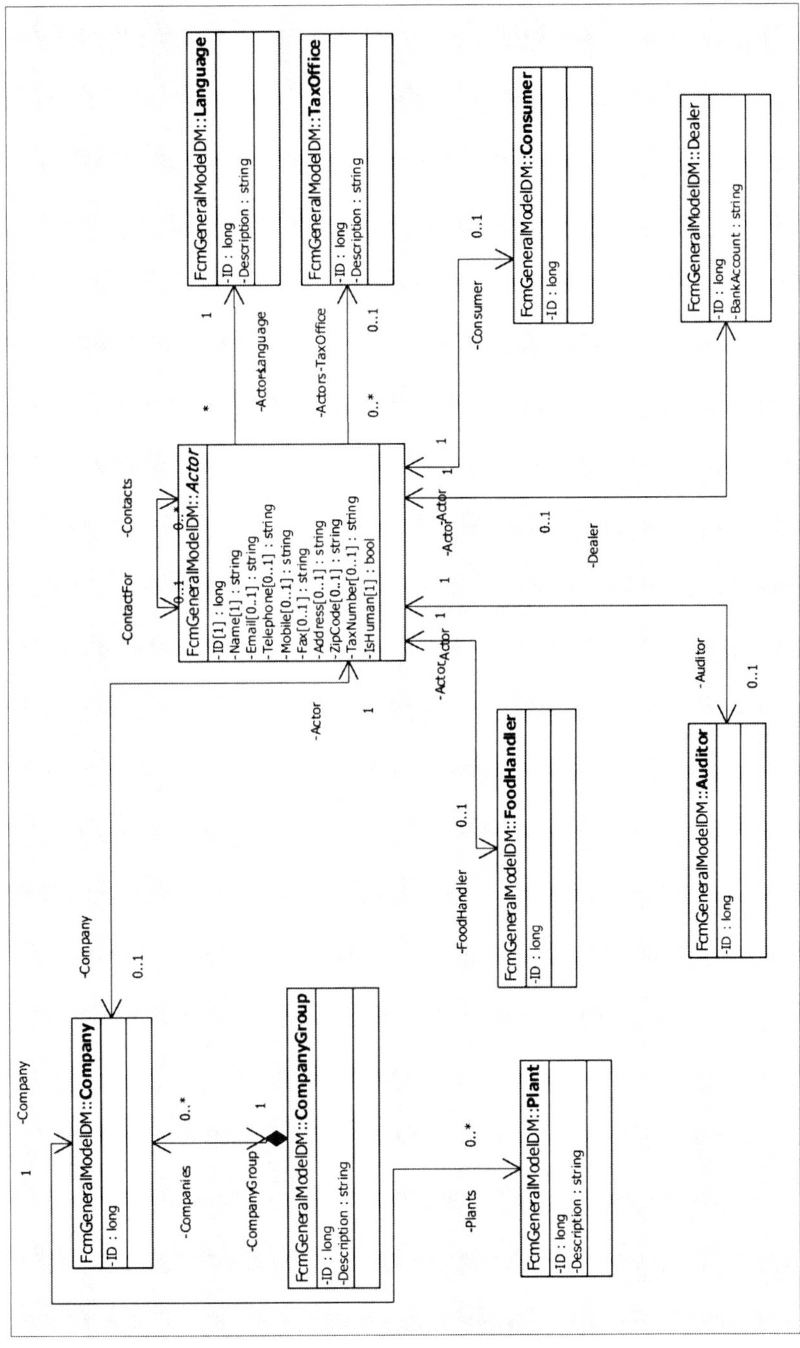

*Figure 6. Class diagram for the entity material*

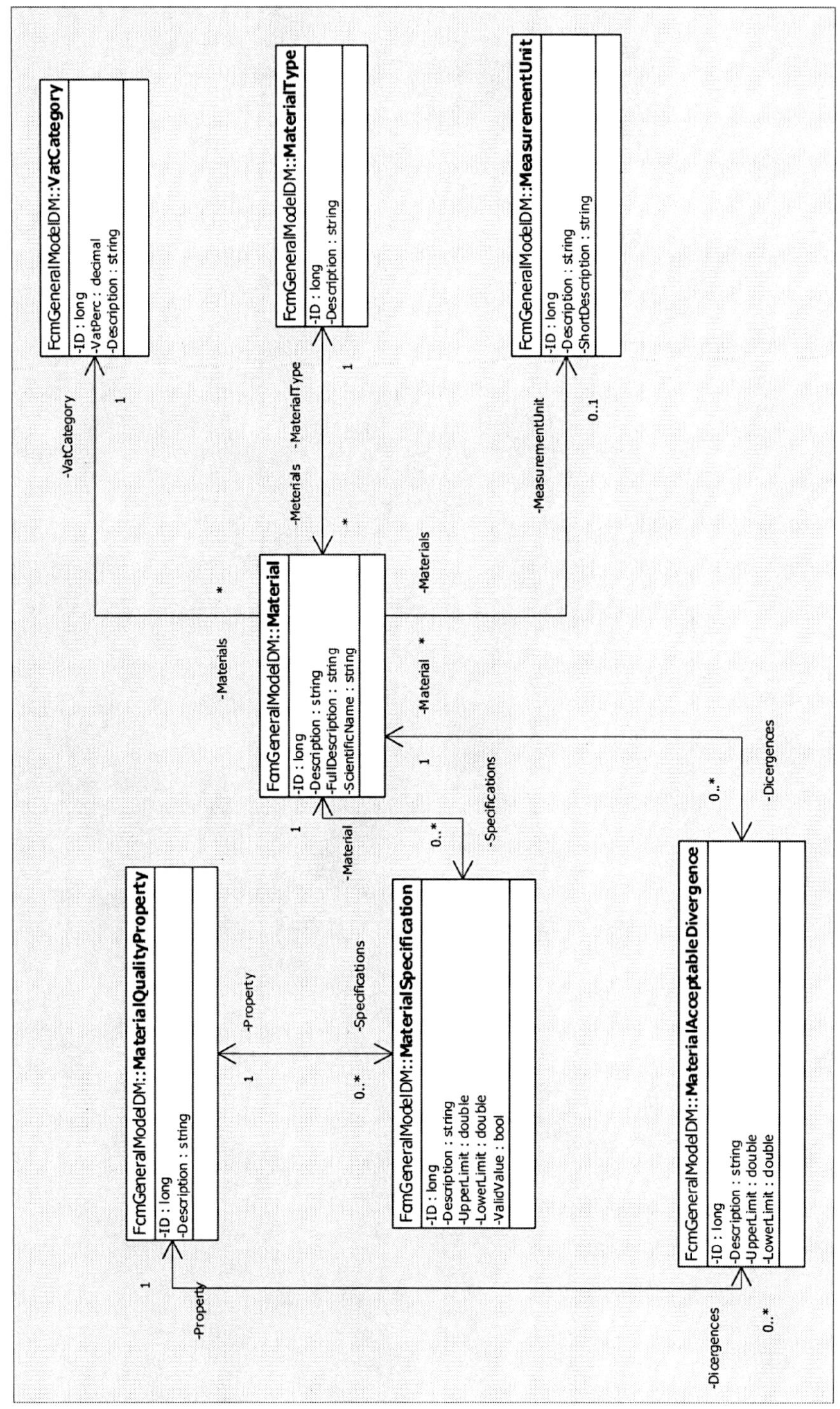

*Figure 7. Class diagram for the entity sample*

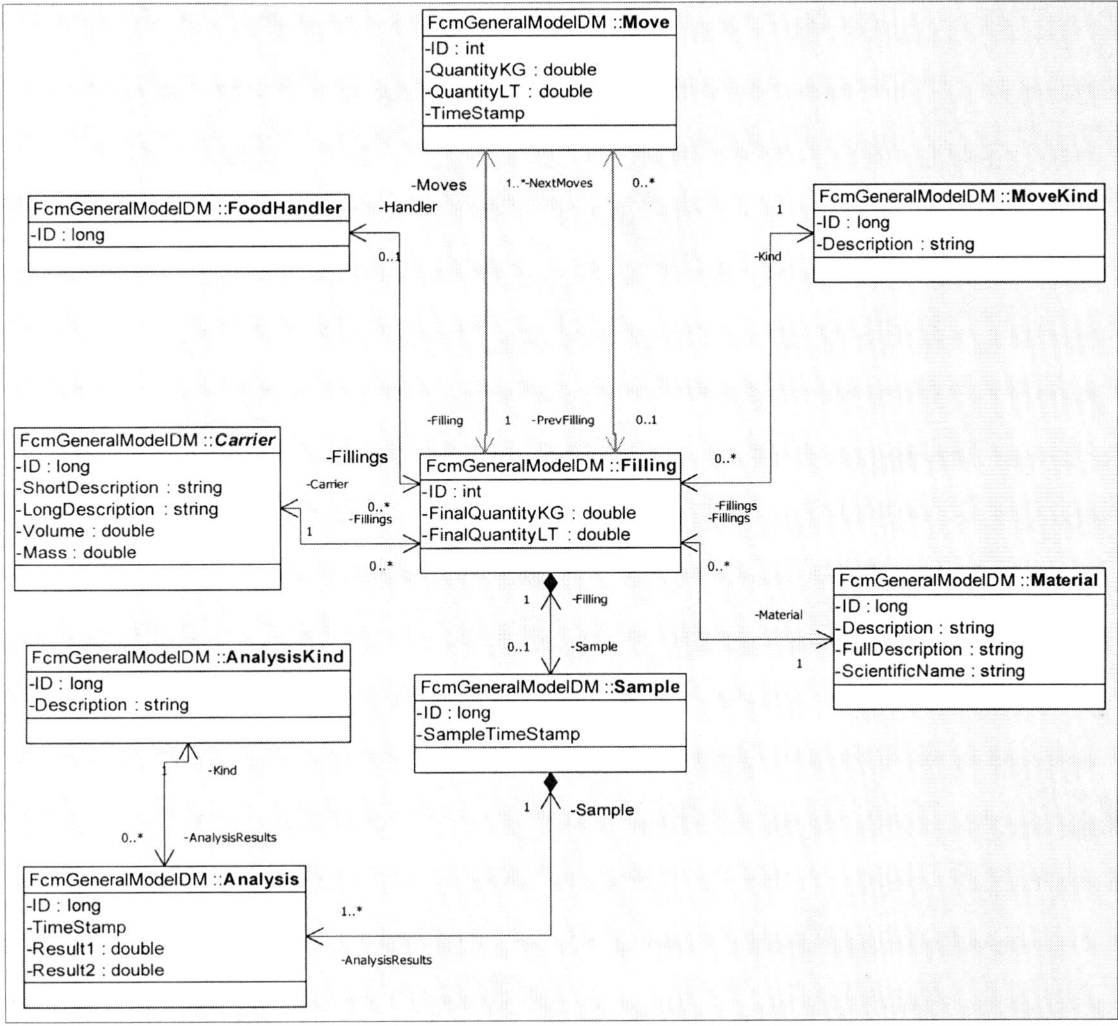


be presented. Another main characteristic is that XML is designed with the Internet in mind. Indeed, one of the main goals in the initial design of XML was to be a web-enabled version of SGML, so as to take advantage of the simple and open transport layers that the Internet provides, such as email (SMTP), web (HTTP), file transfer (FTP), and other mechanisms. It can even take advantage of some of the native security features such as SSL, which are present on the web.

Based on the above, Physical Markup Language (PML) -an XML-based technology- is proposed to be the common "language" for describing physical objects / products in the examined supply chains. PML is intended to be a general, standard means for describing the physical world, by describing physical objects for use in remote monitoring and control of the physical environment. Brock (2001) indicates that the applications include inventory tracking, automatic transaction, supply chain management, machine control and object-to-object communication).

During the final phase the business vocabularies (XML Schemas) from UML diagrams are developed. These XML Schemas are used for the creation and the validation of the messages.

The transformation from diagrams to schemas is the research subject of many projects (Booch et al., 1998; Carlson, 2001; Conrad et al., 2000; Routledge et al., 2002). Practically an intermediate stage between the conceptual models (corresponding classes) and natural models of XML Schemas is applied to all of the approaches. This can be achieved by extending UML diagrams with the use of Stereotypes, which can describe more complicated business rules.

Based on the above research projects the following figure presents the XML Schema that models one of the main entities of the proposed model (Carrier), as well as, its relationships with the other entities (see Figure 8).

## MODELING TRACEABILITY DATA IN PHYSICAL MARKUP LANGUAGE (PML) FORMAT

In the above sections the identification and categorization of the required data for the effective traceability and the modeling of business processes of the agribusiness supply chain were presented and analyzed. The next phases include traceability data modeling in PML format and the development of the web-platform for the monitoring of the required data in order to support the examined traceability process.

The PML is proposed to model the traceability data that were identified and classified in the previous phase of the framework and to provide information about various parameters / elements such as product properties, process properties, tracing properties, business entities properties, properties of means of production used on the product and data measurement properties. In order to provide this information, PML provides a number of XML elements and data types. These include a root element, which captures a "snapshot" of the whole physical environment and sub-elements in a product decomposition tree that hold the above mentioned properties.

PML is proposed to model the data that were identified and classified in the previous phase of the framework and to provide information about various parameters / elements such as: product properties, process properties, tracing properties, business entities properties, properties of means of production used on the product and data measurement properties. In order to provide this information, PML provides a number of XML elements and data types. These include a root element, which captures a snapshot of the whole physical environment and sub-elements in a product decomposition tree that hold the above mentioned properties.

First element presents static and inherent data of the product. It refers to static information that pertains to a smallest homogeneous product unit. For example, this element includes information such as: product ID, product description, supplier name, country of origin, date of pack, quantity, receiving date, dispatching date, etc.

The Second element presents the process properties / characteristics of the product in a node hierarchical syntax, which is simply a nested collection of node elements as the product moves across the various levels of the supply chain. The PML node structure captures the organization of physical objects. Brock (2001) notes that in every node a unique characteristic is rendered to the product referring to the Unified Product Name (UPN), transportation vehicle, pallet or container, lot, etc. by using Electronic Product Code (EPC) format.

The Third element describes all the corresponding information of the tracing process recording the movement history of a product, as it moves through the supply chain. The trace element includes one or more steps that indicate waypoints along the path followed by an object. Each step has a number of optional elements including an owner, date and location. These constitute the history of what has happened to a unit.

Ownership, roles and responsibilities of every business entity (people, companies or organiza-

*Figure 8. XML schema for the entity container*


```
<?xml version="1.0" encoding="utf-8"?>
<xs:schema xmlns:tns="http://schemas.datacontract.org/2004/07/PAVETGeneralModel"elementFormDefault="qualified"
targetNamespace="http://schemas.datacontract.org/2004/07/PAVETGeneralModel" xmlns:xs="http://www.w3.org/2001/XMLSchema">

<!-- Carriers -->
<xs:complexType name="Carrier">
<xs:sequence>
<xs:element minOccurs="0" name="ID" nillable="true" type="xs:long" />
<xs:element minOccurs="0" name="LongDescription" nillable="true" type="xs:string" />
<xs:element minOccurs="0" name="Mass" nillable="true" type="xs:double" />
<xs:element minOccurs="0" name="ShortDescription" nillable="true" type="xs:string" />
<xs:element minOccurs="0" name="Volume" nillable="true" type="xs:double" />
</xs:sequence>
</xs:complexType>
<xs:element name="Carrier" nillable="true" type="tns:Carrier" />

<xs:complexType name="PrimaryProductionCarrier">
<xs:sequence>
</xs:sequence>
</xs:complexType>
<xs:element name="PrimaryProductionCarrier" nillable="true" type="tns:PrimaryProductionCarrier" />

<xs:complexType name="TransportationCarrier">
<xs:sequence>
</xs:sequence>
</xs:complexType>
<xs:element name="TransportationCarrier" nillable="true" type="tns:TransportationCarrier" />

<xs:complexType name="TransformationCarrier">
<xs:sequence>
</xs:sequence>
</xs:complexType>
<xs:element name="TransformationCarrier" nillable="true" type="tns:TransformationCarrier" />

<xs:complexType name="PackagingCarrier">
<xs:sequence>
</xs:sequence>
</xs:complexType>
<xs:element name="PackagingCarrier" nillable="true" type="tns:PackagingCarrier" />

<xs:complexType name="StorageCarrier">
<xs:sequence>
</xs:sequence>
</xs:complexType>
<xs:element name="StorageCarrier" nillable="true" type="tns:StorageCarrier" />

<xs:complexType name="CarrierOperationProperty">
<xs:sequence>
<xs:element minOccurs="0" name="Description" nillable="true" type="xs:string" />
<xs:element minOccurs="0" name="ID" nillable="true" type="xs:long" />
</xs:sequence>
</xs:complexType>
<xs:element name="CarrierOperationProperty" nillable="true" type="tns:CarrierOperationProperty" />

<xs:complexType name="CarrierOperationAcceptableDivergence">
<xs:sequence>
<xs:element minOccurs="0" name="Carrier_ID" nillable="true" type="xs:long" />
<xs:element minOccurs="0" name="Description" nillable="true" type="xs:string" />
<xs:element minOccurs="0" name="ID" nillable="true" type="xs:long" />
<xs:element minOccurs="0" name="LowerLimit" nillable="true" type="xs:double" />
<xs:element minOccurs="0" name="Property_ID" nillable="true" type="xs:long" />
<xs:element minOccurs="0" name="UpperLimit" nillable="true" type="xs:double" />
</xs:sequence>
</xs:complexType>
<xs:element name="CarrierOperationAcceptableDivergence" nillable="true" type="tns:CarrierOperationAcceptableDivergence" />

<xs:complexType name="CarrierOperationSpecification">
<xs:sequence>
<xs:element minOccurs="0" name="Carrier_ID" nillable="true" type="xs:long" />
<xs:element minOccurs="0" name="Description" nillable="true" type="xs:string" />
```

tions) that own or manage a product in the supply chain, are the information that is provided by the Fourth element.

Fifth element provides the properties of means of production or manufacturing that are being used on the product. It includes machines and labor inputs. In the case of machines, contamination issues can arise. For example, if a machine is used for mashing non-Genetically Modified (GM) tomatoes after having been used for GM tomatoes, very rigorous cleaning is necessary. This ensures consumers that the final product (tomato sauce) is GMO free. In the case of labor, ethical issues can arise, as when customers refuse to buy products because they suspect that children have been used to produce it.

Finally, the Sixth element includes information about the data measurement about: location of the product in the supply chain, time / duration measurements (e.g., shipping, delivery, receiving, transport, expiration date, etc.) and also the measurement units of length, mass, temperature, amount, etc.

All the six elements given above provide an integrated description of a product for a specific time, using PML. By establishing and modeling these basic concepts we facilitate the development of systems and applications in the supply chain. The proposed reference model is the base for the development of a web application for event management in the supply chain networks.

## TRACEABILITY PLATFORM: ARCHITECTURE AND MODULES

The final phase of the project was the design and development of the web-platform that supports traceability in the food sector. The proposed traceability application software was developed with Microsoft.NET environment and it is based on 3-tier architecture:

1. Data (base)-tier, which can be any of the market RDMBS, such as Microsoft SQL Server, Oracle, IBM DB2, Postgres, etc.
2. Server-tier, which supports the required services for the effective traceability data management.
3. Client-tier, which is a typical user interface, for the execution of a number of services in the server-tier. The connection among client and server is accomplished with Web Services technology and it is based on Service Oriented Architecture (SOA).

The proposed traceability application software consists of four (4) modules based on the proposed generic traceability framework, as it is depicted in Figure 9.

*Generic entities*. This is the main module of the application. It allows the management of the basic entities that were described in the proposed traceability model. With this module, the final user can add a new entity, modify or delete an existed one, or present it in various views (see Figure 10 and Figure 11).

*Receiving*. The main objective of this module is to monitor the receiving process of the raw milk (first volume). What it follows are the steps of the: 1) Collection of the first volume and its storing into a specialized container (or carrier), 2) Collection of the raw milk from the various collection points (specialized containers), and 3) Moving to central milk processing unit. In each step the quantity of milk is measured and maintained by the application. Furthermore, a number of services are provided: 1) Management of the entities, which are related with the receiving processes, 2) Routing / movement of the first volume, and 3) Importation of the first volume into the specialized containers (see Figure 12).

Figure 9. Application main modules

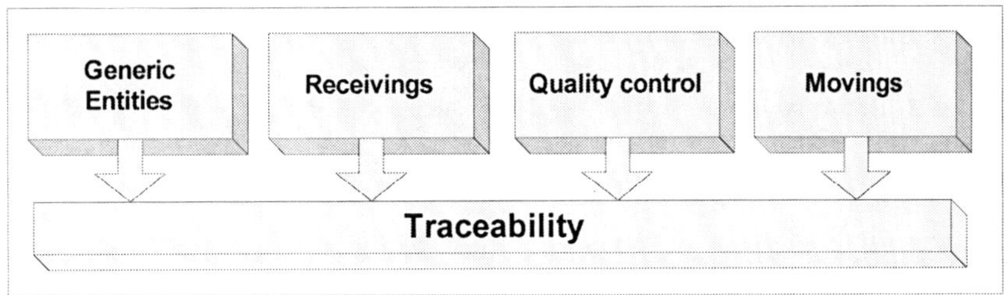


Figure 10. Management of generic entities

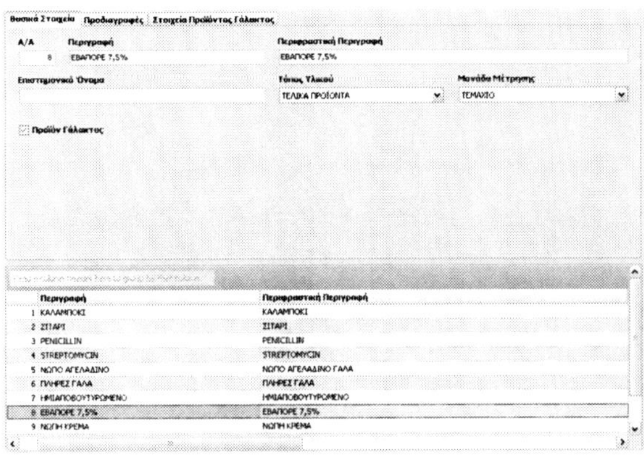

Figure 11. Management of generic entities

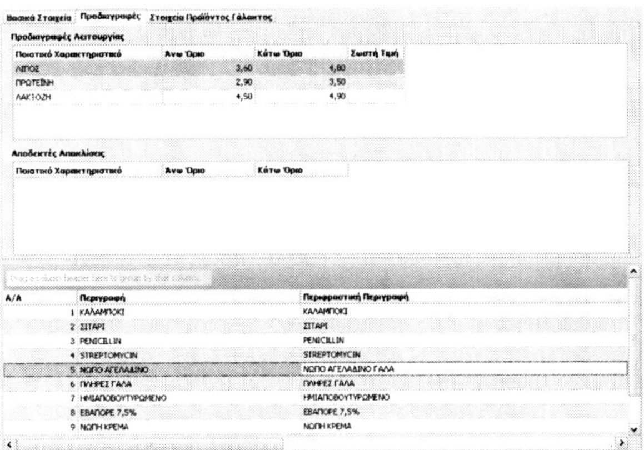

*Figure 12. Receiving module*

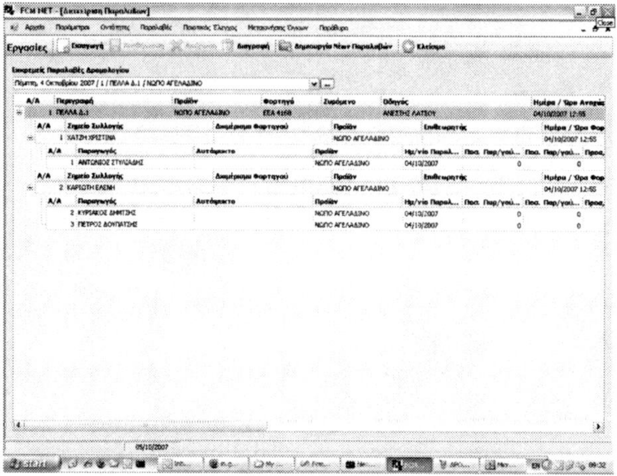

*Quality control.* This module allows the qualitative identification of a specific quantity of products in every stage of their processing. It provides the following capabilities: 1) Management of the entities that refer with the quality control, such as the findings / results of quality analysis process, 2) Management of the samples that are taken from specific quantities of the products, and 3) Management of the results from the samples' analysis (see Figure 13).

*Moving of volumes.* This module allows the examination of milk products inside the process unit (process carrier). The design and development of this module was based in the concept of fillings that is described in previous section. Movement of volumes module provides the following services: 1) Management of the various entities, which refers with the movement of volumes, such as the storing or processing containers or the containers, etc., 2) Management of the fillings of the various containers via the movement

*Figure 13. Quality control module*

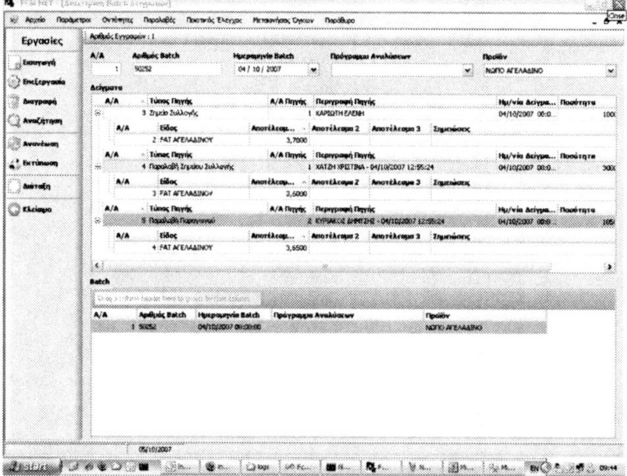

*Figure 14. Various traceability views*

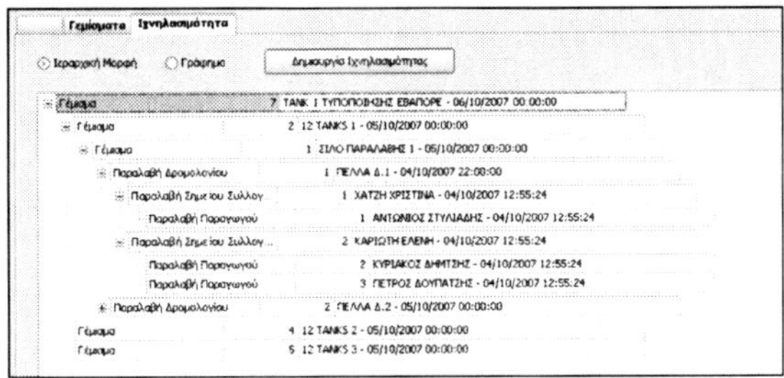

of products from one container to another, and 3) Management of special moving, such as the production of batch and lot final products.

*Traceability module.* Traceability module utilizes data of the previous modules in order to present backwards traceability of the final milk product. This module refers to all of the batch fillings, since that these fillings are the final phase of milk processing procedure. The corresponding info refers to both quantitative identification (usage of receiving and moving modules' data) and qualitative identification (usage of quality control module's data) for all the milk production stages. The client part of the application presents the traceability of the final product in two different formats, in a hierarchical and graphically format (see Figure 14 and Figure 15).

## CONCLUSION

The efficiency of a traceability system depends on the ability to identify uniquely each unit that is produced and distributed, in a way that enables the continuous tracking, from the primary production to the retail point of sale. An efficient traceability system must follow some rules that define which data must be gathered and stored in each stage of the supply chain. This is achieved by standardization of the gathered data and typification of the processes that enable storing and communication of the data.

In this project, by establishing and modeling these basic concepts, a reference model and a

*Figure 15. Various traceability views*

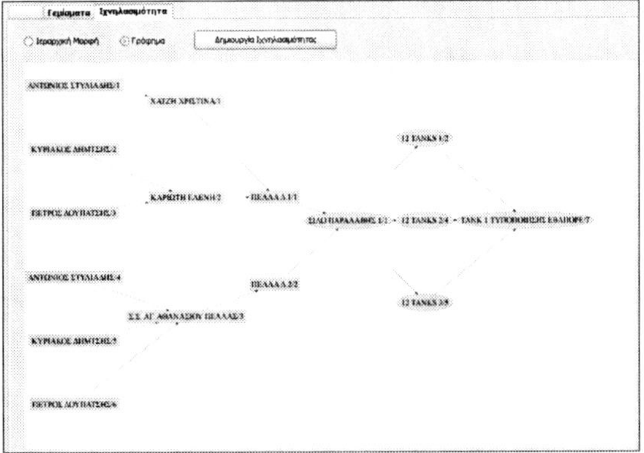

corresponding web application for traceability management for liquid food were developed. The proposed web-platform aims to support efficiently food traceability by monitoring and administering the data gathered from the various logistics processes along the supply chain.

Finally, the use of the proposed web-platform in a specific dairy production line in the Northern Greece (as the final stage of the project), produced the following outcomes. Business users -especially managers from the Production and Quality Departments- noticed that the main benefits derived from applying the proposed web platform were: user friendliness (it required more business than technical background by the users), more realistic identification of requirements of the traceability data and processes, and risk reduction. Moreover, the data utilized in the model already exist in data bases that support HACCP and ISO standards, while data communication tools (RFID, EPC) are based in EAN-UCC standards. Data processing uses XML technologies and information filtering is achieved by implementing the six elements model (PML) presented and analyzed above. On the other hand, users with technical background observed that the main feature of the proposed platform is the simplicity in use and the ability of communicating information through commonly accessible means such as the Internet, e-mail, and cell phones.

Further research though is required in order for the XML Schemas that model the main entities of the proposed framework must have the ability to adapt easily and directly in order to cover the specific needs of a potential supply chain partner (farmers, fishermen, cattle breeders, etc.), or environmental / market changes. Furthermore, advanced applications must be developed for the full integration of enterprises' business systems, as system integration among partners is not implemented only with the modeling and integration of business information, but with the requirements for a full-integrated traceability framework.

## REFERENCES


Bertolini, M., Bevilacqua, M., & Massini, R. (2006). FMECA approach to product traceability in the food industry. *Food Control, 17*(2), 137–145. doi:10.1016/j.foodcont.2004.09.013

Beulens, M., Broens, F., Folstar, P., & Hofstede, J. (2005). Food safety and transparency in food chains and networks, relationships and challenges. *Food Control, 16*(6).

Bevilacqua, M., Ciarapica, F., & Giacchetta, G. (2008). Business process reengineering of a supply chain and a traceability system: a case study. *Journal of Food Engineering, 93*, 13–22. doi:10.1016/j.jfoodeng.2008.12.020

Booch, G., Christerson, M., Fuchs, M., & Koistinen, J. (1998). *UML for XML schema mapping specification*. Retrieved May 10, 2009, from http://xml.coverpages.org/fuchsuml_xmlschema33.pdf

Brock, L. (2001, February). *The physical markup language - a universal language for physical objects, Auto-ID White Paper WH-003*. Retrieved May 10, 2009, from http://xml.coverpages.org/PML-MIT-AUTOID-WH-003.pdf

Carlson, D. (Ed.). (2001). *Modeling XML applications with UML, practical e-business applications*. Reading, MA: Addison-Wesley.

Conrad, R., Scheffner, D., & Freytag, J. (2000). XML conceptual modeling using UML. In *Proceedings of the International Conceptual Modeling Conference,* Salt Lake City, UT (pp. 309-322). New York: Springer Verlag.

Feder, B. J. (2004, December 28). Wal-Mart edict on radio tags hits snag. *The Denver Post,* 6C.

Folinas, D., Vlachopoulou, M., Manthou, V., & Manos, B. (2003, July 5-9). A web-based integration of data and processes in agribusiness supply chain. In *Proceedings of the EFITA 2003 Conference*, Hungary (Vol. 1, pp. 143-150).

Food Standards Agency. (2002). *Traceability in the food chain; a preliminary study*. UK: Food Standards Agency.

Fordice, R. (2004, November). Under control. *Meat Processing,* 34-40.

*GS1 Official Website*. (2006). Retrieved from www.gs1.org

Giraud, G., & Halawany, R. (2006). Consumers perception of food traceability in Europe. In *Proceedings of the 98th EAAE seminar: Marketing dynamics within the global trading system*, Chania, Greece (p. 7).

Jansen-Vullers, M. H., van Dorp, C. A., & Beulens, A. J. M. (2003). Managing traceability information in manufacture. *International Journal of Information Management*, *23*, 395–413. doi:10.1016/S0268-4012(03)00066-5

Karkkainen, M. (2003). Increasing efficiency in the supply chain for short shelf life goods using RFID tagging. *International Journal of Retail and Distribution Management*, *31*, 529–553. doi:10.1108/09590550310497058

Kim, H. M., Fox, M. S., & Gruninger, M. (1995). *Ontology of quality for enterprise modeling*. Los-Albamitos, CA: WET-ICE.

Laurent, S. (Ed.). (1999). *XML™: a primer* (2nd ed.). New York: MIS Press.

Moe, T. (1998). Perspectives on traceability in food manufacture. *Trends in Food Science & Technology*, *9*(5), 211–214. doi:10.1016/S0924-2244(98)00037-5

Morissey, T., & Almonacid, S. (2005). Rethinking technology transfer. *Journal of Food Technology*, *67*, 135–145.

Noy, N. F., & McGuiness, D. L. (2000). *Ontology development 101: a guide to creating your first ontology*. Retrieved May 10, 2009, from http://protege.stanford.edu/publications/ontology_development/ontology101.pdf

Regattieri, A., Gamberi, M., & Manzini, G. (2006). Traceability of food products: general framework and experimental evidence. *Journal of Food Engineering*, *81*, 347–356. doi:10.1016/j.jfoodeng.2006.10.032

Resende-Filho, M. A., & Buhr, B. L. (2008). A principal-agent model for evaluating the economic value of traceability system: a case study with injection. Site lesion control in fed castle. *American Journal of Agricultural Economics*, *90*(4), 1091–1102. doi:10.1111/j.1467-8276.2008.01150.x

Roussos, G. (2006). Enabling RFID in retail. *IEEE Computer*, *39*(3), 25–30.

Routledge, N., Bird, L., & Goodchild, A. (2002). UML and XML Schema. In X. Zhou (Ed.), *Proceedings of the Thirteenth Australasian Database Conference (ADC2002),* Melbourne, Australia (Vol. 5).

Salin, V. (1998). Information technology in agri-food supply chains. *International Food and Agribusiness Review*, *1*, 329–334. doi:10.1016/S1096-7508(99)80003-2

Salomie, J., Dinsoreanu, M., Bianca Pop, C., & Liviu Suciu, S. (2008, November 24-26). Model and SOA solutions for traceability in logistic chains. In *Proceedings of iiWAS2008*, Linz, Austria (pp. 339-344).

Sasazaki, S., Itoh, K., Arimitsu, S., Imada, T., Takasuga, A., & Nagaishi, H. (2004). Development of breed identification markers derived from AFLP in beef cattle. *Meat Science*, *67*, 275–280. doi:10.1016/j.meatsci.2003.10.016

The Extensible Markup Language (XML) Query Specification. (n.d.). *World Wide Web Consortium*. Retrieved from http://www.w3.org/XML/Query/

United Nations Centre for Trade Facilitation and Electronic Business (UN/CEFACT). (n.d.) Retrieved from http://www.unece.org/cefact/

Van Dorp, C. A. (2003). A traceability application based on gozinto graphs. In *Proceedings of the EFITA 2003 Conference,* Debrecen, Hungary.

Wilson, T. P., & Clarke, W. R. (1998). Food safety and traceability in the agricultural supply chain: using the internet to deliver traceability. *Supply Chain Management*, *3*, 127–133. doi:10.1108/13598549810230831

*This work was previously published in International Journal of Information Systems and Supply Chain Management, Volume 3, Issue 3, edited by John Wang, pp. 29-49, copyright 2010 by IGI Publishing (an imprint of IGI Global)*

# Chapter 18
# A Composite Method to Compare Countries to Ascertain Supply Chain Success:
## Case of USA and India

**Mark Gershon**
*Temple University, USA*

**Jagadeesh Rajashekhar**
*SDM Institute for Management Development, India*

## ABSTRACT

*Supply chains are assessed for the contribution they make in improving business processes. Assessment also looks at the return on investment and improves the overall functioning of the entire chain. However, supply chains extend beyond geographical borders and span a wide variety of activities; therefore, a systematic examination of factors required for success of supply chains is essential. This paper proposes a composite method by which supply chains could be assessed at multiple levels to enable a comprehensive comparison. The objective is to first compare at a global level and then narrow down to the firms' level. Although over time a number of measures have been developed to evaluate supply chain performance, this paper provides a methodology involving well-known techniques to assess the supply chain success based on objective considerations. Furthermore, the authors demonstrate how global players can select the partnering countries to reap maximum benefits. Finally, a comprehensive model is provided involving three approaches that look at the issue of comparison from different perspectives and are debated with respect to India and the United States.*

DOI: 10.4018/978-1-4666-0918-1.ch018

## INTRODUCTION

Supply chains have multiple objectives to satisfy and require different measures to ascertain their performance. Considering the fact that today's supply chain extend beyond borders and span a wide range of activities, assessment of supply chains is not a simple task. Further the question is at what level assessments should be made and how different entities can be measured. The entities here could be two countries, two firms or two sectors. Over a period of time several measures have evolved to suit different issues and aspects of supply chains and still some more may be developed considering the type of changes and new developments. This paper develops a composite model to assess the supply chain capabilities and potential that leads to success between say two partnering countries. Considering the huge investments made across the world, the supply chain performance assessment and also countrywide comparisons become crucial in making decisions about investments and types of changes to be made. The model described in this paper allows for comparing different countries or regions with regard to their support for successful international supply chains. Further the model starting from a country-wise level comparison moves down to assessment of specific success factors of supply chain in the second level comparison and finally ends up with comparison across a particular sector or domain selected from the previous level. As the comparison moves from the first level to the third level the scope of comparison reduces and becomes specialized. This multilevel comparison is accomplished with the help of different techniques that allow the comparison to be made smoothly and objectively. This in turn leads to better decision making process enabling higher success rate.

The multilevel comparison proposed in this paper examines the issue of comparison from three different levels as follows:

1. Using global competitive index as a basis for comparison across the countries and using the same factors to decide about the suitability of various countries for establishing supply chains particularly while partnering.
2. Using Analytical Hierarchy Process (AHP) to enable decision makers to compare the partnering countries by making pair-wise comparisons using specific criteria that are considered vital for the success of supply chains. The criteria considered at this level are a subset of the previous level factors.
3. Using Data Envelope Analysis (DEA) to create a series of inputs and outputs that can be used as a basis of comparison for each country. As in the previous step, the factors considered are a subset of the factors considered in the previous level.

In addition it is also possible to establish a fourth level comparison that is at the firms' level which examines all the supply chain success factors at a micro level. This ensures comparison down to operating mechanisms and environment at firms' level. However in this paper the fourth level comparison is kept outside the scope and is suggested as an extension for further work. All the above three levels of comparison are illustrated with a specific case involving comparison between USA and India as partnering countries.

The paper proceeds like this. The next section provides a literature review, at the end of which a discussion of criteria for evaluating supply chain performance is made. Immediately after that, the need for comparison across countries is shown and a new methodology is proposed. To illustrate the applicability of the proposed model a comparison between the USA and India is made. The paper concludes illustrating the benefits of the new model along with further extensions possible for the model.

## BRIEF LITERATURE REVIEW OF SUPPLY CHAIN MONITORING AND ASSESSMENT

Performance assessment of supply chains has aroused considerable interest among researchers who have examined various issues of supply chain for assessment. Stewart (1995) describes how performance assessment can be benchmarked, and Tan, Kannan, and Handfield (1998) comment that assessment can be at different levels. Monitoring supply chain performance is an intriguing new field according to Lee and Whang (2001). Gunasekaran, Patel, and Tirtiroglu (2001) have attempted to develop a framework for measuring the strategic, tactical and operational level performance in a supply chain. Based on trust, terms like Supply Chain Event Management, Supply Chain Process Management, or Supply Chain Execution Management are used interchangeably. Supply chain monitoring must start with tight tracking of the many different processes involved in a supply chain. As products and information flow through different parts of the supply chain, it is necessary to capture the information and ensure that the end users' requirements are satisfied. Supply chain automation is a major trend in this direction that offers a variety of tools and techniques to monitor and improve supply chain performance, (Huhns & Stephens, 2001).

Some other researchers, who have investigated the issue of performance measurement considering various aspects of supply chain include Chan (2003), Bhagwat and Sharma (2007), and Wong and Lee (2008). Brun, Salama, and Gerosa (2009) provide a framework for the selection of the right Performance Measurement System (PMS) for different Supply Chain typologies.

The Internet can be used to facilitate performance measurement across a supply chain and is currently a main platform to share information for the benefit of all the players along the supply chain. Performance can be assessed at different levels across different countries and sectors. For example, supply chain performance can be improved by focusing on five key areas namely:

1. Performance Monitoring - Measuring critical supplier and operator service attributes
2. Trend tracking - Identifying trends over time to aid in decision-making and help fine-tune distribution
3. Gap Analysis - Comparing expectations to actual results, so the team can work on closing gaps
4. Industry Benchmarks - Finding how well a supply chain is performing relative to the expected service levels of superior performers
5. Quality Assurance - Observing deficiencies before they become detrimental

### Supply Chain Performance Measures

Performance measures can be classified broadly into two categories (1) qualitative measures (such as customer satisfaction and product quality) and (2) quantitative measures (such as order-to-delivery lead time, supply chain response time, flexibility, and resource utilization). Improving supply chain performance requires a multi-dimensional strategy that addresses how the organization will service diverse customer needs. While the performance measurements may be similar, the specific performance goals of each segment may be quite different. Further, the quantitative metrics of supply chain performance can be divided into two major groups: (1) Non-financial and (2) Financial (Biswas, 2000).

- Some of the non-financial performance measures include the Cycle time, Customer Service Level, Inventory Levels, and Inventory Turns. Financial measures are used to assess different types of fixed and operational costs associated with a supply chain and ultimately, the aim is to maximize the revenue by keeping the supply chain costs low. Costs arise due to inven-

*Table 1. Some commonly used performance measures*

| Metric | Calculation |
|---|---|
| Number of open orders | Number of order line items that have not been shipped. . |
| Volume of Open Orders | Total volume on order |
| Unavailability of trucks/rail cars | On-Hand + estimated arriving – required by schedule |
| Month to date orders vs. forecast | Shipped orders divided by the sum of estimate the forecast of shipments to date and the shipped orders. The forecast of shipments is calculated using the order profiler. |
| Value of receivables | Total dollar value of receivables |
| Weighted dollar days | SUM(Dollar value of each outstanding invoice * number of days since issues) / total value of outstanding invoices |
| Shipped and not invoiced orders | Number of orders that have been shipped but not invoiced. As above, allow for volume measure in addition to a count. |
| Month-to-date expediting costs | Total accumulated sum of expediting costs |
| Days of Supply | Volume of inventory / average daily use for each critical raw material. Also for Finished Product and Semi-Finished and Totals. Allow basis to be $ or Volume. Allow a basis of history or forecasts, and with varying timeframes (example: based on 60 days of history or based on next 3 months' forecast). Also allow the measure to be at an attribute level of detail, such as Days Supply by prod type, or by color, etc. |
| Schedule adherence Matrix | Cumulative daily scheduled production vs. cumulative closed production orders from ERP system. (need to clarify if lot release happens before or after the production order closes) |
| Total daily production | Volume of daily production |

tories, transportation, facilities, operations, technology, materials, and labor.

Some commonly used performance measures are listed in Table 1.

## Critical View of Different Metrics and Motivation for Present Research

Traditionally supply chain performance has been assessed at firms' level and several researchers have come out with different measures that can be used as performance indicators. Thus it is possible to judge the performance of supply chains based on a variety of metrics as discussed in the previous sections. Most of these metrics are applicable at the firm's level and hence are useful to compare two organizations that form a part of the supply chain. An exclusive survey by Eicher Consultancy Services (2005) that covered India's top 1000 companies over a period of six years ending at 2004, has tried to establish the trends to track improvement in supply chain metrics. The survey comments that more than two thirds of the companies that were analyzed exhibited improvement score higher than 8 (on a scale of 0 to 15) and hence showed ample scope for improvement.

Kim (2006) provides a summary of a variety of measures that have been developed by various researchers during the last several years and categorizes them as measures to judge (a) supply chain operational capability, (b) competitive capability, and (c) firm performance. Honggeng and Benton (2007) discuss many issues of information sharing, but do not go so far as to suggest an approach to use that data for comparison purposes.

The key barriers of Supply Chain Performance Measurement System (SCPMS)' have been discussed by Charan, Shankar, and Baisya (2009).

While all these measures have been successfully used to assess the performance of supply chains considering the operational success at firms' level, the question also is how many of these measures would be necessary for a better assessment or how often these measures be used. In an earlier paper, Fisher (1997) comments that the managers lack a framework for determining which methods are appropriate. Further the questions would be how to start comparison when much broader issues are involved and many related factors need to be considered. These questions assume greater significance as companies are looking for partners outside their countries for variety of reasons including outsourcing, collaborating, on site manufacturing or assembly, or even strategic partnerships for specific purpose like developing a local export hub. This is where the currently available firms' level metrics prove to be only partially helpful because the scale of comparison has to include many "macro" issues going beyond the boundaries of the individual organizations. The authors of this paper have thus proposed a multilevel comparison which is sequential in nature that systematically examines the various macro and micro issues to provide a more meaningful and helpful comparison.

## RESEARCH METHODOLOGY

The authors propose a composite method which involves the following three levels of comparison. It is assumed that such a comparison is narrowed down to two partnering countries which have been identified to be a part of the proposed supply chain.

The development of the composite method proceeds like this:

1. First level comparison using Global Competitiveness Index, (GCI) developed by World Economic Forum, which is reviewed annually and updated. These rankings are not constant but change due to changes taking place within the country and outside.
2. Second level comparison using Analytical Hierarchy Processing (AHP). This uses some of the pertinent criteria required for supply chain success based on well established practices or preferences. These will be a subset of the data used in the first level.
3. The third and final level of comparison using Data Envelope Analysis (DEA) to further narrow down the comparison process to select micro factors to enable a better decision process.

The above three approaches were selected because comparison studies require an approach that can score or rank all of the alternatives, not just limited to develop an optimal solution. These methods chosen are all well known for ranking sets of alternatives. They are also flexible enough to allow different decision makers to reach conclusions relevant to a given context.

### Comparison Using Global Competitive Index

The Global Competitiveness Index (GCI) is a measure of a country's strengths and weaknesses in terms of some factors identified by various researchers and administered by the World Economic Forum. The GCI though simple in nature provides a holistic overview of factors that are critical to driving productivity and competitiveness and groups the factors into nine pillars as follows:

1. Institutions
2. Infrastructure
3. Macro-economy
4. Health and primary education
5. Higher education and training
6. Market efficiency
7. Technological readiness
8. Business sophistication
9. Innovation

As stated by Lopez – Claros (2006) in the executive summary of the Global Competitiveness Report 2006-07, the selection of these nine pillars and the factors underlying them is based on the best theoretical and empirical research. It is important to note that none of these factors alone can ensure competitiveness.

Beyond these pillars, which capture a more comprehensive set of growth factors, the GCI has a number of other important distinguishing features. One is the formal incorporation of the notion that countries around the world are functioning at different stages of economic development. The relative importance of particular factors for improving the competitiveness of a country will be a function of the starting conditions, that is, those institutional and structural features which characterize a country in comparison with others in terms of development, as measured by per capita income. For example, what presently drives productivity in Sweden is necessarily different from what drives it in Ghana. Thus, the GCI separates countries into three specific stages: factor-driven, efficiency-driven, and innovation-driven, each implying a growing degree of complexity in the operation of the economy.

The pillars are organized into three sub-indexes, each critical to a particular stage of development:

A. "The basic requirements sub-index" groups those pillars most critical for countries in the factor-driven stage (institutions, infrastructure, macro-economy, health and primary education)
B. "The efficiency enhancers sub-index" includes those pillars critical for countries in the efficiency-driven stage (higher education and training, market efficiency, technological readiness)
C. "The innovation and sophistication factors sub-index" includes all pillars critical to countries in the innovation-driven stage (business sophistication, innovation).

## Comparison Using Analytical Hierarchy Processing (AHP)

Analytical Hierarchy Processing (AHP) is a well proven technique for decision making under multi-criteria situation. Developed by Thomas L. Saaty the Analytic Hierarchy Process is a procedure designed to quantify managerial judgments of the relative importance of each of several conflicting criteria used in the decision making process (Saaty, 1980).

AHP is based upon the well-established and theoretically sound techniques of (1) structuring problems into hierarchies, (2) reducing complex judgments into a series of pair-wise relative comparisons, (3) using redundant judgments to assess participant consistency, and (4) using an eigenvector method for deriving weights. These techniques are directly applicable to problem of prioritizing and comparing strategic objectives and particularly well-suited to the challenges of implementing a multi-criteria performance management system.

AHP uses hierarchical structures to solve complicated, unstructured decision problems, especially in situations where there are important qualitative aspects that must be considered in conjunction with various measurable quantitative factors (Rangone, 1996). Applications of AHP include wide ranging fields like manufacturing, hospital management, faculty selection, and various support functions of an organization. Vargas (1990) provides an overview of various application of AHP.

As stated by Liedtka (2005), AHP is a popular method for assessing multiple criteria and deriving priorities for decision-making purposes. Major companies (e.g., Ford, General Electric), public accounting firms (e.g., KPMG, Pricewaterhouse Coopers) and government agencies (e.g., United States Treasury Department, United States State Department) already utilize AHP for various purposes. Additionally, academics have employed AHP in over 2,000 studies. In the accounting

*Table 2. The business competitiveness index*

| Country | BCI Ranking | Quality of the national business environment ranking | Company operations and strategy ranking |
|---|---|---|---|
| USA | 1 | 1 | 1 |
| India | 27 | 27 | 25 |

literature, for instance, researchers have applied AHP to a number of complex problems such as analytical review, internal control evaluation, and assessment of management fraud "red flags". Dey, Hariharan, and Clegg (2006) have used AHP to measure the operational performance in health care industry. Xia and Lim (2007) demonstrate the application of AHP for supply chain performance measurement.

### Comparison Using Data Envelope Analyses

In the last two decades Data Envelope Analyses (DEA) has seen a wide range of applications and continues to be used for comparison in those cases where there are multiple inputs and multiple outputs. DEA is extensively covered by Charnes, Cooper, and Rhodes (1978), Banker, Charnes, and Cooper (1984), and Ramanathan (2003). Emrouznejad (2001) provides an extensive bibliography of Data Envelopment Analysis.

DEA is a popular technique to measure the efficiency of units when there are multiple inputs and multiple outputs. In an earlier paper it was observed that Narasimhan, Talluri, and David (2001) have used DEA for supplier evaluation. As the supply chain performance is dependent on multiple inputs and a supply chain is expected to serve several objectives, the choice of DEA for the performance assessment of supply chains is justified in this paper. For example transportation is considered as a vital aspect of supply chain, with movement of goods happening through rail, road, air, and water, which are considered as inputs. At the output side, the multiple outputs can be safe delivery, on time delivery, minimum cost of travel, and minimum resource consumption. Thus DEA can help in the performance analysis.

## INTERNATIONAL COMPARISON: CASE OF USA AND INDIA

The present case discusses the assessment and comparison of factors responsible for supply chain success between USA and India using the composite method developed by the authors. For the first level comparison the data available in the Global Competitiveness Report has been used. The data used is for an overall comparison and indicates the relative strengths and weaknesses of the two countries with respect to various factors. For the next two levels of comparison data collected as a part of the Executive MBA program has been used. One of the authors had the opportunity to supervise the projects carried out by the students of Executive MBA program as a part of the supply chain management course. This enabled the authors to obtain the data from those projects and use the same for analysis in their research work.

### Comparison of USA and India Using GCI

The data used in establishing the GCI for the two countries is reproduced in Table 2, Table 3, Table 4, and Table 5. The numbers in the columns indicate the overall ranking.

Looking at these tables it can be stated that United States is way ahead of India on all the factors considered.

*Table 3. Global competitiveness index: basic requirements*

| Country | Basic Requirements | 1st Pillar | 2nd Pillar | 3rd Pillar | 4th Pillar |
|---|---|---|---|---|---|
| U S A | 27 | 27 | 12 | 69 | 40 |
| India | 60 | 34 | 62 | 88 | 93 |

*Table 4. Global competitiveness index: efficiency enhancers*

| **Country** | **Efficiency Enhancers** | **5th Pillar** | **6th Pillar** | **7th Pillar** |
|---|---|---|---|---|
| **U S A** | 1 | 5 | 2 | **8** |
| **India** | **41** | **49** | **21** | **55** |

*Table 5. Global competitiveness index: innovation factors*

| Country | Innovation factors | 8th Pillar | 9th Pillar |
|---|---|---|---|
| U S A | 4 | 8 | 2 |
| India | 26 | 25 | 26 |

*Table 6. Most problematic factors for doing business*

| **Factor** | **Percent of responses** | **Percent of responses** |
|---|---|---|
| Inadequate supply of infrastructure | 26.30 | 3.62 |
| Inefficient Government bureaucracy | 18.45 | 9.89 |
| Restrictive labor regulations | 15.52 | 7.12 |
| Corruption | 10.50 | 4.03 |
| Tax regulations | 7.00 | 15.43 |
| Tax Rates | 4.45 | 15.46 |
| Policy instability | 4.26 | 4.49 |
| Access to financing | 3.60 | 6.07 |
| Poor work ethic in national labor force | 2.65 | 7.12 |
| Foreign currency regulations | 2.55 | 3.62 |
| Inadequately educated workforce | 2.27 | 9.33 |
| Inflation | 1.04 | 6.71 |
| Crime and theft | 0.76 | 4.43 |
| Government instability/coups | 0.66 | 2.68 |

*Table 7. Overall ranking of India and USA (out of 125) considering supply chain issues*

| Factor | India | U S A |
|---|---|---|
| Extent of marketing | 29 | 2 |
| Local supplier quantity | 9 | 6 |
| Control of international distribution | 25 | 6 |
| Willingness to delegate authority | Not rated | 8 |
| Local supplier quality | 28 | 10 |
| Production process sophistication | 33 | Not rated |
| Value chain presence | 22 | Not rated |

Table 6 lists the most problematic factors for doing business in India compared with United States. This to some extent explains why India's ranking is low in many areas.

With reference to supply chain issues how the two countries are ranked (out of 125) is shown in Table 7.

Observing the overall rank of GCI, Lopez – Claros (2006) state that the United States is sixth during the year of assessment. It remains a world leader in a number of key categories assessed by the GCI, such as market efficiency, innovation, higher education and training, and business sophistication. However, growing imbalances have dented a number of macroeconomic indicators, and the levels of efficiency and transparency underpinning its public institutions do not match those of the most developed industrial countries.

India's overall rank of 43 demonstrates remarkably high scores in capacity for innovation and sophistication of firm operations. This is especially true of the quality of scientific research and the number of scientists and engineers, which are increasingly supplying highly skilled professionals to the private sector. Firm use of technology and rates of technology transfer are high, although penetration rates of the latest technologies are still quite low by international standards, reflecting India's low levels of per capita income and high incidence of poverty. However, weaknesses in the coverage of educational opportunities and poor-quality infrastructure limit the more equitable distribution of the benefits of India's high growth rates. It may be observed that though there appears to be a big gap between the two countries many US firms have made significant investments in India exploring various business opportunities. From the supply chain perspective looking at the essential factors India scores poorly on all the three success factors namely Business Requirements, Efficiency Enhancers, and Innovation Factors. However it may noted that this comparison is broad based and does not reflect the differences that may be favoring one sector and not so for another sector. Hence a deeper examination is necessary.

## Comparison of USA and India Using AHP

The second step is to use the AHP methodology as illustrated by Scott (2006). The purpose is to refine the results of the previous step with regard to a particular set of criteria. So where the previous step looks at published data for many criteria, at this step anyone can refine the process using criteria relevant to their situation. For our purpose here, the basic criteria to evaluate a country's supply chain management efficiency can be divided into four general categories namely, (1) infrastructure, (2) workforce, (3) political and cultural factors, and (4) risks. These four categories make up the first row of the hierarchy. The subcategories are a follows:

*Table 8. Final priority matrix for internal transportation method in USA*

| Goal | Preference |
|---|---|
| Rail | .296 |
| Road | .320 |
| Water | .138 |
| Air | .247 |

1. Infrastructure: Infrastructure is the largest section and the easiest one to quantify. Infrastructure is further broken down into three categories. The first category is the movement of material within a country or internal transportation. The second category is movement of material into or out of a country or external transportation. The third category is communication.
2. Workforce: Workforce is designed to be descriptive of the available labor within the country. It is further refined into the three categories of availability, capability, and cost.
3. Political and cultural: The political and cultural section includes the four categories of legal climate, political climate, business climate, and social acceptance
4. Risks: For the purpose of supply chain management risks are further broken down into political stability, economic stability, and security.

Using the data provided by Kanoi (2006) as an example the supply chain management efficiencies of the U.S. and India are compared. Starting with the internal transportation sub-hierarchy a matrix based on the decision alternatives of rail, road, water, and air is created. This is followed by a priority matrix for cost, time, current load, capacity, reliability, and growth trend. To conserve space only the final prioritizations for the decision point are shown in Table 8.

*Table 9. Priorities for the U.S. and India in the main hierarchy*

| Goal | Preference |
|---|---|
| India | .477 |
| U.S. | .533 |

In this case the preferred internal transportation method for the U.S. is road transportation with a preference of 32%, although rail came in a close second at 29.6% and could be a good alternative. Completing all the priorities for the U.S. and India in the main hierarchy gives the final goal prioritization as given in Table 9.

Further, using the gradient curves (Scott, 2006) one can see how the preference for the U.S. and India changes as the preference for each criteria change. For example, infrastructure has the highest preference at 60.6%. Decreasing this preference will increase the overall preference for India. At 40% India and the U.S. are equally preferred and moving the preference lower will make India preferred over the U.S. The opposite can be said for workforce. The more the workforce preference increases the more India will be preferred. If the workforce preference is increased past 30% then India will be preferred higher than the U.S. Looking at the political cultural and risk criteria we see that the U.S. will be slightly preferred for every preference setting though the gap widens as the preference grows.

This analyses shows that the U.S. has the superior infrastructure rating and our strong preference for infrastructure over the other criteria placed the U.S. higher than India. However, India was close behind because it has a superior workforce rating due to high preference for workforce cost. If workforce preference is increased or infrastructure preference is decreased then India could be preferred over the U.S.

It is to be noted that AHP may not necessarily make a final decision for the stated goal. However it allows to analyze the preferences and see how

| | |
|---|---|
| India: | $124.27T_1 + 411.36T_2$ |
| | $2.08W_1 + .034W_2 + .0003W_3$ |
| U S A: | $497.1T_1 + 1{,}200.701T_2$ |
| | $1.13W_1 + .04W_2 + .0042W_3$ |

they relate to and effect the different options in a complex decision making process.

## Comparison of USA and India Using Data Envelopment Analysis

The most detailed phase of the analysis uses the DEA. After looking at the high level criteria with AHP, the DEA will assess the next level of criteria, the most detailed level. For this model we are going to choose a country's support to supply chain management as a process. Next we need to develop a series of inputs and outputs to describe the process. In the AHP model many subjective values were added. However, in DEA we are going to concentrate on objective values and focus on infrastructure. While this is only one of our four criteria, focusing on one here allows us to demonstrate the approach in a more understandable manner. Again, we are using the data from Kanoi (2006).

The following inputs and outputs are created:

**Inputs**

1. Road Density ($W_1$) measured in miles per square mile
2. Rail Density ($W_2$) measured in miles per square mile
3. Airport Density ($W_3$) measured in airports per square mile

**Outputs**

1. Travel per day ($T_1$) measured in miles
2. Railway Tons per mile ($T_2$) measured in millions of tons

Using these inputs and outputs we can create an output to input ratio for each country.

If the process is 100% efficient the output to input ratio will be equal to 1. It is impossible to get more output than the input you entered and you cannot get less than zero output so all of the output to input ratios must be positive and less than or equal to 1. DEA uses this fact as a way to compare the different processes. Using the output to input ratios we are going to build a linear program for each country. In order to get the comparisons we will use the other countries output to input ratio being less than or equal to 1 as an added constraint. It is written as the inputs minus the outputs being greater than or equal to zero, however, it means the same thing. The rest of the linear program requires setting the inputs to 1 and maximizing the outputs to see how close to 100% we can get it. Setting the inputs to 1 automatically puts the answer in decimal percentage form.

**India:**

Max $z = 124.7T_1 + 411.36T_2$

$124.7T_1 - 411.36T_2 + 2.08W_1 + .034W_2 + .00029W_3 \geq 0$

$-497.1T_1 - 1200.701T_2 + 1.13W_1 + .04W_2 + .0042W_3 \geq 0$

$2.08W_1 + .034W_2 + .00029W_3 = 1$

All Variables $\geq 0$

**U.S.A.**

Max $z = 497.1T_1 + 1200.701T_2$

$-124.7T_1 - 411.36T_2 + 2.08W_1 + .034W_2 + .00029W_3 \geq 0$

$-497.1T_1 - 1200.701T_2 + 1.13W_1 + .04W_2 + .0042W_3 \geq 0$

$1.13W_1 + .04W_2 + .0042W_3 = 1$

All Variables $\geq 0$

Using the computer to solve each of these linear programs we find that the U.S. has an efficiency of 1 and India has an efficiency of .9986. When it comes to travel per day and tons per railroad mile India is almost as efficient as the U.S. In the AHP model the U.S. was significantly better on the infrastructure however we see here that India could easily catch up. By looking at both the DEA and AHP model we see that India is fairly efficient on the functional parts of its infrastructure. To improve India needs to increase its reliability and capacity. If India can do this it will be able to surpass the U.S. because of its already high preference on workforce and the models high preference on infrastructure.

## CONCLUSION

In this paper a sequence of three approaches has been used to compare two countries, namely USA and India to assess the supply chain performance. Considering several factors that promote supply chain efficiency, these approaches develop a quantitative comparison based on various issues. While the superiority of USA is well established in the comparisons, India offers its own benefits in certain matters like availability of skilled manpower and low cost of operations. If the investors or administrators are looking for specific factors like infrastructure, political climate, etc., they can first look at Global Competitive Index and get an overall rating. Then in the next step they can use techniques like AHP and DEA to come out with quantitative assessment of various issues. It may be necessary to look at the various metrics and their usability in any given situation. Based on the observations, certain fine tuning may be necessary to make the metrics more relevant and useful to a given industry. In this regard further research is well justified and would help industries to benchmark their supply chains.

Our main result is not any conclusive opinion about the two countries concerned. Rather it offers a methodology for such comparisons and leaves the final decision to the concerned decision makers.

## REFERENCES


Banker, R. D., Charnes, R. W., & Cooper, W. W. (1984). Some models for estimating technical and scale inefficiencies in data envelopment analysis. *Management Science*, *30*, 1078–1092. doi:10.1287/mnsc.30.9.1078

Bhagwat, R., & Sharma, M. K. (2007). Performance measurement of supply chain management: A balanced scorecard approach. *Computers & Industrial Engineering*, *53*(1), 43–62. doi:10.1016/j.cie.2007.04.001

Biswas, S. (2000). *Supply chain performance measures*. Retrieved December 8, 2008, from http://lcm.csa.iisc.ernet.in/scm/coimbatore/node11.html

Brun, A., Salama, K. F., & Gerosa, M. (2009). Selecting performance measurement systems: matching a supply chain's requirements. *European Journal of Industrial Engineering*, *3*(3), 336–362. doi:10.1504/EJIE.2009.025051

Chan, F. T. S. (2003). Performance measurement in a supply chain. *International Journal of Advanced Manufacturing Technology*, *21*, 534–548. doi:10.1007/s001700300063

Charan, P., Shankar, R., & Baisya, R. (2009). Modeling the barriers of supply chain performance measurement system implementation in the Indian automobile supply chain. *International Journal of Logistics Systems and Management*, *5*(6), 614–630. doi:10.1504/IJLSM.2009.024794

Charnes, A., Cooper, W., & Rhodes, E. (1978). Measuring the efficiency of decision-making units. *European Journal of Operational Research*, *2*, 429–444. doi:10.1016/0377-2217(78)90138-8

Dey, P. K., Hariharan, S., & Clegg, B. (2006). Measuring the operational performance of intensive care units using the analytical hierarchy process approach. *International Journal of Operations & Production Management*, *26*(8), 849–865. doi:10.1108/01443570610678639

Eicher Consultancy Services. (2005). Supply Chain Metrics. *Industry 2.0*, 87-148.

Emrouznejad, A. (2001). *An extensive bibliography of data envelopment analysis (DEA), Volume I – V.* Retrieved from http://www.warwick.ac.uk/~bsrlu

Fisher, M. L. (1997). What is the right supply chain for your product. *Harvard Business Review*, *75*(2), 105–116.

Gunasekaran, A., Patel, C., & Tirtiroglu, E. (2001). Performance measures and metrics in a supply chain environment. *International Journal of Operations & Production Management*, *21*(1-2), 71–87. doi:10.1108/01443570110358468

Huhns, M. N., & Stephens, L. M. (2001). Automating supply chains. *IEE Internet Computing*. Retrieved from http://computer.org/internet/

Kanoi, M. (2006). *Working paper on supply chain management*. Temple University, Philadelphia.

Kim, S. W. (2006). The effect of supply chain integration on the alignment between corporate competitive capability and supply chain operational capability. *International Journal of Operations & Production Management*, *26*(10), 1084–1107. doi:10.1108/01443570610691085

Lee, H. L., & Whang, S. (2001, November). *E-Business and supply chain integration* (Tech. Rep. No. SGSCMF- W2-2001). Stanford, CA: Stanford University, Stanford Global Supply Chain Management Forum.

Liedtka, S. L. (2005). The analytic hierarchy process and multi-criteria performance management systems. *Journal of Cost Management*, *19*(6), 30–38.

Lopez-Claros, A. (2006). *The Global Competitiveness Report 2006-2007* (28th ed.). Basingstoke, UK: Palgrave.

Narasimhan, R., Talluri, S., & Mendez, D. (2001). Supplier evaluation and rationalization via data envelopment analysis: an empirical examination. *Journal of Supply Chain Management*, *37*(3), 28–37. doi:10.1111/j.1745-493X.2001.tb00103.x

Ramanathan, R. (2003). *An Introduction to Data Envelopment Analysis: A tool for Performance Measurement*. Thousand Oaks, CA: Sage.

Rangone, A. (1996). An analytical hierarchy process framework for comparing the overall performance of manufacturing departments. *International Journal of Operations & Production Management*, *16*(8), 104–119. doi:10.1108/01443579610125804

Saaty, T. L. (1980). *The Analytic Hierarchy Process*. New York: McGraw-Hill.

Scott, I. (2006). *Analyses of supply chain management efficiency by country*. Working paper on supply chain management, Temple University, Philadelphia.

Stewart, G. (1995). Supply chain performance benchmarking study reveals keys to supply chain excellence. *Logistics Information Management*, *8*(2), 38–44. doi:10.1108/09576059510085000

Tan, K. C., Kannan, V. R., & Handfield, R. B. (1998). Supply chain management: supplier performance and firm performance. *International Journal of Purchasing and Materials Management*, *34*(3), 2–9.

Vargas, L. G. (1990). An overview of the analytic hierarchy process and its applications. *European Journal of Operational Research, 48*(1), 2–8. doi:10.1016/0377-2217(90)90056-H

Wong, W. P., Jaruphongsa, W., & Lee, L. H. (2008). Supply chain performance measurement system: a monte carlo DEA-based approach. *International Journal of Industrial and Systems Engineering, 3*(2), 162–188. doi:10.1504/IJISE.2008.016743

Xia, L. X. X., Bin, M., & Lim, R. (2007, September 25-28). AHP based supply chain performance measurement system. In *Proceedings of ETFA IEEE Conference on Emerging Technologies and Factory Automation*, Patras, Greece (pp. 1308-1315). Washington, DC: IEEE.

Zhou, H., & Benton, W. C. Jr. (2007). Supply chain practice and information sharing. *Journal of Operations Management, 25*(6), 1348–1365. doi:10.1016/j.jom.2007.01.009

*This work was previously published in International Journal of Information Systems and Supply Chain Management, Volume 3, Issue 3, edited by John Wang, pp. 66-79, copyright 2010 by IGI Publishing (an imprint of IGI Global)*

# Chapter 19
# An Automated Supply Chain Management System and Its Performance Evaluation

**Firat Kart**
*Tibco Software, Inc., USA*

**Louise E. Moser**
*University of California, Santa Barbara, USA*

**P. M. Melliar-Smith**
*University of California, Santa Barbara, USA*

## ABSTRACT

*The MIDAS system is an automated supply chain management system that enables customers, manufacturers, and suppliers to cooperate over the Internet. MIDAS aims to achieve high customer satisfaction by supporting the build-to-order customization model and to reduce inventory carrying costs and logistics administration costs at the manufacturer by supporting the just-in-time manufacturing model. It allows a manufacturer to choose from the MIDAS Registry, suppliers of components, and negotiate based on the prices, availability, and delivery times of those components. The manufacturer can use one of several strategies to aggregate customers' orders before processing them, and one of several strategies to accumulate suppliers' quotes before deciding on a particular supplier. The paper presents an evaluation of these strategies in terms of the customer's satisfaction, as measured by the customer response time, and the manufacturer's gain, as measured by the number of orders aggregated or the best price ratio.*

## INTRODUCTION

Over the last decade, globalization and the Internet have led to a major shift in business management thinking which, in turn, has had a significant impact on how companies do business. Globalization has affected how businesses interact with other businesses, and has resulted in increased consolidation and competition, as many industries have gone through a process of removing restrictions and regulations worldwide (Murch, 2004). New business models have emerged with the Internet,

DOI: 10.4018/978-1-4666-0918-1.ch019

the World Wide Web, and related technologies, such as the Service Oriented Architecture and Web Services (Moser & Melliar-Smith, 2009).

One of the new business models is the build-to-order model (Gunasekarana, 2005), which enables a customer to customize a product by choosing the materials that constitute the product. The build-to-order model is appropriate when the customers attach substantial value to ownership of a customized product, such as mobile, hand-held or wearable computing and communication devices, clothing and jewelry, automobiles, etc. For certain kinds of products, the build-to-order model depends strongly on standardized interfaces that allow the assembly of products from components and the substitution of one component for another.

Supply chains are profoundly challenged by the new business and technology environment (ComputerWorld, 2006). A supply chain moves products or services from the suppliers to the customers. It involves suppliers providing raw materials or services, manufacturers assembling components into products, warehouses storing raw materials and manufactured goods, distributors providing finished products or services to customers, and customers purchasing products and services.

The main objective of supply chain management is to achieve the most efficient use of resources to meet the needs of the customers (Wikipedia, 2008). Among the highest costs incurred by the manufacturers are inventory carrying costs and logistics administration costs (Wilson, 2005). The just-in-time manufacturing model aims to reduce those costs, by manufacturing products on-demand and reducing the physical inventory in warehouses, resulting in a more efficient supply chain. Supply chain management deals with three types of flow:

- **Information flow:** Pertains to placing, transmitting and filling orders, and updating their delivery status
- **Product flow:** Involves movement of goods from a supplier to a customer, as well as customer returns
- **Financial flow:** Relates to credit terms, payments, payment schedules, consignment, and title ownership.

The MIDAS (Managing Integrated Demand and Supply) system that we have developed focuses on information flow. It provides a dynamic environment for customers, manufacturers, and suppliers to cooperate as they have never done before. The MIDAS system in one enterprise interacts with the MIDAS system in other enterprises dynamically over the Internet. MIDAS supports communication between manufacturers and suppliers, even if the manufacturer did not have any prior business with those suppliers and, thus, it increases the ease of collaboration between them. MIDAS makes it easier for small suppliers to get into business with large manufacturers by automating the procurement process. Most importantly, MIDAS aims to meet the needs of the customers on time, and to reduce the costs of the manufacturer by eliminating the need for a large inventory.

At the manufacturer, MIDAS receives orders from the customer, and places orders with the suppliers, automatically and dynamically. MIDAS allows a manufacturer and the suppliers to negotiate a business deal on-line either by accepting a quote as is, or by negotiating. MIDAS allows the manufacturer to use one of several strategies to aggregate customers' orders before processing them, and one of several strategies to accumulate suppliers' quotes before deciding on a particular supplier. This paper presents an evaluation of these strategies in terms of the customer's satisfaction, as measured by the customer response time, and the manufacturer's gain, as measured by the number of orders aggregated or the best price ratio.

## THE MIDAS ARCHITECTURE

The MIDAS system is based on the idea of the Service Oriented Architecture (Erl, 2005; Moser & Melliar-Smith, 2009; Newcomer, 2004; OASIS, 2006) and Web Services (Alonso et al., 2004; Chatterjee & Webber, 2003; W3C, 2004) to achieve its objective of automated supply chain management. The advantages of MIDAS, and of the Service Oriented Architecture in general, are that it increases business flexibility and it lets businesses adapt more quickly to changing business needs.

As a Service Oriented Architecture (SOA), MIDAS leverages existing IT infrastructure to enable users to automate their supply chains. Through its use of Web Services, MIDAS provides interoperability between legacy back-end enterprise software systems. The MIDAS software is modular, which allows it to be re-used at multiple levels of the supply chain and to be modified without disrupting the services provided to the consumers. By applying SOA practices to supply chain applications, MIDAS aims to automate the supply chain and, thereby, to reduce human intervention, errors, and costs.

In this paper we consider a three-level supply chain and a single manufacturer; however, as shown in Figure 1, the MIDAS strategy generalizes to deeper supply chains with $N$ levels, $N \geq 3$, where a manufacturer is a supplier of the products it manufactures and a supplier is a manufacturer of the supplies that it offers. MIDAS can also accommodate other participants within its information flow, such as contract assemblers and shippers. MIDAS is present at the businesses in the supply chain that act as both manufacturer and supplier. By considering the entire supply chain, MIDAS captures supply chain needs more effectively and provides faster adaptation to changing supply and demand.

MIDAS uses the following concepts in supply chain management. A *material* is a product that is sold to a customer and that consists of one or more components. A *component* is a particular category for which there are one or more supplies. A *supply* is a product that is produced by a supplier. A supply can be obtained from one or more suppliers, and the manufacturer can select the supplier of that supply dynamically. A *supply item* is an instance of a supply.

MIDAS also uses the concept of *logical inventory* in supply chain management. Logical inventory is data, stored in the computers and databases of the supply chain management system that are related to the customers' needs and the customers' orders. With logical inventory, the manufacturer does not need to maintain a large physical inventory in its warehouses but, rather, can obtain the supplies that it needs on demand.

Typically, a Service Oriented Architecture is fronted by a client user interface that uses underlying services; the end users see only this user interface. Depending on the business application, a customized user interface is provided for customers that use the underlying services. Although the MIDAS client user interface is customized for the particular application and product, the underlying MIDAS system is general and can be used by manufacturers and suppliers of various kinds of products.

As infrastructure software that supports Web Services, MIDAS uses a Registry that allows a manufacturer to discover suppliers, to select suppliers dynamically, and to find alternate suppliers on demand if an existing supplier becomes unavailable, and to redirect its requests seamlessly to an alternate supplier. MIDAS uses a Reservation Protocol (Zhao et al., 2008) to improve the performance of business transactions that span multiple businesses in the supply chain. Use of the Reservation Protocol decreases the probability of inconsistencies for business transactions between manufacturer and suppliers.

At the manufacturer, the MIDAS system comprises the following modules: Materials Manager, Orders Manager, Database (DB) Monitor, Communication Manager, and Quotes Manager. These modules are shown in Figure 2.

*Figure 1. Use of MIDAS in a supply chain*

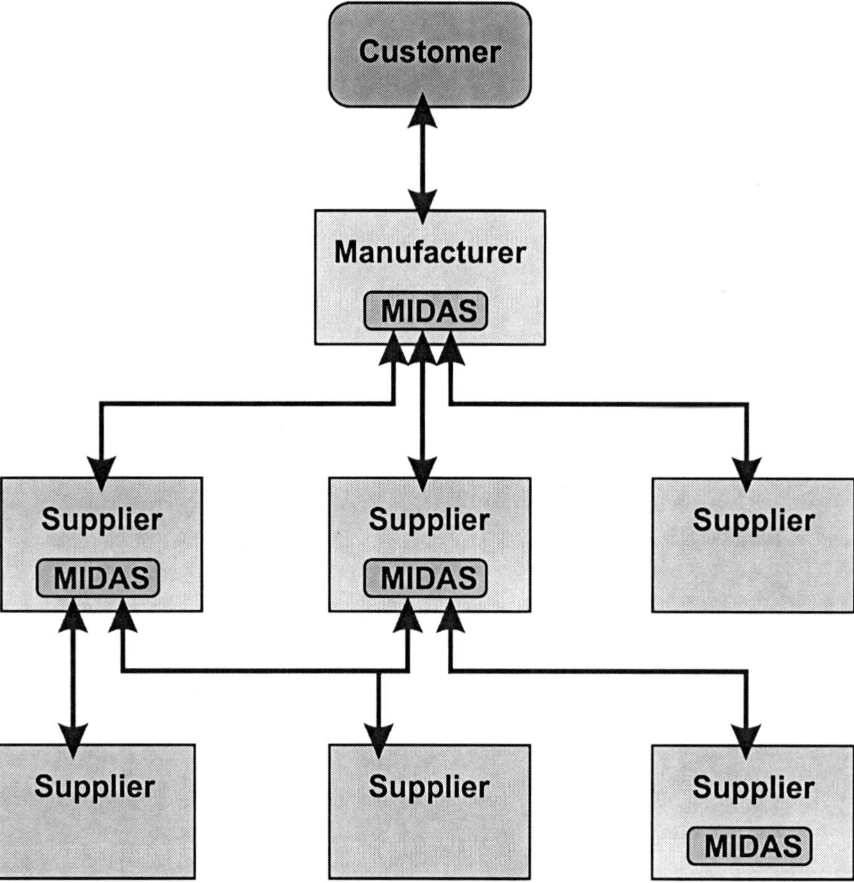


Customers obtain product information from a catalog provided by a Customer Web Service, which retrieves information from the Materials Manager. The Materials Manager relates a material to its components and a component to its supplies. On receiving orders from the customers, the Materials Manager passes the information to the Orders Manager. The Orders Manager inserts, into the Orders Database, information about the customers and the products that the customers are interested in purchasing, and manages the status of the customers' orders. On receiving an order, the Orders Manager informs the DB Monitor, which scans the orders and triggers a business activity that starts purchasing supplies from suppliers. The DB Monitor checks the Orders Database and decides, depending on the particular strategy chosen, whether to inform the Quotes Manager to initiate a search for suppliers and communicate with them. The Quotes Manager handles Quote requests, and relates Quote replies to Quote requests.

Each of these modules plays a role in the two phases of the manufacturer's processing an order from a customer. These two phases are:

- **Waiting phase:** Involves the collection of orders from the customers before making Quote requests for aggregated supply items from the suppliers
- **Quotes phase:** Involves the collection of Quote replies from the suppliers, and making a decision on which supplier will provide the particular supply.

*Figure 2. The modules of MIDAS at the manufacturer*

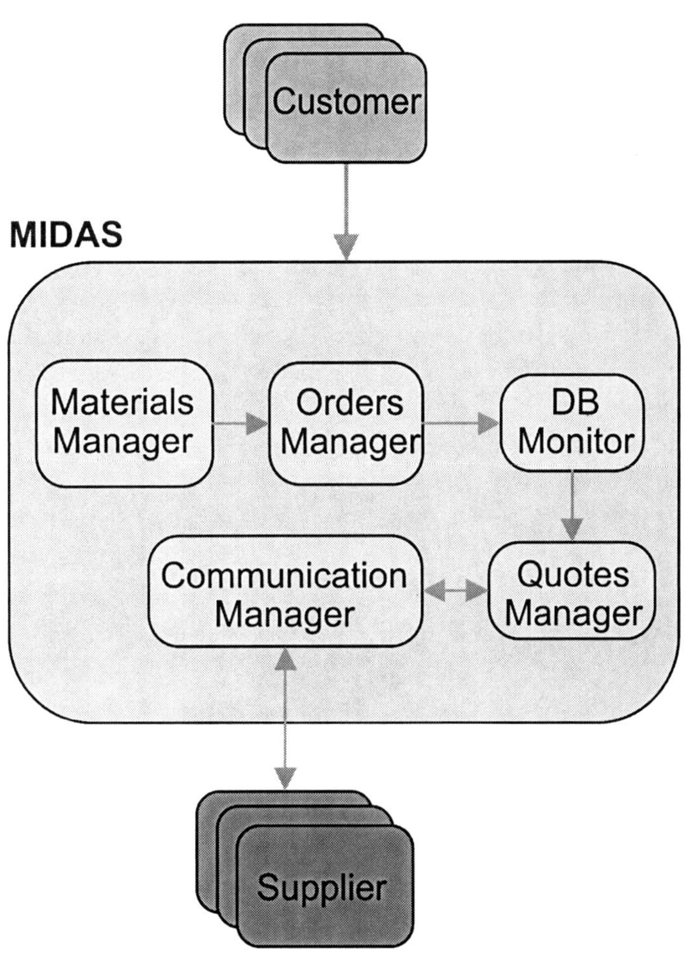


The strategies used by the DB Monitor in the Waiting phase to decide whether to stop collecting customer orders and to inform the Quotes Manager are discussed in the sections on the DB Monitor and the Waiting Phase. The strategies used by the Quotes Manager in the Quotes phase to decide whether to stop collecting Quote replies and to make a decision on a supplier are discussed in the sections on the Quotes Manager and the Quotes Phase.

The MIDAS architecture has interfaces on the customer side and the manufacturer side. Each interface involves different modules of MIDAS, and some modules of MIDAS serve as a bridge between the two interfaces. The Customer Service and the Manufacturer Service are discussed below.

## The Customer Service

The MIDAS architecture is based on the premise that it is the customer's opinion that counts. MIDAS adopts the build-to-order model (Gunasekarana, 2005), which enables a customer to customize a product that the customer purchases by choosing the materials that constitute that product. The customer customizes a product order using a Customer Web Service that communicates with other services to obtain the materials necessary to manufacture the product and to arrange shipping and financing of the product.

The customer side of MIDAS deals with the customer/manufacturer communication and interactions. The Customer Service provides the following functionality:

- Provides catalog information
- Receives information from the customer
- Displays the status of the customer's orders

The customer user interface uses the Customer Web Service to interact with MIDAS. Existing Web Services frameworks ease the process of implementing the user interface that interacts with the services.

## The Manufacturer Service

The customers send their orders for products to the manufacturer, and the manufacturer processes their orders. At the manufacturer, MIDAS does not contact the suppliers each time it receives an order from a customer. Rather, it uses one of several strategies (discussed below) to aggregate orders from different customers. However, the manufacturer must not take too long to confirm a customer's order with the price, product delivery time, etc.

The Manufacturer Service provides the following functionality:

- Monitors orders
- Searches for suppliers
- Contacts relevant suppliers
- Decides on the supplier(s) from which to obtain supplies

This functionality is discussed below in terms of the Orders Manager, DB Monitor, Quotes Manager, and Registry, shown in Figure 2.

## The Orders Manager

MIDAS assumes that, the more orders the manufacturer accumulates, the more gain the manufacturer has. Thousands of supply items have a per item price that is different from the price of a single item.

The Orders Manager collects orders from different customers and aggregates supply items (instances of a particular supply) for those customer orders. It tries to accumulate as many orders for a particular supply as it can. The Orders Manager then forwards the orders to the Database Monitor.

## The Database Monitor and the Orders Database

The Database (DB) Monitor keeps track of the orders for each supply in the Orders Database and triggers order events, based on one of several strategies discussed below. When the DB Monitor triggers an order event for a particular supply, the processing of the order begins. The DB Monitor then informs the Quotes Manager to find appropriate suppliers for that supply and to communicate with them.

The main problem for manufacturers that depend on suppliers is that, if a supplier is not available, the manufacturer cannot make progress. In a dynamic environment, a manufacturer must be able to find new suppliers as the demand arises and to satisfy the customers' needs in a timely manner.

The DB Monitor uses one of the following strategies to trigger an order event and to initiate the processing of an order for a particular supply by informing the Quotes Manager to place an order for that supply.

- **System user decides when to place orders:** The DB Monitor provides information to the system user at the manufacturer about the number of supply items (instances of a particular supply) required for the order of that supply and the wait times for those items. The user triggers manually the Quotes Manager, which initiates communication with the appropriate suppliers.

- **Threshold number of supply items for a particular supply is reached:** On receiving orders from the customers, the Orders Manager directs the DB Monitor to scan the database. If the number of supply items (instances of a particular supply) is greater than the threshold, the DB Monitor informs the Quotes Manager.
- **Timeout occurs at which an order needs to be placed:** The DB Monitor scans the database at defined intervals. For each scan, when a timeout occurs, the DB Monitor informs the Quotes Manager.
- **Hybrid of threshold number of orders and timeout:** This strategy combines the threshold strategy and the timeout strategy, such that whichever occurs first triggers the Quotes Manager.
- **Average wait time for a set of orders is reached:** The Orders Manager accumulates the number of supply items (instances of a particular supply) for the orders from different customers. When a component has an average wait time that is greater than a threshold, the DB Monitor informs the Quotes Manager.

The DB Monitor retrieves information (supply ID, amount to order, average wait time, etc.) from the Orders Database. According to the strategy used, the DB Monitor decides whether it is time to contact the suppliers of a supply and then informs the Quotes Manager to do so.

## The Quotes Manager

The Quotes Manager initiates communication with the suppliers, requesting the aggregated number of supply items. It submits Quote requests to selected suppliers, receives Quote replies with a proposed price, number of items, and proposed delivery time, and then decides on a supplier with which to place an order.

The Quotes Manager retrieves information about the respective suppliers of a supply from its Registry. Having decided on the suppliers with which it will communicate, the Quotes Manager sends Quote requests to those suppliers initiating the second phase, the Quotes phase. After sending Quote requests, the Quotes Manager uses one of the following strategies to decide on a particular supplier with which to place an order. (It could order items for a particular supply from multiple suppliers but, for simplicity, we consider only a single supplier for that supply.)

- **System user decides on a supplier:** The system user has control over the processing of quotes from the suppliers and deciding on a supplier of a particular supply.
- **Threshold percentage of Quote replies:** The Quotes Manager continues to aggregate Quote replies until the total number of Quote replies is above a threshold percentage. For example, when the threshold percentage is 100%, the Quotes Manager waits until it receives Quote replies from all of the respective suppliers.
- **Timeout for Quote replies from the suppliers:** The Quotes Manager continues to wait until a specific time. If the Quotes Manager receives all of the expected Quote replies before the timeout, the Quotes Manager initiates processing of the Quote replies.
- **Hybrid of threshold number of Quote replies and timeout:** This strategy combines the strategies of waiting for a threshold number of Quote replies and waiting for a specific time, whichever occurs first.
- **Average wait time threshold for Quote replies:** The Quotes Manager accumulates Quote replies and when the average wait time for the Quote replies is greater than

*Figure 3. Quote request and quote reply interactions*

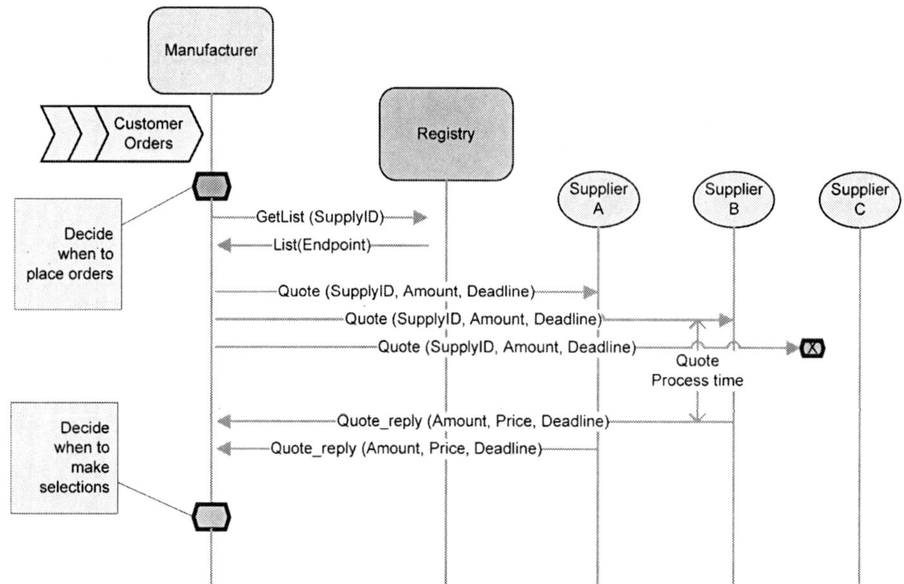


a threshold, the Quotes Manager initiates processing of the Quote replies.

The manufacturer can define the number of items needed, the type of supply, and the delivery time expected. Based on the suppliers' offers, the Quotes Manager then updates its Quote request and sends the updated Quote request to the supplier again, continuing the negotiation.

Figure 3 shows an example sequence diagram for the Quotes phase. In this example, the Quotes Manager could not communicate successfully with Supplier C, perhaps because of a communication failure or unavailability of Supplier C's service. However, the Quotes Manager is aware of the total number of Quote requests that it sent and the number of Quote replies that it must receive. The Quotes Manager decides on the suppliers with which it will do business. Having decided on the status of a quote, the Quotes Manager updates the status of the customer order associated with the quote and informs the Orders Manager. Completion of a quote does not necessarily mean completion of a customer order. The customer order is completed once a decision about all of the different components for that order is made.

As Figure 3 shows, there are two important decision points that affect the time the customer must wait to obtain information about the price and delivery time of the product. First, the Orders Manager accumulates orders for particular supplies, collecting as many as it can, while not increasing the average customer waiting time, before it decides to proceed to the next decision. Next, the Quotes Manager accumulates Quote replies from the suppliers, aiming to make the delay for the customer as short as possible, while not missing better quotes after it decides.

Note that, while waiting for Quote replies, the Quotes Manager is not blocked. All of the messages are sent asynchronously, and no assumption is made about the order of delivery of Quote replies. The time at which the Quotes Manager receives Quote replies from the suppliers varies according to the supplier's quote processing time.

*Figure 4. Communication links and protocols used between the manufacturer, registry, and suppliers*

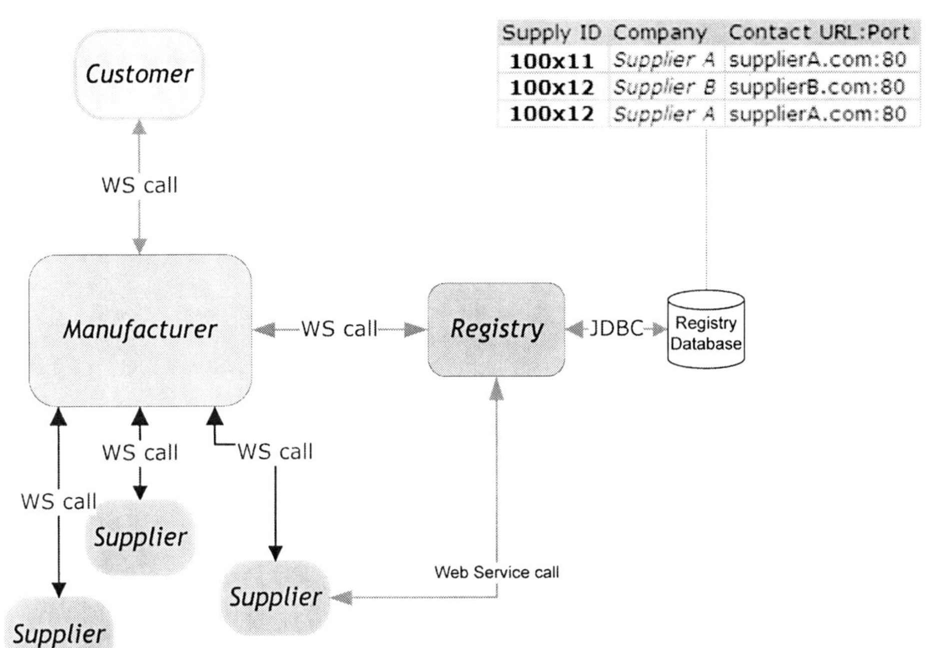


## MIDAS REGISTRY

The MIDAS Registry keeps information about the suppliers, in particular, the contact information of each supplier and the supply IDs of the products that it offers. The manufacturer and the suppliers are assumed to use the same supply IDs, and the supply IDs are assumed to identify, uniquely, the supplies across different manufacturers and different suppliers. The Registry enables the manufacturer to find relevant suppliers of the supplies it needs on demand. The Registry offers suppliers a fair business environment where they have a chance of doing business along with other competing suppliers.

Figure 4 shows the Registry and the communication links and protocols, as well as the structure of the Registry. A supplier registers by communicating with the Registry and passing information about itself to the Registry. The supplier provides its contact information and the products that it offers. If there is a change (such as a change of address, offering of new products, or removing product information), the supplier must update this information in the Registry. After registering, the supplier can receive Quote requests from manufacturers using MIDAS.

### The Reservation Protocol

The Reservation Protocol employed by MIDAS is an extended transaction protocol that is designed for business transactions that span multiple enterprises (Zhao et al., 2008). Business activities between the manufacturer and the supplier are executed as two steps. The first step involves an explicit reservation of resources according to the business logic. The second step involves the confirmation or cancellation of the reservation. For example, a manufacturer that is interested in buying a product from a supplier sends Quote requests to the suppliers in the first step. Once the manufacturer receives Quote replies from the suppliers and makes a decision, the manufacturer sends confirmation or cancellation messages to the suppliers.

Alternative transaction methods suffer in terms of response time and throughput when there are concurrent requests. During the transaction of a request, resources are not available to other requests, increasing the response time and decreasing the throughput. Because the Reservation Protocol executes each step as a separate traditional short-running transaction, resources are classified as available and reserved, which differs from blocking the resources until the current request is complete. Use of the Reservation Protocol decreases the wait time of customers, because MIDAS can complete the business transactions between the manufacturer and a supplier much faster, even though there are many concurrent requests arriving at the suppliers.

The Reservation Protocol improves the performance of business transactions that span multiple enterprises in a multi-level supply chain. A supplier at level *N* receiving a Quote request from a manufacturer at level *N+1* can contact its suppliers at level *N-1* to make a reservation to reply to a Quote request from a manufacturer at level *N+1*. Receiving Quote replies from the suppliers at level *N-1*, a supplier at level *N* can send a Quote reply to a manufacturer at level *N+1*. Depending on the decision of the manufacturer at level *N+1*, the supplier at level *N* sends a decision to its suppliers at level *N-1*. For traditional transaction protocols, the same scenario can result in longer delays and also inconsistencies in the databases of the different enterprises in the supply chain.

## IMPLEMENTATION

The MIDAS automated supply chain system comprises software modules for the customers, the manufacturers, and the suppliers, as described in the MIDAS Architecture section. The Web Services for MIDAS are built on the Apache Axis2 Framework (which is the core engine for Web Services built on Apache Axiom) and the Apache Tomcat Server (Apache, 2008).

### The Customer Side

Customers can access the MIDAS system and the manufacturer's catalog of products using their Web browsers. The customers are represented in the manufacturer's database, and authentication is required before the resources for a particular customer can be accessed. At the manufacturer, servlets use the Customer Web Service interface and make Web Service calls to MIDAS through the API provided. Once the customer is provided with the product catalog, the customer can create his own customized product by selecting a particular supply for each component.

### The Manufacturer Side

#### The Orders Database

The Orders Manager and the Materials Manager at the manufacturer use the Orders Database to insert, query, and update the status of customer orders. The Orders Database includes tables for the materials, components, supplies, Quote requests, Quote replies and sales. It also includes a table that relates materials and components, and a table that relates components and supplies.

#### The Registry and Registry Database

The Registry enables the manufacturer to find relevant suppliers of the supplies it needs on demand. The Registry Database at the manufacturer keeps information about the suppliers, in particular, the contact information of each supplier and the supply IDs of the products that it offers. To register, a supplier communicates with the Registry and provides its contact information and the products that it offers. Once it has registered, the supplier can receive Quote requests from the manufacturer using MIDAS. The Registry supports a fair business environment for suppliers where they have a chance of doing business along with other competing suppliers.

*Figure 5. Waiting phase: Timeline of a customer order*

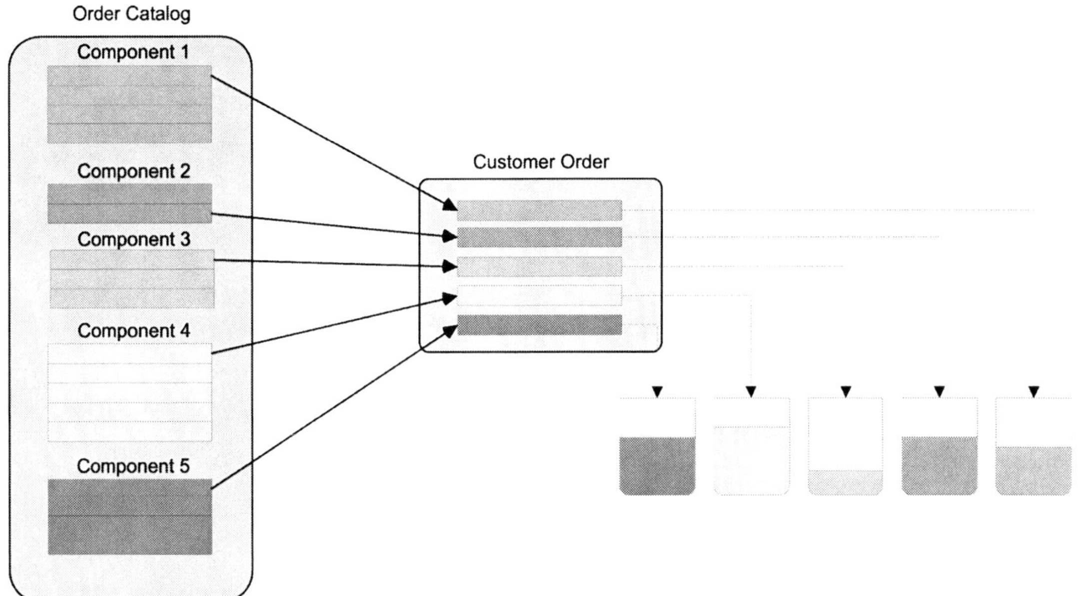


## Order Process

Before presenting the performance evaluation results, we present two customer order timelines to enable better understanding of the evaluation.

Figure 5 depicts the timeline of a customer order submitted to MIDAS in the Waiting phase. The customer retrieves the supply catalog from the manufacturer. Using this catalog, the customer creates his own customized product and submits his order to MIDAS. On receiving orders from the customers, MIDAS first aggregates the supply items ordered, and then contacts the suppliers when the total number of items for a particular supply reaches a threshold or, alternatively, a timeout occurs. Depending on the strategy used, the aggregation time for a supply varies. Frequently chosen supplies are aggregated faster than infrequently chosen supplies. The performance evaluation for the Waiting phase is discussed in the section on the Waiting phase.

Figure 6 depicts the timeline of the customer order for the Quotes phase. After deciding it is time to contact the suppliers for a specific supply, MIDAS retrieves information about the respective suppliers for that supply from the Registry. MIDAS submits a Quote request for that supply to those suppliers. A supplier makes a callback to deliver its Quote reply. Once a supplier has finished the process of making a reservation related to the Quote request, it sends a Quote reply using the callback Web Service. MIDAS collects these Quote replies, and makes a decision about the status of a quote, which affects the status of different supply items. The choice of quotation might depend on price, availability, delivery schedule, or other factors. The selected supplier is notified with a confirmation of the purchase. Suppliers that were not selected are contacted with a cancellation of the reservation. Once a decision for all order items for a customer order is made, the customer order is complete. The performance evaluation for the Quotes phase is discussed in the section on the Quotes phase.

*Figure 6. Quotes phase: Timeline of a customer order*

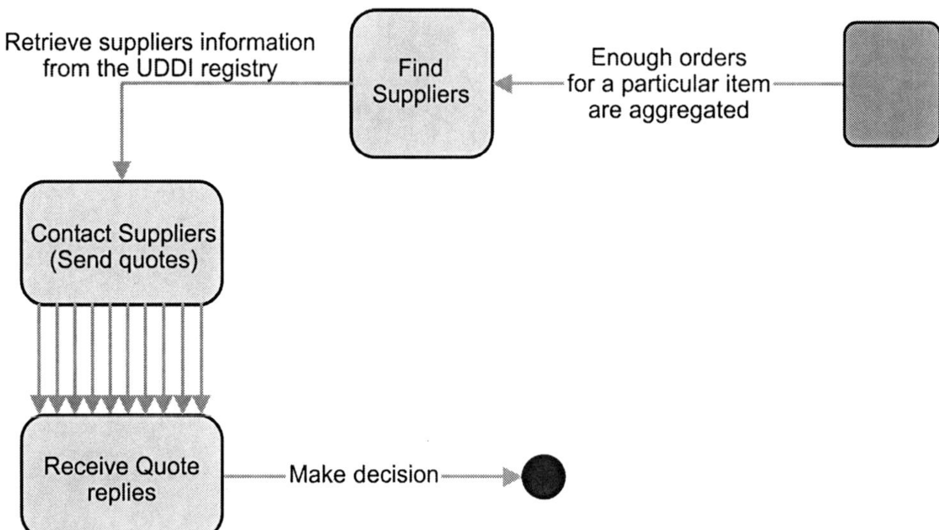


## PERFORMANCE EVALUATION

In the DB Monitor section we discussed different strategies that the DB Manager can use in the Waiting phase to aggregate orders from customers before it triggers the Quotes Manager to communicate with suppliers. In the Quotes Manager section we discussed different strategies that the Quotes Manager can use in the Quotes phase when it accumulates Quote replies from suppliers. The delay for aggregating orders, $delay_{AO}$, affects the delay for accumulating quotes, $delay_{AQ}$, particularly if the supplier processes and responds to Quote replies sequentially.

To evaluate the different strategies that the DB Manager and the Quotes Manager can use, we used the PlanetLab global research network (PlanetLab, 2008).

For the different strategies, we evaluated the customer's satisfaction, as measured by the customer response time, and the manufacturer's gain, as measured by the number of orders aggregated or the best price ratio (probability of hitting the best quote among the possible quotes). Of course, the overall customer response time is not when the manufacturer says the product is ready to be shipped but when the customer actually receives it. However, the customer usually wants to know when the product is ready to be shipped and to track its status. Moreover, in this paper, we are concerned with information flow (rather than product flow), so the customer response time that we're using here is appropriate.

We evaluated the time spent in both the Waiting phase and the Quotes phase of MIDAS, i.e., the time spent to aggregate orders and the time spent to aggregate Quote replies. In addition, we investigated the probability that MIDAS makes a decision for the best quote possible. Finally, we calculated the average time needed to complete a customer order for the different strategies.

### The Waiting Phase

As discussed in the DB Monitor section, the Waiting phase of MIDAS is based on one of several strategies. Here we consider:

- Aggregating a threshold number of orders
- Timeout for a specific amount of time
- Hybrid of threshold number of orders and timeout

*Figure 7. Order aggregation for different customer order arrival rates and different numbers of component alternatives*

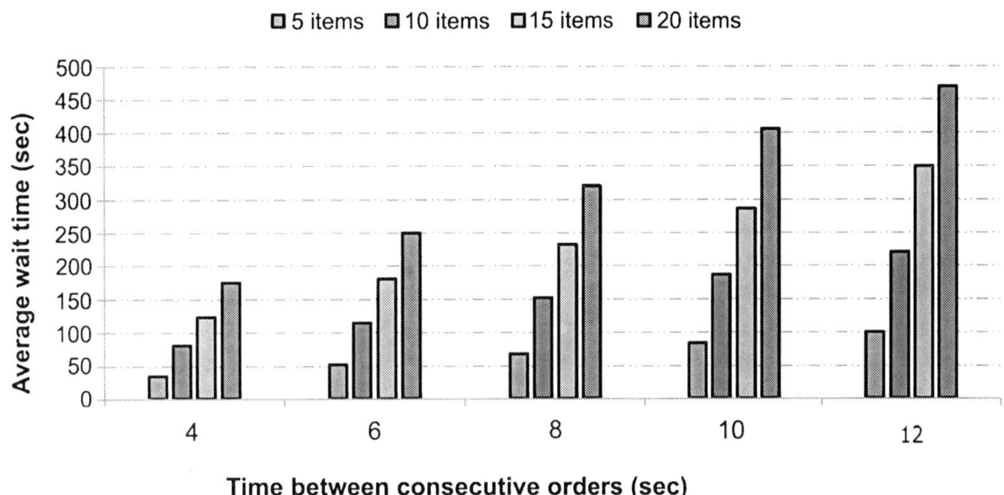


We evaluated these strategies under different customer order arrival rates. In our evaluation, the manufacturer receives a total of 1000 customer orders.

## Aggregating a Threshold Number of Orders

This strategy aggregates orders for a specific number of items for a particular supply before the Quotes Manager issues Quote requests to the suppliers.

We evaluated the time spent to aggregate orders for 5, 10, 15, and 20 supply items with different customer order arrival rates (customer order arrives every 4, 6, 8, 10, and 12 seconds) with the same number (four) alternatives for each component and with different numbers of alternatives for each component. The results in the two cases are very similar.

Figure 7 shows the results for different numbers of alternatives (2, 3, 4, 5, and 6) for the different components. Having fewer alternatives for a component makes aggregation of the supply item faster than if there are more alternatives, but the weighted average of the wait time during order aggregation is the same as for the same number of alternatives. This behavior is expected because a supply that is aggregated faster has a larger number of occurrences but a smaller wait time. A supply that appears in fewer customer orders has a higher wait time. Taking the weighted average provides the same average wait time as the aggregation of orders.

As we see from Figure 7, the average wait time to aggregate enough supply items increases when there is a lower customer order arrival rate. Moreover, the increased threshold delays the time to complete the Waiting phase. The threshold number of orders aggregation strategy keeps the number of supply items at a certain limit; however, a low customer order arrival rate might delay the order of a particular supply if that supply is not popular.

Customers will be satisfied if their orders are completed earlier. However, contacting suppliers with orders for more supply items is more favorable for the manufacturer. The more supply items purchased from the supplier, the more gain the manufacturer has. Thus, the manufacturer faces a conflict as to whether to be concerned about the number of supply items to aggregate (manufacturer

*Figure 8. Order aggregation for different timeout values*

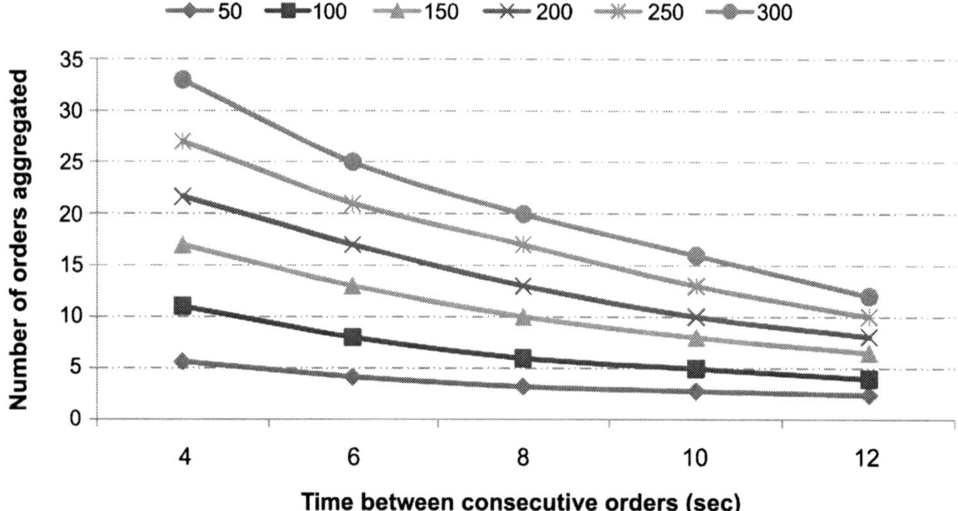


gain) or the wait time for aggregation of supply items (customer satisfaction). The appropriate balance between manufacturer gain and customer satisfaction is a sensitive business decision that is beyond the scope of this paper.

## Timeout for a Specific Amount of Time

The timeout strategy constrains the wait time for orders by placing an upper bound on it. Therefore, we consider the number of orders aggregated as a function of the customer order arrival rate and the timeout value. The results for the timeout strategy are presented in Figure 8.

As Figure 8 shows, an increased timeout value results in the aggregation of a larger number of items for a particular supply. For the same timeout value, the customer order arrival rate affects the number of orders aggregated before the timeout occurs.

## Hybrid of Threshold Number of Orders and Timeout

For the threshold number of orders strategy, when the customer order arrival rate is low, the wait time for the aggregation of orders increases and the strategy suffers. If the customer order arrival rate is very low, the Quotes phase will not start unless a timeout occurs. On the other hand, the timeout strategy limits the wait time for the aggregation of orders but, if the customer order arrival rate is high (e.g., when customer orders come in bursts), it might wait longer than necessary. Therefore, a hybrid strategy might be able to obtain the threshold number of orders in less than the maximum wait time of the Waiting phase but still bound the wait time so that the Quotes phase can proceed.

## The Quotes Phase

As discussed in the section on the Quotes Manager, the Quotes phase of MIDAS is based on one of several strategies. Here we consider:

*Figure 9. Processing time for the suppliers*

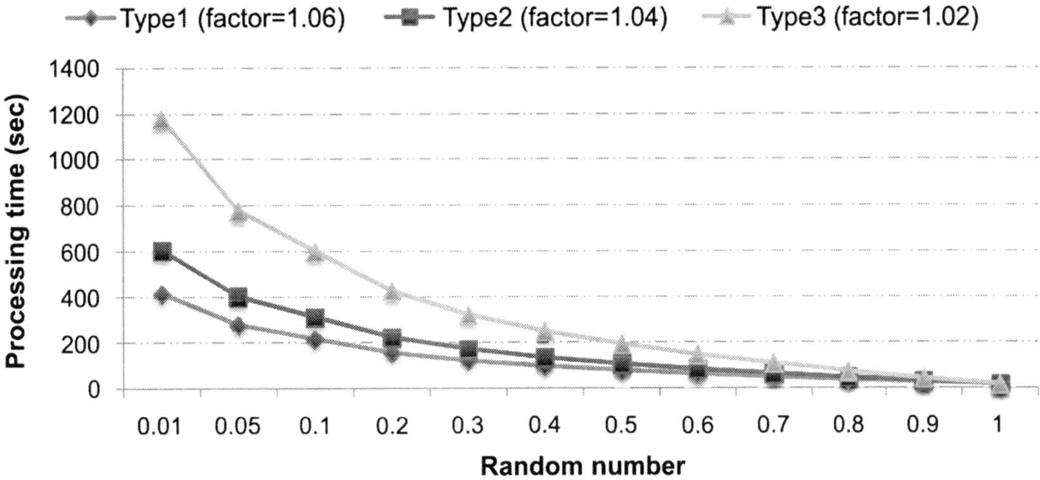


- Aggregating a threshold percentage of Quote replies
- Timeout for a specific amount of time
- Hybrid of threshold percentage of Quote replies and timeout

In the Quotes phase, MIDAS contacts the relevant suppliers and submits a Quote request to each of them. On receiving a Quote request, a supplier incurs some processing time before it replies to MIDAS using the callback Web Service endpoint passed in the Quote request. The manufacturer collects the Quote replies concurrently in an asynchronous manner, and the suppliers respond to Quote requests after a processing time that depends on the type of supplier.

The processing time for the suppliers, shown in Figure 9, is given by the following formula:

*Processing time = 20 - 5 * log (random) / log (factor)*

The processing time depends on a random number between 0 and 1, and corresponds to processing times between 20 seconds and ∞. With a high probability, the processing time is close to 20 seconds, and with a very low probability, the processing time is ∞, which represents a supplier that crashes after receiving a Quote request from MIDAS.

The processing time also depends on a scaling factor to represent different types of suppliers with different processing times. The factors 1.06, 1.04, and 1.02 correspond to processing times in the ranges 20-400, 20-600, and 20-1200. Type 1 suppliers (factor=1.06) respond to Quote requests earlier than Type 3 suppliers (factor=1.02), and Type 2 suppliers respond to Quote requests in between the other two.

The time spent before determination of the status of a quote depends on the strategy and the type of supplier (1, 2, or 3). Depending on the strategy, it is possible to decide the status of a quote before receiving all of the Quote replies. Aggregation of more Quote replies increases the chance of selecting the best quote.

For each of the above strategies in the Quotes phase, we used the threshold strategy in the Waiting phase to evaluate the average wait time and the best quote ratio. The best quote ratio is the probability that MIDAS decides the best possible Quote reply.

## Aggregating a Specific Threshold Percentage of Quote Replies

In this strategy, if the number of Quote replies exceeds a certain percentage, the Quotes Manager

Figure 10. Average wait time for the threshold percentage strategy

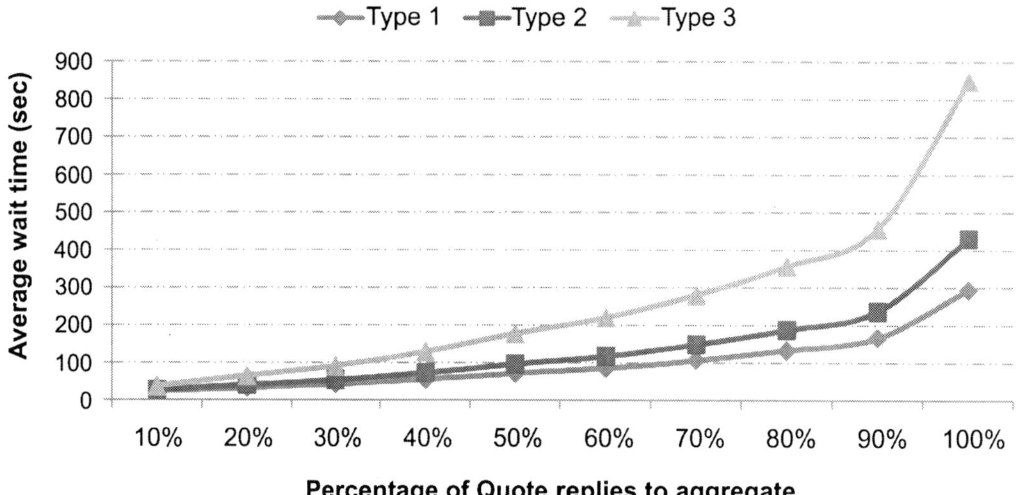


stops collecting Quote replies from the suppliers and completes the customer orders. Compared to the strategy of collecting all Quote replies, this strategy decreases the customer's response time (increases the customer's satisfaction), at the cost of reducing the best price ratio.

We evaluated the average wait time for the Quotes phase using the threshold percentage strategy. Figure 10 shows the average wait time for the threshold percentage strategy in the Quotes phase. Depending on the number of suppliers that the Quotes Manager contacted, the Quotes Manager waits for the number of Quote replies given by:

*Expected replies = Requests sent * Percentage to aggregate*

The results show a linear increase and then an exponential increase for the case that MIDAS waits to collect all Quote replies. The average wait time is dominated by the maximum processing time from one of the suppliers for that quote. The average wait time can be kept reasonable, although MIDAS must wait for 90% of the expected Quote replies.

We also evaluated the best quote ratio against the percentage of expected Quote replies for the threshold percentage strategy. The results indicate that the best quote ratio is independent of the processing time at the suppliers, i.e., the type of supplier.

## Timeout for a Specific Amount of Time

When the Quotes Manager collects Quote replies from the suppliers, it waits for a specific amount of time, $delay_{AQ}$. When the timeout occurs, the Quotes Manager decides on a supplier for a specific supply for the customer's order and ignores late Quote replies. Because the Quotes Manager doesn't need to collect Quote replies from all of the suppliers, this strategy decreases the customer's response time (increases the customer's satisfaction), compared to the strategy of collecting all Quote replies. On the other hand, the Quotes Manager cannot always obtain the best price from the suppliers, because the Quotes Manager might receive the Quote reply with the best price too late and thus ignore it.

First, we evaluated the average wait time, i.e., $delay_{AQ}$, in the Quotes phase using the timeout strategy. Figure 11 shows that the average wait time increases as the time spent aggregating Quote replies increases, as expected. The timeout strategy decides the status of a quote, either when

*Figure 11. Average wait time for the timeout strategy*

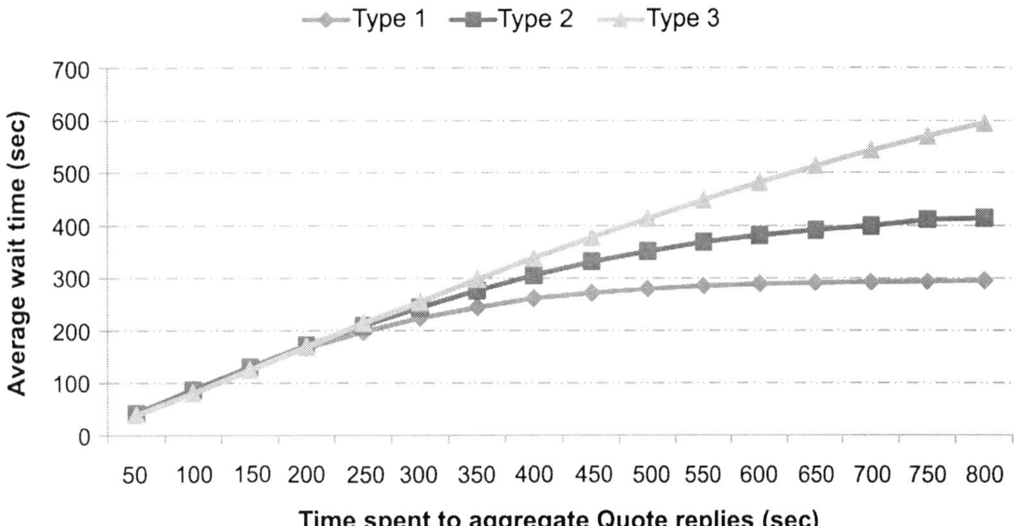


a timeout occurs, or before the timeout when all of the expected Quote replies are received. The average wait time stabilizes around the maximum processing time of the suppliers. After stabilization, increasing the timeout does not affect the time taken to make a decision for the quote.

Next, we evaluated the probability that MIDAS catches the best quote depending on the timeout value. Figure 12 shows the best quote ratio for the timeout strategy. When the Quotes Manager decides the status of a quote early, the probability of catching the best quote is low because a better Quote reply might arrive at a later time. Type 1 suppliers that have a small processing time send their Quote replies early, which increases the number of Quote replies aggregated in the Quotes phase. Similarly, once the best quote ratio reaches a maximum value, increasing the timeout does not affect the best quote ratio. Analysis of the best quote ratio allows the selection of timeouts that can achieve high manufacturer gain without adversely affecting customer satisfaction. Note that this strategy provides an incentive for prompt responses by suppliers because late responses might not be considered.

## Hybrid of Treshold Percentage of Quote Replies and Timeout

For the threshold percentage of Quote replies strategy, when the supplier's processing time is high (supplier's response rate is low), the aggregation time for Quote replies increases and the threshold percentage strategy suffers. On the other hand, the timeout strategy limits the wait time for the aggregation of Quote replies but, if the supplier's response rate is high (such as when the supplier's Quote replies come in bursts), it might wait longer than necessary. Therefore, a strategy that takes both factors into account might obtain the threshold percentage of Quote replies in less than the maximum wait time but still bound the wait time of the Quotes phase of MIDAS.

## Customer Order Analysis

Finally, we provide a customer order analysis in terms of the wait time that elapses before completion of a customer order, i.e., all of the supply items needed for the customer order have been ordered. Thus, the order completion time is the maximum of the completion times for all supply

Figure 12. Best quote ratio for the timeout strategy

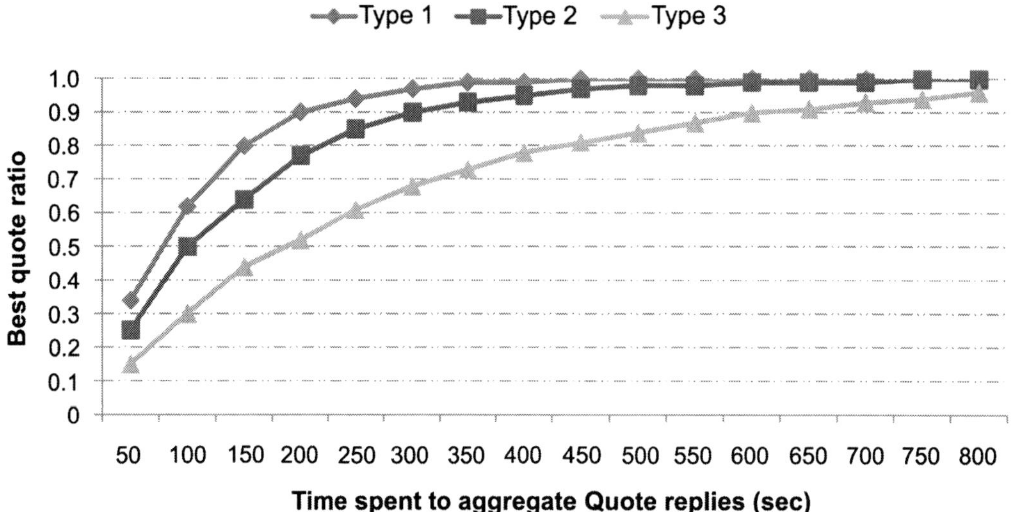


items $i$ needed to fill the customer order, which is given by:

*Order completion time* $= max_{\forall i} (delay_{AO}(i) + delay_{AQ}(i))$

The results of the analysis for the threshold percentage strategy in the Quotes phase are shown in Figure 13 for the case in which MIDAS receives orders every 10 seconds and 5 items are aggregated before MIDAS enters the Quotes phase. Note that the order completion time is similar to the behavior for the wait time using the threshold percentage strategy in the Quotes phase shown in Figure 10.

Under the same conditions, we investigated the customer order completion time using the timeout strategy in the Quotes phase. The results are shown in Figure 14. Note that the order completion time is similar to the behavior for the wait time using the threshold percentage strategy

Figure 13. Order completion time for the threshold percentage strategy

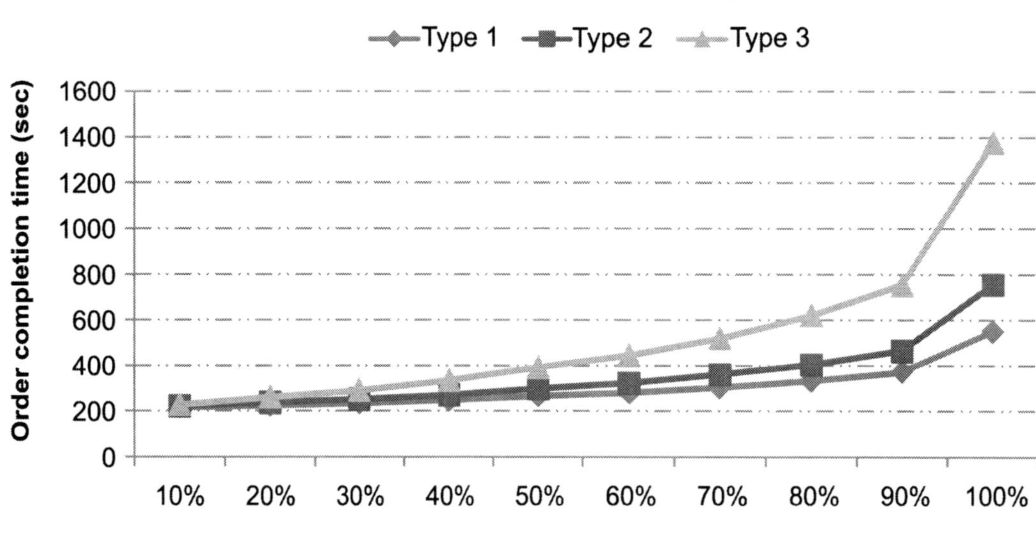

*Figure 14. Order completion time for the timeout strategy*

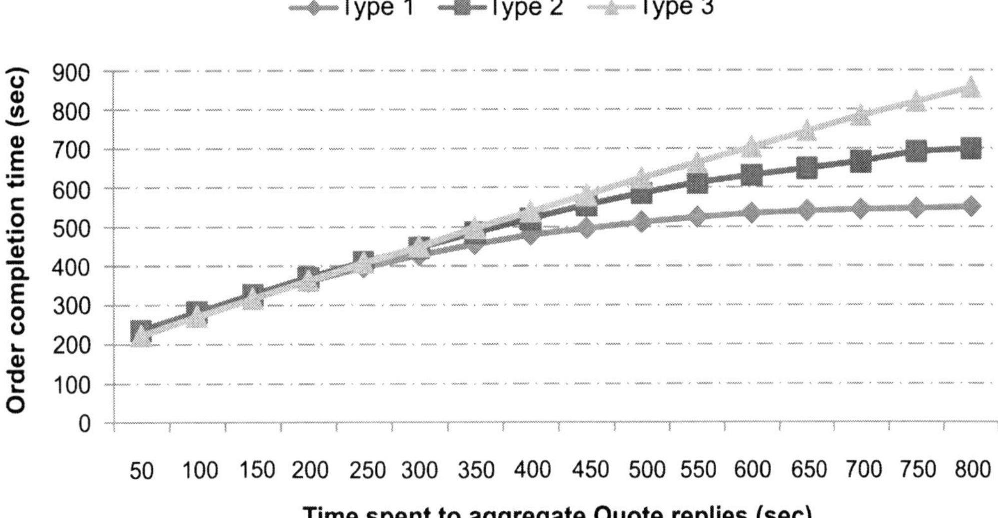


in the Quotes phase shown in Figure 11. The customer order completion time is important because it determines:

- When a customer order can be accepted and an anticipated delivery time can be provided to the customer
- When a response can be provided to a request for quotation in a multi-level supply chain.

## RELATED WORK

The MIDAS system is an automated supply chain management system that implements the *build-to-order* model. Several researchers have presented discussions of this model at a conceptual level, but have not presented an implemented automated supply chain management system and its performance evaluation, as we do in this paper. Gunasekarana and Ngai (2005) present a comprehensive review of build-to-order supply chain management and a framework for development. They note that the build-to-order strategy can meet the requirements of customers with diverse needs by leveraging outsourcing and Internet technologies.

Rodriguez and Montoya-Torres (2009) consider a build-to-order supply chain without finished product inventory. They study various scheduling algorithms with different levels of information sharing among the supply chain members and their effects on the performance of the supply chain. In particular, they measure the impact of supplier-customer information sharing on production scheduling, in order to obtain a more reactive and flexible supply chain. MIDAS could be augmented, so that it also addresses the issue of production scheduling.

Graham and Hardaker (2000) highlight the role of the Internet in building flexible, agile, on-demand supply chains based on the build-to-order model. They note that the build-to-order model not only addresses diverse customer needs, but also lowers inventory stock, because parts are pulled from the suppliers as needed. Fontanella (2000) observes that an integrated I-commerce model can strengthen the relationship between global supply chain management and network orientation. Moreover, he notes that Internet technologies make foreign markets more accessible and make it easier to integrate foreign customers, suppliers and intermediate companies into the supply chain, increasing savings and providing innovation.

Lancioni et al. (2000) also discuss the use of the Internet for managing the major aspects of supply chains, particularly order processing, purchasing/procurement and transportation. They note that enterprises use the build-to-order model to achieve lower inventory costs (with somewhat higher production costs), track inventory more accurately, and report the status of orders. Gomez-Padilla (2009) analyzes the importance of contracts for supply chain coordination, where inventory holding costs are shared among the companies in the supply chain. In MIDAS, we aim to minimize the inventory holding costs at the manufacturer.

Forza and Salvador (2002) discuss the difficulties in managing build-to-order systems. They focus particularly on the challenges due to custom customer orders, which must be maintained and processed carefully. Kolish (2000) also discusses the difficulties of the build-to-order model with the production of ships, airplanes and large-scale machine tools. He addresses the coordination of fabrication and assembly with respect to scarce capacities.

In this paper, we focus on information flow in supply chains, as realized by MIDAS. Wu et al. (2005) also consider information flow in supply chains. They observe that, in a competitive e-business environment, an enterprise must be involved in managing the supply chain from both the upstream side (the suppliers) and the downstream side (the customers). They propose a novel approach that brings together business processes and services for supply chain management. Nunez-Munoz and Montoya-Torres (2009) analyze the impact of coordinated decisions with information sharing in a three-echelon supply chain. They model demand as a geometric Brownian process, and measure the impact of the degree of coordination among supply chain companies. In contrast, in MIDAS, we consider different strategies in the Waiting phase and the Quotes phase, and also different types of suppliers with different processing times.

As pointed out previously, the MIDAS system aims to automate the supply chain and, thus, to reduce human intervention, errors and costs, resulting in a more efficient supply chain. Dong and O'Brien (1999) have a similar objective, but they base their business model on the four criteria: profit, lead time, performance, and promptness of delivery. They analyze supply chain performance at two levels: the chain level and the operation level. At the chain level, they set objectives associated with the criteria for each supply chain stage in order to satisfy customer service targets and to select the best supply chain management strategy. At the operation level, they optimize manufacturing and logistics activities for the given targets.

Jinho and Rogers (2005) investigate the use of UML for building a flexible supply chain business model. They regard a supply chain as five view models with four business domains, where each domain consists of functions, resources, processes, interactions, and business rules. However, they do not describe a system that demonstrates their approach, as we do for MIDAS.

The MIDAS system enables customers to build their own customized products before they order. Zhou et al. (2003) also investigate customization in supply chains. They address mass customization, which supports customer innovation and which integrates mass production and customized production. Yang et al. (2005) investigate customization and postponement in supply chains to reduce costs, realize diversity, and improve agility. They consider four kinds of postponement in supply chains: supply, manufacture, delivery, and service. Ghiassi and Spera (2003) also discuss industry solutions, and best practices, for Internet-based supply chains that support mass customization.

MIDAS realizes the concept of on-demand, just-in-time manufacturing to reduce inventory carrying costs and business logistics costs, by maintaining logical inventory, rather than maintaining physical inventory in warehouses. Tanik et al. (2001) propose a zero-time framework based on what they call the *T-strategy*, which allows enterprises to adapt in a timely manner to changing market conditions.

In our performance evaluation of MIDAS, we have considered the customer's satisfaction (measured by the average customer response time) and the manufacturer's gain (measured by the number of orders aggregated or the best price ratio). In contrast, Li et al. (2005) consider the average supplier response time in their evaluation of supply chains. Giglio and Minciardi (2003) investigate the modeling and optimization of supply chains that involve multiple production sites and multiple suppliers, using mathematical programming techniques to minimize the costs in the network. Zhao and Jin (2005) investigate optimized coordination of supply chains based on relationships and dependencies. Levy et al. (2009) present a value-satisfaction taxonomy of the effectiveness of information systems, where user satisfaction is the key measure, and also a case study of their value-satisfaction taxonomy.

Sharma et al. (2008) present a review of the literature on *quality function deployment*, which provides a means of translating the needs of the customers through the various stages of product planning, engineering and manufacturing into a final product. MIDAS is likewise concerned with satisfying the needs of the customers, and delivering the greatest value to the customers.

Wang et al. (2004) observe that appropriate supply chain partners have a large effect on the output value of a supply chain system. They propose an Internet-driven electronic marketplace that can provide an effective platform to select the right partners. In some ways, such a marketplace is like the MIDAS Registry. They present a model of procurement strategies, and discuss factors for success of such a marketplace.

## CONCLUSION AND FUTURE WORK

We have presented the MIDAS system for automated supply chain management, which supports the build-to-order model to increase customer satisfaction and the just-in-time manufacturing model to reduce inventory carrying costs and logistics administration costs, yielding a more efficient supply chain. At the manufacturer, MIDAS uses two phases in processing orders from the customers: the Waiting phase in which it aggregates orders from the customers before it makes Quote requests from the suppliers, and the Quotes phase in which it collects Quote replies from the suppliers before deciding on the supplier that will provide the particular supply. The manufacturer can use one of several threshold and timeout strategies in deciding how long to wait to collect orders from customers and how long to wait to collect quotes from suppliers.

Based on our experimental evaluation of these strategies, we draw the following conclusions. To enhance the satisfaction of the customers by reducing the customer response time, the system should process orders as soon as possible. On the other hand, aggregating orders can benefit the manufacturer, by reducing the manufacturer's costs. Thus, a conflict exists between the customer and the manufacturer, so a balance point must be found. The results show that the Waiting phase could be adjusted to provide the minimum response time by using a hybrid scheme based on both the threshold percentage and timeout strategies. The analysis of the Quotes phase shows that the manufacturer does not really need to aggregate all possible Quote replies. Above the best quote ratio, increasing the threshold percentage of Quote replies aggregated or the timeout does not affect the manufacturer's gain considerably, but delays the customer's order completion time. The strategies used in the Quotes phase try to reduce the response time by controlling the number of Quote replies received. The best price from the suppliers might not be captured because, again, not all of the Quote replies are taken into account. The threshold and timeout strategies in the Quotes phase reduce the customer response time at the cost of decreasing the best price ratio. Again, a hybrid strategy is preferable.

This paper has addressed the services of MIDAS related to information flow. In the future, we plan to design and implement the services related to product flow and financial flow. We also plan to provide estimated delivery times for customer orders. In particular, we plan to augment MIDAS with a workflow component that handles relationships and dependencies between the components of a material. In a typical build-to-order system, the customer is given a limited set of components from which to choose and is provided feedback if he makes an incompatible choice. The build-to-order system incorporates configuration rules that are specific to the components and product being produced that limit these choices.

The MIDAS system, as presented here, deals with business processes up to the point where the decision to do business with a specific supplier is completed. However, the status of an order might change after the manufacturer has placed an order with a supplier. For example, if the delivery time changes, this information needs to be updated and the customer needs to be informed. MIDAS might use historical data to allow the manufacturer to make decisions at two decision points, the decision to issue Quote requests after collecting orders from the customers and the decision to stop collecting Quote replies and place an order with a particular supplier. Depending on the rates at which the customers place orders, the DB Monitor can adjust its threshold or timeout values accordingly. Moreover, the Quotes Manager can select better suppliers dynamically by waiting until it receives the Quote replies that it ignored previously because they were late even though they provide a better price.

The design and performance of the MIDAS Registry (which allows the manufacturer to find suppliers) needs to be evaluated. If the Registry contains the information the Quotes Manager needs to make a decision, the Quotes Manager can skip the Quotes phase and initiate purchases from the suppliers immediately, yielding a faster response time for the customer.

## ACKNOWLEDGMENT

This research was supported in part by the UC Discovery Grant Program and QAD, Inc, grant number COM05-10194.

## REFERENCES

W3C. (2004). *Web Services Architecture*. Retrieved April 2009 from http://www.w3.org/TR/ws-arch

Alonso, G., Casati, F., Kuno, H., & Machiraju, V. (2004). *Web Services concepts: Architectures and applications*. Berlin: Springer-Verlag.

Apache Web Services Project. (2008). *Home*. Retrieved from http://ws.apache.org/

Chatterjee, S., & Webber, J. (2003). *Developing enterprise Web Services: An architect's guide*. Englewood Cliffs, NJ: Prentice Hall.

Computer World. (2006). *Supply chain management*. Retrieved April 2008, from http://www.computerworld.com/softwaretopics/erp/story/0,10801,66625,00.html

Dong, L., & O'Brien, C. (1999). Integrated decision modeling of supply chain efficiency. *International Journal of Production Economics*, *59*(1-3), 147–157. doi:10.1016/S0925-5273(98)00097-8

Erl, T. (2005). *Service-Oriented Architecture (SOA): Concepts, technology, and design*. Englewood Cliffs, NJ: Prentice Hall.

Fontanella, J. (2000). The Web-based supply chain. *Supply Chain Management Review*, *3*(4), 17–20.

Forza, C., & Salvador, F. (2002). Managing for variety in the order acquisition and fulfillment process: The contribution of product configuration systems. *International Journal of Production Economics*, *76*(1), 87–98. doi:10.1016/S0925-5273(01)00157-8

Friedman, T. (2006). *Globalism is good*. Retrieved April 2009, from http://www.impactlab.com/modules.php?sid=5338

Ghiassi, M., & Spera, C. (2003). Defining the Internet-based supply chain system for mass customized markets. *Computers & Industrial Engineering*, *45*(1), 17–41. doi:10.1016/S0360-8352(03)00017-2

Giglio, D., & Minciardi, R. (2003). Modelling and optimization of multi-site production systems in supply chain networks. In *Proceedings of the IEEE International Conference on Systems, Man and Cybernetics, 3* (pp. 2678-2683).

Gomez-Padilla, A. (2009). Supply chain coordination by contracts with inventory holding cost share. *International Journal of Information Systems and Supply Chain Management*, *2*(2), 36–47.

Graham, G., & Hardaker, G. (2000). Supply-chain management across the Internet. *International Journal of Physical Distribution and Logistics Management*, *30*(3/4), 286–295. doi:10.1108/09600030010326055

Gunasekarana, A., & Ngai, E. W. T. (2005). Build-to-order supply chain management: A literature review and framework for development. *Journal of Operations Management*, *23*(5), 423–451. doi:10.1016/j.jom.2004.10.005

Jinho, K., & Rogers, K. J. (2005). An object-oriented approach for building a flexible supply chain model. *International Journal of Physical Distribution and Logistics Management*, *35*(7), 481–502. doi:10.1108/09600030510615815

Kolisch, R. (2000). Integration of assembly and fabrication for make-to-order production. *International Journal of Production Economics*, *68*(3), 287–306. doi:10.1016/S0925-5273(99)00011-0

Lancioni, R. A., Smith, M. F., & Oliva, T. A. (2000). The role of the Internet in supply chain management: Logistics catches up with strategy. *Industrial Marketing Management*, *29*(1), 45–56. doi:10.1016/S0019-8501(99)00111-X

Levy, Y., Murphy, K. E., & Zanakis, S. H. (2009). A value-satisfaction taxonomy of IS effectiveness (VSTISE): A case study of user satisfaction with IS and user-perceived value of IS. *International Journal of Information Systems in the Service Sector*, *1*(1), 93–118.

Li, Y. S., Ye, F. F., Fang, Z. M., & Yang, J. G. (2005). Flexible supply chain optimization and its SRT analysis. *Industrial Engineering and Management*, *10*(1), 89–93.

Moser, L. E., & Melliar-Smith, P. M. (2009). *Service Oriented Architecture and Web Services. Encyclopedia of Computer Science and Engineering* (pp. 2504–2511). Hoboken, NJ: John Wiley & Sons.

Murch, R. (2004). *Autonomic computing*. Armonk, NY: IBM Press.

Newcomer, E., & Lomow, G. (2004). *Understanding SOA with Web Services*. Independent Technology Guides.

Nunez-Munoz, M., & Montoya-Torres, J. R. (2009). Analyzing the impact of coordinated decisions with a three-echelon supply chain. *International Journal of Information Systems and Supply Chain Management*, *2*(2), 1–15.

OASIS. (2006). *Reference model for Service Oriented Architecture*. Retrieved April 2009 from http://www.oasis-open.org/committees/tc_home.php?wg_abbrev=soa-rm

PlanetLab. (2009). *Home*. Retrieved April 2009 from http://www.planet-lab.org/

Rodriguez, G., & Montoya-Torres, J. R. (2009). Measuring the impact of supplier-customer information sharing on production scheduling. *International Journal of Information Systems and Supply Chain Management*, *2*(2), 48–61.

Sharma, J. R., Rawani, A. M., & Barahate, M. (2008). Quality function deployment: A comprehensive literature review. *International Journal of Data Analysis Techniques and Strategies*, *1*(1), 78–103. doi:10.1504/IJDATS.2008.020024

Tanik, U., Tanik, M. M., & Jololian, L. (2001). Internet enterprise engineering. A zero-time framework based on the T-strategy. In *Proceedings of the IEEE Southeast Conference* (pp. 263-270).

Wang, L. P., Qiu, F. Y., Dai, H. R., & Chen, Z. C. (2004). Relationships of electronic marketplace and partner selection of supply chain. *Computer Integrated Manufacturing Systems*, *10*(5), 550–555.

Wikipedia. (2006). *Supply chain.* Retrieved April 2009 from http://en.wikipedia.org/wiki/supply_chain

*Wilson, R. (2005).* Council of Supply Chain Management Professionals, 16th Annual State of Logistics Report.

Wu, B., Dewan, M., Li, L., & Yang, Y. (2005). Supply chain protocolling. In *Proceedings of the 7th IEEE International Conference on E-Commerce Technology* (pp. 314-321).

Yang, J., Zhao, S. Z., & Wang, J. R. (2005). Study on postponement in customization supply chain. *Industrial Engineering and Management*, *10*(4), 35–44.

Zhao, T. Z., & Jin, Y. H. (2005). Optimized coordination of supply chains based on dependency. *Computer Integrated Manufacturing Systems*, *10*(8), 929–933.

Zhao, W., Kart, F., Moser, L. E., & Melliar-Smith, P. M. (2008). A reservation-based extended transaction protocol for coordination of Web Services. *Journal of Web Services Research*, *5*(3), 64–95.

Zhou, D., Xiang, B. H., & Zou, G. S. (2003). The strategy of mass customization and the countermeasure of Chinese enterprises. *Industrial Engineering and Management*, *8*(5), 12–16.

*This work was previously published in International Journal of Information Systems and Supply Chain Management, Volume 3, Issue 2, edited by John Wang, pp. 84-107, copyright2010 by IGI Publishing (an imprint of IGI Global)*

# Chapter 20
# Supplier Selection by the Pair of AR-NF-IDEA Models

**Reza Farzipoor Saen**
*Islamic Azad University - Karaj Branch, Iran*

**Mark Gershon**
*Temple University, USA*

## ABSTRACT

*Supplier selection is the process by which suppliers are reviewed, evaluated, and chosen to become part of a company's supply chain. To select the best suppliers in the presence of cardinal data, ordinal data, nondiscretionary factors, and weight restrictions, this paper proposes a new model considering all of these assumptions. A numerical example demonstrates the application of the proposed method.*

## INTRODUCTION

Supplier selection is the process by which suppliers are reviewed, evaluated, and chosen to become part of the company's supply chain. Shin et al. (2000) argue that several important factors have caused the current shift to single sourcing or a reduced supplier base. First, multiple sourcing prevents suppliers from achieving the economies of scale based on order volume and learning curve effect. Second, multiple supplier system can be more expensive than a reduced supplier base in terms of the labor and order processing costs to managing multiple source inventories. Third, multiple sourcing lowers the overall quality level because of the increased variation in incoming quality among suppliers. Fourth, a reduced supplier base helps eliminate mistrust between buyers and suppliers due to lack of communication. Fifth, worldwide competition forces firms to find the best suppliers in the world. Farzipoor Saen (2009c) proposes a method for ranking suppliers in the presence of nondiscretionary factors. Farzipoor Saen (in press) demonstrates the use of advanced Data Envelopment Analysis (DEA) modeling for measuring how well suppliers perform on multiple criteria relative to other suppliers competing in the same marketplace. The approach allows the buyer to

DOI: 10.4018/978-1-4666-0918-1.ch020

evaluate effectively each supplier's performance relative to the performance of the 'best suppliers' in the marketplace, through calculation of DEA efficiency measures.

One of the uses of DEA is supplier selection. In original DEA formulations the assessed Decision Making Units (DMUs) can freely choose the weights or values to be assigned to each input and output in a way that maximizes its efficiency, subject to this system of weights being feasible for all other DMUs. In supplier selection problems, Decision Makers (DMs) may have value judgments that can be formalized *a priori*, and therefore should be taken into account in supplier selection. For example, in the supplier selection problem in general, one input (material price) usually overwhelms all other inputs, and ignoring this aspect may lead to biased efficiency results. Suppliers might also supply some outputs that require considerably more resources than others and this marginal rate of substitution between outputs should somehow be taken into account when selecting a supplier.

Traditionally, supplier selection models are based on cardinal data with less emphasis on ordinal data. However, with the widespread use of manufacturing philosophies such as Just-In-Time (JIT), emphasis has shifted to the simultaneous consideration of cardinal and ordinal data in supplier selection decisions.

In any realistic situation, there may exist nondiscretionary criteria that are beyond the control of management. Banker and Morey (1986) illustrate the impact of exogenously determined inputs that are not controllable in an analysis of a network of fast food restaurants. In their study, each of the 60 restaurants in the fast food chain consumes six inputs to produce three outputs. The three outputs (all controllable) correspond to breakfast, lunch, and dinner sales. Only two of the six inputs, expenditures for supplies and expenditures for labor, are discretionary. The other four inputs (age of store, advertising level, urban/rural location, and presence/absence of drive-in capability) are beyond the control of the individual restaurant manager. Their analysis clearly demonstrates the value of accounting for the nondiscretionary character of these inputs explicitly in the DEA models they employ; the result is identification of a considerably enhanced opportunity for targeted savings in the controllable inputs and targeted increases in the outputs. In the case of supplier selection, distance and supply variety are generally considered nondiscretionary criteria.

In supplier selection context, the objective of this paper is to use a model that considers imprecise data, weight restrictions, and nondiscretionary factors.

This paper proceeds as follows. First, literature review is presented. Next, the method which selects the suppliers is introduced. Then, numerical example and concluding remarks are discussed respectively.

## LITERATURE REVIEW

Some mathematical programming approaches have been used for supplier selection in the past. Zeng et al. (2006) considered a simplified partner selection problem which takes into account only the bid cost and the bid completion time of subprojects, and the due date the project. They modeled the problem as a nonlinear integer programming problem and proved that the decision problem of the partner selection problem is NP-complete. Then they analyzed some properties of the partner selection problem and constructed a branch and bound algorithm. Ghodsypour and O'Brien (2001) developed a mixed integer nonlinear programming model to solve the multiple sourcing problems, which takes into account the total cost of logistics, including net price, storage, and transportation and ordering costs. The model is burdensome because it must be run $2^n$ times for $n$ suppliers. Dahel (2003) presented a multiobjective mixed integer programming approach to simultaneously determine the number of vendors to employ and the order quantities to allocate to these vendors in a multiple-product, multiple-supplier competitive sourcing environment. Talluri and Baker (2002) presented a multi-phase mathematical programming approach for effective supply chain design. More specifically, they developed and applied a combination of multi-criteria efficiency models, based on game theory concepts, and linear and integer programming methods. Ip et al. (2004) described the sub-contractor selection problem by a 0-1 integer programming with non-analytical objective function. Kanagaraj and Jawahar (2009) introduced a Reliability-Based Total Cost of Ownership (RBTCO) model, which incorporates procurement, maintenance and downtime costs along with the practical constraints on product reliability and weight limitation for supplier selection decisions. The mathematical formulation of the RBTCO model belongs to Nonlinear Integer Programming (NLIP). A Simulated Annealing Algorithm (SAA) was developed to arrive at the optimal or near-optimal solutions.

Talluri and Narasimhan (2003) proposed a max-min productivity based approach that derives variability measures of vendor performance, which are then utilized in a nonparametric statistical technique in identifying vendor groups for effective selection. To solve the vendor selection problem with multiple objectives, Kumar et al. (2004) applied fuzzy goal programming approach. To incorporate the imprecise aspiration levels of the goals, they formulated a vendor selection problem as a fuzzy mixed integer goal programming that includes three primary goals: minimizing the net cost, minimizing the net rejections, and minimizing the net late deliveries subject to realistic constraints regarding buyer's demand, vendor's capacity, vendor's quota flexibility, purchasing value of items, budget allocation to individual vendor. Hajidimitriou and Georgiou (2002) presented a quantitative model, based on the Goal Programming (GP) technique, which uses appropriate criteria to evaluate potential candidates and leads to the selection of the optimal partner (supplier). Cebi and Bayraktar (2003) proposed an integrated model for supplier selection. In their model, the supplier selection problem has been structured as an integrated Lexicographic Goal Program (LGP) and Analytic Hierarchy Process (AHP) model including both quantitative and qualitative conflicting factors. Cakravastia and Takahashi (2004) proposed a multi-objective model to support the process of supplier selection and negotiation that considers the effect of these decisions on the manufacturing plan. The model also takes into account several theoretical concepts in the negotiation process: concession force, resistance force and effective alternatives. Arunkumar et al. (2006) proposed a GP model for supplier selection with quantity discounts. They converted the piecewise linear problem into an easier linear problem, thereby decreasing the complexity of the problem. Karpak et al. (2001) presented one of the "user-friendly" multiple criteria decision support systems-Visual Interactive Goal programming (VIG). VIG facilitates the introduction of a decision support vehicle that helps improve the supplier selection decisions. Kameshwaran et al. (2007) provided a multiattribute e-procurement system for procuring a large volume of a single item. Their system is formulated as a mixed linear integer multiple criteria optimization problem and GP is used as the solution technique.

To take into account both cardinal and ordinal data in supplier selection, Wang et al. (2004) developed an integrated Analytic Hierarchy Process (AHP) and Preemptive Goal Programming (PGP) based methodology. However, one of the GP problems arises from a specific technical requirement. After the purchasing managers specify the goals for each selected criterion (e.g., amount of price and quality level), they must decide on a preemptive priority order of these goals, i.e., determining in which order the goals will be attained. Frequently such a priori input might not produce an acceptable solution and the priority structure may be altered to resolve the problem once more. In this fashion, it may be possible to generate a solution iteratively that finally satisfies the DM. Unfortunately, the number of potential priority reorderings may be very large. A supplier selection problem with five factors has up to 120 priority reorderings. Going through such a laborious process would be costly and inefficient.

Because of the complexity of the decision making process involved in supplier selection, all the aforementioned literature relied on some form of procedures that assigns weights to various performance measures. The primary problem associated with arbitrary weights is that they are subjective, and it is often a difficult task for the DM to accurately assign numbers to preferences. It is a daunting task for the DM to assess weighting information as the number of performance criteria increases. Therefore, a more robust mathematical technique that does not demand too much and too precise information, i.e., ordinal preferences instead of cardinal weights, from the DM can strengthen the supplier evaluation process. To this end, Weber (1996) demonstrated how DEA can be used to evaluate vendors on multiple criteria and identified benchmark values which can then be used for this purpose. Weber et al. (2000) presented an approach for evaluating the number of vendors to employ in a procurement situation using Multi-Objective Programming (MOP) and DEA. The approach advocates developing vendor-order quantity solutions (referred to as supervendors) using MOP and then evaluating the efficiency of these supervendors on multiple criteria using DEA. Braglia and Petroni (2000) described a MAUT based on the use of DEA, aimed at helping purchasing managers to formulate viable sourcing strategies in the changing market place. To evaluate the aggregate performances of suppliers, Liu et al. (2000) proposed to employ DEA. This extends Weber's (1996) research in using DEA in supplier evaluation for an individual product. Forker and Mendez (2001) proposed an analytical method for benchmarking using DEA that can help companies identify their most efficient suppliers, the suppliers among the most efficient with the most widely applicable Total Quality Management (TQM) programs, and those suppliers who are not on the efficient frontier but who could move toward it by emulating the practices of their "best peer" supplier(s). Talluri and Sarkis (2002) focused upon the supplier performance evaluation and monitoring process, which assist in maintaining effective customer-supplier linkages. They tried to improve the discriminatory power of BCC model proposed by Banker et al. (1984). To select appropriate suppliers, Talluri et al. (2006) suggested a Chance-Constrained Data Envelopment Analysis (CCDEA) approach in the presence of multiple performance measures that are uncertain. Recently, to select the best suppliers in the presence of both cardinal and ordinal data, Farzipoor Saen (2007) proposed an innovative method, which is based on Imprecise Data Envelopment Analysis (IDEA). However, he did not consider the weights restrictions. To select the best suppliers in volume discount environments in the presence of both cardinal and ordinal data, Farzipoor Saen (2009a) proposed an innovative approach.

Liu et al. (2000) proposed to employ DEA for selecting the best suppliers in the existence of nondiscretionary factors, but they did not introduce a model which selects the suppliers in the presence of weight restrictions, nondiscretionary factors, and imprecise data. Farzipoor Saen (2009b) proposed an approach that considers weight restrictions, imprecise data, and nondiscretionary factors for technology selection. However, he did not apply it for supplier selection problem.

To the best of the author's knowledge, there is not any reference that deals with supplier selection in the conditions that weight restrictions, nondiscretionary factors and imprecise data are present simultaneously. The approach presented in this paper has some distinctive features.

- Supplier selection is a straightforward process carried out by the proposed model.

- The proposed model considers imprecise data, weight restrictions, and nondiscretionary factors simultaneously.

## PROPOSED MODEL FOR SUPPLIER SELECTION

One serious drawback of DEA applications in supplier selection has been the absence of DM judgment, allowing total freedom when allocating weights to input and output data of supplier under analysis. This allows suppliers to achieve artificially high efficiency scores by indulging in inappropriate input and output weights. The most widespread method for considering judgments in DEA models is, perhaps, the weight restrictions inclusion.

DEA, as proposed by Charnes et al. (1978) (CCR model) and developed by Banker et al. (1984) (BCC model) is an approach for evaluating the efficiencies of DMUs. This evaluation is generally assumed to be based on a set of cardinal (quantitative) output and input factors. In many real world applications (especially supplier selection problems), however, it is essential to take into account the existence of imprecise data (ordinal and interval factors) when rendering a decision on the performance of a DMU.

The only papers that discuss IDEA in the presence of weight restrictions are the works of Cooper et al. (1999), Cooper et al. (2001a), and Cooper et al. (2001b). In these papers, the Assurance Region[1]-IDEA (AR-IDEA) model was developed to deal not only with imprecise data and cardinal data but also with weight restrictions. Recently, Wang et al. (2005) developed a new pair of interval DEA models for dealing with imprecise data.

In this section, a new pair of Assurance Region-Nondiscretionary Factors-IDEA (AR-NF-IDEA) models is proposed. The final efficiency score for each DMU will be characterized by an interval bounded by the best lower bound efficiency and the best upper bound efficiency of each DMU.

Suppose that there are $n$ DMUs to be evaluated. Each DMU consumes $m$ inputs to produce $s$ outputs. In particular, $DMU_j$ consumes amounts $X_j = \{x_{ij}\}$ of inputs ($i$=1, ..., $m$) and produces amounts $Y_j = \{y_{rj}\}$ of outputs ($r$=1, ..., $s$). Without loss of generality, it is assumed that all the input and output data $x_{ij}$ and $y_{rj}$ ($i$=1, ..., $m$; $r$=1, ..., $s$; $j$=1, ..., $n$) cannot be exactly obtained due to the existence of uncertainty. They are only known to lie within the upper and lower bounds represented by the intervals $\left[x_{ij}^L, x_{ij}^U\right]$ and $\left[y_{rj}^L, y_{rj}^U\right]$, where $x_{ij}^L > 0$ and $y_{rj}^L > 0$.

In order to deal with such an uncertain situation, the following pair of linear programming models has been developed to generate the upper (1) and lower (2) bounds of interval efficiency for each DMU (Wang et al., 2005; Farzipoor Saen, 2009b):

Model 1:

$$Max\theta_{jo}^U = \sum_{r=1}^{s} u_r y_{rj_o}^U$$

s.t.

$$\sum_{i=1}^{m} v_i x_{ij_o}^L = 1 \tag{1}$$

$$\sum_{r=1}^{s} u_r y_{rj}^U - \sum_{i=1}^{m} v_i x_{ij}^L \leq 0, j = 1, \cdots, n, \tag{2}$$

$$u_r, v_i \geq \varepsilon \qquad r, i.$$

Model 2

$$Max\theta^L_{jo} = \sum_{r=1}^{s} u_r y^L_{rj_o}$$

s.t.

$$\sum_{i=1}^{m} v_i x^U_{ij_o} = 1, \tag{3}$$

$$\sum_{r=1}^{s} u_r y^U_{rj} - \sum_{i=1}^{m} v_i x^L_{ij} \leq 0, j = 1, \cdots, n, \tag{4}$$

$$u_r, v_i \geq \varepsilon \qquad r, i.$$

where $j_o$ is the DMU under evaluation (usually denoted by $DMU_o$); $u_r$ and $v_i$ are the weights assigned to the outputs and inputs; $\theta^U_{jo}$ stands for the best possible relative efficiency achieved by $DMU_o$ when all the DMUs are in the state of best production activity, while $\theta^L_{jo}$ stands for the lower bound of the best possible relative efficiency of $DMU_o$. They constitute a possible best relative efficiency interval $\left[\theta^L_{jo}, \theta^U_{jo}\right]$. $\varepsilon$ is the non-Archimedean infinitesimal[2].

In order to judge whether a DMU is DEA efficient or not, the following definition is given.

**Definition 1.** A DMU, $DMU_o$, is said to be DEA efficient if its best possible upper bound efficiency $\theta^{U*}_{jo} = 1$; otherwise, it is said to be DEA inefficient if $\theta^{U*}_{jo} < 1$.

In (5) the various types of weight restriction that can be applied to multiplier models are shown.

$$\left.\begin{array}{l}
\text{Absolute weight restrictions} \\
\delta_i \leq v_i \leq \tau_i \quad (g_i) \qquad \rho_r \leq u_r \leq \eta_r \quad (g_o) \\
\text{Assurance region of} \\
\text{type I (relative weight restrictions)} \\
\alpha_i \leq \dfrac{v_i}{v_{i+1}} \leq \beta_i \quad (h_i) \qquad \theta_r \leq \dfrac{u_r}{u_{r+1}} \leq \zeta_r \quad (h_o) \\
\text{Assurance regions of} \\
\text{type II (input-output weight restrictions)} \\
\gamma_i v_i \geq u_r \qquad (l)
\end{array}\right\} \tag{5}$$

The Greek letters ($\delta_i, \tau_i, \rho_r, \eta_r, \alpha_i, \beta_i, \theta_r, \zeta_r, \gamma_i$) are user-specified constants to reflect value judgments the DM wishes to incorporate in the assessment. They may relate to the perceived importance or worth of input and output factors. The restrictions *(g)* and *(h)* in (5) relate on the left hand side to input weights and on the right hand side to output weights. Constraint *(l)* links directly input and output weights. Absolute weight restrictions are the most immediate form of placing restrictions on the weights as they simply restrict them to vary within a specific range. For an assurance region of type I, link either only input weights ($h_i$) or only output weights ($h_o$). The relationship between input and output weights are termed assurance region of type II.

Restrictions on virtual inputs/virtual outputs assume the form in (6), where the proportion of the total virtual output of $DMU_j$ accounted for by output $r$ is restricted to lie in the range $[a_r, b_r]$ and the proportion of the total virtual input of $DMU_j$ accounted for by input $i$ is restricted to lie in the range $[c_i, d_i]$.

$$a_r \le \frac{u_r y_{rj}}{\sum_{r=1}^{s} u_r y_{rj}} \le b_r, \qquad r = 1,\ldots,s$$

$$c_i \le \frac{v_i x_{ij}}{\sum_{i=1}^{m} v_i x_{ij}} \le d_i, \qquad i = 1,\ldots,m$$

(6)

All proposed virtual weights[3] restrictions can be described by the general set of $w=1,\ldots,t$ weight restrictions, applying to the DMU$_o$:

$$\sum_{i=1}^{m} a_{iw} x_{io} v_i + \sum_{r=1}^{s} b_{rw} y_{ro} u_r \ge k_w \qquad w \qquad (7)$$

where $a_{iw}$ is the weight of $i$th input in $w$th weight restriction and $b_{rw}$ is the weight of *the* $r$th output in the $w$th weight restriction.

This expression encapsulates the three different kinds of virtual weights restrictions, of the classification presented below.

## ABSOLUTE VIRTUAL WEIGHTS RESTRICTIONS

Absolute virtual weights restrictions involve constraining the virtual weight of a single factor. This approach is equivalent to using proportional virtual weights restrictions applied to output virtuals in an output oriented model, as $\sum_{r=1}^{s} y_{ro} u_r = 1$. If applied also to input virtuals in an output oriented model, it will only be equivalent for the DMUs that are efficient (i.e., $\sum_{i=1}^{m} x_{io} v_i = 1$), and therefore define the frontier. They are of the form:

$$\begin{aligned} a_{iw} x_{io} &\ge k_w, & i &= i', \\ a_{iw} &= 0 & \forall i &\ne i', \\ b_{rw} &= 0 & \forall r & \end{aligned}$$

for restricting the virtual input $i'$; and

$$\begin{aligned} b_{rw} y_{ro} &\ge k_w, & r &= r', \\ b_{rw} &= 0 & \forall r &\ne r', \\ a_{iw} &= 0 & \forall i & \end{aligned}$$

for restricting the virtual output $r'$.

These restrictions are useful when the DM is able to specify particular bounds, or wants to assure that a certain factor attains a threshold value, for instance.

## VIRTUAL ASSURANCE REGIONS OF TYPE I

Assurance regions of type I virtual restrictions link virtual inputs or outputs to translate into an ordering of preference. They are

$$\begin{aligned} \sum_{i=1}^{m} a_{iw} x_{io} v_i &\ge 0, \\ b_{rw} &= 0 \qquad \forall r \end{aligned}$$

to link virtual inputs, and

$$\begin{aligned} \sum_{r=1}^{s} b_{rw} y_{ro} u_r &\ge 0, \\ a_{iw} &= 0 \qquad \forall i \end{aligned}$$

to link virtual outputs.

These restrictions are useful when the DM cannot assign particular bounds to the factors, but is able to decide that a factor is more important than another, twice as important, etc.

## VIRTUAL ASSURANCE REGIONS OF TYPE II

Finally, assurance regions of type II virtual restrictions, link the input-output divide. They can be translated by

$$\sum_{i=1}^{m} a_{iw} x_{io} v_i + \sum_{r=1}^{s} b_{rw} y_{ro} u_r \geq 0,$$

where at least one $a_{iw} \neq 0$ and one $b_{rw} \neq 0$.

These restrictions are useful when there is a known relationship between an input and an output. For instance, it is known that to produce a certain output, one needs to have a certain level of a certain input.

The multipliers formulation, with the virtual weights restrictions applying to all DMUs, is as below

Model 3

$$Max\theta_{jo}^{U} = \sum_{r=1}^{s} u_r y_{rj_o}^{U}$$

s.t.

$$\sum_{i=1}^{m} v_i x_{ij_o}^{L} = 1 \tag{8}$$

$$\sum_{r=1}^{s} u_r y_{rj}^{U} - \sum_{i=1}^{m} v_i x_{ij}^{L} \leq 0, j = 1,\cdots,n, \tag{9}$$

$$\sum_{r=1}^{s} b_{rw} y_{rj}^{U} u_r + \sum_{i=1}^{m} a_{iw} x_{ij}^{L} v_i \geq k_w \qquad w, j \tag{10}$$

$$u_r, v_i \geq \varepsilon \qquad r, i$$

Model 4

$$Max\theta_{jo}^{L} = \sum_{r=1}^{s} u_r y_{rj_o}^{L}$$

s.t.

$$\sum_{i=1}^{m} v_i x_{ij_o}^{U} = 1 \tag{11}$$

$$\sum_{r=1}^{s} u_r y_{rj}^{U} - \sum_{i=1}^{m} v_i x_{ij}^{L} \leq 0, j = 1,\cdots,n, \tag{12}$$

$$\sum_{r=1}^{s} b_{rw} y_{rj}^{U} u + \sum_{i=1}^{m} a_{iw} x_{ij}^{L} v_i \geq k_w \qquad w, j \tag{13}$$

$$u_r, v_i \geq \varepsilon \qquad r, i.$$

To determine the peer groups for using in seeking improvements for inefficient suppliers, the dual forms of the Models 3 and 4 are required. The envelopment formulation (dual problem) of Models 3 and 4 becomes

Model 5

$$Minc_{jo}^{U} = \theta_o - \sum_{j=1}^{n} \sum_{w=1}^{t} k_w \rho_{wj} - \varepsilon \sum_{i=1}^{m} s_i^{-} - \sum_{r=1}^{s} s_r^{+}$$

s.t.

$$\sum_{j=1}^{n} y_{rj}^{U} \lambda_j - \sum_{j=1}^{n} \sum_{w=1}^{t} b_{rw} y_{rj}^{U} \rho_{wj} - s_r^{+} = y_{rjo}^{U} \quad r = 1,\cdots,s, \tag{14}$$

$$x_{ijo}^{L}\theta_{o} - \sum_{j=1}^{n} x_{ij}^{L}\lambda_{j} - \sum_{j=1}^{n}\sum_{w=1}^{t} a_{iw}x_{ij}^{L}\rho_{wj} - s_{i}^{-} = 0 \quad i = 1,\cdots,m, \tag{15}$$

$$\begin{array}{ll} \theta_{o} & \text{free}, \\ \rho_{wj} \geq 0 & w, j, \\ \lambda_{j} \geq 0 & j = 1,\cdots,n, \\ s_{i}^{-} \geq 0 & i = 1,\cdots,m, \\ s_{r}^{+} \geq 0 & r = 1,\cdots,s. \end{array}$$

Model 6

$$Minc_{jo}^{L} = \theta_{o} - \sum_{j=1}^{n}\sum_{w=1}^{t} k_{w}\rho_{wj} - \varepsilon\sum_{i=1}^{m} s_{i}^{-} - \varepsilon\sum_{r=1}^{s} s_{r}^{+},$$

s.t.

$$\sum_{j=1}^{n} y_{rj}^{U}\lambda_{j} - \sum_{j=1}^{n}\sum_{w=1}^{t} b_{rw}y_{rj}^{U}\rho_{wj} - s_{r}^{+} = y_{rjo}^{L} \quad r = 1,\cdots,s, \tag{16}$$

$$x_{ijo}^{U}\theta_{o} - \sum_{j=1}^{n} x_{ij}^{L}\lambda_{j} - \sum_{j=1}^{n}\sum_{w=1}^{t} a_{iw}x_{ij}^{L}\rho_{wj} - s_{i}^{-} = 0 \quad i = 1,\cdots,m, \tag{17}$$

$$\begin{array}{ll} \theta_{o} & \text{free}, \\ \rho_{wj} \geq 0 & w, j, \\ \lambda_{j} \geq 0 & j = 1,\cdots,n, \\ s_{i}^{-} \geq 0 & i = 1,\cdots,m, \\ s_{r}^{+} \geq 0 & r = 1,\cdots,s. \end{array}$$

where $\theta_{o}, \lambda_{j}, \rho_{wj}, s_{i}^{-}$, and $s_{r}^{+}$ are the dual variables. $\theta_{o}$ is the radial input shrinkage factor (eventually to become the efficiency measure) and $\lambda = \{\lambda_{j}\}$ is vector of DMU loadings, determining "best practice" for the DMU being evaluated. $c_{jo}^{U}$ stands for the upper bound (best possible) of the relative efficiency achieved by DMU$_o$ when all the DMUs are in the state of best production activity, while $c_{jo}^{L}$ stands for the lower bound of the best possible relative efficiency of DMU$_o$. They constitute a possible best relative efficiency interval $\left[c_{jo}^{L}, c_{jo}^{U}\right]$. The variable $s_{r}^{+}$ is shortfall amount of output *r* and $s_{i}^{-}$ is excess amount of input *i*. From the duality theory in linear programming, for an inefficient DMU$_o$, $\lambda_{j}^{*} > 0$ in the optimal dual solution implies that DMU$_j$ is a unit of the peer group. A peer group of an inefficient DMU$_o$ is defined as the set of DMUs that reach the efficiency score of 1 using the same set of weights that result in the efficiency score of DMU$_o$. It is the existence of this collection of DMUs that forces the DMU$_o$ to be inefficient. In other words, the peer group or reference set is the set of efficient units to which an inefficient unit has been most directly compared with when calculating its efficient score. It contains the efficient units which have the most similar input/output orientation to the inefficient unit and they should therefore provide examples of good operating practice for it to emulate.

Now, suppose that the input variables may be partitioned into subsets of discretionary (D) and nondiscretionary (N) variables. Thus,

$$I = \{1,2,...,m\} = I_{D} \cup I_{N}, \quad I_{D} \cap I_{N} = \Phi$$

The pair of Assurance Region-Nondiscretionary Factors-Imprecise Data Envelopment Analysis (AR-NF-IDEA) models is then finally given by

Model 7

$$Minc_{jo}^{U} = \theta_{o} - \sum_{j=1}^{n}\sum_{w=1}^{t} k_{w}\rho_{wj} - \varepsilon\sum_{1\in D}^{m} s_{i}^{-} - \varepsilon\sum_{r=1}^{s} s_{r}^{+},$$

s.t.

$$\sum_{j=1}^{n} y_{rj}^{U}\lambda_j - \sum_{j=1}^{n}\sum_{w=1}^{t} b_{rw} y_{rj}^{U}\rho_{wj} - s_r^{+} = y_{rjo}^{U}, r = 1,\cdots,s, \tag{18}$$

$$x_{ijo}^{L}\theta_o - \sum_{j=1}^{n} x_{ij}^{L}\lambda_j - \sum_{j=1}^{n}\sum_{w=1}^{t} a_{iw} x_{ij}^{L}\rho_{wj} - s_i^{-} = 0 \quad i \in D, \tag{19}$$

$$x_{ijo}^{L} - \sum_{j=1}^{n} x_{ij}^{L}\lambda_j - \sum_{j=1}^{n}\sum_{w=1}^{t} a_{iw} x_{ij}^{L}\rho_{wj} - s_i^{-} = 0 \quad i \in N, \tag{20}$$

$$\begin{array}{ll} \theta_o & \text{free}, \\ \rho_{wj} \geq 0 & w, j, \\ \lambda_j \geq 0 & j = 1,\cdots,n, \\ s_i^{-} \geq 0 & i \in D, \\ s_i^{-} = 0 & i \in N, \\ s_r^{+} \geq 0 & r = 1,\cdots,s. \end{array}$$

Model 8

$$Min c_{jo}^{L} = \theta_o - \sum_{j=1}^{n}\sum_{w=1}^{t} k_w\rho_{wj} - \varepsilon\sum_{i\in D}^{m} s_i^{-} - \varepsilon\sum_{r=1}^{s} s_r^{+}$$

s.t.

$$\sum_{j=1}^{n} y_{rj}^{U}\lambda_j - \sum_{j=1}^{n}\sum_{w=1}^{t} b_{rw} y_{rj}^{U}\rho_{wj} - s_r^{+} = y_{rjo}^{L} \quad r = 1,\cdots,s, \tag{21}$$

$$x_{ijo}^{U}\theta_o - \sum_{j=1}^{n} x_{ij}^{L}\lambda_j - \sum_{j=1}^{n}\sum_{w=1}^{t} a_{iw} x_{ij}^{L}\rho_{wj} - s_i^{-} = 0 \quad i \in D \tag{22}$$

$$x_{ijo}^{U} - \sum_{j=1}^{n} x_{ij}^{L}\lambda_j - \sum_{j=1}^{n}\sum_{w=1}^{t} a_{iw} x_{ij}^{L}\rho_{wj} - s_i^{-} = 0 \quad i \in N, \tag{23}$$

$$\begin{array}{ll} \theta_o & \text{free}, \\ \rho_{wj} \geq 0 & w, j, \\ \lambda_j \geq 0 & j = 1,\cdots,n, \\ s_i^{-} \geq 0 & i \in D, \\ s_i^{-} = 0 & i \in N, \\ s_r^{+} \geq 0 & r = 1,\cdots,s. \end{array}$$

It is to be noted that the $\theta_o$ to be minimized appears only in the constraints for which $i \in D$, whereas the constraints for which $i \in N$ operate only indirectly (as they should) because the input levels $x_{ijo}$ are not subject to managerial control. Therefore this is recognized by entering all $x_{ijo}$, $i \in N$ at their fixed (observed) value. Note that the slacks $s_i^{-}$, $i \in N$ are omitted from the objective function. Hence these nondiscretionary inputs do not enter directly into the efficiency measures being optimized in Models 7 and 8. They can, nevertheless, affect the efficiency evaluations by virtue of their presence in the constraints. For Models 7 and 8, it is not relevant to minimize the proportional decrease in the entire input vector. Such minimization should be determined only with respect to the subvector that is composed of discretionary inputs.

In order to judge whether a DMU is DEA efficient or not, the following definition is given.

**Definition 2.** A DMU, $DMU_o$, is said to be DEA efficient if its best possible upper bound efficiency $c_{jo}^{U*} = 1$; otherwise, it is said to be DEA inefficient if $c_{jo}^{U*} < 1$.

Comparing definition 1 and definition 2, we have provided a definition for the new problem that is equivalent to the same definitions provided earlier for the base problem. Therefore, one unified model that deals with all aspects of the weights restrictions, imprecise data and nondiscretionary factors in a direct manner has been introduced.

Now, the method of transforming ordinal preference information into interval data is discussed, so that the pair of AR-NF-IDEA models presented in this paper can still work properly even in these situations.

Suppose some input and/or output data for DMUs are given in the form of ordinal preference information. Usually, there may exist three types of ordinal preference information: (1) strong ordinal preference information such as $y_{rj}>y_{rk}$ or $x_{ij}>x_{ik}$, which can be further expressed as $y_{rj} \ge \chi_r y_{rk}$ and $x_{ij} \ge \eta_i x_{ik}$, where $\chi_r > 1$ and $\eta_i > 1$ are the parameters on the degree of preference intensity provided by decision maker; (2) weak ordinal preference information such as $y_{rp} \ge y_{rq}$ or $x_{ip} \ge x_{iq}$; (3) indifference relationship such as $y_{rl} = y_{rt}$ or $x_{il} = x_{it}$. Since the DEA model has the property of unit-invariance, the use of scale transformation to ordinal preference information does not change the original ordinal relationships and has no effect on the efficiencies of DMUs. Therefore, it is possible to conduct a scale transformation to every ordinal input and output index so that its best ordinal datum is less than or equal to unity and then give an interval estimate for each ordinal datum.

Now, consider the transformation of ordinal preference information about the output $y_{rj}$ $(j=1,\ldots, n)$ for example. The ordinal preference information about input and other output data can be converted in the same way.

For weak ordinal preference information $y_{r1} \ge y_{r2} \ge \cdots \ge y_{rn}$, we have the following ordinal relationships after scale transformation:

$$1 \ge \hat{y}_{r1} \ge \hat{y}_{r2} \ge \cdots \ge \hat{y}_{rn} \ge \sigma_r,$$

where $\sigma_r$ is a small positive number reflecting the ratio of the possible minimum of $\{y_{rj} | j=1,\ldots, n\}$ to its possible maximum. It can be approximately estimated by the decision maker. It is referred as the ratio parameter for convenience. The resultant permissible interval for each $\hat{y}_{rj}$ is given by

$$\hat{y}_{rj} \in [\sigma_r, 1], \qquad j = 1,\cdots,n.$$

For strong ordinal preference information $y_{r1} > y_{r2} > \cdots > y_{rn}$, there is the following ordinal relationships after scale transformation:

$$1 \ge \hat{y}_{r1}, \hat{y}_{rj} \ge \chi_r \hat{y}_{r,j+1}$$
$$(j = 1,\cdots,n-1) \text{ and } \hat{y}_{rn} \ge \sigma_r,$$

where $\chi_r$ is a preference intensity parameter satisfying $\chi_r > 1$ provided by the decision maker and $\sigma_r$ is the ratio parameter also provided by the decision maker. The resultant permissible interval for each $\hat{y}_{rj}$ can be derived as follows:

$$\hat{y}_{rj} \in [\sigma_r \chi_r^{n-j}, \chi_r^{1-j}],$$
$$j = 1,\cdots,n \text{ with } \sigma_r \le \chi_r^{1-n}.$$

Finally, for the indifference relationship, the permissible intervals are the same as those obtained for the weak ordinal preference information.

Through the scale transformation above and the estimation of permissible intervals, all the ordinal preference information is converted into interval data and can thus be incorporated into the pair of AR-NF-IDEA models. This completes the proposed model developed in this paper.

In the next section, a numerical example is presented.

## NUMERICAL EXAMPLE

The data set for this example is partially taken from Farzipoor Saen (2007) and contains specifications on 18 suppliers (DMUs). Originally, the supplier data set is selected from a case study that utilized DEA for monitoring customer–supplier relationship. For more details please see Talluri and Baker (2002) and Kleinsorge et al. (1992).

In particular, this example is used to show how ordinal data, bounded data, and weight restriction as well as nondiscretionary factors, can be combined into the one unified approach provided by AR-NF-IDEA. The cardinal inputs considered are Total Cost of shipments (TC) and Distance (D). D is generally considered as a nondiscretionary input variable. Supplier Reputation (SR) is included as a qualitative input while Number of Bills received from the supplier without errors (NB) will serve as the bounded data output. SR is an intangible factor that is not usually included explicitly in evaluation models for supplier selection. This qualitative variable is measured on an ordinal scale so that, for instance, reputation of supplier 18 is given the highest rank, and supplier 17, the lowest. Note that, the measures selected in this paper are not exhaustive by any means, but are some general measures that can be utilized to evaluate suppliers. In an application of this methodology, DMs must carefully identify appropriate inputs and outputs to be used in the decision making process. Table 1 depicts the supplier's attributes.

Suppose the preference intensity parameter and the ratio parameter about the strong ordinal preference information are given (or estimated) as $\eta_3 = 1.12$ and $\sigma_3 = 0.01$, respectively. Us-

*Table 1. Related attributes for 18 suppliers*

| Supplier No. (DMU) | Inputs | | | Output |
|---|---|---|---|---|
| | TC $x_{1j}$ | D (km) $x_{2j}$ | SR* $x_{3j}$ | NB $y_{1j}$ |
| 1 | 253 | 249 | 5 | [50, 65] |
| 2 | 268 | 643 | 10 | [60, 70] |
| 3 | 259 | 714 | 3 | [40, 50] |
| 4 | 180 | 1809 | 6 | [100, 160] |
| 5 | 257 | 238 | 4 | [45, 55] |
| 6 | 248 | 241 | 2 | [85, 115] |
| 7 | 272 | 1404 | 8 | [70, 95] |
| 8 | 330 | 984 | 11 | [100, 180] |
| 9 | 327 | 641 | 9 | [90, 120] |
| 10 | 330 | 588 | 7 | [50, 80] |
| 11 | 321 | 241 | 16 | [250, 300] |
| 12 | 329 | 567 | 14 | [100, 150] |
| 13 | 281 | 567 | 15 | [80, 120] |
| 14 | 309 | 967 | 13 | [200, 350] |
| 15 | 291 | 635 | 12 | [40, 55] |
| 16 | 334 | 795 | 17 | [75, 85] |
| 17 | 249 | 689 | 1 | [90, 180] |
| 18 | 216 | 913 | 18 | [90, 150] |

* Ranking such that 18 ≡ highest rank,…, 1 ≡ lowest rank ($x_{3,18} > x_{3,16} \cdots > x_{3,17}$)

ing the transformation technique described in previous section, an interval estimate for SR of each supplier can be derived, which is shown in Table 2. Therefore, all the input and output data are now transformed into interval numbers and can be evaluated using the pair of AR-NF-IDEA models.

According to the decision of DM, the importance of TC must be greater than SR. Assume that TC is twice as important as SR (*virtual assurance regions of type I*). Applying Models 7 and 8, the efficiency scores of suppliers (DMUs) and peer groups of suppliers have been presented in Table 3. To solve the problem, Lingo software was used. A snapshot of the software for upper bound of efficiency of supplier 1 is presented in Appendix. The positive non-Archimedean infinitesimal, $\varepsilon$, has been set to 0.0001.

Based on the definition 2, suppliers 4, 6, 11, 14, and 17 have the possibility to be DEA efficient. If they are able to use the minimum inputs to produce the maximum outputs, they are DEA efficient (efficient in scale); otherwise, they are not DEA efficient. The remaining 13 suppliers with relative efficiency scores of less than 1 are considered to be inefficient. Therefore, DM can choose one or more of these efficient suppliers. Also, the last column of Table 3 provides peer groups for inefficient suppliers. The peer groups serve as a benchmark to use in seeking improvements for inefficient suppliers.

*Table 2. Interval estimate for the 18 suppliers after the transformation of ordinal preference information*

| Supplier No. (DMU) | SR |
|---|---|
| 1 | [.01574, .22917] |
| 2 | [.02773, .40388] |
| 3 | [.01254, .1827] |
| 4 | [.01762, .25668] |
| 5 | [.01405, .20462] |
| 6 | [.0112, .16312] |
| 7 | [.02211, .32197] |
| 8 | [.03106, .45235] |
| 9 | [.02476, .36061] |
| 10 | [.01974, .28748] |
| 11 | [.05474, .79719] |
| 12 | [.04363, .63552] |
| 13 | [.04887, .71178] |
| 14 | [.03896, .56743] |
| 15 | [.03479, .50663] |
| 16 | [.0613, .89286] |
| 17 | [.01, .14564] |
| 18 | [.06866, 1] |

## CONCLUSION

When suppliers are compared for their overall performances, an aggregate evaluation relevant to the considerations of a purchasing firm needs to be conducted. Such an overall performance evaluation of suppliers should be based on performance measures for all part types supplied to the purchasing company. A potential use of an overall performance evaluation of suppliers is to provide benchmarking data for reducing the number of suppliers, which in turn results in benefits including reduction in costs of parts and order processing, and better partnership with suppliers. This paper has introduced a new pair of AR-NF-IDEA models and employed it for supplier selection.

The supplier-selection approach developed in this paper includes a number of attractive features for managers, as below:

- This paper proposed an approach capable of treating nondiscretionary factors, which are beyond the control of a DMU manager.
- This paper proposed an approach capable of treating weight restrictions.
- This paper proposed an approach capable of treating imprecise data.

*Table 3. The efficiency interval and peer group for the 18 suppliers*

| Supplier No. (DMU) | Efficiency Interval[4] | Peer group |
|---|---|---|
| 1 | [.186893, .3122656] | 4, 14, 17 |
| 2 | [.248115, .278422] | 4, 14, 17 |
| 3 | [.271985, .546011] | 4 |
| 4 | [.993976, 1] | N/A |
| 5 | [.167119, .27319] | 4, 14, 17 |
| 6 | [.303292, 1] | N/A |
| 7 | [.508175, .612737] | 4 |
| 8 | [.328302, .576344] | 4, 14, 17 |
| 9 | [.276986, .423783] | 4, 14, 17 |
| 10 | [.17709, .328534] | 4, 17 |
| 11 | [.821322, 1] | N/A |
| 12 | [.293373, .415662] | 4, 14 |
| 13 | [.286224, .400769] | 4, 14 |
| 14 | [.609409, 1] | N/A |
| 15 | [.215501, .217011] | 4 |
| 16 | [.248104, .272213] | 4, 14 |
| 17 | [.369223, 1] | N/A |
| 18 | [.454985, .678513] | 4, 14 |

- The proposed approach considers multiple criteria. This helps managers to select suppliers using a comprehensive approach that goes beyond just purchase costs.
- The proposed approach is computationally efficient and can be solved in a few seconds on a personal computer.

Further research can be done based on the results of this paper. One of them is as follows:

Consider the case that some of the suppliers are slightly non-homogeneous. One of the assumptions of all the classical models of DEA is based on complete homogeneity of DMUs (suppliers), whereas this assumption in many real applications cannot be generalized. In other words, some inputs and/or outputs are not common for all the DMUs occasionally. Therefore, there is a need to develop a model that deals with these conditions.

## ACKNOWLEDGMENT

The authors wish to thank the anonymous reviewers for their valuable suggestions and comments.

## REFERENCES

Arunkumar, N., Karunamoorthy, L., Anand, S., & Ramesh Babu, T. (2006). Linear approach for solving a piecewise linear vendor selection problem of quantity discounts using lexicographic method. *International Journal of Advanced Manufacturing Technology*, *28*, 1254–1260. doi:10.1007/s00170-004-2474-z

Banker, R. D., Charnes, A., & Cooper, W. W. (1984). Some methods for estimating technical and scale inefficiencies in data envelopment analysis. *Management Science*, *30*, 1078–1092. doi:10.1287/mnsc.30.9.1078

Banker, R. D., & Morey, R. C. (1986). Efficiency analysis for exogenously fixed inputs and outputs. *Operations Research*, *34*, 513–521. doi:10.1287/opre.34.4.513

Braglia, M., & Petroni, A. (2000). A quality assurance-oriented methodology for handling trade-offs in supplier selection. *International Journal of Physical Distribution & Logistics Management*, *30*, 96–111. doi:10.1108/09600030010318829

Cakravastia, A., & Takahashi, K. (2004). Integrated model for supplier selection and negotiation in a make-to-order environment. *International Journal of Production Research*, *42*, 4457–4474. doi:10.1080/00207540410001727622

Cebi, F., & Bayraktar, D. (2003). An integrated approach for supplier selection. *Logistics Information Management, 16*, 395–400. doi:10.1108/09576050310503376

Charnes, A., Cooper, W. W., & Rhodes, E. (1978). Measuring the efficiency of decision making units. *European Journal of Operational Research, 2*, 429–444. doi:10.1016/0377-2217(78)90138-8

Cooper, W. W., Park, K. S., & Yu, G. (1999). IDEA and AR-IDEA: models for dealing with imprecise data in DEA. *Management Science, 45*, 597–607. doi:10.1287/mnsc.45.4.597

Cooper, W. W., Park, K. S., & Yu, G. (2001a). An illustrative application of IDEA (imprecise data envelopment analysis) to a Korean mobile telecommunication company. *Operations Research, 49*, 807–820. doi:10.1287/opre.49.6.807.10022

Cooper, W. W., Park, K. S., & Yu, G. (2001b). IDEA (imprecise data envelopment analysis) with CMDs (column maximum decision making units). *The Journal of the Operational Research Society, 52*, 176–181. doi:10.1057/palgrave.jors.2601070

Cooper, W. W., Seiford, L. M., & Tone, K. (2007). *Data envelopment analysis: a comprehensive text with models, applications, references and DEA-solver software* (2nd ed.). New York: Springer.

Dahel, N. E. (2003). Vendor selection and order quantity allocation in volume discount environments. *Supply Chain Management: An International Journal, 8*, 335–342. doi:10.1108/13598540310490099

Farzipoor Saen, R. (2007). Suppliers selection in the presence of both cardinal and ordinal data. *European Journal of Operational Research, 183*, 741–747. doi:10.1016/j.ejor.2006.10.022

Farzipoor Saen, R. (2009a). Suppliers selection in volume discount environments in the presence of both cardinal and ordinal data. *International Journal of Information Systems and Supply Chain Management, 2*, 69–80.

Farzipoor Saen, R. (2009b). Technology selection in the presence of imprecise data, weight restrictions, and nondiscretionary factors. *International Journal of Advanced Manufacturing Technology, 41*, 827–838. doi:10.1007/s00170-008-1514-5

Farzipoor Saen, R. (2009c). Using data envelopment analysis for ranking suppliers in the presence of nondiscretionary factors. *International Journal of Procurement Management, 2*, 229–243. doi:10.1504/IJPM.2009.024807

Farzipoor Saen, R. (in press). Supplier selection by the pair of nondiscretionary factors-imprecise data envelopment analysis models. *The Journal of the Operational Research Society.*

Forker, L. B., & Mendez, D. (2001). An analytical method for benchmarking best peer suppliers. *International Journal of Operations & Production Management, 21*, 195–209. doi:10.1108/01443570110358530

Ghodsypour, S. H., & O'Brien, C. (2001). The total cost of logistics in supplier selection, under conditions of multiple sourcing, multiple criteria and capacity constraint. *International Journal of Production Economics, 73*, 15–27. doi:10.1016/S0925-5273(01)00093-7

Hajidimitriou, Y. A., & Georgiou, A. C. (2002). A goal programming model for partner selection decisions in international joint ventures. *European Journal of Operational Research, 138*, 649–662. doi:10.1016/S0377-2217(01)00161-8

Ip, W. H., Yung, K. L., & Wang, D. (2004). A branch and bound algorithm for sub-contractor selection in agile manufacturing environment. *International Journal of Production Economics, 87*, 195–205. doi:10.1016/S0925-5273(03)00125-7

Kameshwaran, S., Narahari, Y., Rosa, C. H., Kulkarni, D. M., & Tew, J. D. (2007). Multiattribute electronic procurement using goal programming. *European Journal of Operational Research, 179*, 518–536. doi:10.1016/j.ejor.2006.01.010

Kanagaraj, G., & Jawahar, N. (2009). A simplified annealing algorithm for optimal supplier selection using the reliability-based total cost of ownership model. *International Journal of Procurement Management, 2*, 244–263. doi:10.1504/IJPM.2009.024809

Karpak, B., Kumcu, E., & Kasuganti, R. R. (2001). Purchasing materials in the supply chain: managing a multi-objective task. *European Journal of Purchasing & Supply Management, 7*, 209–216. doi:10.1016/S0969-7012(01)00002-8

Kleinsorge, I. K., Schary, P., & Tanner, R. (1992). Data envelopment analysis for monitoring customer–supplier relationships. *Journal of Public Policy, 11*, 357–372.

Kumar, M., Vrat, P., & Shankar, R. (2004). A fuzzy goal programming approach for vendor selection problem in a supply chain. *Computers & Industrial Engineering, 46*, 69–85. doi:10.1016/j.cie.2003.09.010

Liu, J., Ding, F. Y., & Lall, V. (2000). Using data envelopment analysis to compare suppliers for supplier selection and performance improvement. *Supply Chain Management: An International Journal, 5*, 143–150. doi:10.1108/13598540010338893

Shin, H., Collier, D. A., & Wilson, D. D. (2000). Supply management orientation and supplier/buyer performance. *Journal of Operations Management, 18*, 317–333. doi:10.1016/S0272-6963(99)00031-5

Talluri, S., & Baker, R. C. (2002). A multi-phase mathematical programming approach for effective supply chain design. *European Journal of Operational Research, 141*, 544–558. doi:10.1016/S0377-2217(01)00277-6

Talluri, S., & Narasimhan, R. (2003). Vendor evaluation with performance variability: a max-min approach. *European Journal of Operational Research, 146*, 543–552. doi:10.1016/S0377-2217(02)00230-8

Talluri, S., Narasimhan, R., & Nair, A. (2006). Vendor performance with supply risk: a chance-constrained DEA approach. *International Journal of Production Economics, 100*, 212–222. doi:10.1016/j.ijpe.2004.11.012

Talluri, S., & Sarkis, J. (2002). A model for performance monitoring of suppliers. *International Journal of Production Research, 40*, 4257–4269. doi:10.1080/00207540210152894

Wang, G., Huang, S. H., & Dismukes, J. P. (2004). Product-driven supply chain selection using integrated multi-criteria decision-making methodology. *International Journal of Production Economics, 91*, 1–15. doi:10.1016/S0925-5273(03)00221-4

Wang, Y. M., Greatbanks, R., & Yang, J. B. (2005). Interval efficiency assessment using data envelopment analysis. *Fuzzy Sets and Systems, 153*, 347–370.

Weber, C. A. (1996). A data envelopment analysis approach to measuring vendor performance. *Supply Chain Management, 1*, 28–39. doi:10.1108/13598549610155242

Weber, C. A., Current, J., & Desai, A. (2000). An optimization approach to determining the number of vendors to employ. *Supply Chain Management: An International Journal, 5*, 90–98. doi:10.1108/13598540010320009

Wong, Y. H. B., & Beasley, J. E. (1990). Restricting weight flexibility in data envelopment analysis. *The Journal of the Operational Research Society, 39*, 829–835.

Zeng, Z. B., Li, W., & Zhu, W. (2006). Partner selection with a due date constraint in virtual enterprises. *Applied Mathematics and Computation, 175*, 1353–1365. doi:10.1016/j.amc.2005.08.022

## APPENDIX

The snapshot of Lingo for upper bound of efficiency for supplier 1 (Figures 1 and 2).

1 In the optimal weight $\left(v_i^*, u_j^*\right)$ of DEA models for inefficient DMUs, we may see many zeros — showing that the DMU has a weakness in the corresponding items compared with other (efficient) DMUs. Large differences in weights from item to item may also be a concern. It was concerns like these that led to the development of the assurance region approach which imposes constraints on the relative magnitude of the weights for special items. For example, we may add a constraint on the ratio of weights for Input 1 and Input 2 as follows:

$$L_{1,2} \leq \frac{v_2}{v_1} \leq U_{1,2}$$

*Figure 1.*

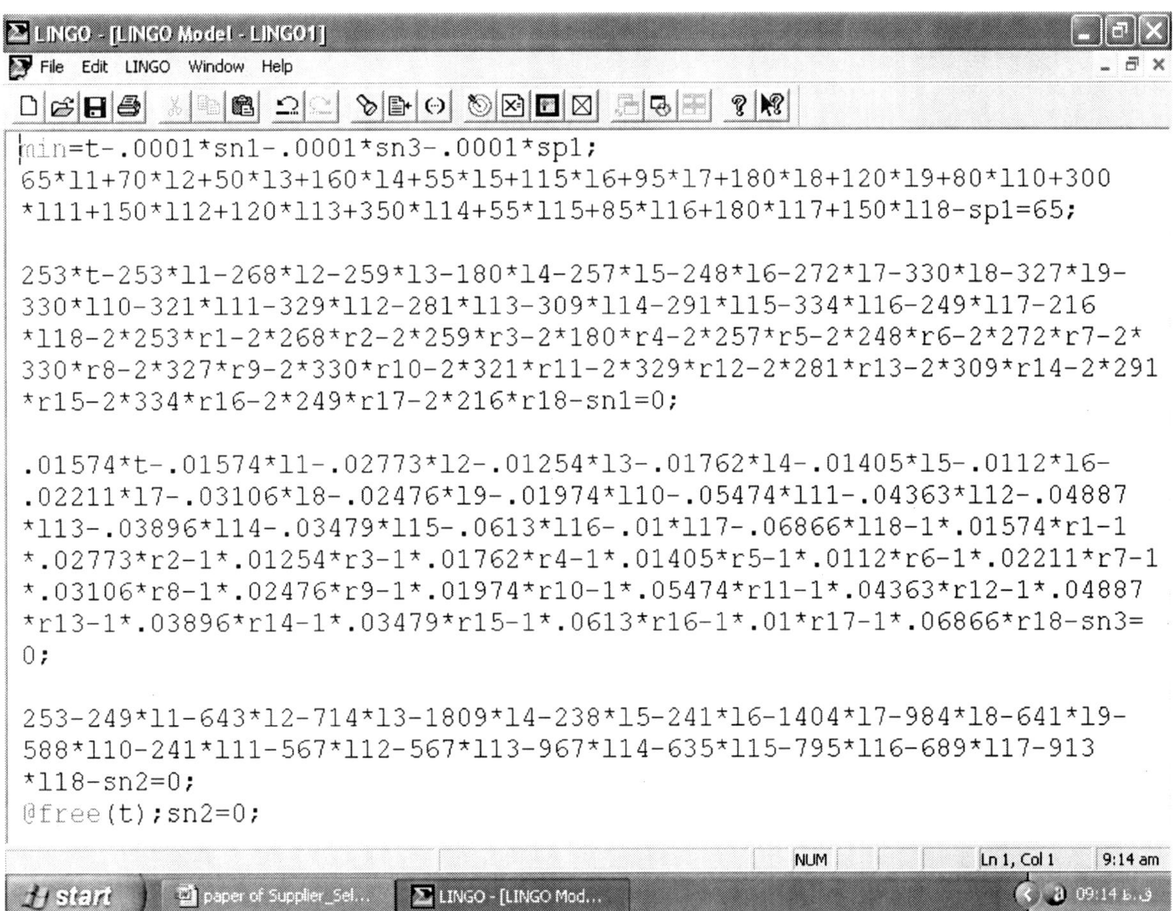

*Figure 2.*

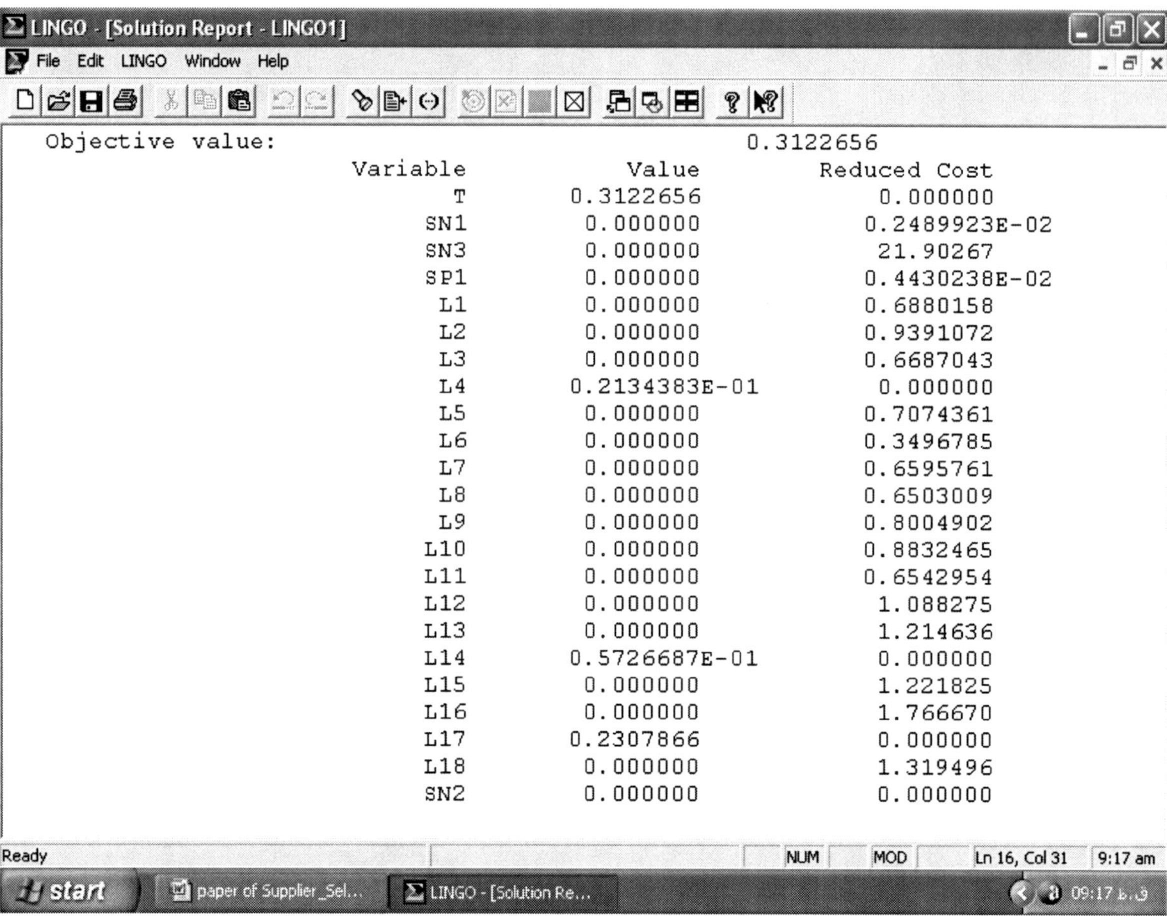

LINGO - [Solution Report - LINGO1]

File Edit LINGO Window Help

Objective value: 0.3122656

| Variable | Value | Reduced Cost |
|---|---|---|
| T | 0.3122656 | 0.000000 |
| SN1 | 0.000000 | 0.2489923E-02 |
| SN3 | 0.000000 | 21.90267 |
| SP1 | 0.000000 | 0.4430238E-02 |
| L1 | 0.000000 | 0.6880158 |
| L2 | 0.000000 | 0.9391072 |
| L3 | 0.000000 | 0.6687043 |
| L4 | 0.2134383E-01 | 0.000000 |
| L5 | 0.000000 | 0.7074361 |
| L6 | 0.000000 | 0.3496785 |
| L7 | 0.000000 | 0.6595761 |
| L8 | 0.000000 | 0.6503009 |
| L9 | 0.000000 | 0.8004902 |
| L10 | 0.000000 | 0.8832465 |
| L11 | 0.000000 | 0.6542954 |
| L12 | 0.000000 | 1.088275 |
| L13 | 0.000000 | 1.214636 |
| L14 | 0.5726687E-01 | 0.000000 |
| L15 | 0.000000 | 1.221825 |
| L16 | 0.000000 | 1.766670 |
| L17 | 0.2307866 | 0.000000 |
| L18 | 0.000000 | 1.319496 |
| SN2 | 0.000000 | 0.000000 |

Ready NUM MOD Ln 16, Col 31 9:17 am

start paper of Supplier_Sel... LINGO - [Solution Re... 09:17

where $L_{1,2}$ and $U_{1,2}$ are lower and upper bounds that the ratio $\frac{v_2}{v_1}$ may assume. The name assurance region (AR) comes from this constraint, which limits the region of weights to some special area. For more details please see Cooper et al. (2007).

2 In abstract algebra, the Archimedean property, named after the ancient Greek mathematician Archimedes of Syracuse, is a property held by some groups, fields, and other algebraic structures. Roughly speaking, it is the property of having no infinitely large or infinitely small elements (i.e. no nontrivial infinitesimals). An algebraic structure in which any two non-zero elements are comparable, in the sense that neither of them is infinitesimal with respect to the other, is called Archimedean. A structure which has a pair of non-zero elements, one of which is infinitesimal with respect to the other, is called non-Archimedean. For example, a linearly ordered group that

is Archimedean is an Archimedean group, and a field with a non-Archimedean absolute value is a non-Archimedean field.

3 Virtual weights are restrictions on weighted inputs/outputs. The virtual weights are based on the use of proportions, and are dimensionless. For more details please see Wong and Beasley (1990).

4 Consider suppliers 1 and 2. While supplier 1 may be very close in efficiency to supplier 2, but if you look at the data set of these suppliers, you will find out that since the inputs of supplier 1 is less than supplier 2, so the upper bound of efficiency of supplier 1 is more than supplier 2.

*This work was previously published in International Journal of Information Systems and Supply Chain Management, Volume 3, Issue 4, edited by John Wang, pp. 25-41, copyright 2010 by IGI Publishing (an imprint of IGI Global)*

# Compilation of References

Ackerman, M. S. (1996). Definitional and contextual issues in organizational and group memories. *Information Technology & People*, *9*(1), 10–24. doi:10.1108/09593849610111553

Adelberg, A. H. (1975). Management information systems and their implications. *Management Accounting*, *53*(9), 328–328.

Agrawal, D. K. (2001). 7 Ms to make supply chain relationships harmonious. In *Proceeding of National Conference on People, Processes and Organizations: Emerging Realities.* New Delhi, India: Excel books.

Aguinis, H. (2004). *Regression Analysis for Categorical Moderators*. New York: Guilford Press.

Al Kattan, I., & Al Khudairi, T. (2007). Improving supply chain management effectiveness using RFID. *In IEEE International Engineering Management Conference,* Austin, Texas.

Al Khudairi, T. (2007). *Integrating RFID into supply chain management in process inventory control.* Unpublished Master thesis dissertation, American University of Sharjah, UAE.

Alaranta, M., & Henningsson, S. (2008). An approach to analyzing and planning post-merger IS integration: Insights from two field studies. *Information Systems Frontiers*, *10*(3), 307–319. doi:10.1007/s10796-008-9079-2

Albright, B. (2006). *Ready or not RFID's coming, frontline solutions website*. Retrieved January 2007. from http://www.findarticles.com/p/articles/mi_m0dis/is_12_4/ai_112366601

Alonso, G., Casati, F., Kuno, H., & Machiraju, V. (2004). *Web Services concepts: Architectures and applications*. Berlin: Springer-Verlag.

Alsene, E. (1999). The computer integration of the enterprise. *IEEE Transactions on Engineering Management*, *46*(1), 26–35. doi:10.1109/17.740033

Altekar, R. V. (2005). *Supply chain management concept and cases*. Washington, DC: IEEE.

Alter, S. (1999). A general, yet useful theory of Information Systems. *Communications of the AIS, 1*(13).

Altiok, T., et al. (2001). *Simulation Modeling and Analysis with ARENA*. Cyber Research Inc.

American Arbitration Association. (2001). E-commerce dispute risk management services offered by AAA. *Dispute Resolution Journal*, *56*(3), 5.

Anand, V., Manz, C. C., & Glick, W. H. (1998). An organizational memory approach to information management. *Academy of Management Review*, *23*(4), 796–809. doi:10.2307/259063

Anbazhagan, N., & Arivarignan, G. (2000). Two-commodity continuous review inventory system with coordinated reorder policy. *International Journal of Information and Management Sciences*, *11*(3), 19–30.

Anbazhagan, N., & Arivarignan, G. (2001). Analysis of two-commodity markovian inventory system with lead time. *Journal of Applied Mathematics & Computing*, *8*(2), 519–530.

Andel, T. (1997). Information supply chain: set and get your goals. *Transportation and Distribution*, *38*(2), 33.

Andraski, J. C. (1998). Leadership and the realization of supply chain collaboration. *Journal of Business Logistics*, *19*(2), 9.

Angeles, R. (2005). RFID technologies: supply-chain applications and implementation issues. *Information Systems Management*, *22*(1), 51–65. doi:10.1201/1078/44912.22.1.20051201/85739.7

Anonymous. (2004, September). *Using ADR in Supply Chain Disputes*. Paper presented at the Purchasing b2b, PMAC-AGM conference, Halifax, Nova Scotia.

Apache Web Services Project. (2008). *Home*. Retrieved from http://ws.apache.org/

Arbin, K., & Essler, U. (2005). Covisint in Europe: Analysing the auto B2B e-marketplace. *International Journal of Automotive Technology & Management*, *5*(1).

Argyris, C. (1990). *Overcoming organizational defenses: Facilitating organizational learning*. Boston, Allyn & Bacon.

Ariba. (2008). *Air Products and Chemicals customer success story*. Retrieved August 2008, from http://www.ariba.com//pdf/Air_Products.pdf

Arnold, L. (1974). *Stochastic differential equations: Theory and applications*. New York: John Willey & Sons.

Arunkumar, N., Karunamoorthy, L., Anand, S., & Ramesh Babu, T. (2006). Linear approach for solving a piecewise linear vendor selection problem of quantity discounts using lexicographic method. *International Journal of Advanced Manufacturing Technology*, *28*, 1254–1260. doi:10.1007/s00170-004-2474-z

Ashenbaum, B., Maltz, A. B., & Rabinovich, E. (2005). Studies of trends in third party logistics usage: What can we conclude? *Transportation Journal*, *44*(3), 39–50.

Ash, R. B. (1972). *Real analysis and probability*. New York: Academic Press.

Attaran, M. (2004). Exploring the relationship between information technology and business process reengineering. *Information & Management*, *41*(5), 585–596. doi:10.1016/S0378-7206(03)00098-3

Auramo, J., Kauremaa, J., & Tanskanen, K. (2005). Benefits of IT in supply chain management: an explorative study of progressive companies. *International Journal of Physical Distribution & Logistics Management*, *35*(2), 82–100. doi:10.1108/09600030510590282

Austin, T. A., Lee, H. L., & Kopczak, L. (1997). *Unlocking Hidden Value in the Personal Computer Supply Chain*. Chicago, IL: Andersen Consulting.

Aviv, Y. (2001). The effect of collaborative forecasting on supply chain performance. *Management Science*, *47*(10), 1326–1343. doi:10.1287/mnsc.47.10.1326.10260

Bacheldor, B. (2006). Process improvement drives manufacturers' RFID implementations. *RFID Journal*. Retrieved July 1, 2008, from http://www.rfidjournal.com/article/articleview/2903/1/1/

Bakos, Y. (1998). The emerging role of electronic marketplaces on the Internet. *Communications of the ACM*, *41*(8).

Balakrishnan, B., & Bowen, F. (2008). A strategic framework for managing failure in JIT supply chains. *Journal of Information Systems and Supply Chain Management*, *1*(4), 20–38.

Baldi, S., & Borgman, H. (2001). Consortium-based B2B e-marketplaces: A case study of the automotive industry. *14th Bled Electronic Commerce Conference,* Slovenia.

Bal, J., & Teo, P. K. (2000). Implementing virtual team working: Part 1 - a literature review of best practice. *Logistics Information Management*, *13*(6), 346. doi:10.1108/09576050010355644

Bal, J., & Teo, P. K. (2001). Implementing virtual team working: Part 2 - a literature review. *Logistics Information Management*, *14*(3), 208. doi:10.1108/09576050110390248

Bal, J., & Teo, P. K. (2001a). Implementing virtual team working: Part 3 – a methodology for introducing virtual team working. *Logistics Information Management*, *14*(4), 276. doi:10.1108/EUM0000000005722

Ballintify, J. L. (1964). On a basic class of inventory problems. *Management Science*, *10*, 287–297. doi:10.1287/mnsc.10.2.287

Ballou, R. H. (2004). *Business logistics management: Planning, organizing, and controlling the supply chain* (5th ed.). Upper Saddle River, NJ: Pearson.

Ballou, R. H. (2007). The evolution and future of logistics and supply chain management. *European Business Review*, *19*(4), 332–348. doi:10.1108/09555340710760152

Balsmeier, P. W., & Voisin, W. J. (1996). Supply chain management: a time based strategy. *Industrial Management (Des Plaines)*, *38*(5), 24–27.

Banker, R. D., Charnes, A., & Cooper, W. W. (1984). Some methods for estimating technical and scale inefficiencies in data envelopment analysis. *Management Science*, *30*, 1078–1092. doi:10.1287/mnsc.30.9.1078

Banker, R. D., & Morey, R. C. (1986). Efficiency analysis for exogenously fixed inputs and outputs . *Operations Research*, *34*(4), 513–521. doi:10.1287/opre.34.4.513

Bardhan, I., Whitaker, J., & Mithas, S. (2006). Information technology, production process outsourcing, and manufacturing plant performance. *Journal of Management Information Systems*, *23*(2), 13–40. doi:10.2753/MIS0742-1222230202

Barrat, M. (2004). Understanding the meaning of collaboration in the supply chain. *Supply Chain Management*, *9*(1), 30. doi:10.1108/13598540410517566

Barratt, M., & Rosdhal, K. (2002). Exploring business-to-business marketsites. *International Journal of Purchasing and Supply Management*, 8.

Barratt, M., & Oliveira, A. (2001). Exploring the experiences of collaborative planning initiatives. *International Journal of Physical Distribution & Logistics Management*, *31*(4).

Bask, A. H. (2001). Relationships between 3PL providers and members of supply chains – a strategic perspective. *Journal of Business and Industrial Marketing*, *16*(6), 470–486. doi:10.1108/EUM0000000006021

Basu, A. K. (1993). *Introduction to stochastic process*. Oxford, UK: Alpha Science International Ltd.

Bayles, D. (2000). Send it back! The role of reverse logistics. *InformIT*. Retrieved from www.informit.com/articles/printerfriendly.aspx?p=164926

Bechtel, C., & Jayaram, J. (1997). Supply chain management: a strategic perspective. *The International Journal of Logistics Management*, 15-34.

Bendoly, E., Citurs, A., & Konsynski, B. (2007). Internal infrastructural impacts on rfid perceptions and commitment: knowledge, operational procedures, and information processing standards. *Decision Sciences*, *38*(3), 423–449. doi:10.1111/j.1540-5915.2007.00165.x

Benjamin, R., & Blunt, J. (1994). IS plans in context: The IS planning environment: Critical IT issues in the year 2000. In R. Galliers, D. Leidner, & B. Baker (Eds.), *Strategic Information Management*. Oxford: Butterworth.

Bensoussan, A., Hurst, E. G. Jr, & Naslund, B. (1974). *Management Application of Modern Control Theory*. New York: Elsevier.

Bertolini, M., Bevilacqua, M., & Massini, R. (2006). FMECA approach to product traceability in the food industry. *Food Control*, *17*(2), 137–145. doi:10.1016/j.foodcont.2004.09.013

Beulens, M., Broens, F., Folstar, P., & Hofstede, J. (2005). Food safety and transparency in food chains and networks, relationships and challenges. *Food Control*, *16*(6).

Bevilacqua, M., Ciarapica, F., & Giacchetta, G. (2008). Business process reengineering of a supply chain and a traceability system: a case study. *Journal of Food Engineering*, *93*, 13–22. doi:10.1016/j.jfoodeng.2008.12.020

Bhagwat, R., & Sharma, M. K. (2007). Performance measurement of supply chain management: A balanced scorecard approach. *Computers & Industrial Engineering*, *53*(1), 43–62. doi:10.1016/j.cie.2007.04.001

Bharadwaj, A. S. (2000). A resource-based perspective on information technology capability and firm performance: an empirical investigation. *Management Information Systems Quarterly*, *24*(1), 169–196. doi:10.2307/3250983

Bhargava, H. (2007). Building smart RFID networks. *RFID Journal*. Retrieved July 7, 2008, from http://www.rfidjournal.com/article/articleview/3387/1/82/

Bhatnagar, R., Sohal, A., & Millen, R. A. (1999). Third party logistics services: A Singapore perspective. *International Journal of Physical Distribution and Logistics Management*, *29*(9), 569–587. doi:10.1108/09600039910287529

Bilbao, J. M., Jiménez-Losada, A., Lebrón, E., & Tijs, S. H. (2002). The τ-value for games on Matroids. *Top (Madrid)*, *10*(1), 67–81. doi:10.1007/BF02578941

Biller, S., Muriel, A., & Zhang, Y. (2006). Impact of price postponement on capacity and flexible investment. *Production and Operations Management*, *15*(2), 198–214. doi:10.1111/j.1937-5956.2006.tb00240.x

Birge, J. J., Drogosz, I., & Duenyas, I. (1998). Setting single-period optimal capacity levels and prices for substitutable products. *International Journal of Flexible Manufacturing Systems*, *10*, 407–430. doi:10.1023/A:1008061605260

Biswas, S. (2000). *Supply chain performance measures*. Retrieved December 8, 2008, from http://lcm.csa.iisc.ernet.in/scm/coimbatore/node11.html

Bitran, G., & Rene, C. (2003). An overview of pricing models for revenue management. *Manufacturing & Service Operations Management*, *5*(3), 203–229. doi:10.1287/msom.5.3.203.16031

Blackhurst, J., Wu, T., & Graighead, C. (2006). A systematic approach to supply chain conflict detection with a hierarchical Petri Net extension. *Omega . International Journal of Management Science*, *36*(5), 680.

Blossom, P. (2005). Levels of RFID maturity, part 2. *RFID Journal.* Retrieved July 5, 2008, from http://www.rfidjournal.com/article/articleview/1347/1/82/

Blumenthal, S. (1969). *Management Information Systems: A Framework for Planning and Development.* Englewood Cliffs, NJ: Prentice-Hall.

Boland, J., & Hirschheim, R. (1987). *Critical issues in information systems research*, Wiley, New York.

Bonfill, A., Espuña, A., & Puigjaner, L. (2007). Decision support framework for coordinated production and transport scheduling in SCM. *Computers & Chemical Engineering*. doi:.doi:10.1016/j.compchemeng.2007.04.020

Booch, G., Christerson, M., Fuchs, M., & Koistinen, J. (1998). *UML for XML schema mapping specification.* Retrieved May 10, 2009, from http://xml.coverpages.org/fuchsuml_xmlschema33.pdf

Boot, R., & Butler, J. (2001). *Momentum in the automotive industry*. KPMG report.

Bose, I., & Lam, C. W. (2008). Facing the challenges of RFID data management. *International Journal of Information Systems and Supply Chain Management*, *1*(4), 1–19.

Bottani, E., & Rizzi, A. (2006). A fuzzy TOPSIS methodology to support outsourcing of logistics services. *Supply Chain Management: An International Journal*, *11*(4), 294–308. doi:10.1108/13598540610671743

Boubekri, N. (2001). Technology enablers for supply chain. *Integrated Manufacturing Systems*, *12*(6), 394–399. doi:10.1108/EUM0000000006104

Bowersox, D. J., Closs, D.J., & Drayer, R.W. (2005, January 1). The digital transformation: Technology and beyond. *Supply Chain Management Review*.

Bowersox, D. J., Daugherty, P. J., Droge, C. L., Rogers, D. S., & Wardlow, D. L. (1989). Leading-edge logistics: Competitive positioning for the 90s. *Council of Logistics Management*, Oak Brook, IL.

Bowersox, D. J., & Closs, D. C. (1996). *Logistical management: the integrated supply chain process*. New York: McGraw-Hill.

Braglia, M., & Petroni, A. (2000). A quality assurance-oriented methodology for handling trade-offs in supplier selection. *International Journal of Physical Distribution & Logistics Management*, *30*, 96–111. doi:10.1108/09600030010318829

Brain, M. (2006). How UPC bar codes work. *Howstuffworks website*. Retrieved May 2009 from http://electronics.howstuffwworks.comupc

Branzei, R., Dimitrov, D., & Tijs, S. (2005). *Models in Cooperative Game Theory – Crisp, Fuzzy, and Multi-Choice Games*. Berlin: Springer.

Brock, L. (2001, February). *The physical markup language - a universal language for physical objects, Auto-ID White Paper WH-003*. Retrieved May 10, 2009, from http://xml.coverpages.org/PML-MIT-AUTOID-WH-003.pdf

Bronder, C., & Pritzl, R. (1992). Developing strategic alliances: A successful framework for cooperation. *European Management Journal*, *10*(4), 412–420. doi:10.1016/0263-2373(92)90005-O

Brown, A. E., & Grant, G. G. (2005). Framing the frameworks: a review of IT governance research. *Communication of the AIS*, *15*, 696–712.

Browne, J., & Jiangang, Z. (1999). Extended and virtual enterprises - similarities and differences. *International Journal of Agile Management Systems*, *1*(1), 30–36. doi:10.1108/14654659910266691

Browne, J., Sockett, P. J., & Wortmann, J. C. (1995). Future manufacturing systems - Towards the extended enterprise. *Computers in Industry*, *25*, 235–254. doi:10.1016/0166-3615(94)00035-O

Brown, J., Dev, C., & Lee, D. (2000). Managing marketing channel opportunism: The efficacy of alternative governance mechanisms. *Journal of Marketing*, *64*(2), 51–66. doi:10.1509/jmkg.64.2.51.17995

Brun, A., Salama, K. F., & Gerosa, M. (2009). Selecting performance measurement systems: matching a supply chain's requirements. *European Journal of Industrial Engineering*, *3*(3), 336–362. doi:10.1504/EJIE.2009.025051

Brunsson, N., & Adler, N. (1989). *The Organization of Hypocrisy: Talk, Decisions, and Actions in Organizations*. New York: Wiley.

Brunsson, N., & Jacobsson, B. (2000). *A World of Standards*. Oxford: Oxford University Press.

Buckley, P. J., & Casson, M. (1988). A theory of cooperation in international business . In Casson, M. (Ed.), *The Economics of Business Culture*. Oxford, UK: Oxford University Press.

Burt, D. N., Dobler, D. W., & Starling, S. L. (2003). *World class supply management: The key to supply chain management* (7th ed.) Boston: McGraw-Hill Irwin.

Byrd, T. A., & Davidson, N. W. (2003). Examining possible antecedents of IT impact on the supply chain and its effect on firm performance. *Information & Management*, *41*(3), 243–255. doi:10.1016/S0378-7206(03)00051-X

Cabdoi, C. (2003). *Fourth party logistics market: A European perspective. 1 Dec 2003.* Retrieved March 3, 2009, from http://www.frost.com/prod/servlet/market-insight-top.pag?docid=8341069

Cakravastia, A., & Takahashi, K. (2004). Integrated model for supplier selection and negotiation in a make-to-order environment. *International Journal of Production Research*, *42*, 4457–4474. doi:10.1080/00207540410001727622

Caldwell, B. (1999). Reverse logistics. *Information Week Online*. Retrieved from http://www.informationweek.com/729/logistics.htm

Carlson, D. (Ed.). (2001). *Modeling XML applications with UML, practical e-business applications*. Reading, MA: Addison-Wesley.

Carroll, J. M., & Swatman, P. A. (2000). Structured-case: A methodological framework for building theory in information systems research. *European Journal of Information Systems*, *9*(4), 235–242. doi:10.1057/palgrave/ejis/3000374

Carr, P., Rainbird, M., & Walters, D. (2004). Mcasuring the implications of virtual integration in the new economy: a process-led approach. *International Journal of Physical Distribution & Logistics Management*, *34*(3), 358–372. doi:10.1108/09600030410533646

Carter, J. R., Ferrin, B. G., & Carter, C. R. (1995). The effect of less-than-truckload rates on the purchase order lot size decision. *Transportation Journal*, *34*(4), 35–44.

Carter-Steel, A. (2009). IT service departments struggle to adopt a service-oriented philosophy. *International Journal of Information Systems in the Service Sector*, *1*(2), 69–77.

Cebi, F., & Bayraktar, D. (2003). An integrated approach for supplier selection. *Logistics Information Management*, *16*, 395–400. doi:10.1108/09576050310503376

Cederlund, J., Kohli, R., Sherer, S., & Yao, Y. (2007). How Motorola put CPFR into action. *Supply Chain Management Review*.

Cellini, R., & Lambertini, L. (2002). A differential game approach to investment in product differentiation. *Journal of Economic Dynamics & Control*, *27*, 51–62. doi:10.1016/S0165-1889(01)00026-4

Chalasani, S., & Sounderpandian, J. (2004). Performance benchmarks and cost sharing models for B2B supply chain information systems. *Benchmarking*, *11*(5), 447–464. doi:10.1108/14635770410557690doi:10.1108/14635770410557690

Chan, H. (2004, February). The supply-chain squeeze. *Optimize*.

Chandra, C., & Kumar, S. (2000). Supply chain management in theory and practice: a passing fad or a fundamental change. *Industrial Management & Data Systems*, *100*(3), 100–113. doi:10.1108/02635570010286168

Chandrashekar, A., & Schary, P. B. (1999). Toward the virtual supply chain: the convergence of IT and organization. *International Journal of Logistics Management*, *10*(2), 27–39. doi:10.1108/09574099910805978

Chan, F. T. S. (2003). Performance measurement in a supply chain. *International Journal of Advanced Manufacturing Technology*, *21*, 534–548. doi:10.1007/s001700300063

Charan, P., Shankar, R., & Baisya, R. (2009). Modeling the barriers of supply chain performance measurement system implementation in the Indian automobile supply chain. *International Journal of Logistics Systems and Management*, *5*(6), 614–630. doi:10.1504/IJLSM.2009.024794

Charnes, A., Cooper, W. W., & Rhodes, E. (1978). Measuring the efficiency of decision making units. *European Journal of Operational Research*, *2*(6), 429–444. doi:10.1016/0377-2217(78)90138-8

Charnes, A., Cooper, W. W., Wei, Q. L., & Huang, Z. M. (1989). Cone-ratio data envelopment analysis and multi-objective programming. *International Journal of Systems Science*, *20*(7), 1099–1118. doi:10.1080/00207728908910197

Charnes, A., Cooper, W., & Rhodes, E. (1978). Measuring the efficiency of decision-making units. *European Journal of Operational Research*, *2*, 429–444. doi:10.1016/0377-2217(78)90138-8

Chase, R., Jacobs, R., & Aquilano, J. (2004). *Operations Management*. New York: McGraw Hill Irwin.

Chatterjee, S., & Webber, J. (2003). *Developing enterprise Web Services: An architect's guide*. Englewood Cliffs, NJ: Prentice Hall.

Checkland, P. (1991). From framework through experience to learning: the essential nature of action research. In H. Nissen, H. Klein, & Hirschheim (Eds.), *Information Systems Research*. Elsevier Science.

Chen, C. J. (2004). The effects of knowledge attribute, alliance characteristics, and absorptive capacity on knowledge transfer performance. *R & D Management*, *34*(3), 311–321. doi:10.1111/j.1467-9310.2004.00341.x

Chen, C., & Lee, W. (2004). Multi-objective optimization of multi-echelon supply chain networks with uncertain product demands and prices. *Computers & Chemical Engineering*, *28*, 1131–1144. doi:10.1016/j.compchemeng.2003.09.014

Childerhouse, P., & Towill, D. R. (2006). Enabling seamless market-orientated supply chains. *International Journal of Logistics Systems and Management*, *2*(4).

Child, J., & Faulkner, D. (1998). *Strategies of Cooperation: Managing Alliances, Networks, and Joint Ventures*. New York, NY: Oxford University Press.

Choi, B., Tsai, N., & Jones, T. (2008). Building enterprise network infrastructure for a supermarket store chain. *Journal of Cases on Information Technology*, 31–46.

Chopra, S., & ManMohan, S. (2004, fall). Managing risk to avoid supply chain breakdown. *MIT Sloan Management Review*, *3*, 53-60.

Chopra, S., & Meindl, P. (2001). *Supply chain management: Strategy, planning and operation*. Upper Saddle River: Prentice Hall.

Chopra, S., & Meindl, P. (2001). *Supply chain management – strategy, planning and operation*. Upper Saddle River, NJ: Prentice Hall.

Chowdhury, N., Sherer, S. A., & Ray, M. (2001). Realizing IT value at Air Products and Chemicals. *Communications of the AIS*, *7*(23).

Chow, H. K. H., Choy, K. L., & Lee, W. B. (2007). Knowledge management approach in build-to-order supply chains. *Industrial Management & Data Systems*, *107*(6), 882–919. doi:10.1108/02635570710758770

Choy, K. L., Lee, W. B., & Lo, V. (2004). Development of a case based intelligent supplier relationship management system - linking supplier rating system and product coding system. *Supply Chain Management*, *9*(1), 86–101. doi:10.1108/13598540410517601doi:10.1108/13598540410517601

Christopher, M. (1998). *Logistics and supply chain management* (2nd ed.). Essex, UK: Pearson.

Christopher, M. G., & Towill, D. R. (2002). An Integrated Model for the Design of Agile Supply Chains. *International Journal of Physical Distribution & Logistics, 31*(4), 262–264.

Ciborra, C. (1993). *Teams, markets and systems: Business innovation and information technology.* Cambridge University Press.

Cleland, D. I., Bidanda, B., & Chung, C. A. (1995). Human issues in technology integration – Part 1. *Industrial Management (Des Plaines), 37*(4), 22.

Clemhout, S., Leitmann, G., & Wan, H. Y. Jr. (1971). A differential game model of duopoly. *Econometrica, 39*(6), 911–938. doi:10.2307/1909667

Clemons, E. K., & Row, M. C. (1991). Sustaining IT Advantage: The Role of Structural Differences. *MIS Quarterly, 15*(3), 275–293. doi:10.2307/249639

Closs, D. J., & Goldsby, T. J. (1997). Information technology influences on world class logistics capability. *International Journal of Physical Distribution & Materials Management, 27*(2), 4–17.

Closs, D. J., & Savitskie, K. (2003). Internal and external logistics information technology integration. *The International Journal of Logistics Management, 14*(1), 63–76. doi:10.1108/09574090310806549

Cohen, S., & Roussel, J. (2005). *Strategic supply chain management.* New York: McGraw-Hill.

Cohen, W. M., & Levinthal, D. A. (1990). Absorptive capacity: a new perspective on learning and innovation. *Administrative Science Quarterly, 35*(1), 128–152. doi:10.2307/2393553

Collins, J. (2001). *Good to great.* New York: Harper Business.

Collins, J. (2004). Soap maker cleans up with RFID. *RFID Journal.* Retrieved July 21, 2008, from http://www.rfidjournal.com/article/articleview/1101

Collis, D. J. (1994). Research note: how valuable are organizational capabilities. *Strategic Management Journal, 15*, 143–152. doi:10.1002/smj.4250150910

Colombo, L. (2002). *Product differentiation and cartel stability with stochastic demand market.* Societa Italiana di Economia Publica.

Computer World. (2006). *Supply chain management.* Retrieved April 2008, from http://www.computerworld.com/softwaretopics/erp/story/0,10801,66625,00.html

Connelly, M. (2001). Where Jacques Nasser went wrong. *Automotive News.*

Conrad, R., Scheffner, D., & Freytag, J. (2000). XML conceptual modeling using UML. In *Proceedings of the International Conceptual Modeling Conference,* Salt Lake City, UT (pp. 309-322). New York: Springer Verlag.

Cooper, B. L., Watson, H. J., Wixom, B. H., & Goodhue, D. L. (2000). Data warehousing supports corporate strategy at first American corporation. *Management Information Systems Quarterly, 24*(4), 547–567. doi:10.2307/3250947

Cooper, M. C., Lambert, D. M., & Pagh, J. D. (1997). Supply chain management: more than new name for logistics. *International Journal of Logistics Management, 8*(1), 1–14. doi:10.1108/09574099710805556

Cooper, W. W., Park, K. S., & Yu, G. (1999). IDEA and AR-IDEA: models for dealing with imprecise data in DEA. *Management Science, 45*, 597–607. doi:10.1287/mnsc.45.4.597

Cooper, W. W., Park, K. S., & Yu, G. (2001a). An illustrative application of IDEA (imprecise data envelopment analysis) to a Korean mobile telecommunication company. *Operations Research, 49*, 807–820. doi:10.1287/opre.49.6.807.10022

Cooper, W. W., Park, K. S., & Yu, G. (2001b). IDEA (imprecise data envelopment analysis) with CMDs (column maximum decision making units). *The Journal of the Operational Research Society, 52*, 176–181. doi:10.1057/palgrave.jors.2601070

Cooper, W. W., Seiford, L. M., & Tone, K. (2007). *Data envelopment analysis: a comprehensive text with models, applications, references and DEA-solver software* (2nd ed.). New York: Springer.

Council of Supply Chain Management Professionals. (2005). *Glossary of supply chain and logistics terms and glossary.* http://www.cscmp.org, (February 2005).

Cournot, A. (1960). *Researches into the mathematical principtes of the theory of weatth* (N. T. Bacon trans., English ed. of Coumot, 1838). New York: Kelley.

Crain, K. (2008). Global market data book. *Automotive News Europe*. Crain communications.

Croasdell, D. T. (2001). IT's role in organizational memory and learning. *Information Systems Management*, (Winter): 8–11.

Croom, S. R. (2001). The dyadic capabilities concept: examining the processes of key supplier involvement in collaborative product development. *European Journal of Purchasing and Supply Management, 6*.

Croson, R., & Donohue, K. (2006). Behavioral causes of the bullwhip effect and the observed value of inventory information. *Management Science*, *52*(3), 323–336. doi:10.1287/mnsc.1050.0436

Croxton, K. L., Garcia-Dastugue, S. J., Lambert, D. M., & Rodgers, D. S. (2001). The supply chain management processes. *International Journal of Logistics Management*, *12*(2), 13–36. doi:10.1108/09574090110806271

CSCMP. (2007). *Council of Supply Chain Management Professionals' home page*. Retrieved November 1, 2007, from http://cscmp.org/

Culliton, J. W., Lewis, H. T., & Steele, J. D. (1956). *The role of air freight in physical distribution*. Boston: Harvard University.

Curiel, I. (1997). *Cooperative Game Theory and Applications – Cooperative Games Arising from Combinatorial Optimization Problems*. Boston: Kluwer Academic Publishers.

Dahel, N. E. (2003). Vendor selection and order quantity allocation in volume discount environments. *Supply Chain Management: An International Journal*, *8*, 335–342. doi:10.1108/13598540310490099

Damsgaard, J., & Lyytinen, K. (1998). Contours of diffusion of electronic data interchange in Finland: Overcoming technological barriers and collaborating to make it happen. *The Journal of Strategic Information Systems*, 7.

Dapiran, P., Lieb, R. C., Millen, R. A., & Sohal, A. (1996). Third party logistics service usage by large Australian firms. *International Journal of Physical Distribution and Logistics Management*, *26*(10), 36–45. doi:10.1108/09600039610150442

Davenport, T. (2006). Competing on analytics. *Harvard Business Review*, *84*(1), 98–106.

Davenport, T. H. (2005). The coming commoditization of processes. *Harvard Business Review*(June).

Davenport, T., Eccles, R., & Prusak, L. (1992). Information politics. *Sloan Management Review*. Fall.

Davidow, W. H., & Malone, M. S. (1992). *The virtual corporation*. HarperBusiness.

Davidson, C., & Deneckere, R. (1986). Long-run competition in capacity, short-run competition in price, and the cournot model. *The Rand Journal of Economics*, *17*, 404–415. doi:10.2307/2555720

Davis, E. W., & Spekman, R. E. (2004). *The extended enterprise*. Upper Saddle River: Prentice Hall.

De Boer, L., Harink, J., & Heijboer, G. (2002). A conceptual model for assessing the impact of electronic procurement. *European Journal of Purchasing & Supply Management, 8*.

De Kok, A. G., Van Donselaar, K. H., & Van Woensel, T. (2008). A break-even analysis of RFID technology for inventory sensitive to shrinkage. *International Journal of Production Economics*, *112*, 521–531. doi:10.1016/j.ijpe.2007.05.005

Denning, L. (2008, December 3). Ships ahoy: New world's supply chain. *The Wall Street Journal*.

Derks, J., & Tijs, S. (2000). On merge properties of the Shapley value. *International Game Theory Review*, *2*(4), 249–257. doi:10.1142/S0219198900000214

Desanti, S. (2000, June). Evolution of electronic B2B marketplaces. *Federal Trade Commission Public Workshop*.

Dey, P. K., Hariharan, S., & Clegg, B. (2006). Measuring the operational performance of intensive care units using the analytical hierarchy process approach. *International Journal of Operations & Production Management*, *26*(8), 849–865. doi:10.1108/01443570610678639

Dillard, S. (2006, November 25). Litigation nation. *The Wall Street Journal*, A9.

Disney, S. M., Naim, M. M., & Potter, A. (2004). Assessing the impact of e-business on supply chain dynamics. *International Journal of Production Economics*, *89*(2), 109–118. doi:10.1016/S0925-5273(02)00464-4

Dixit, A., & Pindyck, R. (1993). *Investment under uncertainty*. Princeton, NJ: Princeton University Press.

Dong, S., Xu, S. X., & Zhu, K. X. (2009). Information technology in supply chains; the value of IT-enabled resources under competition. *Information Systems Research*, *20*(1), 18–32. doi:10.1287/isre.1080.0195doi:10.1287/isre.1080.0195

Dong, L., & O'Brien, C. (1999). Integrated decision modeling of supply chain efficiency. *International Journal of Production Economics*, *59*(1-3), 147–157. doi:10.1016/S0925-5273(98)00097-8

Dos Santos, B. L., & Smith, L. S. (2008). RFID in the supply chain: panacea or pandora's box? *Communications of the ACM*, *51*(10), 127–131. doi:10.1145/1400181.1400209

Driessen, T. S. H. (1985). *Contributions to the Theory of Cooperative Games: The τ-Value and k-Convex Games.* Unpublished doctoral dissertation, University of Nijmegen, The Netherlands.

Driessen, T. S. H., & Tijs, S. H. (1983). *Extensions and Modifications of the τ-Value for Cooperative Games* (Report No. 8325). Nijmegen, The Netherlands: University of Nijmegen, Department of Mathematics.

Driessen, T. (1987). The τ-value: a survey . In Peters, H. J. M., & Vrieze, O. J. (Eds.), *Surveys in Game Theory and Related Topics* (pp. 209–213). Amsterdam, The Netherlands: Stichting Mathematisch Centrum.

Driessen, T. S. H., & Tijs, S. H. (1985). The τ-value, the core and semiconvex games. *International Journal of Game Theory*, *14*(4), 229–247. doi:10.1007/BF01769310

Driessen, T., & Tijs, S. (1982). The τ-value, the nucleolus and the core for a subclass of games . In Loeffel, H., & Stähly, P. (Eds.), *Methods of Operations Research 46* (pp. 395–406). Königstein, Germany: Verlagsgruppe Athenäum Hain Hanstein.

Drucker, P. (1962, April). The economy's dark continent. *Fortune Magazine*, 103.

Dudey, M. (1992). Dynamic Edgeworth-Bertrand competition. *The Quarterly Journal of Economics*, *107*(4), 1461–1477. doi:10.2307/2118397

Dudley, L., & Lasserre, P. (1998). Information as a substitute for inventories. *European Economic Review*, *33*(1), 67–88. doi:10.1016/0014-2921(89)90038-X

Dvorak, P. (2009, May 18). Ups and downs whipsaw supply chains. *The Wall Street Journal*, A1.

Dyer, J. H. (2000). *Collaborative advantage.* Oxford University Press.

Earl, M., & Feeney, D. (2000). How to be a CEO for the information age. *Sloan Management Review.*

Eaton, B. C., & Eaton, D. F. (1991). *Microeconomics* (2nd ed.). New York: W. H. Freeman and Company.

Economist. (2009, 17 January). The big chill (pp. 65-67).

Eicher Consultancy Services. (2005). Supply Chain Metrics. *Industry 2.0*, 87-148.

Eisenhardt, K. (1989). Building theories from case study research. *Academy of Management Review*, *14*(4).

Eisenhardt, K. M. (1989). Building Theories from Case Study Research. *Academy of Management Review*, *14*(4), 532–550. doi:10.2307/258557

Eisenhardt, K. M., & Martin, J. A. (2000). Dynamic capabilities: what are they? *Strategic Management Journal*, *21*(10/11), 1105–1121. doi:10.1002/1097-0266(200010/11)21:10/11<1105::AID-SMJ133>3.0.CO;2-E

Ellram, L. M., & Cooper, M. C. (1990). Supply chain management, partnerships and the shipper-third party relationship. *International Journal of Logistics Management*, *1*(2), 1–10.

Emiliani, M. L., & Stec, D. J. (2004). Aerospace parts suppliers' reaction to online reverse auctions. *Supply Chain Management*, *9*(2), 139. doi:10.1108/13598540410527042

Emmons, H., & Gilbert, S. M. (1998). Note: The role of return policies in pricing and inventory decisions for catalogue goods. *Management Science*, *44*(2), 276–283. doi:10.1287/mnsc.44.2.276

Emrouznejad, A. (2001). *An extensive bibliography of data envelopment analysis (DEA), Volume I – V.* Retrieved from http://www.warwick.ac.uk/~bsrlu

Erlenkotter, D. (1990). Ford Whitman Harris and the Economic Order Quantity Model of 1913. *Management Science*, *38*(6), 937–846.

Erl, T. (2005). *Service-Oriented Architecture (SOA): Concepts, technology, and design*. Englewood Cliffs, NJ: Prentice Hall.

Ernst & Whinney (1987). *Corporate profitability & logistics.* Oak Brook: Council of Logistics Management.

Estrin, L., Foreman, J. T., & Garcia, S. (2003). *Overcoming barriers to technology adoption in small manufacturing enterprises (SMEs).* White paper Carnegie Mellon University, Software Engineering Institute (June).

Euclides, F. K. (2004). Supply chain approach to sustainable beef production from a Brazilian perspective. *Livestock Production Science*, *90*(1), 53–61. doi:10.1016/j.livprodsci.2004.07.006

European Logistics Association (ELA) & A. T. Kearney Management Consultants. (2004). *Differentiation for Performance. Results of the Fifth Quinquennial European Logistics Study, Excellence in logistics 2003/2004, ELA*. GER.

Evans, M. G. (1985). A monte carlo study of the effects of correlated method variance in moderated multiple regression analysis. *Organizational Behavior and Human Decision Processes*, *36*, 302–323. doi:10.1016/0749-5978(85)90002-0

Evans, P., & Wolf, B. (2005, July-August). Collaboration rules. *Harvard Business Review*, *83*(7/8), 96.

Evgeniou, T. (2002). Information Integration and Information Strategies for Adaptive Enterprises. *European Management Journal*, *20*(5), 486–494. doi:10.1016/S0263-2373(02)00092-0

Fan, Y. Y., Bhargava, H. K., & Natsuyama, H. (2005). Dynamic pricing via dynamic programming. *Journal of Optimization Theory and Applications*, *127*(3), 565–577. doi:10.1007/s10957-005-7503-z

Farzipoor Saen, R. (2007). Suppliers selection in the presence of both cardinal and ordinal data. *European Journal of Operational Research*, *183*, 741–747. doi:10.1016/j.ejor.2006.10.022

Farzipoor Saen, R. (2009). A mathematical model for selecting third-party reverse logistics providers. *International Journal of Procurement Management*, *2*(2), 180–190. doi:10.1504/IJPM.2009.023406

Farzipoor Saen, R. (2009a). Suppliers selection in volume discount environments in the presence of both cardinal and ordinal data. *International Journal of Information Systems and Supply Chain Management*, *2*, 69–80.

Farzipoor Saen, R. (2009b). Technology selection in the presence of imprecise data, weight restrictions, and nondiscretionary factors. *International Journal of Advanced Manufacturing Technology*, *41*, 827–838. doi:10.1007/s00170-008-1514-5

Farzipoor Saen, R. (2009c). Using data envelopment analysis for ranking suppliers in the presence of nondiscretionary factors. *International Journal of Procurement Management*, *2*, 229–243. doi:10.1504/IJPM.2009.024807

Farzipoor Saen, R. (in press). Supplier selection by the pair of nondiscretionary factors-imprecise data envelopment analysis models. *The Journal of the Operational Research Society*.

Fawcett, S. E., & Magnan, G. M. (2005). *Achieving world-class supply chain alignment: Benefits, barriers, and bridges*. White paper Center for Advanced Purchasing Studies, *http://*.www.capsresearch.org.

Fawcett, S. E., Magnan, G. M., & McCarter, M. W. (2008). A three-stage implementation model for supply chain collaboration. *Journal of Business Logistics*, *29*(1), 93.

Fearne, A. (1998). The evolution of partnerships in the meat supply chain: insights from the British beef industry. *Supply Chain Management, Bradford*, *3*(4), 214. doi:10.1108/13598549810244296

Feder, B. J. (2004, December 28). Wal-Mart edict on radio tags hits snag. *The Denver Post,* 6C.

Federgruen, A., Groenvelt, H., & Tijms, H. C. (1984). Coordinated replenishment in a multi-item inventory system with compound Poisson demands. *Management Science, 30*, 344–357. doi:10.1287/mnsc.30.3.344

Fehr, E., & Schmidt, K. M. (1999). A theory of fairness, competition, and cooperation. *The Quarterly Journal of Economics, 114*(3), 817–868. doi:10.1162/003355399556151

Feichtinger, G., & Hartl, R. F. (1985). Optimal pricing and production in an inventory model. *European Journal of Operational Research, 19*, 45–56. doi:10.1016/0377-2217(85)90307-8

Felix, T. S., Chan, K. N., Tiwari, M. K., Lau, H. C. W., & Choy, K. L. (2008). Global supplier selection: a fuzzy-AHP approach. *International Journal of Production Research,* [REMOVED HYPERLINK FIELD]*46*(14), 3825-3857.

Feng, C. M., & Yuan, C.-Y. (2006). The impact of information and communication technologies on logistics management. *International Journal of Management, 23*(4), 909–944.

Fershtman, C., & Kamien, M. I. (1987). Dynamic duopolistic competition with sticky prices. *Econometrica, 55*(2), 1151–1164. doi:10.2307/1911265

Fine, C., & Freund, R. (1986). Optimal investment in product-flexible manufacturing capacity. *Management Science, 36*(4), 449–466. doi:10.1287/mnsc.36.4.449

Finney, S., & Corbett, M. (2007). ERP implementation: a compilation and analysis of critical success factors. *Business Process Management Journal, 13*(3), 329–347. doi:10.1108/14637150710752272

Fisher, M. F. (1997). What is the right supply chain for your product? A simple framework can help you figure out the answer. *Harvard Business Review*, 105–116.

Fisher, M. L. (1997). What is the right supply chain for your product. *Harvard Business Review, 75*(2), 105–116.

Fites, D. V. (1996). Make your dealers your partners. *Harvard Business Review, 74*(2), 84–97.

Folinas, D., Vlachopoulou, M., Manthou, V., & Manos, B. (2003, July 5-9). A web-based integration of data and processes in agribusiness supply chain. In *Proceedings of the EFITA 2003 Conference*, Hungary (Vol. 1, pp. 143-150).

Fontanella, J., & Klein, E. (2008). Supply chain technology spending outlook. *Supply Chain Management Review, 12*(4), 14.

Fontanella, J. (2000). The Web-based supply chain. *Supply Chain Management Review, 3*(4), 17–20.

Food Standards Agency. (2002). *Traceability in the food chain; a preliminary study*. UK: Food Standards Agency.

Fordice, R. (2004, November). Under control. *Meat Processing,* 34-40.

Forker, L. B., & Mendez, D. (2001). An analytical method for benchmarking best peer suppliers. *International Journal of Operations & Production Management, 21*, 195–209. doi:10.1108/01443570110358530

Forrester (2005, September 19). *APAC study chain apps spending outlook*. http://www.forrester.com.

Forza, C., & Salvador, F. (2002). Managing for variety in the order acquisition and fulfillment process: The contribution of product configuration systems. *International Journal of Production Economics, 76*(1), 87–98. doi:10.1016/S0925-5273(01)00157-8

Frasquet, M., Cervera, A., & Gil, I. (2008). The impact of IT and customer orientation on building trust and commitment in the supply chain. *International Review of Retail, Distribution and Consumer Research, 18*(3), 343. doi:10.1080/09593960802114164

Friedman, T. (2006). *Globalism is good.* Retrieved April 2009, from http://www.impactlab.com/modules.php?sid=5338

Friedman, T. L. (2004). *The world is flat*. New York: Farrar, Straus and Giroux.

Fromen, B. (2004). *Fair Distribution in Enterprise Networks – Solution Concepts Stemming from Cooperative Game Theory*. Doctoral dissertation, University of Duisburg-Essen, Germany. Wiesbaden, Germany: Gabler.

Frost & Sullivan. (2004). F*ourth-party logistics: Turning a cost into a value proposition.* Retrieved March 3, 2009, from http//www.ebizq.net/topics/scm/features/3881.html

Fu, M. C. (2007). Are we there yet? The marriage between simulation & optimization. *OR/MS Today, 34*(3), 16-17.

Fujiwara, K. (2006). A Stackelberg game model of dynamic duopolistic competition with sticky prices. *Economic Bulletin, 12*(12), 1–9.

Fukuyama, F. (1996). *Trust: The social virtues and the creation of prosperity.* New York: Free Press.

Fung, V. K., Fung, W. K., & Wind, Y. (2007). *Competing in a flat world.* Upper Saddle River: Wharton School Publishing.

Fung, R. Y. K., & Ma, X. (2001). A new method for joint replenishment problems. *The Journal of the Operational Research Society, 52*, 358–362. doi:10.1057/palgrave.jors.2601091

Gaimon, C. (1988). Simultaneous and dynamic price, production, inventory, and capacity decisions. *European Journal of Operational Research, 35*, 426–441. doi:10.1016/0377-2217(88)90232-9

Gain, S. (2005, September 19). *Perfect projects.* ITP Technology. http://.www.itp.net.

Galliers, R., Leidner, D., & Baker, B. (1999). *Strategic information management.* Oxford: Butterworth Heinemann.

Gal, T., & Greenberg, H. J. (1997). *Advances in sensitivity analysis and parametric programming.* Boston: Kluwer Academic.

Galunic, D. C., & Rodan, S. (1998). Resource recombinations in the firm: knowledge structures and the potential for schumpeterian innovation. *Strategic Management Journal, 19*, 1193–1201. doi:10.1002/(SICI)1097-0266(1998120)19:12<1193::AID-SMJ5>3.0.CO;2-F

García-Arca, J., & Prado-Prado, J. C. (2007). The implementation of new technologies through a participative approach. *Creativity and Innovation Management, 16*(4), 386. doi:10.1111/j.1467-8691.2007.00450.x

Gartner (2005, March 24). *IBM transforms its supply chain to drive growth.* http://www.gartner.com.

Ghiani, G., Laporte, G., & Musmanno, R. (2004). *Introduction to logistics systems planning and control.* New York: John Wiley & Sons.

Ghiassi, M., & Spera, C. (2003). Defining the Internet-based supply chain system for mass customized markets. *Computers & Industrial Engineering, 45*(1), 17–41. doi:10.1016/S0360-8352(03)00017-2

Ghodsypour, S. H., & O'Brien, C. (2001). The total cost of logistics in supplier selection, under conditions of multiple sourcing, multiple criteria and capacity constraint. *International Journal of Production Economics, 73*, 15–27. doi:10.1016/S0925-5273(01)00093-7

Giannoccaro, L., & Pontrandolfo, P. (2004, February). Supply chain coordination by revenue sharing contracts. *International Journal of Production Economics, 89*(2), 131. doi:10.1016/S0925-5273(03)00047-1

Gibbons, M., Limoges, C., Nowotny, H., Schwartzman, S., Scott, P., & Trow, M. (1994). *The New Production of Knowledge: the Dynamics of Science and Research in Contemporary Societies.* London: Sage.

Gibbons, R. (1992). *A Primer in Game Theory.* Upper Saddle River, NJ: Pearson Education Limited.

Giglio, D., & Minciardi, R. (2003). Modelling and optimization of multi-site production systems in supply chain networks. In *Proceedings of the IEEE International Conference on Systems, Man and Cybernetics, 3* (pp. 2678-2683).

Giraud, G., & Halawany, R. (2006). Consumers perception of food traceability in Europe. In *Proceedings of the 98th EAAE seminar: Marketing dynamics within the global trading system*, Chania, Greece (p. 7).

Godon, D., Visich, J. K., & Li, S. (2007). An exploratory study of RFID implementation benefits and challenges in the supply chain. In *Proceedings of the 38th Annual Meeting of the Decision Sciences Institute* (pp. 5261-5266).

Goldman, S. L., Nagel, R. N., & Preiss, K. (1995). *Agile competitors and virtual organizations.* New York: Van Nostrand Reinhold.

Gomez-Padilla, A. (2009). Supply chain coordination by contracts with inventory holding cost share. *International Journal of Information Systems and Supply Chain Management, 2*(2), 36–47.

Goranson, H. T. (1999). *The agile virtual enterprise.* Westport: Quorum Books.

Gottschalk, P. (2000). Studies of key issues in IS management around the world. *International Journal of Information Management*, *20*(3), 169–180. doi:10.1016/S0268-4012(00)00003-7

Gowen, C. R., & Tallon, W. J. (2003). Enhancing supply chain practices through human resource management. *Journal of Management Development*, *22*(1-2), 32.

Goyal, S. K. (1973). Determination of economic packaging frequency of items jointly replenished. *Management Science*, *20*, 232–235. doi:10.1287/mnsc.20.2.232

Goyal, S. K. (1974). Determination of optimal packaging frequency of jointly replenished items. *Management Science*, *21*, 436–443. doi:10.1287/mnsc.21.4.436

Goyal, S. K. (1988). Economic ordering policy for jointly replenished items. *International Journal of Production Research*, *26*, 1237–1240. doi:10.1080/00207548808947937

Goyal, S. K., & Belton, A. S. (1979). On a simple method of determining order quantities in joint replenishments for deterministic demand. *Management Science*, *25*, 604.

Goyal, S. K., & Satir, T. (1989). Joint replenishment inventory control: Deterministic and stochastic models. *European Journal of Operational Research*, *38*, 2–13. doi:10.1016/0377-2217(89)90463-3

Graham, G., & Hardaker, G. (2000). Supply-chain management across the Internet. *International Journal of Physical Distribution and Logistics Management*, *30*(3/4), 286–295. doi:10.1108/09600030010326055

Greenberg, H. (2007). Representing uncertainty in decision support. *OR/MS Today*, *34*(3), 14-16.

Greenfield, A., Patel, J., & Fenner, J. (2001). Online invoicing for business-to-business users. *Information Week, November, 863,* 80-82.

Grenier, R., & Metes, G. (1995). *Going virtual.* Upper Saddle River: Prentice Hall.

Grey, B. (1989). *Negotiations: Arenas for reconstructing meaning.* Unpublished working paper, Pennsylvania State University, Center for Research in Conflict and Negotiation, University Park, PA.

Grimsley, M., & Meehan, A. (2007). e-Government information systems: Evaluation-led design for public value and client trust. *European Journal of Information Systems*, *16*(2), 134–148. doi:10.1057/palgrave.ejis.3000674

*GS1 Official Website*. (2006). Retrieved from www.gs1.org

Gulati, R., Sytch, M., & Mehrotra, P. (2007, March 3). Preparing for the exit. *The Wall Street Journal*, R11. *ISO/IEC- 20000*. (2009). Retrieved from www.isoiec-20000certification.com

Gunasekaran, A., Patel, C., & Tirtiroglu, E. (2001). Performance measures and metrics in a supply chain environment. *International Journal of Operations & Production Management*, *21*(1-2), 71–87. doi:10.1108/01443570110358468

Gunasekarana, A., & Ngai, E. W. T. (2005). Build-to-order supply chain management: A literature review and framework for development. *Journal of Operations Management*, *23*(5), 423–451. doi:10.1016/j.jom.2004.10.005

Guo, J.-y., Liu,, J., & Qiu, L. (2006). Research on supply chain performance evaluation based on DEA/AHP model. In *Proceedings of the 2006 IEEE Asia-Pacific Conference on Services Computing* (pp. 609-612).

Gupta, A., & Maranas, C. D. (2000a). A two-stage modeling and solution framework for multisite midterm planning under demand uncertainty. *Industrial & Engineering Chemistry Research*, *39*(10), 3799–3813. doi:10.1021/ie9909284

Gupta, A., & Maranas, C. D. (2003). Managing demand uncertainty in supply chain planning. *Computers & Chemical Engineering*, *27*, 1219–1227. doi:10.1016/S0098-1354(03)00048-6

Gupta, A., Maranas, C. D., & McDonald, C. M. (2000b). Midterm supply chain planning under demand: Customer demand satisfaction and inventory management. *Computers & Chemical Engineering*, *24*, 2613–2621. doi:10.1016/S0098-1354(00)00617-7

Gustin, C. M., Daugherty, P. J., & Stank, T. P. (1995). The effect of information availability on logistics integration. *Journal of Business Logistics*, *16*(1), 1–21.

Gustin, C. M., Daughery, P. J., & Stank, T. P. (1995). The effect of information availability on logistics integration. *Journal of Business Logistics*, *16*(1), 1–21.

Hage, J., & Aiken, M. (1967). Program change and organizational properties: a comparative analysis. *American Journal of Sociology*, *72*, 503–519. doi:10.1086/224380

Hajidimitriou, Y. A., & Georgiou, A. C. (2002). A goal programming model for partner selection decisions in international joint ventures. *European Journal of Operational Research*, *138*, 649–662. doi:10.1016/S0377-2217(01)00161-8

Hall, R. (1992). The strategic analysis of intangible resources. *Strategic Management Journal*, *13*, 135–144. doi:10.1002/smj.4250130205

Hammant, J. (1995). Information technology trends in logistics. *Logistics Information Management*, *8*(1), 32–38. doi:10.1108/09576059510102235

Hanfield, R. B., & Nichols, E. L., Jr. (2002). *Supply chain redesign*. Upper Saddle River: Prentice Hall.

Hanseth, O., & Braa, K. (2001). Hunting for the treasure at the end of the rainbow. Standardizing corporate IT infrastructure. *Journal of Collaborative Computing*, *10*(3-4), 261–292. doi:10.1023/A:1012637309336

Harland, C. (1996). Supply chain management: relationships, chains and networks. *British Journal of Management*, 7.

Harney, J. (2005). Enterprise content management for SMBs. *AIIM E-Doc Magazine*, *19*(3), 59.

Harps, L. H. (2003). Revving up returns. *Inbound Logistics*. Retrieved from www.inboundlogistics.com/articles/featurs/1103_feature02.shtml

Harrington, H. J., & Harrington, J. S. (1995). *Total quality management*. New York: McGraw-Hill, Inc.

Harrington, H. J., Hefner, M. B., & Cox, C. K. (1995). Environmental change plans: Best practices for improvement planning and implementation. *Harrington and Harrington's Total Improvement Management*. New York: McGraw-Hill, Inc.

Hartl, R. F., & Sethi, S. P. (1984). Optimal control problem with differential inclusions: Sufficiency conditions and an application to a production-inventory model. *Optimal Control Applications & Methods*, *5*(4), 289–307. doi:10.1002/oca.4660050403

Hart, P., & Saunders, C. (1997, Jan/Feb). Power and trust: Critical factors in the adoption and use of electronic data interchange. *Organization Science*, *8*(1).

Hedman, J., & Kalling, T. (2003). The business model concept: theoretical underpinnings and empirical illustrations. *European Journal of Information Systems*, *12*, 49–59. doi:10.1057/palgrave.ejis.3000446

Heeks, R., & Davies, A. (1999). Different approaches to information age reform. In Heeks (Ed.), *Reinventing Government in the Information Age – International practice in IT-enabled public sector reform.* Routledge, London.

Heikkila, J., & Holmstrom, J. (2006). The impact of RFID and agent technology innovation on cost efficiency and agility. *International Journal of Agile Manufacturing*, *9*(1), 1–6.

Helo, P., & Szekely, B. (2005). Logistics information systems –An analysis of software solutions for supply chain co-ordination. *Industrial Management + Data Systems*, *105*(1), 5-18.

Helper, S., & MacDuffie, J. (2000). *Evolving the auto industry: E-business effects on consumer and supplier relationships.* Fischer Center, UC Berkeley.

Helper, S., & MacDuffie, J. P. (2002). B2B and modes of exchange: Evolutionary and transformative effects. In B. Kogut (Ed.), *The Global Internet Economy.*

Helper, S. (1991). How much has really changed between US automakers and their Suppliers? *Sloan Management Review*, *32*(4).

Henriksen, H. Z. (2006). Motivators for IOS adoption in Denmark. *Journal of Electronic Commerce in Organizations*, *4*(2), 25–39.

Hertz, S., & Alfredsson, M. (2003). Strategic development of third party logistics providers. *Industrial Marketing Management*, *32*, 139–149. doi:10.1016/S0019-8501(02)00228-6

Hewitt, F. (1994). Supply chain redesign. *The International Journal of Logistics Management*, 1-9.

Higle, J., & Wallace, S. W. (2003). Sensitivity analysis and uncertainty in linear programming. *Interfaces*, *33*(4), 53–60. doi:10.1287/inte.33.4.53.16370

Holloway, S. (2006). *Potential of RFID in the Supply Chain*. Chicago: Solidsoft Ltd.

Holweg, M., & Pil. F. (2001). Successful build-to-order strategies: Start with the customer. *MIT Sloan Management Review*.

Hong, J., Chin, A., & Lin, B. (2004). Logistics outsourcing by manufacturers in China: A survey of the industry. *Transportation Journal*, *43*(1), 17–25.

Howard, M., Vidgen, R., Powell, P., & Graves, A. (2001). Planning for IS-related industry transformation: The case of the 3DayCar. *Procs of 9th European Conference on Information Systems*, Slovenia.

Howard, M., Vidgen, R., Powell, P., & Graves, A. (2002). Are hubs the centre of things? E-procurement in the automotive industry. *Procs of 10th European Conference on IS*, Poland.

Howard, M., Vidgen, R., & Powell, P. (2006). Automotive e-hubs: exploring motivations and barriers to collaboration and interaction. *The Journal of Strategic Information Systems*, *15*(1), 51–75.

Hsu, L. (2005). SCM system effects on performance for interaction between suppliers and buyers. *Industrial Management (Des Plaines)*, *105*(7), 857.

Huber, N., Ward, H., Goodwin, B., & Simons, M. (2000, July). The e-procurement dilemma. *Computer Weekly*.

Huber, G. P. (1991). Organizational learning: the contributing processes and literature. *Organization Science*, *2*, 88–115. doi:10.1287/orsc.2.1.88

Huber, G. P. (1996). Organizational learning: a guide for executives in technology-critical organizations. *International Journal of Technology Management*, *11*(7), 821–832.

Huchzermeier, A., & Cohen, M. (1996). Valuing operational flexibility under exchange rate risk. *Operations Research*, *44*, 100–113. doi:10.1287/opre.44.1.100

Hughes, D. (1994). *Breaking with tradition: building partnerships and alliances in the European food industry*. Ashford, UK: Wye College Press.

Huhns, M. N., & Stephens, L. M. (2001). Automating supply chains. *IEE Internet Computing*. Retrieved from http://computer.org/internet/

Hwang, C. L., Fan, L. T., & Erickson, L. E. (1967). Optimal production planning by the maximum Principle. *Management Science*, *13*, 750–755. doi:10.1287/mnsc.13.9.751

Hwang, C. L., & Yoon, K. (1981). *Multiple attribute decision making methods and applications*. Berlin: Springer Verlag.

IBM (2006). *Expanding the innovation horizon*. Somers: IBM Global Services.

Ip, W. H., Yung, K. L., & Wang, D. (2004). A branch and bound algorithm for sub-contractor selection in agile manufacturing environment. *International Journal of Production Economics*, *87*, 195–205. doi:10.1016/S0925-5273(03)00125-7

Iyer, A. V., & Bergen, M. E. (1997). Quick response in manufacturer-retailer channels. *Management Science*, *43*, 559–570. doi:10.1287/mnsc.43.4.559

Jähn, H., Zimmermann, M., Fischer, M., & Käschel, J. (2006). Performance evaluation as an influence factor for the determination of profit shares of competence cells in non-hierarchical regional networks. *Robotics and Computer-integrated Manufacturing*, (22): 526–535. doi:10.1016/j.rcim.2005.11.011

Jansen, J. J. P., Van Den Bosch, F. A. J., & Volberda, H. W. (2005). Managing potential and realized absorptive capacity: how do organizational antecedents matter? *Academy of Management Journal*, *48*(6), 999–1015.

Jansen-Vullers, M. H., van Dorp, C. A., & Beulens, A. J. M. (2003). Managing traceability information in manufacture. *International Journal of Information Management*, *23*, 395–413. doi:10.1016/S0268-4012(03)00066-5

Jessup, L., & Valacich, J. (2008). *Information systems today – Managing in the digital world* (3rd ed.). Upper Saddle River, NJ: Pearson/Prentice Hall.

Jinho, K., & Rogers, K. J. (2005). An object-oriented approach for building a flexible supply chain model. *International Journal of Physical Distribution and Logistics Management*, *35*(7), 481–502. doi:10.1108/09600030510615815

Jones, P., Clarke-Hill, C., Hiller, D., & Comfort, D. (2005). The benefits, challenges and impacts of radio frequency identification technology (RFID) for retailers in the UK. *Marketing Intelligence & Planning*, *23*(4), 395–402. doi:10.1108/02634500510603492

Jonsson, P., & Zineldin, M. (2003). Achieving high satisfaction in supplier-dealer working relationships. *Supply Chain Management*, *8*(3-4), 224.

Journal, R. F. I. D. (2006). The cost of RFID equipment. *RFID Journal website*. Retrieved May 2009 from http://www.rfidjournal.com/faq/20: RFID Research Center. (2001). *The myths and realities of RFID*. University of Arkansas, USA: Information Technology Research Institute.

Journal, W. S. (2009, May 19). Sarbox and the constitution. *Review and Outlook*, A16.

Jung, J. Y., Blau, G., & Pekny, J. F. (2004). A simulation based optimization approach to supply chain management under demand uncertainty. *Computers & Chemical Engineering*, *28*, 2087–2106. doi:10.1016/j.compchemeng.2004.06.006

Jun, M., Cai, S., & Kim, D. (2008). The strategic implications of e-network integration and transformation paths for synchronizing supply chains. *International Journal of Information Systems and Supply Chain Management*, *1*(4), 39–59.

Jun, M., Cai, S., & Kim, D. (2008). The strategic implications of e-network integration and transformation paths for synchronizing supply chains. *International Journal of Information Systems and Supply Chain Management*, *1*(4), 39–59.

Kahneman, D., & Tversky, A. (1982). *Judgment under uncertainty: Heuristics and biases* (Kahneman, D., Slovic, P., & Tversky, A., Eds.). Cambridge, UK: Cambridge University Press.

Kale, P., Dyer, J. H., & Singh, H. (2002). Alliance capability, stock market response, and long-term alliance success: The role of the alliance function. *Strategic Management Journal*, (23): 747–767. doi:10.1002/smj.248

Kalling, T. (2003). ERP systems and the strategic management processes that lead to competitive advantage. *Information Resources Management Journal*, *16*(4), 46–67.

Kalpakam, S., & Arivarignan, G. (1993). A coordinated multicommodity (s,S) inventory system. *Mathematical and Computer Modelling*, *18*, 69–73. doi:10.1016/0895-7177(93)90206-E

Kameshwaran, S., Narahari, Y., Rosa, C. H., Kulkarni, D. M., & Tew, J. D. (2007). Multiattribute electronic procurement using goal programming. *European Journal of Operational Research*, *179*, 518–536. doi:10.1016/j.ejor.2006.01.010

Kanagaraj, G., & Jawahar, N. (2009). A simplified annealing algorithm for optimal supplier selection using the reliability-based total cost of ownership model. *International Journal of Procurement Management*, *2*, 244–263. doi:10.1504/IJPM.2009.024809

Kanakamedala, K., Ramsdell, G., & Srivatsan, V. (2003). Getting supply chain software right. *The McKinsey Quarterly*.

Kanoi, M. (2006). *Working paper on supply chain management*. Temple University, Philadelphia.

Kanter, R. (1994, July/August). Collaborative advantage: The art of alliances. *Harvard Business Review*.

Kaplan, S., & Sawhney, M. (2000, May-June). E-hubs: The new B2B marketplaces. *Harvard Business Review*.

Karimi, J., Somers, T.M., & Bhattacherjee, A. (2007). The impact of erp implementation on business process outcomes: a factor-based study. *Journal of Management Information Systems, Summer, 24*(1), 101-134.

Karkkainen, M. (2003). Increasing efficiency in the supply chain for short shelf life goods using RFID tagging. *International Journal of Retail and Distribution Management*, *31*, 529–553. doi:10.1108/09590550310497058

Karpak, B., Kumcu, E., & Kasuganti, R. R. (2001). Purchasing materials in the supply chain: managing a multi-objective task. *European Journal of Purchasing & Supply Management*, *7*, 209–216. doi:10.1016/S0969-7012(01)00002-8

Karuppan, C. M., & Karuppan, M. (2008). Resilience of super users' mental models of enterprise-wide systems. *European Journal of Information Systems*, *17*(1), 29–46. doi:10.1057/palgrave.ejis.3000728

Kashyap, R. N. (1972). Management information systems for corporate planning and control. *Long Range Planning*, *5*(2), 25–31. doi:10.1016/0024-6301(72)90042-8

Kaspi, M., & Rosenblatt, M. J. (1991). On the economic ordering quantity for jointly replenished items. *International Journal of Production Research*, *29*(1), 107–114. doi:10.1080/00207549108930051

Katz, M. L., & Shapiro, C. (1985). Network externalities, competition, and compatibility. *The American Economic Review*, *75*(3), 424–440.

Kearns, G. S., & Lederer, A. L. (2004). The impact of industry contextual factors on IT focus and the use of IT for competitive advantage. *Information & Management*, *41*(2), 899–919. doi:10.1016/j.im.2003.08.018

Khan, A., & Kurnia, S. (2005). *Explaining the Potential of RFID*. Department of Information Systems, University of Melbourne, Australia.

Khan, A., & Kurnia, S. (2006). *Exploring the potential benefits of RFID: a literature-based study.* Department of Information Systems, University of Melbourne, Australia (pp. 1-12).

Killingsworth, K. (2008). Improving fleet management performance. *Inbound Logistics, 32*. Retrieved from www.inboundlogistics.com/articles/itmatters/itmatters0508.shtml

Kilton, D., Sadowski, R., & Sturrock, D. (2003). *Simulation with Arena* (3rd ed.). McGraw-Hill, Series Industrial Engineering and Management Science.

Kim, H. M., Fox, M. S., & Gruninger, M. (1995). *Ontology of quality for enterprise modeling*. Los-Albamitos, CA: WET-ICE.

Kim, L. (1997a). The dynamics of samsung's technological learning in semiconductors. *California Management Review*, *39*(3), 86–100.

Kim, L. (1997b). *From imitation to innovation: The dynamics of Korea's technological learning*. Cambridge, MA: Harvard Business School Press.

Kim, S. W. (2006). The effect of supply chain integration on the alignment between corporate competitive capability and supply chain operational capability. *International Journal of Operations & Production Management*, *26*(10), 1084–1107. doi:10.1108/01443570610691085

Kirkman, B.L., Rosen, B., Gibson, C.B., & Tesluk, P.E., & McPherson, Simon O. (2002). Five challenges to virtual team success: Lessons from Sabre, Inc. *The Academy of Management Executive*, *16*(3), 67.

Kirvennummi, M., Hirvo, H., & Eriksson, I. (1998). Framework for barriers to IS-related change: development and evaluation of a theoretical model. In De Gross et al., (Eds.), *Proceedings of Joint IFIP Working Conference*, Finland.

Kisiel, R. (2002, April 29). Online trade exchanges: What went wrong? *Automotive News Europe, 2*.

Kisiel, R. (2002a, July 8). Covisint gets one last chance. *Automotive News Europe, 3*.

Kisiel, R., & Whitbread, C. (2000). E-business. *Automotive News Europe* (pp. 37–40).

Klaus, P, Henning, H., Muller-Steinfahr, U., & Stein. (1993). The promise of interdisciplinary research in logistics. In *Proceedings of the Twenty second Annual Transportation and Logistics Educators Conference* (pp. 161-87).

Kleinsorge, I. K., Schary, P., & Tanner, R. (1992). Data envelopment analysis for monitoring customer–supplier relationships. *Journal of Public Policy*, *11*, 357–372.

Knudsen, T., & Madsen, T. K. (2002). Export strategy: a dynamic capabilities perspective. *Scandinavian Journal of Management*, *18*, 475–501. doi:10.1016/S0956-5221(01)00019-7

Ko, H.-C., Tseng, F.-C., Yin, C.-P., & Huang, L.-C. (2008). The factors influence suppliers satisfaction of green supply chain management systems in Taiwan. *International Journal of Information Systems and Supply Chain Management*, *1*(1), 66–79.

Kogut, B. (1985). Designing global strategies: profiting from operational flexibility. *Sloan Management Review*, 27–38.

Kolisch, R. (2000). Integration of assembly and fabrication for make-to-order production. *International Journal of Production Economics*, *68*(3), 287–306. doi:10.1016/S0925-5273(99)00011-0

Konsynski, B. R. (1993). Strategic control in the extended enterprise. *IBM Systems Journal*, *30*(1), 111–142.

Korpela, J., Lehmusvaara, A., & Tuominen, M. (2001). An analytic approach to supply chain development. *International Journal of Production Economics*, *71*(1-3), 145–155. doi:10.1016/S0925-5273(00)00114-6

Koulikoff-Souviron, M., & Pascal, V. (2005). A dose of collaboration. [London.]. *European Business Forum*, *22*, 59.

Krcmar, H., Bjørn-Andersen, N., & O'Callaghan, R. (Eds.). (1995). *EDI in Europe: How it Works in Practice*. Chichester: John Wiley & Sons.

Kreps, D., & Scheinkman, J. (1983). Quantity precommitment and Bertrand competition yield Cournot outcomes. *The Bell Journal of Economics*, *14*, 326–337. doi:10.2307/3003636

Kristianto, Y., & Helo, P. (2009). Strategic thinking in supply and innovation in dual sourcing Procurement. *Int. J. Applied Management Science*, *1*(4), 401–419.

Kristianto, Y., & Helo, P. (2010). Built-to-order supply chain: response analysis with control model. *Int. J. Procurement Management*, *3*(2), 181–198. doi:10.1504/IJPM.2010.030734

Krol, B., Keller, S., & Zelewski, S. (2005). E-logistics ocercome the bullwhip effect. *International Journal of Operations and Quantitative Management*, *11*(4), 281–289.

Kuhn, H. W., Harsanyi, J. C., Selten, R., Weibull, J. W., Damme, E., van Nash, J. F. Jr, & Hammerstein, P. (1996). The work of John Nash in game theory – Nobel seminar, december 8, 1994. *Journal of Economic Theory*, *69*(1), 153–185. doi:10.1006/jeth.1996.0042

Kumar, N. (2000). The power of trust in manufacturer-retailer relationships. *Harvard Business Review's Managing The Value Chain.* Boston: HBS Publishing.

Kumar, M., Vrat, P., & Shankar, R. (2004). A fuzzy goal programming approach for vendor selection problem in a supply chain. *Computers & Industrial Engineering*, *46*, 69–85. doi:10.1016/j.cie.2003.09.010

Kumar, S. (2007). Connective technology as a strategic tool for building effective supply chain. *International Journal of Manufacturing Technology and Management*, *10*(1), 41–56. doi:10.1504/IJMTM.2007.011400

Kuroiwa, S. (1999). Growing an open business environment from CALS, JIT & supply chain. *Logistics in the Information Age, 4th Conference Proceedings ISL*, Florence, Italy.

Kwon, K., & Zmud, R. (1987). Unifying the fragmented models of information systems implementation. In R.J. Boland,& R.A. Hirschheim (Eds.), *Critical Issues in Information Systems Research.* Chicester: Wiley.

Lambert, D. M., Cooper, M. C., & Pagh, J. D. (1998). Supply chain management: implementation issues and research opportunities. *International Journal of Logistics Management*, *9*(2), 1–19. doi:10.1108/09574099810805807

Lambertini, L., & Mantovani, A. (2004). *Process and product innovation. A differential game approach to product life cycle.* Retrieved January 5, 2008, from http://www2.dse.unibo.it/seminari/Mantovani

Lamming, R. (1993). *Beyond partnership: Strategies for innovation and lean supply.* Prentice Hall.

Lam, W. (2005). Investigating success factors in enterprise application integration: a case-driven analysis. *European Journal of Information Systems*, *14*(2), 175–187. doi:10.1057/palgrave.ejis.3000530

Lancioni, R. A., Smith, M. F., & Oliva, T. A. (2000). The role of the Internet in supply chain management: Logistics catches up with strategy. *Industrial Marketing Management*, *29*(1), 45–56. doi:10.1016/S0019-8501(99)00111-X

Lane, M., & Koronios, A. (2000). Using stakeholder salience theory to facilitate management of stakeholder requirements in business-to-customer web information systems. *13th International Bled Electronic Commerce Conference*, Slovenia.

Lane, J. P., & Lubatkin, M. (1998). Relative absorptive capacity and interorganizational learning. *Strategic Management*, *19*, 461–477. doi:10.1002/(SICI)1097-0266(199805)19:5<461::AID-SMJ953>3.0.CO;2-L

Langley, C. J., & Holcomb, M. C. (1992). Creating logistics customer value. *Journal of Business Logistics*, *13*(2), 1–27.

Larson, A. (1991). Partner networks leveraging external ties to improve entrepreneurial performance. *Journal of Business Venturing*, *6*, 173–188. doi:10.1016/0883-9026(91)90008-2

Larson, P., & Gammelgaard, B. (2001). Logistics in Denmark: A survey of the industry. *International Journal of Logistics: Research and Applications*, *4*(2), 191–205.

Lau, A., Lau, H., & Zhou, Y. (2006). Considering asymmetrical manufacturing cost information in a two-echelon system that uses price-only contracts. *IIE Transactions*, *38*, 253–271. doi:10.1080/07408170590961148

Lau, Kwok, H., & Ma, W. L. (2008). A supplementary framework for evaluation of integrated logistics service provider. *International Journal of Information Systems and Supply Chain Management*, *1*(3), 49–69.

Laurent, S. (Ed.). (1999). *XML™: a primer* (2nd ed.). New York: MIS Press.

Law, C. C. H., & Ngai, E. W. T. (2007). ERP systems adoption: An exploratory study of the organizational factors and impacts of erp success. *Information & Management*, *44*, 418–432. doi:10.1016/j.im.2007.03.004

Lawrence, C. (2006). Friend or foe. *Farmers Weekly*, *144*(25), 18–19.

Lee, H. L., & Whang, S. (2001, November). *E-Business and supply chain integration* (Tech. Rep. No. SGSCMF-W2-2001). Stanford, CA: Stanford University, Stanford Global Supply Chain Management Forum.

Lee, H., & Whang, S. (1998). Information sharing in the supply chain. *Research Paper Series*, working paper, Stanford University.

Lee, Y., Cheng, F., & Leung, Y. (2005). A quantitative view on how RFID will improve a supply chain. *IBM Research Report* (pp. 1-45).

Lee, H. L., Padmanabhan, V., & Whang, S. (1997). Information distortion in a supply chain: The bullwhip effect. *Management Science*, *43*(4), 546–558. doi:10.1287/mnsc.43.4.546

Lee, H. L., Padmanabhan, V., & Whang, S. (1997). Information distortion in supply chain: the bullwhip effect. *Management Science*, *43*(4), 546–558. doi:10.1287/mnsc.43.4.546

Lee, H. L., & Tang, C. S. (1998b). Managing supply chains with contract manufacturing . In Lee, H. L., & Ng, S. M. (Eds.), *Global supply chain and technology management*. Orlando, FL: Production and Operations Management Society Publishers.

Lee, H., & Tang, C. S. (1998a). Variability reduction through operations reversal. *Management Science*, *44*, 162–173. doi:10.1287/mnsc.44.2.162

Leenders, M. R., Fraser, J. P., Flynn, A. E., & Fearson, H. E. (2006). *Purchasing and supply management.* Boston: McGraw-Hill Irwin.

Lee, T. (2008). Supply chain risk management. *International Journal of Information and Decision Sciences*, *1*(1), 98–111. doi:10.1504/IJIDS.2008.020050

Lee, Y. M. (2008). Balancing accuracy of promised ship date and IT cost. *International Journal of Information Systems and Supply Chain Management*, *1*(1), 1–14.

Lee, Y. M., Cheng, F., & Leung, Y. T. (2009). A quantitative view on how RFID can improve inventory management in a supply chain. *International Journal of Logistics Research and Applications*, *12*(1), 23–43. doi:10.1080/13675560802141788

Lee, Y. W., & Strong, D. M. (2004). Knowing-why about data processing and data quality. *Journal of Management Information Systems*, *20*(3), 13–39.

Levy, D. (1995). International sourcing and supply chain stability. *Journal of International Business Studies*, *26*, 343–360. doi:10.1057/palgrave.jibs.8490177

Levy, M., & Grewal, D. (2000). Overview of the issues of supply chain management in a networked economy. *Journal of Retailing*, *76*(4), 415–429. doi:10.1016/S0022-4359(00)00043-9

Levy, Y., Murphy, K. E., & Zanakis, S. H. (2009). A value-satisfaction taxonomy of IS effectiveness (VSTISE): A case study of user satisfaction with IS and user-perceived value of IS. *International Journal of Information Systems in the Service Sector*, *1*(1), 93–118.

Lewis, I., & Talalayersky, A. (2000). Third party logistics: Leveraging information technology. *Journal of Business Logistics*, *21*(2), 173–185.

Lewis, I., & Talalayevsky, A. (1997). Logistics and information technology: a coordination perspective. *Journal of Business Logistics*, *18*(1), 141–157.

Lewis, I., & Talalayevsky, A. (2004). Improving the inter-organizational supply chain through optimization of information flows. *Journal of Enterprise Information Management*, *17*(3), 229–237. doi:10.1108/17410390410531470

Liao, S. H., Chem, Y. M., & Liu, F. H. (2004). Information technology and relationship management: a case study of Taiwan's small manufacturing firm. *Technovation*, *24*(2), 97–108. doi:10.1016/S0166-4972(02)00037-8

Li, C. L., & Kouvelis, P. (1999). Flexible and risk-sharing supply contracts under price uncertainty. *Management Science*, *45*(10), 1378–1398. doi:10.1287/mnsc.45.10.1378

Lidd, S. J. (1979). The Pressing Need for Management Information Systems. *Management Focus*, *26*(5), 44–44.

Li, E., Du, T., & Wong, J. (2005). Access control in collaborative commerce. *Decision Support Systems*, 1–11. Retrieved from www.elsevier.com/locate/dsw.

Lieb, R. C. (1992). The use of third-party logistics services by large American manufacturers. *Journal of Business Logistics*, *13*(2), 29–42.

Lieb, R. C., & Bentz, B. A. (2004). The use of 3PL services by large American manufacturers: The 2003 survey. *Transportation Journal*, *43*(3), 24–33.

Lieb, R. C., & Bentz, B. A. (2005). The use of 3PL services by large American manufacturers: The 2004 survey. *Transportation Journal*, *44*(2), 5–15.

Lieb, R. C., Millen, R. A., & Van Wassenhove, L. N. (1993). Third party logistics: A comparison of experienced American and European manufacturers. *International Journal of Physical Distribution and Logistics Management*, *23*(6), 35–44. doi:10.1108/09600039310044894

Lieb, R. C., & Miller, J. (2002). The use of 3PL services by large American manufacturers: The 2000 survey. *International Journal of Logistics: Research and Applications*, *5*(1), 1–12.

Lieb, R. C., & Randall, H. L. (1996). A comparison of the use of third-party logistics services by large American manufacturers, 1991, 1994 and 1995. *Journal of Business Logistics*, *17*(1), 305–320.

Liedtka, S. L. (2005). The analytic hierarchy process and multi-criteria performance management systems. *Journal of Cost Management*, *19*(6), 30–38.

Li, F., & Williams, H. (1999). Interfirm collaboration through interfirm networks. *Information Systems Journal*, 9.

Lim, D., & Palvia Prashant, C. (2001). EDI in strategic supply chain: Impact on customer service. *International Journal of Information Management*, *21*(3), 193–211. doi:10.1016/S0268-4012(01)00010-X

Lindsay, P., & Norman, D. (1977). *Human information processing*. Orlando, FL: Academic Press.

Lin, F., Sheng, O., & Wu, S. (2005). An integrated framework for e-chain bank accounting systems. *Industrial Management & Data Systems*, *105*(3), 291–306. doi:10.1108/02635570510590129

Lipnack, J., & Stamps, J. (1997). *Virtual teams: reaching across space, time, and organizations with technology*. New York: John Wiley and Sons.

Liu, J., Ding, F. Y., & Lall, V. (2000). Using data envelopment analysis to compare suppliers for supplier selection and performance improvement. *Supply Chain Management: An International Journal*, *5*, 143–150. doi:10.1108/13598540010338893

Liu, M. L., & Sahinidis, M. V. (1998). Robust process planning under uncertainty. *Industrial & Engineering Chemistry Research*, *37*, 1883–1899. doi:10.1021/ie970694t

Lium, A.-G., Crainic, T. G., & Wallace, S. W. (2007). Correlations in stochastic programming: a case from stochastic service network design. *Asia-Pacific Journal of Operational Research*, *24*(2), 161–179. doi:10.1142/S0217595907001206

Li, X., & Wang, Q. (2006). Coordination mechanisms of supply chain systems. *European Journal of Operational Research*, *179*(1), 1–16. doi:10.1016/j.ejor.2006.06.023

Li, Y. S., Ye, F. F., Fang, Z. M., & Yang, J. G. (2005). Flexible supply chain optimization and its SRT analysis. *Industrial Engineering and Management*, *10*(1), 89–93.

Lochamy, A. III, & McCormack, K. (2004). Linking SCOR planning practices to supply chain performance. *International Journal of Operations & Production Management, 24*(11/12), 1192.

Lombardo, S. (2001). *AHP reference listing*. Retrieved January 2, 2003, from http://www.expertchoice.com/ahp/default.htm

Lopez-Claros, A. (2006). *The Global Competitiveness Report 2006-2007* (28th ed.). Basingstoke, UK: Palgrave.

Lummus, R. R., Vokurka, R. J., & Krumwiede, D. (2008). Supply chain integration and organizational success. *S.A.M. Advanced Management Journal, 73*(1), 56–62.

Lumsden, K. R., & Hulthén, L. A. R. (1998). Outline for a conceptual framework on complexity in logistics systems. In *Proceedings of NOFOMA 98,* Helsinki, Finland.

Lundsford, J., & Glader, P. (2007, June 9). Boeing's nuts-and-bolts problem. *The Wall Street Journal,* A8.

Mabert, V. A., & Venkatraman, M. A. (1998). Special research focus on supply chain linkages: challenges for design and management in the 21st century. *Decision Sciences, 29*(3), 537–550. doi:10.1111/j.1540-5915.1998.tb01353.x

Mabisan, S., & Nambisan, P. (2008). How to profit from a better "virtual customer environment". *MIT Sloan Management Review, 49*(3), 53–61.

Mahdavi, I., Mohebbi, S., Cho, N., Paydar, M. M., & Mahdavi-Amiri, N. (2008). Designing a dynamic buyer-supplier coordination model in electronic markets using stochastic Petri nets. *International Journal of Information Systems and Supply Chain Management, 1*(3), 1–20.

Maku, T. C., & Collins, T. R. (2005). The impact of human interaction on supply chain management practices. *Performance Improvement, 44*(7), 26. doi:10.1002/pfi.4140440708

Malhotra, A., Gosain, S., & El Sawy, O. A. (2005). Absorptive capacity configurations in supply chains: gearing for partner-enabled market knowledge creation. *MIS Quarterly, March, 29*(1), 145-187.

Malone, T. W., Crowston, K., & Herman, G. A. (2003). *Organizing Business Knowledge: The MIT Process Handbook*: MIT Press.

Maloni, M. (2006). *Understanding Radio Frequency Identification (RFID) and its impact on the supply chain.* Pen State Behrend: Black School of Business, PA, USA

Maltz, A. B., Riley, L., & Boberg, K. (1993). Purchasing logistics in the US-Mexico transborder situation: Logistics outsourcing in the US-Mexico co-production. *International Journal of Physical Distribution and Logistics Management, 23*(8), 46–55. doi:10.1108/09600039310049835

Manetti, J. (2001). How technology is transforming manufacturing. *Production and InventoryManagement Journal, 1st Quarter.*

Mangan, J., Lalwani, C., & Butcher, T. (2008). *Global Logistics and Supply Chain Management*. Hoboken, NJ: John Wiley & Sons.

Manhattan Associates. (2006). *A fine-tuned slotting strategy keeps Pep Boys rolling*. Retrieved from www.manh.com

Markus, M. L. (2000). Paradigm shifts - E-Business and business/systems integration. *Communication of the AIS, 4*(10), 1–45.

Martin, S. (1995). R & D joint ventures and tacit product market collusion. *European Journal of Political Economy, 11*, 733–741. doi:10.1016/0176-2680(95)00026-7

Masetti, B., & Zmud, R. (1996). Measuring the extent of EDI usage in complex organizations: Strategies and illustrative examples. *MIS Quarterly, 20*(3), 331–345. doi:10.2307/249659

Mata, F. J., Fuerst, W. L., & Barney, J. B. (1995). Information technology and sustained competitive advantage: A resource-based analysis. *MIS Quarterly, 19*(4), 487–506. doi:10.2307/249630

McAdam, R., & McCormack, D. (2001). Integrating business processes for global alignment and supply chain management. *Business Process Management, 7*(2), 113. doi:10.1108/14637150110389696

McClellan, M. (2003). *Collaborative manufacturing.* New York: St. Lucie Press.

McCormack, K. P., & Johnson, W. C. (2003). *Supply Chain Networks and Business Process Orientation.* Boca Raton, FL: St. Lucie Press.

McCullen, P., & Towill, D. (2002). Diagnostics and reduction of bullwhip in supply chains. *Supply Chain Management, 7*(3), 164–179. doi:10.1108/13598540210436612

McLaren, T. S., & Vuong, D. C. H. (2008). A "genomic" classification scheme for supply chain management information systems. *Journal of Enterprise Information Management, 21*(4), 409–423. doi:10.1108/17410390810888688doi:10.1108/17410390810888688

Meertens, M. A., & Potters, J. A. M. (2006). The nucleolus of trees with revenues. *Mathematical Methods of Operations Research, 64*(2), 363–382. doi:10.1007/s00186-006-0088-y

Mentzer, J. T., DeWitt, W., Keebler, J. S., Min, S., Nix, N. W., Smith, C. D., & Zacharia, Z. G. (2001). Defining supply chain management. *Journal of Business Logistics, 22*(2), 1–25.

Mentzer, J. T., DeWitt, W., Keebler, J. S., Min, S., Nix, N. W., & Zacharia, Z. G. (2001). Defining supply chain management. *Journal of Business Logistics, 22*(2), 1–25.

Mentzer, J., Dewitt, W., & Keebler, J. (2001). Defining supply chain management. *Journal of Business Logistics, 22*(2), 1–25.

Merton, R. (1968). *Social theory and social structure.* Free Press, NY.

Metters, R. (1997). Quantifying the bullwhip effect in supply chains. *Journal of Operations Management, 15*(2), 89–100. doi:10.1016/S0272-6963(96)00098-8

Meyer, N. D. (2004). Systemic IS governance: An introduction. *Information Systems Management, 21*(4), 23–34. doi:10.1201/1078/44705.21.4.20040901/84184.3

Michael, K. (2005). The pros and Cons of RFID in supply chain management. In *Proceedings of the International Conference on Mobile Business, IEEE Computer Society* (pp. 1-7).

Mieghem, J. A. V. (1998). Investment strategies for flexible resources. *Management Science, 44*(8), 1071–1078. doi:10.1287/mnsc.44.8.1071

Mieghem, J. A. V., & Dada, M. (1999). Price versus production postponement: Capacity and competition. *Management Science, 45*(12), 1631–1649.

Mighell, R. L., & Jones, L. A. (1963). Vertical coordination in agriculture. *USDA-ERS Agricultural Economics Report*, 19.

Miles, M., & Huberman, M. (1994). *Qualitative data analysis: An expanded sourcebook.* Sage: London.

Miles, R. E., & Snow, C. C. (2007). Organization theory and supply chain management: An evolving research perspective. *Journal of Operations Management, 25*(2), 459–463. doi:10.1016/j.jom.2006.05.002

Mills, J., Schmitz, J., & Frivol, G. (2004). A strategic view of supply networks. *International Journal of Operations & Production Management, 24*(9/10), 1012. doi:10.1108/01443570410558058

Min, H., & Galle, W. (1999). Electronic commerce usage in business-to-business purchasing. *International Journal of Operations & Production Management, 19*(9).

Min, H., & Zhou, G. (2002). Supply chain modeling: past, present, and future. *Computers & Industrial Engineering, 43*(2), 231–249. doi:10.1016/S0360-8352(02)00066-9

Ministry of Transportation and Communication. (2006). *Finland state of logistics 2006.* Helsinki, Finland: Edita Publishing Ltd.

Min, S., & Mentzer, J. T. (2000). The role of marketing in supply chain management. *International Journal of Physical Distribution & Logistics Management, 30*(9), 765–787. doi:10.1108/09600030010351462

Mintzberg, H. (1979). *The structure of organisations.* New York: Prentice Hall.

Min, W. (2007). Topsis-AHP simulation model and its application to supply chain management. *World Journal of Modelling and Simulation, 3*(3), 196–201.

Moberg, C. R., Seph, T. W., & Freese, T. L. (2003). SCM: making the vision a reality. *Supply Chain Management Review, 7*(5), 34–39.

Moe, T. (1998). Perspectives on traceability in food manufacture. *Trends in Food Science & Technology, 9*(5), 211–214. doi:10.1016/S0924-2244(98)00037-5

Mohtadi, H. (2008). Information sharing in food supply chains. *Canadian Journal of Agricultural Economics, 56*(2), 163. doi:10.1111/j.1744-7976.2008.00123.x

Monczka, R. M., Trent, R. J., & Petersen, K. J. (2006). *Effective global sourcing and supply for superior results*. Tempe, AZ: CAPS Research.

Moon, K. L., & Ngai, E. W. T. (2008). The adoption of RFID in fashion retailing; a business value-added framework. *Industrial Management & Data Systems, 108*(5), 596–612. doi:10.1108/02635570810876732

Morissey, T., & Almonacid, S. (2005). Rethinking technology transfer. *Journal of Food Technology, 67*, 135–145.

Morrell, M., & Ezinheard, J. (2001). Revisiting adoption factors of inter-organizational information systems in SMEs. *Logistics Information Management, 15*(1-2), 46.

Moser, L. E., & Melliar-Smith, P. M. (2009). *Service Oriented Architecture and Web Services. Encyclopedia of Computer Science and Engineering* (pp. 2504–2511). Hoboken, NJ: John Wiley & Sons.

Moyaux, T., & Chaib-draa, B. (2007). Information sharing as a coordination mechanism for reducing the bullwhip effect in supply chain. *IEEE Transaction on Systems, Man, and Cybernetics, May, 37*(3), 396-409.

Mukhopadhyay, T., & Cooper, R. B. (1993). A microeconomic production assessment of the business value of management information systems: The case of inventory control. *Journal of Management Information Systems, 10*(1), 33–56.

Mulani, N. (2008). Business intelligence enters the supply chain. *Logistics management, 47*(4), 27.

Murch, R. (2004). *Autonomic computing*. Armonk, NY: IBM Press.

Murphy, J. V. (2008). CPG maker that bought forecasting software 10 years ago still benefits. *Global logistics & supply chain strategies*. Retrieved from www.sup錯pychainbrain.com

Murphy, S. (2008). The supply chain in 2008. *Supply Chain Management Review, 12*(1).

Murphy, P. R., & Poist, R. F. (1998). Third-party logistics usage: An assessment of propositions based on previous research. *Transportation Journal, 37*(4), 26–35.

Nachiappan, S. P., Gunasekaran, A., & Jawahar, N. N. (2007). Knowledge management system for operating parameters in two-echelon vmi supply chains. *International Journal of Production Research, 45*(11), 2479-2505.

Nunnally, J.C. (1978). *Psychometric Theory*. New York, NY: McGraw-Hill.

Nahmias, S. (2005). *Production & operations analysis* (5th ed.). New York: McGraw-Hill/Irwin.

Nam, S. H., Vitton, J., & Kurata, H. (2009). Determining the optimal number of suppliers utilized by contractors. *International Journal of Production Economics*.

Narasimhan, R., & Jayaram, J. (1998). Causal linkages in supply chain management: An exploratory study of North American manufacturing firms. *Decision Sciences, 29*(3), 579–605. doi:10.1111/j.1540-5915.1998.tb01355.x

Narasimhan, R., & Kim, S. W. (2001). Information system utilization strategy for supply chain integration. *Journal of Business Logistics, 22*(2), 51–75.

Narasimhan, R., Talluri, S., & Mendez, D. (2001). Supplier evaluation and rationalization via data envelopment analysis: an empirical examination. *Journal of Supply Chain Management, 37*(3), 28–37. doi:10.1111/j.1745-493X.2001.tb00103.x

Narayanan, V. G., & Raman, A. (2004). *Aligning incentives in supply chains*. Boston: Harvard Business School Publishing.

National Research Council. (2000). *Surviving supply chain integration*. Washington: National Academy Press.

Nelson, D., Moody, P. E., & Stegner, J. (2001). *The purchasing machine*. New York: The Free Press.

Neureuther, B. D., & Kenyon, G. N. (2008). The impact of information technologies on the US beef industry's supply chain. *International Journal of Information Systems and Supply Chain Management, 1*(1), 48–65.

Neureuther, B. D., & Kenyon, G. N. (2008). The impact of information technologies on the US beef industry's supply chain. *International Journal of Information Systems and Supply Chain Management, 1*(1), 48–65.

Neureuther, B. D., & Kenyon, G. N. (2008). The impact of information technologies on the US beef industry's supply chain. *International Journal of Information Systems and Supply Chain Management*, *1*(1), 48–65.

Newcomer, E., & Lomow, G. (2004). *Understanding SOA with Web Services*. Independent Technology Guides.

Nilsson, A., Segerstedt, A., & van der Sluis, E. (2007). A new iterative heuristic to solve the joint replenishment problem using a spreadsheet technique. *International Journal of Production Economics*, *108*(1-2), 399–405. doi:10.1016/j.ijpe.2006.12.022

Nilsson, A., & Silver, E. A. (2008). A simple improvement on Silver's heuristic for the joint replenishment problem. *The Journal of the Operational Research Society*, *59*(10), 1415–1421. doi:10.1057/palgrave.jors.2602446

Nitschke, T., & O'Keefe, M. (1997). Managing the linkage with primary producers: experiences in the Australian grain industry. *Supply Chain Management, Bradford*, *2*(1), 4. doi:10.1108/13598549710156295

Nonaka, I., & Takeuchi, H. (1995). *The knowledge-creating company: How japanese companies create the dynamics of innovation*. Oxford University Press.

Novack, R.A., Langly, Jr., C.J., & Rinehart, L.M. (1995). *Creating logistics value: themes for the future*. Oak Brook: Council on Logistics Management.

Nowotny, H., Scott, P., & Gibbons, M. (2001). *Re-thinking Science. Knowledge and the Public in the Age of Uncertainty*. Oxford: Polity Press.

Noy, N. F., & McGuiness, D. L. (2000). *Ontology development 101: a guide to creating your first ontology*. Retrieved May 10, 2009, from http://protege.stanford.edu/publications/ontology_development/ontology101.pdf

Núñez-Muñoz, M., & Montoya-Torres, J. (2009). Analyzing the impact of coordinated decisions within a three-echelon supply chain. *International Journal of Information Systems and Supply Chain Management*, *2*(2), 1–15.

Nunez-Munoz, M., & Montoya-Torres, J. R. (2009). Analyzing the impact of coordinated decisions with a three-echelon supply chain. *International Journal of Information Systems and Supply Chain Management*, *2*(2), 1–15.

O'Brien, J. A. (2001). *Management Information Systems: Managing Information Technology in the Internetworked Enterprise*. New York, NY: Mc-Graw Hill.

O'Connor, M. C. (2006). LEGO puts the RFID pieces together. *RFID Journal, February 12*. Retrieved August 1, 2008, from http://www.rfidjournal.com/article/articleview/2145

OASIS. (2006). *Reference model for Service Oriented Architecture*. Retrieved April 2009 from http://www.oasis-open.org/committees/tc_home.php?wg_abbrev=soa-rm

Ohno, T. (1988). *Toyota production system*. Portland: Productivity Press.

Olsen, A. L. (2005). An evolutionary algorithm to solve the joint replenishment problem using direct grouping. *Computers & Industrial Engineering*, *48*, 223–235. doi:10.1016/j.cie.2005.01.010

*OR/MS Today*. (2007). *34*(1-6).

Ozelkan, E. (2008). When does RFID make business sense for managing supply chains? *International Journal of Information Systems and Supply Chain Management*, *1*(1), 15–47.

Padmanabhan, B., & Tuzhilin, A. (2003). On the use of optimization for data mining: theoretical interactions and ecrm opportunities. *Management Science*, *49*(10), 1327–1343. doi:10.1287/mnsc.49.10.1327.17310

Pagarkar, M., Natesan, M., & Prakash, B. (2005). *RFID in Integrated Order Management Systems*. Chennai, India: Tata Consultancy Services.

Panayides, P., & So, M. (2005). The impact of integrated logistics relationships on third party quality performance. *Maritime Economics & Logistics*, *7*(1), 36–48. doi:10.1057/palgrave.mel.9100123

Panchal, J. H., Fernandez, M. G., Paredis, C. J. J., Allen, J. K., & Mistree, F. (2007). An interval-based constraint satisfaction (IBCS) method for decentralized, collaborative multifunctional design. *Concurrent Engineering . Research and Application*, *15*(3), 309–323.

Pant, S., Sethi, R., & Bhandari, M. (2003). Making sense of the e-supply chain landscape: An implementation framework. *International Journal of Information Management*, *23*(3), 201–221.

Parkin, M. (1997). *Microeconomics* (3rd ed.). Reading, MA: Addison-Wesley.

Patel, V. (2006). Contract management: The new competitive edge. *Supply Chain Management Review*, *10*(3), 42–44.

Patni Americas, Inc. (2008). *Thought Paper: Global Data Synchronization: A Foundation Block for Realizing RFID Potential.* Patni Americas, Inc., Cincinnati, Ohio. Retrieved July 24, 2008, from http://www.patni.com/resource-center/collateral/RFID/tp_RFID_Global-Data-Synchronization.html

Patterson, J. (1996). An investigation of conflict resolution between buyers and suppliers in strategic long-term cooperative relationships. In *Proceedings of the ISM 81st Annual International Conference*, Chicago.

Pazner, E. A. (1977). Pitfalls in the theory of fairness. *Journal of Economic Theory*, *14*(2), 458–466. doi:10.1016/0022-0531(77)90146-6

Pekelman, D. (1974). Simultaneous price-production decision. *Operations Research*, *22*, 788–794. doi:10.1287/opre.22.4.788

Perloff, J. M., & Salop, S. C. (1985). Equilibrium with product differentiation. *The Review of Economic Studies*, *52*(1), 107–120. doi:10.2307/2297473

Perrow, C. (1967). A framework for the comparative analysis of organizations. *American Sociological Review*, *32*, 194–208. doi:10.2307/2091811

Peypoch, R. (1998). The case for electronic business communities. *Business Horizons*, *4*(6), 17–20. doi:10.1016/S0007-6813(98)90073-8

Pfeffer, J., & Salancik, G. (1978). *The external control of organizations: A resource dependent perspective.* Harper and Row.

Pfeffer, J., & Salancik, G. R. (2003). *The External Control of Organizations: A Resource Dependence* Stanford: Stanford University Press.

Phios. (1999). *New Tools for Managing Business Processes*. Cambridge, MA: Phios Corporation.

PlanetLab. (2009). *Home*. Retrieved April 2009 from http://www.planet-lab.org/

Posey, C., & Bari, A. (2009). Information sharing and supply chain performance: Understanding complexity, compatibility, and processing. *International Journal of Information Systems and Supply Chain Management*, *2*(3), 67–76.

Pradhan, S. et al. (2005). RFID and sensing in the supply chain: challenges and opportunities. *HP Laboratories Palo Alto HPL* (pp. 1-11).

Prahalad, C. K., & Hamel, G. (1990). The core competence of the corporation. *Harvard Business Review*, *68*(3).

Prater, E. et al. (2005). Future impact of RFID on e-supply chains in grocery retailing. *Supply Chain Management.* ABI/INFORM Global (pp. 134-142).

Pyburn, P. J. (1983). Linking the MIS plan with corporate strategy: an exploratory study. *MIS Quarterly*, *7*(2), 1–14. doi:10.2307/248909

Raghuram, G., & Rangraj, N. (2000). *Logistics and supply chain management, concepts and cases.* New Delhi, India: Macmillan India Ltd.

Rahman, Z. (2004). Use of internet in supply chain management: a study of Indian companies. *Industrial Management & Data Systems*, *104*(1), 31–41. doi:10.1108/02635570410514070

Rahman, Z. (2006). Integrating the physical, information and financial flows - the next corporate paradigm. *The ICFAI Journal of Supply Chain Management*, *3*(3), 12–22.

Rai, A., Patnayakuni, R., & Seth, N. (2006). Firm performance impacts of digitally enabled supply chain integration capabilities. *Management Information Systems Quarterly*, *30*(2), 225–246.

Rainbird, M. (2004). Demand and supply chain: the value catalyst. *International Journal of Physical Distribution & Logistic Management*, *34*(3), 230–250. doi:10.1108/09600030410533565

Raisinghani, M. S., & Meade, L. L. (2005). Strategic decisions in supply-chain intelligence using knowledge management: an analytic-network-process framework. *Supply Chain Management*, *10*(2), 114–121. doi:10.1108/13598540510589188

Ramanathan, R. (2003). *An Introduction to Data Envelopment Analysis: A tool for Performance Measurement*. Thousand Oaks, CA: Sage.

Ranganathan, C., & Dhaliwal, J. S. (2001). A survey of business process reengineering practices in Singapore. *Information & Management*, *39*(2), 125–134. doi:10.1016/S0378-7206(01)00087-8

Rangone, A. (1996). An analytical hierarchy process framework for comparing the overall performance of manufacturing departments. *International Journal of Operations & Production Management*, *16*(8), 104–119. doi:10.1108/01443579610125804

Regattieri, A., Gamberi, M., & Manzini, G. (2006). Traceability of food products: general framework and experimental evidence. *Journal of Food Engineering*, *81*, 347–356. doi:10.1016/j.jfoodeng.2006.10.032

Rekik, Y., Sahin, E., & Dallery, Y. (2009). Inventory inaccuracy in retail stores due to theft: An analysis of the benefits of RFID. *International Journal of Production Economics*, *118*, 189–198. doi:10.1016/j.ijpe.2008.08.048

Resende-Filho, M. A., & Buhr, B. L. (2008). A principal-agent model for evaluating the economic value of traceability system: a case study with injection. Site lesion control in fed castle. *American Journal of Agricultural Economics*, *90*(4), 1091–1102. doi:10.1111/j.1467-8276.2008.01150.x

Reyes, P. M., Frazier, G., Prater, E., & Cannon, A. (2007). RFID: the state of the union between promise and practice. *International Journal of Integrated Supply Management*, *3*(2), 192–206. doi:10.1504/IJISM.2007.011976

Reynolds, J. (2000). Supply chain, distribution and fulfilment. *International Journal of Retail and Distribution Management*, *28*(10).

Richmond, B., Burns, A., Maybe, J., Nutthsll, L., & Toole, R. (1998). Supply chain management tools: minimizing the benefits . In Gattoma, J. (Ed.), *Strategic Supply Chain Alignment: Best Practices in Supply Chain Management* (pp. 509–520). Aldershot, UK: Gower.

Rindfleisch, A., & Heide, J. B. (1997). Transaction Cost Analysis: past, present and future applications. *Journal of Marketing*, *61*(4), 30–54. doi:10.2307/1252085

Roberti, M. (2008). Laying the foundation for RFID. *RFID Journal*. Retrieved August 10, 2008, from http://www.rfidjournal.com/article/articleview/3524/1/435/

Rockstroh, J. (2002). Achieving quality ROI across the supply chain. *Quality*, *41*(6), 54.

Rodriguez, G., & Montoya-Torres, J. R. (2009). Measuring the impact of supplier-customer information sharing on production scheduling. *International Journal of Information Systems and Supply Chain Management*, *2*(2), 48–61.

Rogers, S. (2004). *Supply chain management: Six elements of superior design*. http://www.manufacturing.net.

Romeo, P. (2006). Buyers embrace more sophisticated supply-purchasing procedures. *Nations Restaurant News*, *40*(21), 92–93.

Ross, C. F. (1925). A mathematical theory of competition. *American Journal of Mathematics*, *47*(3), 163–175. doi:10.2307/2370550

Roussos, G. (2006). Enabling RFID in retail. *IEEE Computer*, *39*(3), 25–30.

Routledge, N., Bird, L., & Goodchild, A. (2002). UML and XML Schema. In X. Zhou (Ed.), *Proceedings of the Thirteenth Australasian Database Conference (ADC2002)*, Melbourne, Australia (Vol. 5).

Rubin, S., & Fogarty, T. (2000). *Automotive B2B: The creation of B2B enterprises should add value for GM and Ford shareholders*. Warburg Dillon Read, Global Equity Research.

Runde, C., & Flannegan, T. (2008). *Building Conflict Competent Teams* (p. 238). San Francisco, CA: Jossey-Bass.

Rupnick R., Kukar, M., & Krisper, M. (2007). Integrating data mining and decision support through data mining based decision support system. *Journal of Computer Information Systems, Spring*, *47*(3), 89-104.

Russell, D. M., & Hoag, A. M. (2004). People and information technology in the supply chain: Social and organizational influences on adoption. *International Journal of Physical Distribution & Logistics Management*, *34*(1-2), 102. doi:10.1108/09600030410526914

Rutherford, T. D. (2000). Re-embedding, Japanese investment and the restructuring buyer-supplier relations in the Canadian automotive components industry during the 1990s. *Regional Studies*, *34*(8), 739. doi:10.1080/00343400050192838

Rutner, S. M., Gibson, B. J., & Gustin, C. M. (2001). Longitudinal study of supply chain information systems. *Production and Inventory Management Journal*, *42*(2), 49–56.

Ryan, T. P. (2000). *Statistical methods for quality improvement.* New York: John Wiley & Sons.

Ryssel, R., Ritter, T., & Gemunden, H. G. (2004). The impact of information technology deployment on trust, commitment and value creation in business relationships. *Journal of Business and Industrial Marketing*, *19*(3), 197. doi:10.1108/08858620410531333

Saaty, T. L. (1980). *The analytic hierarchy process*. New York: McGraw-Hill.

Saeed, K. A., Malhotra, M. K., & Grover, V. (2005). Examining the impact of interorganizational systems on process efficiency and sourcing leverage in buyer-supplier dyads. *Decision Sciences*, *36*(3), 365–396. doi:10.1111/j.1540-5414.2005.00077.x

Saharidis, K. D., Georgios, K., Vassilis, S., & Dallery, Y. (2009). Centralized and decentralized control polices for a two-stage stochastic supply chain with subcontracting. *International Journal of Production Economics;*,(n.d) 117–126. doi:10.1016/j.ijpe.2008.10.001

Sahay, B. S., & Ranjan, J. (2008). Real time business intelligence in supply chain analytics. *Information Management & Computer Security*, *16*(1), 28–48. doi:10.1108/09685220810862733doi:10.1108/09685220810862733

Sahay, B. S., & Mohan, R. (2006). 3PL practices: An Indian perspective. *International Journal of Physical Distribution and Logistics Management*, *36*(9), 666. doi:10.1108/09600030610710845

Sakamura, K. (2001). Radio frequency identification and non-contact smart cards. *IEEE Micro*, 4–6. doi:10.1109/MM.2001.977752

Sako, M. (1992). *Prices, quality and trust: Inter-firm relations in Britain and Japan*. Cambridge University Press.

Salin, V. (1998). Information technology in Agri-Food supply chains. *International Food and Agribusiness Management Review*, *1*(3), 329–334. doi:10.1016/S1096-7508(99)80003-2

Salin, V. (2000). Information technology and cattle-beef supply chains. *American Journal of Agricultural Economics*, *82*(5), 1105–1111. doi:10.1111/0002-9092.00107

Salomie, J., Dinsoreanu, M., Bianca Pop, C., & Liviu Suciu, S. (2008, November 24-26). Model and SOA solutions for traceability in logistic chains. In *Proceedings of iiWAS2008*, Linz, Austria (pp. 339-344).

Sanchanta, M. (2007, February 2). Boeing suppliers face production hitches. *Financial Times (North American Edition)*, 22.

Sanders, N. R., & Premus, R. (2002). IT applications in supply chain organizations: a link between competitive priorities and organizational benefits. *Journal of Business Logistics*, *23*(1), 65–83.

Sanders, N. R., & Premus, R. (2005). Modeling the relationship between firm IT capability, collaboration, and performance. *Journal of Business Logistics*, *26*(1), 1.

Sankaran, J., Mun, D., & Charman, Z. (2002). Effective logistics outsourcing in New Zealand: An inductive empirical investigation. *International Journal of Physical Distribution and Logistics Management*, *32*(8), 682–702. doi:10.1108/09600030210444926

SAP. (2007). Air Products becomes one company with SAP software. *Business transformation study*. Retrieved from http://www.sap.com/usa/solutions/business-suite/erp/customersuccess/index.epx

SAP. (2007). *SAP's homepage*. Retrieved November 5, 2007 from http://www.sap.com/solutions/business-suite/scm/index.epx

Sarrico, C. S., & Dyson, R. G. (2004). Restricting virtual weights in data envelopment analysis. *European Journal of Operational Research*, *159*(1), 17–34. doi:10.1016/S0377-2217(03)00402-8

Sasazaki, S., Itoh, K., Arimitsu, S., Imada, T., Takasuga, A., & Nagaishi, H. (2004). Development of breed identification markers derived from AFLP in beef cattle. *Meat Science*, *67*, 275–280. doi:10.1016/j.meatsci.2003.10.016

Sauvage, T. (2003). The relationship between technology and logistics third-party providers. *International Journal of Physical Distribution and Logistics Management*, *33*(3), 236–253. doi:10.1108/09600030310471989

Schein, E. H. (1985). *Organizational culture and leadership.* San Francisco: Jossey-Baas Publishers.

Schein, E. H. (1994). The role of the CEO in the management of change: the case of information technology. In R. Galliers, D. Leidner, & B. Baker (Eds.), *Strategic information management.* Oxford: Butterworth Heinemann.

Scherer, F. (1980). *Industrial market structure and economic performance*. Rand McNally.

Schmeidler, D. (1969). The nucleolus of a characteristic function game. *SIAM Journal on Applied Mathematics*, *17*(6), 1163–1170. doi:10.1137/0117107

Schultz, M. (1994). *On studying organizational culture.* New York: Walter de Gruyter.

Scott Morton, M. S. (1991). *The corporation of the 1990s.* Oxford University Press, NY.

Scott, I. (2006). *Analyses of supply chain management efficiency by country*. Working paper on supply chain management, Temple University, Philadelphia.

Segars, A. H., & Grover, V. (1999). Profiles of Strategic Information Systems Planning. *Information Systems Research*, *10*(3), 199–233. doi:10.1287/isre.10.3.199

Seppanen, M. (2000). Developing industrial strength simulation models using visual basic for applications (VBA). *Proceedings of the 2000 Winter Simulation Conference* (pp. 77-82).

Sevkli, M., Koh, S. C., Zaim, L., Selim, D. M., & Tatoglu, E. (2007). An application of data envelopment analytic hierarchy process for supplier selection: a case study of BEKO in Turkey. *International Journal of Production Research*, *45*(9), 1973–2003. doi:10.1080/00207540600957399

Shang, S., & Seddon, P. B. (2002). Assessing and managing the benefits of enterprise systems: the business manager's perspective. *Information Systems Journal*, *12*(4), 271–299. doi:10.1046/j.1365-2575.2002.00132.x

Shao, B. B. M., & Lin, W. T. (2001). Measuring the value of information technology in technical efficiency with stochastic production frontiers. *Information and Software Technology*, *43*(7), 447–456. doi:10.1016/S0950-5849(01)00150-1

Shapiro, J. F. (2007). *Modeling the supply chain* (2nd ed.). Belmont, CA: Thomson Higher Education.

Shapley, L. S. (1953). A value for n-person games. In H. W. Kuhn & A. W. Tucker (Eds.), *Contributions to the Theory of Games – Volume II, Annals of Mathematics Studies 28* (pp. 307-317). Princeton, NJ: Princeton University Press.

Sharma, J. R., Rawani, A. M., & Barahate, M. (2008). Quality function deployment: A comprehensive literature review. *International Journal of Data Analysis Techniques and Strategies*, *1*(1), 78–103. doi:10.1504/IJDATS.2008.020024

Sharma, M. K., & Rajat, B. (2007). An integrated BSC-AHP approach for supply chain, management evaluation. *Measuring Business Excellence*, *11*(3), 57–68. doi:10.1108/13683040710820755

Shere, S. A. (2005). From supply chain management to value network advocacy: implication for e-supply chains. *Supply Chain Management: An International Journal*, *10*(2), 77–83. doi:10.1108/13598540510589151

Shi, K., & Xiao, T. (2008). Coordination of a supply chain with a loss-averse retailer under two types of contracts. *Int. J. Information and Decision Sciences*, *1*(1), 5–25. doi:10.1504/IJIDS.2008.020033

Shin, H., Collier, D. A., & Wilson, D. D. (2000). Supply management orientation and supplier/buyer performance. *Journal of Operations Management*, *18*, 317–333. doi:10.1016/S0272-6963(99)00031-5

Shutzberg, L. (2004). *Radio Frequency Identification (RFID) in the Consumer Goods Supply Chain: Mandated Compliance or Remarkable Innovation?* Norcross, GA: Rock-Tenn Company.

Siegel, J. G., & Shim, J. K. (1991). *Financial management.* Hauppauge, NY: Barron's Business Library.

Silver, E. A. (1974). A control system of coordinated inventory replenishment. *International Journal of Production Research*, *12*, 647–671. doi:10.1080/00207547408919583

Silver, E. A. (1976). A simple method of determining order quantities in joint replenishments under deterministic demand. *Management Science*, *22*, 1351–1361. doi:10.1287/mnsc.22.12.1351

Silver, E. A., Pyke, D. F., & Peterson, R. (1998). *Inventory management and production planning and scheduling* (3rd ed.). New York: Wiley & Sons.

Simatupang, T. M., & Sridharan, R. (2004). Benchmarking supply chain collaboration: An empirical study. *Benchmarking*, *11*(5), 484. doi:10.1108/14635770410557717

Simchi-Levi, D., Kaminsky, P., & Simchi-Levi, E. (2003). *Managing the supply chain: The definitive guide for the business professional*. New York: McGraw-Hill.

Simchi-Levi, D., Kaminsky, P., & Simchi-Levi, E. (2004). *Managing the Supply Chain: The Definitive Guide for the Business Professional*. New York: McGraw-Hill.

Simon, H. A. (1957). *Administrative behavior* (4th ed.). New York: Free Press.

Singh, N., Lai, K.-H., & Cheng, T. C. E. (2007). Intra-organizational perspectives on IT-enabled supply chains. *Communications of the ACM*, *50*(1), 59–65. doi:10.1145/1188913.1188918

Singh, N., & Vives, X. (1984). Price and quantity competition in a differentiated duopoly. *The Rand Journal of Economics*, *15*, 546–554. doi:10.2307/2555525

Sink, H. L., & Langley, C. J. (1997). A managerial framework for the acquisition of third-party logistics services. *Journal of Business Logistics*, *18*(2), 63–89.

Sivakumar, B., Anbazhagan, N., & Arivarignan, G. (2007). Two commodity inventory system with individual and joint ordering polices and Renewal demands. *Stochastic Analysis and Applications*, *25*(6), 1217–1241. doi:10.1080/07362990701567314

Smaros, J., Holmstrom, J., & Kamarainen, V. (2000). New service opportunities in the e-grocery business. *International Journal of Logistics Management*, *11*(1).

Smith, R. (2008). Reality of supply chain models. *APICS-Extra online, APICS*. Retrieved May 28, 2008, from www.apics.com

Smithies, A., & Savage, L. J. (1940). A dynamic problem in duopoly. *Econometrica*, *8*(2), 130–143. doi:10.2307/1907032

Sodhi, M., & Tang, C. (2009). Rethinking links in the global supply chain. [insert]. *Financial Times (North American Edition)*, 11.

Sohail, M. S., & Sohal, A. (2003). The use of 3PL services: A Malaysian perspective. *Technovation*, *23*, 401–408. doi:10.1016/S0166-4972(02)00003-2

Sohal, A., Millen, R. A., & Moss, S. (2002). A comparison of the use of 3PL services by Australian firms between 1995-1999. *International Journal of Physical Distribution and Logistics Management*, *32*(1), 59–68. doi:10.1108/09600030210415306

Spangler, W. E., Gal-Or, M., & May, J. H. (2003). Using data mining to profile tv viewers. *Communications of the ACM*, *46*(2), 66–72. doi:10.1145/953460.953461

Speckman, R., & Davis, E. (2004, May). Risky business: expanding the discussion on risk and the extended enterprise. *International Journal of Physical Distribution & Logistics Management*, *34*(5), 414–429. doi:10.1108/09600030410545454

Spence, M. (1976). Product differentiation and welfare. *The American Economic Review*, *66*(2).

Stanley, T. (2006, February 1). High-stakes analytics. *Optimize*.

Stein, E. W. (1995). Organizational memory: review of concepts and recommendations for management. *International Journal of Information Management*, *15*(2), 17–32. doi:10.1016/0268-4012(94)00003-C

Stevens, G. C. (1990). Successful supply chain management. *Management Decision*, *28*(8), 25–30. doi:10.1108/00251749010140790

Stewart, G. (1995). Supply chain performance benchmarking study reveals keys to supply chain excellence. *Logistics Information Management*, *8*(2), 38–44. doi:10.1108/09576059510085000

Stock, J. R. (1997). Applying theories from other disciplines of logistics. *International Journal of Physical Distribution & Logistics Management*, *27*(9/10), 515–539. doi:10.1108/09600039710188576

Stock, J. R., & Lambert, D. M. (2001). *Strategic Logistics Management* (4th ed.). New York: McGraw-Hill.

Stonebraker, P. W., & Afifi, R. (2004). Toward a contingency theory of supply chains. *Management Decision*, *42*(9), 1131. doi:10.1108/00251740410565163

Supply Chain Digest. (2008). Supply chain software: AMR Research remains bullish on supply chain software spend. *Supply Chain Digest*. Retrieved from http://www.scdigest.com/assets/On_Target/08-05-19-2.php?cid=1688

Svensson, G. (2003). Holistic and cross-disciplinary deficiencies in the theory generation of supply chain management. *Supply Chain Management: An International Journal*, *8*(4), 303–316. doi:10.1108/13598540310490062

Swatman, P., M., & Swatman, P., A. (1992). EDI system integration: A definition and literature survey. *The Information Society*, 8.

Szmerekovsky, J. G., & Zhang, J. (2008). Coordination and adoption of item-level RFID with vendor managed inventory. *International Journal of Production Economics*, *114*, 388–398. doi:10.1016/j.ijpe.2008.03.002

Szulanski, G. (1996). Exploring internal stickiness: impediments to the transfer of best practice within the firm. *Strategic Management Journal*, *17*, 27–43.

Tait, N., & Peel, M. (2006, August 8). Tool for dispute resolution loses its certainty. *Financial Times (North American Edition)*, 14.

Talluri, S., & Baker, R. C. (2002). A multi-phase mathematical programming approach for effective supply chain design . *European Journal of Operational Research*, *141*(3), 544–558. doi:10.1016/S0377-2217(01)00277-6

Talluri, S., & Narasimhan, R. (2003). Vendor evaluation with performance variability: a max-min approach. *European Journal of Operational Research*, *146*, 543–552. doi:10.1016/S0377-2217(02)00230-8

Talluri, S., Narasimhan, R., & Nair, A. (2006). Vendor performance with supply risk: a chance-constrained DEA approach. *International Journal of Production Economics*, *100*, 212–222. doi:10.1016/j.ijpe.2004.11.012

Talluri, S., & Sarkis, J. (2002). A model for performance monitoring of suppliers. *International Journal of Production Research*, *40*, 4257–4269. doi:10.1080/00207540210152894

Tan (1999). The use of information technology to enhance supply chain management. *Production and Inventory Management Journal*, 40.

Tang, C. H. (2006). Perspectives in supply chain risk management. *International Journal of Production Economics*, *103*, 451–488. doi:10.1016/j.ijpe.2005.12.006

Tanik, U., Tanik, M. M., & Jololian, L. (2001). Internet enterprise engineering. A zero-time framework based on the T-strategy. In *Proceedings of the IEEE Southeast Conference* (pp. 263-270).

Tan, K. C., Kannan, V. R., & Handfield, R. B. (1998). Supply chain management: supplier performance and firm performance. *International Journal of Purchasing and Materials Management*, *34*(3), 2–9.

Tapscott, D., Ticoll, D., & Lowy, A. (2000). *Digital capital: Harnessing the power of business webs*. Nicholas Brealey.

Teece, D. J., Pisano, G., & Shuen, A. (1997). Dynamic capabilities and strategic management. *Strategic Management Journal*, *18*(7), 509–533. doi:10.1002/(SICI)1097-0266(199708)18:7<509::AID-SMJ882>3.0.CO;2-Z

Tellkamp, C. (2006). *The impact of Auto-ID technology on process performance RFID in the FMCG supply chain*. Unpublished doctoral dissertation, Graduate School of Business Administration, Economics, Law and Social Sciences (HSG) University of St. Gallen, Switzerland.

The Extensible Markup Language (XML) Query Specification. (n.d.). *World Wide Web Consortium*. Retrieved from http://www.w3.org/XML/Query/

Thompson, P. (1998). Bank lending and the environment: policies and opportunities. *International Journal of Bank Marketing*, *16*(6), 243–252. doi:10.1108/02652329810241384

Thompson, R. G., Langemeier, L. N., Lee, C. T., Lee, E., & Thrall, R. M. (1990). The role of multiplier bounds in efficiency analysis with application to Kansas farming . *Journal of Econometrics*, *46*(1/2), 93–108. doi:10.1016/0304-4076(90)90049-Y

Threlkel, M., & Kavan, B. (1999). From traditional EDI to Internet-based EDI: managerial considerations. *Journal of Information Technology*, 14.

Thurstone, L. L. (1931, September). Multiple factor analysis. *Psychological Review*, *38*(5), 406–427. doi:10.1037/h0069792

Tijs, S. H. (1981). Bounds for the core and the τ-value . In Moeschlin, O., & Pallaschke, D. (Eds.), *Game Theory and Mathematical Economics* (pp. 123–132). Amsterdam, The Netherlands: North Holland Publishing Company.

Tijs, S. H. (1987). An axiomatization of the tau-value. *Mathematical Social Sciences*, *13*(2), 177–181. doi:10.1016/0165-4896(87)90054-0

Tijs, S. H., & Driessen, T. S. H. (1983). *The τ-Value as a Feasible Compromise Between Utopia and Disagreement (Report 8312)*. Nijmegen, The Netherlands: University of Nijmegen, Department of Mathematics.

Tijs, S. H., & Driessen, T. S. H. (1986). Game theory and cost allocation problems. *Management Science*, *32*(8), 1015–1028. doi:10.1287/mnsc.32.8.1015

Tiliquist, J., King, J. L., & Woo, C. (2002). Analyzing IT and organizational dependency. *MIS Quarterly*, *26*(2), 91–118. doi:10.2307/4132322

Timmers, P. (1998). Business models for electronic markets. *Electronic Markets*, *8*(2).

Transportation and Logistics Institute (TLI). (2003). *Report on the working group on logistics: Developing Singapore into a global integrated logistics hub.*

Trappey, A. J. C., Trappey, C. V., Hou, J. L., & Chen, B. J. G. (2004). Mobile agent technology and application for online global logistic services. *Industrial Management & Data Systems*, *104*(1/2), 169–184. doi:10.1108/02635570410522143

Trent, R. J. (2008). *End-to-end lean management: a guide to complete supply chain improvement.* Ft. Lauderdale, FL: J. Ross Publishing.

Tsiakis, P., Shah, N., & Pantelides, C. (2001). Design of multi-echelon supply chain networks under demand uncertainty. *Industrial & Engineering Chemistry Research*, *40*, 35–85. doi:10.1021/ie0100030

Turban, E., McLean, E., & Wetherbe, J. (2001). *Information technology for management* (2nd ed.). New York: John Wiley and Sons.

UNCITRAL. (1976). *United Nations Commission on International Trade Law*. Retrieved from http://www.united.nations.org

UNCITRAL. (1994). *Model Law on International Commercial Arbitration.* Retrieved from http://www.uncitral.org

United Nations Centre for Trade Facilitation and Electronic Business (UN/CEFACT). (n.d.) Retrieved from http://www.unece.org/cefact/

Uzzi, B., & Lancaster, R. (2003). Relational embeddedness and learning: the case of bank loan managers and their clients. *Management Science, April, 49*(4), 383-399.

van Damme, D. A., & Ploos van Amstel, M. J. (1996). Outsourcing logistics management activities. *International Journal of Logistics Management*, *7*(2), 85–95. doi:10.1108/09574099610805548

Van Den Hoven, J. (2004). Data architecture standards for the effective enterprise. *Information Systems Management, Summer, 21*(3), 61-64.

Van Dorp, C. A. (2003). A traceability application based on gozinto graphs. In *Proceedings of the EFITA 2003 Conference,* Debrecen, Hungary.

Van Eijs, M. J. G. (1993). A note of the joint replenishment problem under constant demand. *The Journal of the Operational Research Society*, *44*(2), 185–191.

Van Grembergen, W. (2005). Introduction to the Minitrack "IT Governance and its Mechanisms". *System Sciences, 2005. HICSS '05. Proceedings of the 38th Annual Hawaii International Conference on* (pp. 235-235).

van Hoek, R. (2002). Using information technology to leverage transport and logistics service operations in the supply chain: An empirical assessment of the interrelation between technology and operation management. *International Journal of Information Technology and Management*, *1*(1), 115–130. doi:10.1504/IJITM.2002.001191

Vanany, I., & Zailani, S. (2009). Supply chain risk management: A literature review and future research. *International Journal of Information Systems and Supply Chain Management, 2*(1), 16–33.

Vanany, I., Zailani, S., & Pujawan, N. (2009). Supply chain risk management: Literature review and future research. *International Journal of Information Systems and Supply Chain Management, 2*(1).

Vargas, L. G. (1990). An overview of the analytic hierarchy process and its applications. *European Journal of Operational Research, 48*(1), 2–8. doi:10.1016/0377-2217(90)90056-H

Varian, H. L. (1976). Two problems in the theory of fairness. *Journal of Public Economics, 5*(3/4), 249–260. doi:10.1016/0047-2727(76)90018-9

Vassiliadis, B. (2009). The grid as a virtual enterprise enabler. *International Journal of Information Systems in the Service Sector, 1*(1), 78–92.

Vassiliadis, W. (2009). The grid as a virtual enterprise enabler. *International Journal of Information Systems in the Service Sector, 1*(1), 78–92.

Vidgen, R., & Goodwin, S. (2000). XML: What is it good for? *Computing and Control Engineering Journal, 11*(3), 119–124.

Vieira, G. (2004). Ideas for modeling and simulation of supply chains with arena. *Proceedings of the 2004 Winter Simulation Conference* (pp. 1418-1427).

Vijayaraman, B., & Osyk, B. (2006). An empirical study of RFID implementation in the warehousing industry. *The International Journal of Logistics Management, 17*(1), 6–20. doi:10.1108/09574090610663400

Viswanathan, S. (1996). A new optimal algorithm for the joint replenishment problem. *The Journal of the Operational Research Society, 47*, 936–944.

Viswanathan, S. (2002). On optimal algorithms for the joint replenishment problem. *The Journal of the Operational Research Society, 53*, 1286–1290. doi:10.1057/palgrave.jors.2601445

Viswanathan, S. (2007). An algorithm for determining the best lower bound for the stochastic joint replenishment problem. *Operations Research, 55*(5), 992–996. doi:10.1287/opre.1070.0401

W3C. (2004). *Web Services Architecture*. Retrieved April 2009 from http://www.w3.org/TR/ws-arch

Wagner, H. (2002). And then there were none. *Operations Research, 50*(1), 217–226. doi:10.1287/opre.50.1.217.17777

Wallace, S. W. (2000). Decision making under uncertainty: Is sensitivity analysis of any use? *Operations Research, 48*(1), 20–25. doi:10.1287/opre.48.1.20.12441

Wallace, S. W. (2009). *Stochastic programming and the option of doing it differently*. Operations Research.

Walsh, J. P., & Ungson, G. R. (1991). Organizational memory. *Academy of Management Review, 16*(1), 57–91. doi:10.2307/258607

Wang, Z., Yan, R., Hollister, K., & Xing, R. (2009). A relative comparison of leading supply chain management software packages. *International Journal of Information Systems and Supply Chain Management, 2*(1), 81–96.

Wang, G., Huang, S. H., & Dismukes, J. P. (2004). Product-driven supply chain selection using integrated multi-criteria decision-making methodology. *International Journal of Production Economics, 91*, 1–15. doi:10.1016/S0925-5273(03)00221-4

Wang, L. P., Qiu, F. Y., Dai, H. R., & Chen, Z. C. (2004). Relationships of electronic marketplace and partner selection of supply chain. *Computer Integrated Manufacturing Systems, 10*(5), 550–555.

Wang, Y. (2008, March). Strategic management of international subcontracting. *International Journal of Information Systems and Supply Chain Management, 1*(3), 21–32.

Wang, Y. M., Greatbanks, R., & Yang, J. B. (2005). Interval efficiency assessment using data envelopment analysis. *Fuzzy Sets and Systems, 153*, 347–370.

Wasserman, E. (2005). Accenture outlines secrets of RFID success. *RFID Journal*. Retrieved August 5, 2008, from http://www.rfidjournal.com/article/articleview/1506/1/1/

Wasserman, E. (2007). Beaver street fisheries automates RFID tagging. *RFID Journal.* Retrieved August 15, 2008, from http://www.rfidjournal.com/article/articleview/3060

Watson, H., & Wixom, B. (2007). Enterprise agility and mature BI capabilities. *Business Intelligence Journal, 12*(3), 4–6.

Webb, P., Pollard, C., & Ridley, G. (2006). Attempting to define IT governance: Wisdom or folly? *System Sciences, 2006. HICSS '06. Proceedings of the 39th Annual Hawaii International Conference on, 8*, 194-199.

Weber, C. A. (1996). A data envelopment analysis approach to measuring vendor performance. *Supply Chain Management, 1*, 28–39. doi:10.1108/13598549610155242

Weber, C. A., Current, J., & Desai, A. (2000). An optimization approach to determining the number of vendors to employ. *Supply Chain Management: An International Journal, 5*, 90–98. doi:10.1108/13598540010320009

Webster, J. (1995). Networks of collaboration or conflict? Electronic data interchange and power in the supply chain. *The Journal of Strategic Information Systems, 4*(1).

Weill, P., & Broadbent, M. (1998). *Leveraging the New Infrastructure* Boston: Harvard Business School Press.

Weill, P. (2004). Don't just lead govern: How top-performing firms govern IT. *MIS Quarterly Executive, 3*(1), 1–17.

Welty, B., & Becerra-Frenandez, I. (2001). Managing trust and commitment in collaborative supply chain relationships. *Communications of the ACM, 44*(6), 67. doi:10.1145/376134.376170

West, L. A. Jr. (1994). Researching the costs of information systems. *Journal of Management Inquiry, 11*(2), 75–108.

West, L. A. Jr, & Courtney, J. F. (1993). The information problems in organizations: a research model for the value of information and information systems. *Decision Sciences, 24*(2), 229–252. doi:10.1111/j.1540-5915.1993.tb00473.x

Whitaker, J., Mithas, S., & Krishnan, M. (2007). A field study of RFID deployment and return expectations. *Production and Operations Management, 16*(5), 599–612.

White, A., Daniel, E. M., & Mohdzain, M. (2005). The role of emergent information technologies and systems in enabling supply chain agility. *International Journal of Information Management, 25*(5), 396–410. doi:10.1016/j.ijinfomgt.2005.06.009

Wijnhoven, F., Spil, T., Stegwee, R., & Fa, R. T. A. (2006). Post-merger IT integration strategies: An IT alignment perspective. *The Journal of Strategic Information Systems, 15*(1), 5–28. doi:10.1016/j.jsis.2005.07.002

Wikipedia. (2006). *Supply chain.* Retrieved April 2009 from http://en.wikipedia.org/wiki/supply_chain

Wildeman, R. E., Frenk, J. B. G., & Dekker, R. (1997). An efficient optimal solution method for the joint replenished problem. *European Journal of Operational Research, 99*, 433–444. doi:10.1016/S0377-2217(96)00072-0

Wilding, R., & Juriado, R. (2004). Customer perceptions on logistics outsourcing in the European consumer goods industry. *International Journal of Physical Distribution and Logistics Management, 34*(8), 628–644. doi:10.1108/09600030410557767

Willcocks, L., Feeny, D., & Olson, N. (2006). Implementing core IS capabilities: Feeny-Willcocks IT governance and management framework revisited. *European Management Journal, 24*(1), 28–37. doi:10.1016/j.emj.2005.12.005

Williams, J. R., Haka, S. F., Bettner, M. S., & Carcello, J. V. (2008). *Financial accounting* (13th ed.). New York: McGraw-Hill Publishing.

*Wilson , R. ( 2005 ).* Council of Supply Chain Management Professionals, 16th Annual State of Logistics Report.

Wilson, T. P., & Clarke, W. R. (1998). Food safety and traceability in the agricultural supply chain: using the internet to deliver traceability. *Supply Chain Management, 3*, 127–133. doi:10.1108/13598549810230831

Wise, R., & Morrison, D. (2000). Beyond the exchange: The future of B2B. *Harvard Business Review, 78*(6), 86–96.

Withey, M., Daft, R. L., & Cooper, W. H. (1983). Measures of perrow's work unit technology: an empirical assessment and a new scale. *Academy of Management Journal, 26*, 45–63. doi:10.2307/256134

Wong, W. P., Jaruphongsa, W., & Lee, L. H. (2008). Supply chain performance measurement system: a monte carlo DEA-based approach. *International Journal of Industrial and Systems Engineering*, *3*(2), 162–188. doi:10.1504/IJISE.2008.016743

Wong, Y. H. B., & Beasley, J. E. (1990). Restricting weight flexibility in data envelopment analysis . *The Journal of the Operational Research Society*, *41*(9), 829–835.

Wong, Y. H. B., & Beasley, J. E. (1990). Restricting weight flexibility in data envelopment analysis. *The Journal of the Operational Research Society*, *39*, 829–835.

Woodbury, R. S. (1960). The legend of Eli Whitney and interchangeable parts. *Technology and Culture*, *1*(3), 235–253. doi:10.2307/3101392

Wu, B., Dewan, M., Li, L., & Yang, Y. (2005). Supply chain protocolling. In *Proceedings of the 7th IEEE International Conference on E-Commerce Technology* (pp. 314-321).

Wu, F., Zsidisin, G. A., & Ross, A. D. (2007). Antecedents and outcomes of e-procurement adoption: An integrative model. *IEEE Transactions on Engineering Management*, *54*(3), 576–587. doi:10.1109/TEM.2007.900786

Wyld, D. (2006). RFID 101: the next big thing for management. *Management and Research News, 29*(4).

Xia, L. X. X., Bin, M., & Lim, R. (2007, September 25-28). AHP based supply chain performance measurement system. In *Proceedings of ETFA IEEE Conference on Emerging Technologies and Factory Automation*, Patras, Greece (pp. 1308-1315). Washington, DC: IEEE.

Xiao, T., Luo, J., & Jin, J. (2007). Coordination of a supply chain with demand stimulation and random demand disruption. *International Journal of Information Systems and Supply Chain Management*, *2*(1), 1–15.

Xiao, T., Luo, J., & Jin, J. (2009). Coordination of a supply chain with demand stimulation and random demand disruption. *International Journal of Information Systems and Supply Chain Management*, *2*(1), 1–15.

Xu, K., & Dong, Y. (2004). Information gaming in demand collaboration and supply chain performance. *Journal of Business Logistics*, *25*(1), 121–144.

Yadavalli, V. S. S., Arivarignan, G., & Anbazhagan, N. (2006). Two commodity coordinated inventory system with Markovian demand. *Asia - Pacific Journal of Operational Research, 23*(4), 497-508.

Yadavalli, V. S. S., Anbazhagan, N., & Arivarignan, G. (2004). A two-commodity stochastic inventory system with lost sales. *Stochastic Analysis and Applications*, *22*(2), 479–497. doi:10.1081/SAP-120028606

Yang, J., Zhao, S. Z., & Wang, J. R. (2005). Study on postponement in customization supply chain. *Industrial Engineering and Management*, *10*(4), 35–44.

Yao, Y., Palmer, J., & Dresner, M. (2007). An interorganizational perspective on the use of electronically-enabled supply chains. *Decision Support Systems*, *43*(3), 884. doi:10.1016/j.dss.2007.01.002

Yin, R. (1994). *Case Study Research, design and methods*. Sage: Thousand Oaks, CA.

Zahra, S. A., & George, G. (2002). Absorptive capacity: a review, reconceptualization, and extension. *Academy of Management Review*, *27*(2), 185–203. doi:10.2307/4134351

Zeng, Z. B., Li, W., & Zhu, W. (2006). Partner selection with a due date constraint in virtual enterprises. *Applied Mathematics and Computation*, *175*, 1353–1365. doi:10.1016/j.amc.2005.08.022

Zhang, D. (2006). A network economic model for supply chain versus supply chain competition. *Omega – The International Journal of Management Science, 34*(3), 283-295.

Zhao, T. Z., & Jin, Y. H. (2005). Optimized coordination of supply chains based on dependency. *Computer Integrated Manufacturing Systems*, *10*(8), 929–933.

Zhao, W., Kart, F., Moser, L. E., & Melliar-Smith, P. M. (2008). A reservation-based extended transaction protocol for coordination of Web Services. *Journal of Web Services Research*, *5*(3), 64–95.

Zhou, D., Xiang, B. H., & Zou, G. S. (2003). The strategy of mass customization and the countermeasure of Chinese enterprises. *Industrial Engineering and Management*, *8*(5), 12–16.

Zhou, H., & Benton, W. C. Jr. (2007). Supply chain practice and information sharing. *Journal of Operations Management, 25*(6), 1348–1365. doi:10.1016/j.jom.2007.01.009

Zokaei, A. K., & Simons, D. W. (2006). Value chain analysis in consumer focus improvement – A case study of the UK red meat industry. *International Journal of Logistics Management, 17*(2), 141–162. doi:10.1108/09574090610689934

# About the Contributors

**Zhongxian Wang** is a professor at Montclair State University, New Jersey, USA. Professor Wang teaches Operations Analysis, Production/Operations Management, Decision Support & Expert Systems, Business Statistics, and Management Sciences. He is a member of Institute for Operations Research and the Management Sciences (INFORMS), Information Resources Management Association (IRMA), The Decision Sciences Institute (DSI), The Production and Operations Management Society (POMS).

* * *

**Bjørnar Aas** is Assistant Professor of Logistics at Molde University College, The Norwegian School of Logistics, Molde, Norway. He obtained his MSc (in 2003) and PhD (in 2008) in Logistics from Molde University College. His research focuses on upstream logistics in the oil and gas industry and much of the research has been done in collaboration with the Norwegian oil company Statoil ASA. The results have been published in several refereed journals. His current research interests include supply chain management, logistics planning under uncertainty, information systems, and routing of supply vessels.

**N. Anbazhagan** is currently Reader in Alagappa University, Karaikudi, India. He received his M. Phil and Ph.D in Mathematics from Madurai Kamaraj University, Madurai, India and M.Sc in Mathematics from Cardamom Planters Association College, Bodinayakanur, India. He has received Young Scientist Award (2004) from DST, New Delhi, India, Young Scientist Fellowship (2005) from TNSCST, Chennai, India and Career Award for Young Teachers(2005) from AICTE, India. He has successfully completed one research project, funded by DST, India. His research interests include Stochastic modeling, Optimization Techniques, Inventory and Queueing Systems. He has published the research articles in several journals, including Stochastic analysis and applications, APJOR and ORiON.

**Rebecca Angeles** is Full Professor, Management Information Systems Area, Faculty of Business Administration, University of New Brunswick Fredericton, Canada. Her research publications have appeared in such publications as *Information & Management, Decision Support Systems, Supply Chain Management: An International Journal, Industrial Management & Data Systems, International Journal of Integrated Supply Management, International Journal of Management and Enterprise Development, International Journal of Value Chain Management, International Journal of Physical Distribution & Logistics Management, Logistics Information Management, Journal of Business Logistics,* among others. Her research interests are in the areas of radio frequency identification, supply chain management issues, outsourcing and its consequences on supply chains, electronic trading partnership management issues, electronic business, business-to-business exchanges, electronic trading partnerships, electronic data interchange (EDI), Internet-EDI, and interorganizational systems, and innovative education approaches in Management Information Systems.

**Kritchanchai Duangpun** was engaged as the industrial consultant for several large-scale projects, such as the supply chain management project for Thailand Textile Institute in 2001 and the Logistics for FTA China ASEAN, Thailand Research Fund in 2005. Her research interests are in supply chain management, logistics policy for FTA issues and performance measurement in supply chains. She is currently serving in Mahidol University, in the capacity of Associate Professor in the Centre of Logistics Management, as well as holding the position of Coordinating Chair of Logistics Research Group, Thailand Research Fund.

**Reza Farzipoor Saen** is an Associate Professor at the Department of Industrial Management of Islamic Azad University – Karaj Branch (Iran). He completed his Ph.D. in Industrial Management from Islamic Azad University – Science & Research Branch in 2002. He has published over 26 refereed papers in many prestigious journals such as Applied Mathematics and Computation, Journal of Operational Research Society, European Journal of Operational Research, Applied Mathematical Modelling, the International Journal of Advanced Manufacturing Technology, International Journal of Procurement Management, *International Journal of Physical Distribution & Logistics Management,* World Applied Sciences Journal, WSEAS Transactions on Mathematics, Asia Pacific Management Review and the International Journal of Information Systems and Supply Chain Management.

**Dimitris Folinas**, possess a Ph.D in e-Logistics from University of Macedonia, Thessaloniki, Greece, and a Master of Information Systems from the same Institution. He has held various teaching posts with the University of Macedonia (Department of Marketing and Operations Management), and ATEI Thessaloniki (Department of Logistics). His research interests include Logistics and Supply Chain Management, Supply chain information systems, Logistics Information Systems and Technologies, Enterprise Information Systems and e-Logistics / e-Business.

**Mark Gershon** is the Chairman of the Management Science and Operations Management department at Fox Business School, Temple University, Philadelphia, USA. An industrial engineer by training, he began his career as a quality engineer with the Project Management office at the Army Armaments Development Command. He then worked as a project manager for Control Data Corporation, responsible for software development applications and consulting projects for the minerals industry before coming to Temple in 1983. Since then, Dr. Gershon has focused on developing new ideas for project management and quality management. Besides earning a Professional Engineer's license, he holds a certification as a Project management Professional and a Six Sigma Master Black Belt. He consults and gives seminars on these topics around the world. Dr. Gershon has published widely, including two text books and over thirty refereed journal articles. He has been honored with the Lindback Award by Temple for excellence in teaching.

**Jonas Hedman** is an associate professor at Center for Applied Information Communication Technology (CAICT) at Copenhagen Business School, Denmark. He earned his Ph. D. from Lund University,

Sweden. His research interest spans over several areas including, business models, ERP systems, integration of information systems, information systems development, nero-informatics, and mobile services and technologies. He is currently working on two projects. The DREAMS project aims at understanding new emergent mobile business models and designing new mobile services. The second project is related towards building knowledge in the area of information systems integration and in particular how to measure integration in business and industries. He is co-author of the book "IT and Business Models: Concepts and Theories", editor of two books, and has written 50 plus refereed journal and conference papers.

**Stefan Henningsson** is currently employed as senior researcher at Copenhagen Business School, Center for Applied Information Communication Technology (CAICT). His current research addresses IS integration in various contexts, including mergers & acquisitions, industry-wide supply chains, and international trade processes. He received his Ph.D. and masters degree in Information Systems at Lund University, Sweden. Henningsson's Ph. D. thesis addressed IS integration in the context of M&A and developed managerial tools for meeting the IS integration challenge. Previous publications include more than 20 refereed papers on IS in contexts such as organizational integration, mergers & acquisitions, and organizational learning.

**Mickey Howard** is a Senior Lecturer in Operations & Supply Management at the School of Management, University of Bath. He worked as an industrial designer for the retail and auto industry for 10 years. His research explores managing innovation across the auto, defence, and ICT sectors. He is currently investigating product-service innovation and servitization strategy in terms of the impact on procurement policy and practice. He regularly publishes in the *International Journal of Operations & Production Management*, the *Journal of Purchasing & Supply Management*, and *Supply Chain Management: an International Journal*. In 2008 he was awarded the Dean's prize for research excellence and a Chartered Institute of Purchasing & Supply research fellow scholarship.

**Peter Hosie** BEd WAIT, BA (Hons) Murd, MBus (Dist) Curtin, PhD UWA is Associate Professor in Management with the Faculty of Business and Management at the University of Wollongong in Dubai and an Adjunct Senior Fellow with Curtin Business School, Curtin University of Technology. Before joining UOWD in 2007, he held positions at Curtin University, Edith Cowan University and the University of Western Australia. Peter has taught most aspects of Human Resource Management in Asia Pacific and Dubai. He has published over 100 widely cited internationally refereed articles, and conference papers, practitioner articles and reports. Peter's research interests include the predictors of manager's performance (especially job related psychological wellbeing and intrinsic job satisfaction), crisis and security management, Technologically Mediated Learning, and logistics and information technology.

**Seong-Hyun Nam** is an associate professor in the College of Business and Public Administration at the University of North Dakota. He earned his Ph.D. at the University of Wisconsin-Milwaukee in POM. He also holds two master's degrees in Mathematics and Statistics from the University of Wisconsin-Madison. His interests are in Supply Chain Management, Stochastic Optimal Control and Model, Operations Research, and Operations Management. He has published in journals such as Management Science, IIE-Transactions, International Journal of Production Economics, and Decision Sciences Journal of Innovative Education.

**Firat Kart** received the Ph.D. degree in Electrical and Computer Engineering at the University of California, Santa Barbara. He received the B.S. degree in Computer Science from Bilkent University in Ankara, Turkey. Currently, he is a principal engineer / architect at Tibco in Palo Alto, California. His research interests include distributed systems, computer networks, Service Oriented Architectures, Web Services, database transactions, and supply chain and healthcare applications.

**Ibrahim Al Kattan** is an Associate Professor of Industrial Engineering in Graduate Program in Engineering Systems Management at American University of Sharjah, United Arab Emirates. He obtained his Ph.D. in Industrial and Manufacturing Engineering from Tennessee Technological University, an MS in Industrial Engineering & Management from Oklahoma State University, and another MS in Engineering Production and Management from the Birmingham University, England. He is the founder and Ex-Director of one of the most successful Graduate program in the UAE. He has over twenty five years of experience as a teacher, researcher, consultant in engineering management, simulation and modeling, quality management, and supply chain management in USA, UK, UAE and Iraq.

**George Kenyon is** an Associate Professor of Supply Chain and Operations Management at Lamar University. He received his Ph.D. in Business Administration from Texas Tech University in 1997, his M.S. in Management Science from Florida Institute of Technology in 1993, and his B.S. in Technology from the University of Houston in 1982. His research interests are in the fields of supply chain management, quality management, and operations management. He has published in *Journal of Marketing Channels, Quality Management Journal, International Journal of Production Economics, International Journal of Information Systems and Supply Chain Management, and Journal of Case Studies in Accreditation and Assessment.* His professional experience includes fifteen years of industry experience, were knowledge of in various aspects of engineering including systems testing, systems design, and manufacturing, as well as, business planning and supply chain management was acquired. The base of this experience was gained from companies such as Texas Instruments, Rockwell International, The Boeing Company, Aspen Technologies, and Hewlett Packard. As a consultant, his assignments have included scheduling and planning projects with Gulf States Steel, the Westlake Corporation, Miller Brewing, and Phillips Chemical.

**Taha Al-Khudairi** is Lab Instructor in the Department of Computer Engineering at American University of Sharjah, UAE. He obtained his BS and MS in Electrical Engineering in 1992 and 1996 respectively. In May 2007 he received his second Master of Science in Engineering Systems Management degree from American University of Sharjah, United Arab Emirates. In 2002 and 2003, Mr. Al-Khudairi has received an Engineering Student Council Award for the best lab instructor in the Department of Computer Engineering at American University of Sharjah, UAE.

**Yohanes Kristianto** obtained an undergraduate degree in Chemical Engineering and a master degree in Industrial Engineering from Sepuluh Nopember Institute of Technology, Surabaya, Indonesia, Prior to his academic career, he worked for a Quality function of a multinational company. He is now a PhD student in Industrial Management at University of Vaasa, Finland. His research interests are in the area of supply-chain strategy/management and production/operations management.

**Hisashi Kurata** is an assistant professor of Operations Management at the Graduate School of International Management, International University of Japan. He has a Ph.D. from the University of Wisconsin-Milwaukee, and M.S. from the Pennsylvania State University, and a B.A. from the University of Tokyo. His research interest includes supply chain management, interface studies between marketing and operations, and inventory management. His recent teaching interest includes operations management, supply chain management, and service management. He has published in journals such as European Journal of Operational Research, International Journal of Production Economics, Journal of Revenue Management, and International Journal of Operations and Quantitative Management.

**Debendra Kumar Mahalik** is presently working as a Lecturer at PG Department of Business Administration, Sambalpur University. He has around seven years of teaching and six years of Industry experience. He is also pursuing his Ph.D in the area of management. He has published around twelve papers in international and national journals and conference proceedings. His areas of interest include application of Information technology in business, efficiency analysis and multi-criteria decision making.

**Ioannis Manikas** holds a Bachelor in Agriculture and a Master of Science in the field of logistics, at Cranfield University, United Kingdom. He is a PhD fellow at the Department of Agricultural Economics in AUTH and his primary interest includes Supply Chain Management, Logistics and Agribusiness Management. He holds a teaching post as an external lecturer at ATEI Thessaloniki (Department of Logistics) and has a wide experience as a self employed project manager and consultant in the Agrifood sector.

**John Mawhinney** is Executive Assistant Professor of Supply Chain Management in the Palumbo/Donahue Schools of Business, Duquesne University. He posses 35 years of supply chain industry, consulting, and academic experience. Mr. Mawhinney earned a B.S. in Logistics and Marketing at Ohio State University and an MBA from the University of Pittsburgh. He is currently pursuing a Doctorate in Education with a focus on Supply Chain Management outcomes assessment. Mr. Mawhinney is also working on the U.S. Air Force Research Labs.

**P. M. Melliar-Smith** is a professor in the Department of Electrical and Computer Engineering at the University of California, Santa Barbara. Previously, he worked as a research scientist at SRI International in Menlo Park. His research interests encompass the fields of distributed systems and applications, and network architectures and protocols. He has published more than 250 conference and journal papers in computer science and engineering. Dr. Melliar-Smith is a pioneer in the field of fault-tolerant distributed computing. He received the Ph.D. in Computer Science from the University of Cambridge, England.

**Louise E. Moser** is a professor in the Department of Electrical and Computer Engineering at the University of California, Santa Barbara. Her research interests span the fields of computer networks, distributed systems and software engineering. Dr. Moser has authored or coauthored more than 250 conference and journal publications. She has served as an associate editor for the IEEE Transactions on Services Computing and the IEEE Transactions on Computers and as an area editor for IEEE Computer in the area of networks. She received the Ph.D. in Mathematics from the University of Wisconsin, Madison.

**Brian D. Neureuther** is an Associate Professor of Supply Chain and Operations Management at the State University of New York, College at Plattsburgh. He received his Ph.D. in Production and Operations Management from Texas Tech University, his M.B.A. degree from Wright State University in Dayton, Ohio, with a concentration in management science and his B.A. in mathematics from the State University of New York, College at Geneseo. His research interests include supply chain management, supply chain disruption, information technology in supply chains, simulation for production planning and control, and quality control. He has published over 30 peer reviewed journal articles and his work has appeared in journals such as the Journal of Integrated Design and Process Science, the International Journal of Production Economics, IEEE Transactions on Semiconductor Manufacturing, Production Planning and Control, the International Journal of Information Systems in the Service Sector, the Quality Management Journal, the International Journal of Information Systems and Supply Chain Management, and the Journal of Marketing Channels. He has been guest editor of the Journal of Marketing Channels and is on the editorial advisory board of the International Journal of Information Systems and Supply Management and the Journal of Marketing Channels. He has presented at over 32 international and national conferences on topic ranging from teaching pedagogy to managing supply chain risk and has consulted with companies such as Rider University, Neoteric Hovercraft, EDI Telecommunications, Southwestern Wire Cloth, and the Cleveland County Chamber of Commerce (North Carolina). He is a member of the Production and Operations Management Society and APICS, the Society of Operations Management.

**Jaesun Park** is a Professor in the College of Business and Public Administration at the University of North Dakota. He earned his Ph.D. at Northwestern University. His research and teaching interest includes Operations Research and Supply Chain Management. He has published in journals like Management Science, IEEE and Decision Sciences Journal of Innovative Education.

**Gokulananda Patel** is currently Senior Professor at Birla Institute of Management Technology, Greater Noida, in the areas of Operations and Quantitative methods. Prior to joining BIMTECH he was with Post Graduate Department of Business Administration, Sambalpur University. Professor Patel is known for quantitative applications in management decision making. He has published around thirty papers in international and national journals and conference proceedings. He has around thirty years of teaching experience and his field of Specializations is Quantitative Techniques, Operational Research, E-Governance and Efficiency Measurement. He has guided number of scholars for their Ph.D.

**Malte L. Peters** is senior lecturer and research assistant at the Institute for Production and Industrial Information Management, University of Duisburg-Essen (Campus Essen). He was involved in several research and practice projects. His current research interests include operations research especially efficiency analysis, multicriteria decision making, knowledge management and process management.

**Lee Pickler** teaches International Marketing and Management in the Systems MBA and International MBA programs, and has a research interest in globalization. Other research work includes on-going studies of the delivery of healthcare in the United States and Japan through his work with Socio-Medical Research of Tokyo. Dr. Pickler has been engaged in a wide range of management consulting assignments, to include projects relating to outsourcing to China, the planning for a distribution center for serving the automobile manufacturing in Brazil, as well as global marketing research studying the aircraft turbine engine aftermarket maintenance services.

**Philip Powell** is Professor of Information Management at the University of Bath. Formerly, Professor of Information Systems, University of London, and Director of the Information Systems Research Unit at Warwick Business School, he has worked and taught in Australia, Africa, US and Europe. Prior to becoming an academic he worked in insurance, accounting and systems analysis. He is the author of eight books on information systems and financial modeling. He has published numerous book chapters and his work has appeared in over ninety international journals and at over 100 conferences. He is Managing Editor of the Information Systems Journal and associate editor and editorial board member of a number of other journals. He is a past President of the UK Academy for Information Systems. His research concerns the role and use of IS in organisations especially issues of strategy and evaluation in the context of small firms, e-business and knowledge management.

**Jagadeesh Rajashekhar** is a Professor at SDM Institute for Management Development, Mysore, India, in the area of Operations Management and Quantitative Techniques, with over 25 years of experience in academics. His qualifications include B.E. (Mysore University), M.E. (Madras University), Diploma in Production Management (Annamalai University), Diploma in Software Capability Maturity Model, and Ph.D. (I I T, Bombay, India), and also certifications in Six Sigma Green Belt and Project Management. Since 2006, he is an Adjunct Faculty at Richard J Fox School of Business and Management at Temple University, Philadelphia, USA, where, he teaches a course on Operations Management. He has authored the Indian adaptation of the book "Operations Management" by Heizer and Render, and published more than 70 papers in reputed national and international journals and conferences. He is also a referee for several international journals. He has successfully conducted training programs for people from industries, educational institutions, and various Government departments.

**Kenneth A. Saban** is Associate Professor of Marketing in the Palumbo/Donahue Schools of Business, Duquesne University. He posses 36 years of manufacturing, consulting, and academic experience. Dr. Saban earned a B.S.B.A. from Youngstown State University, a M.S.J. from Northwestern University, and a Ph.D. from the University of Pittsburgh. He has developed a particular expertise in the areas of *business collaboration* and *technology management* through a wide-range of research projects, grants and industry partnerships. He has published over 20 articles and given numerous academic and industry presentations. He is also working with the U.S. Air Force Research Labs to explore the impact of "human collaboration" on supply chain performance.

**Susan A. Sherer** is the Kenan Professor of IT Management and Chair of the Department of Management at Lehigh University. Her research interests include IT investment and risk management, supply chain management information systems, and cross cultural issues in information systems. She is the author of Software Failure Risk: Measurement and Management. Her research has been published in Information and Management, Information Systems Frontiers, Journal of Global Information Management, Journal of Information Systems, International Journal of Electronic Commerce, Supply Chain Management, International Journal of Electronic Business, and Communications of AIS. Sherer received her Ph.D. in Decision Sciences from the Wharton School of the University of Pennsylvania, M.S. Industrial Engineering from SUNY Buffalo, and B.S. Mathematics from SUNY Albany.

**Sreekumar** is Ph.D in Management and is presently working as an Associate Professor at Rourkela Institute of Management Studies, Rourkela 769015, INDIA. His areas of interest include application of Data Envelopment Analysis for efficiency analysis, Multi-criteria Decision-Making. He has around sixteen years of teaching experience in the areas of Quantitative Techniques for management decision making. He has published around twenty five research papers in various international and national conferences and journals. He has also authored two text books.

**Albert Tan Wee Kwan** is an associate director at The Logistics Institute, National University of Singapore. His research interests are in reverse logistics, process modelling and reengineering, and information technology to coordinate supply chain. He holds an MBS from the National University of Ireland and a PhD in Supply Chain Management from the Nanyang Technological University. His research works have been published in the International Journal of Physical Distribution & Logistics Management, the International Journal of Logistics Systems and Management, the International Journal of Logistics Management and the Asia Pacific Journal of Marketing and Logistics.

**Richard Vidgen** is Professor of Information Systems in the School of Management, University of Bath. He worked in information systems development in industry for 15 years. He holds a first degree in Computer Science and Accounting, an MSc in Accounting. He left industry to join the University of Salford, where he completed a PhD in systems thinking and information system quality. His current research interests include complex adaptive systems theory, agile project management, and e-commerce quality. He has published the books *Data Modelling for Information Systems* (1996) and *Developing Web Information Systems* (2002) as well as more than 100 chapters and journal papers.

**B. Vigneshwaran** is a Lecturer in Mathematics in the Department of Mathematics, Thiagarajar College of Engineering, Madurai, India. He has M.Sc(Applied Mathematics), MPhil(Mathematics) degrees with five years of teaching experience. He is currently pursuing his PhD in Mathematics at the Alagappa University, Karaikudi, India. His current research interests include Applied Probability, Inventory and Queueing Systems.

**John J. Vitton** is an Associate Professor in the College of Business and Public Administration at the University of North Dakota. He served as Management Department Chairman from 2001 through 2006. He earned his Ph.D. at the University of Nebraska-Lincoln in Business Administration, his MBA at Ohio State University, and his B.S. in Metallurgical Engineering at Michigan Technological University. His interests are in Supply Chain Management, Strategic Management, and Human Resource Management. He is a member of the Academy of management, the Society For Case Research, the North American Case Research Association, and Beta Gamma Sigma Honor Society.

**Stein W. Wallace** is Professor of Operational Research at Lancaster University Management School, UK, He has earlier held professorships in Norway (Trondheim and Molde) and Hong Kong. He obtained his PhD in informatics at The University of Bergen in 1984. His main interest is decision-making under uncertainty. He is on the editorial board of several journals, including INFORMS Journal on Computing (since 1991) and the new Energy Systems. He co-authored (with Peter Kall) Stochastic Programming in 1994 and co-edited (with William T. Ziemba) Applications of Sochastic Programming in 2005, and has published more than 60 articles in refereed journals. He chaired COSP – the International community for stochastic programming - from 1992 to 1995.

**Frank Wolf** is on the faculty of Nova Southeastern University's H. Wayne Huizenga School of Business and Entrepreneurship where he teaches MBA courses in Operations Management and General Management best practices. He has a research interest in international supply chains, arbitration and mediation. Dr. Wolf entered academic life after an industry career with major US corporations working in engineering, operations research, and general management as general manager of a wholly owned IT subsidiary. He also held an appointment to the White House Fellowship (PIE) and served at the United Nations as a non-governmental representative (NGO), prior to starting two entrepreneurial firms in information technology services. He is currently an active arbitrator and also serves on the mediator roster with FINRA, the Financial Industry Regulatory Authority, formerly NASD and NYSE.

**Ruiliang Yan** is an assistant professor of marketing at Indiana University Northwest. He received his PhD in marketing from the University of Wisconsin, Milwaukee. He has published one book and a number of articles in the different refereed journals. He also is serving as referee for many highly prestigious journals. He specializes in marketing modeling, retailing and supply chain management.

**Stephan Zelewski** teaches business management with focus on production Management at the University of Duisburg-Essen. He holds the chair for Production and Industrial Information Management at the faculty of economics at the campus of Essen. He also is member of the Institute of Business and Economic Studies (IBES), the Institute for Computer Science and Business Information Systems (ICB) as well as the Centre for Logistics & Traffic (ZLV). His work scope contains production management, especially logistics and supply chain management, the use of modern computer technologies in the area of production management, operations research and game theory, knowledge management and artificial intelligence and their operational applications (especially knowledge based systems and multi-agent systems), production theory as well as philosophy of science (especially constructions of economic theories from the perspective of the non statement view). For further information please visit: http://www.pim.wiwi.uni-due.de/team/stephan-zelewski.

# Index

## U



## V



## W

CPSIA information can be obtained at www.ICGtesting.com
Printed in the USA
BVOW050222170212
282949BV00007BA/12/P

9 781466 609181